Crop Ecology

Productivity and Management in Agricultural Systems

Second Edition

Food security and environmental conservation are two of the greatest challenges facing the world today. It is predicted that food production must increase by at least 70% before 2050 to support continued population growth, although the size of the world's agricultural area will remain essentially unchanged.

This updated and thoroughly revised second edition provides in-depth coverage of the impact of environmental conditions and management on crops, resource requirements for productivity, and effects on soil resources. The approach is explanatory and integrative, with a firm basis in environmental physics, soils, physiology, and morphology. System concepts are explored in detail throughout the book, giving emphasis to quantitative approaches, management strategies and tactics employed by farmers, and associated environmental issues.

Drawing on key examples and highlighting the role of science, technology, and economic conditions in determining management strategies, this book is suitable for agriculturalists, ecologists, and environmental scientists.

David J. Connor is Emeritus Professor of Agriculture at the University of Melbourne, Australia. His research programs deal with land and environmental relationships of a range of irrigated and rainfed cropping systems. In 2003 he was awarded the Donald Medal for outstanding contributions by the Australian Society of Agronomy.

Robert S. Loomis is Emeritus Professor in the Department of Plant and Environmental Sciences at the University of California, Davis, USA. His research interests include photosynthetic productivity, nutrient and water management, and integrated simulation models. He holds numerous honors from scientific societies and universities worldwide. Most recently, in 2001, he was awarded Douter honoris causa, Universidade Técnica, Lisboa.

Kenneth G. Cassman is Professor of Agronomy at the University of Nebraska, USA. His research focuses on nutrient cycling and crop nutrient requirements, crop yield potential, and water productivity of irrigated crops. In 2006 he received the Agronomic Research Award from the American Society of Agronomy.

Crop Ecology

Productivity and Management in Agricultural Systems

Second Edition

DAVID J. CONNOR
University of Melbourne, Australia

ROBERT S. LOOMIS
University of California, Davis, USA

KENNETH G. CASSMAN
University of Nebraska, Lincoln, USA

CAMBRIDGE
UNIVERSITY PRESS

CAMBRIDGE UNIVERSITY PRESS
Cambridge, New York, Melbourne, Madrid, Cape Town,
Singapore, São Paulo, Delhi, Tokyo, Mexico City

Cambridge University Press
The Edinburgh Building, Cambridge CB2 8RU, UK

Published in the United States of America by Cambridge University Press, New York

www.cambridge.org
Information on this title: www.cambridge.org/9780521761277

First published 2011

Printed in the United Kingdom at the University Press, Cambridge

A catalogue record for this publication is available from the British Library

Library of Congress Cataloguing in Publication data
Connor, D. J.
Crop ecology : productivity and management in agricultural systems / David J. Connor, Robert S. Loomis,
Kenneth G. Cassman. – 2nd ed.
    p.   cm.
Rev. ed. of: Crop ecology / R.S. Loomis, D.J. Connor. 1992.
Includes bibliographical references and index.
ISBN 978-0-521-76127-7 (hardback) – ISBN 978-0-521-74403-4 (paperback)
1. Agricultural ecology.   2. Agricultural systems.   I. Loomis, R. S.   II. Cassman, Kenneth G.
III. Loomis, R. S. Crop ecology.   IV. Title.
S589.7.L66   2011
630.2′77 – dc22   2010053580

ISBN 978-0-521-76127-7 Hardback
ISBN 978-0-521-74403-4 Paperback

Contents

Preface

Humans make extensive use of land, water, energy, labor, and other resources in the production of crops and pastures. We do this because it is essential to our survival and well-being. As world population grows, so does demand for continuing success in agriculture. And as more land is used in agriculture, concerns for loss of natural ecosystems and biodiversity increase as well. The conflict between production and conservation can only be resolved with cropping systems that are highly productive, efficient, and sustainable.

Agricultural management involves plant communities and areas of land. It requires knowledge of individual plant behavior under crowded conditions and interactions of plant communities with aerial and soil environments. These organismal and higher levels of biological organization are the subjects of ecology at different spatial scales, but explanation of these behaviors depends upon integration of relevant knowledge spanning lower levels from molecules and cells to organs. Ecology can thus be characterized as an integration of other disciplines. In turn, however, it provides specialist disciplines with context and relevance and, further, explains that in isolation they rarely affect system outcome. Crop ecology has additional dimensions in agricultural technology that interface with engineering, information and social sciences, and perspectives provided through history.

The tools of crop ecology (strong basic physics, chemistry, and mathematics) are not different from those of other biological disciplines. Mathematical models are especially useful in integration and are generally appropriate to crop ecology. In essence, ecological thinking derives from an eagerness to understand the whole and a willingness to maintain a broad appreciation of component disciplines.

We designed this book as a text and reference for advanced undergraduate and post-graduate students and for practicing educators and industry professionals. It derives from our experience in teaching over many years and our frustration with the great breadth and diffuse nature of appropriate readings. We especially want to encourage young scientists to use information in orderly ways to expand our understanding of crop ecology, and to develop new ways in which it can be applied to the changing problems of plant production. We do not, however, see the book limited to agriculturalists. It can also provide ecological context for courses in environmental sciences that would benefit from an agricultural perspective.

Our approach is explanatory and integrative. Although we review many topics, and introduce some new topics slowly, the text generally builds quickly on basic plant biology, soil science, environmental physics, and chemistry. Integration is apparent in system themes introduced at the outset and brought to a focus in several case studies (Chapters 16 and 17) that can serve as models for analysis of evolution and management in other farming systems. The final chapter seeks a vision and analysis of the challenges facing agriculture to 2050.

We wish to record our appreciation to colleagues and friends who have provided data, figures, or helped in discussion and by critical evaluation of various chapters.

Australia: John Angus and Tony Fischer – CSIRO, Canberra. Rob Norton – The University of Melbourne. Garry O' Leary, Victorian Institute for Dryland Agriculture. Des Whitfield, Mark O'Connell, and Ian Goodwin – Institute for Sustainable Irrigated Agriculture, Tatura. Mark Johns – farmer, Horsham. Victor Sadras – Research and Development Institute, South Australia.

Spain: María Inés Mínguez-Tudela, María Gómez del Campo, Miguel Quemada, Carlos Gregorio Hernández and Margarita Ruiz-Ramos – Universidad Politécnica de Madrid. Santiago Bonachela – Universidad de Almeria. Luciano Mateos – Instituto de Agricultura Sostenible (CSIC), Córdoba.

The Philippines: Achim Dobermann, Shaobing Peng, Grace Centeno, and K. L. Heong – International Rice Research Institute.

The USA: Patricio Grassini, Maribeth Milner, Justin van Wart, Dan Walters, Viacheslav (Slava) Adamuchuk, Don Lee, Dennis McCallister, Tom Hoegemeyer, and Richard Ferguson – University of Nebraska. R. Ford Denison – University of Minnesota. Jerry Hatfield and Daniel Olk – USDA National Laboratory for Agriculture and the Environment. Michele Wander – University of Illinois, and Haishun Yang – Monsanto Company.

Acknowledgments to sources of all figures and tables are given in their legends. Chapter 16 is an extended version of a paper (Connor 2004) included with permission of the publisher.

Finally, we thank the Universities of Melbourne, California, and Nebraska and our wives, Inés, Ann, and Susie, for their support and patience during this project.

Part I

Farming systems and their biological components

A view from space gives emphasis to areal dimensions of vegetation and agriculture at the thin interface between atmosphere and solid earth. It is only by spreading plants across the landscape that they can efficiently intercept fluxes of limiting resources such as CO_2, water, and sunlight. Farmers make strategic and tactical decisions about planting and management to optimize rates of crop growth and accumulation of yield. In farming, land is divided into individual fields as the units of management and production. In ecological terms, plants that occupy those fields constitute a community of cohabiting organisms. A community, considered together with the chemical and physical features of the environment, forms a further fundamental grouping, the ecosystem.

Farmers' efforts in crop and pasture management aim at beneficial control over the structure of crop communities and physical and chemical aspects of their environment. These issues are introduced in Chapter 1, which also presents five major themes that recur throughout the book. Chapter 2 presents concepts of trophic chains seminal to understanding the role of animals in agriculture and the nutritional requirements of humans. Establishment and productivity of plant communities dominated by agricultural species is presented in Chapter 3, and their genetic resources in Chapter 4. This part terminates, in Chapter 5, with a discussion of plant phenological development as the primary basis for adaptability to environment and determination of reproductive yield.

Part II Farming systems

Farming systems and their biological components

1   Agricultural systems

Humans depend on agriculture to provide food, feed, fiber, and fuel. Production of these organic materials in individual fields depends upon the physiological abilities of plants and the soil and aerial environments in which they grow. What crops are grown, and how, are human decisions that depend upon usefulness and value of products, costs of production, and risks involved. At the farm level, those considerations are rationalized with need for animal feeds, availability of labor, and requirements for crop rotation to raise fertility and control disease, weeds, or erosion. Additional constraints are imposed by market forces and availability of capital and technology.

Within these socioeconomic considerations, crop response to environment and management follows the laws of thermodyamics and conservation of energy and mass. Therefore, we can understand and predict crop performance using ecological analyses in terms of biological, chemical, and physical principles. This is the context and content of crop ecology.

In this chapter, we introduce ideas about the nature, objectives, and management of farming to provide a foundation for detailed analyses of crop performance in agricultural systems. We also present the guiding principles upon which this book is constructed.

1.1   On the nature of agriculture

Agriculture can be studied at various organizational and geospatial levels, from individual fields, to their grouping in farms, and to grouping of farms within regions. This is illustrated in Fig. 1.1 and identifies the need to establish a coherent terminology.

1.1.1   Terminology

Individual production fields are the fundamental units for studies in crop ecology. Field sizes range from 10 to 100 ha in large-scale mechanized agriculture typical of Argentina, Australia, Brazil, Canada, Europe, and the USA, to 1 ha or less in small-scale, labor-intensive agriculture in densely populated countries such as Bangladesh, China, India, Indonesia, Malawi, the Philippines, and Vietnam. A **cropping system (or livestock–pasture system)** is the temporal sequence of crops and management practices in individual fields. At this level we can examine the production processes of plants, their

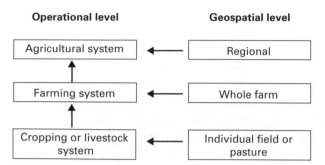

Fig. 1.1 Relationship between operational and geospatial scales in agriculture.

dependence on environmental conditions, and the role of soil processes that support plant growth. By observing individual fields over years, we can understand the effects of crop rotation, tillage practices, soil amendments, and removal of harvested material on soil properties. Yields can be analyzed as a function of resource requirements for nutrients, water, and energy. Economic budgets of costs and returns and assessment of labor requirements are also useful at the field level.

At a higher level, fields are components of farms under management of particular farmers. The principal crops, livestock, and management practices employed on a particular farm constitute a **farming system**. Ruthenberg (1980) noted that because farms are organized to produce net economic return they are the fundamental units for economic and sociological analyses. Other goals such as supplying food for the farm family (subsistence agriculture) and perpetuating the farm for later generations are also important. A farm, then, is a goal-oriented system in which goals dictate how capital and labor are used in production activities. Availability of capital and labor, plus the need to manage risk, impose constraints on which crops are grown, sequences employed, and intensity of farming. Some aspects of agricultural ecology are therefore analyzed best at the farm level. Examples include integration of livestock with crops and pastures and cycling of nutrients from one field, through manure, to another.

The range of combinations of species is broad, but there are structural typologies that are repeated in many parts of the world. Thus in Table 1.1, which describes the major cultivated systems of the world, one can see clear relationships between farming activities and the environment. Irrigation is practiced across the range of climates but is most common in humid tropics and sub-tropics where water is most readily available and the thermal regime allows year-round cropping. Rainfed systems concentrate on cropping in humid areas, plantation crops in highlands subject to erosion hazard, and integration of crops and livestock in semi-arid/arid areas. Shifting cultivation is generally restricted to humid tropical areas under increasing pressure from encroachment for conversion to continuous cropping to provide food, fiber, and energy for an expanding world population (Sections 15.5 and 17.4).

Attempts to analyze farming systems encounter the problem that no two farms are exactly alike. Soil types, crops, field size, financial resources, and farmer's skills and opinions about best practices vary from farm to farm. It is logical, however, to assume farmers behave rationally, which means there are good reasons why farms are similar or different. When most farms within a region are organized in a similar way, we can apply a descriptive phrase that characterizes regional farming practices. Chapters 16 and 17, for example, deal with technological change and evolution of various cropping systems in Australia, the USA, Brazil, and Asian countries, including examples with both crops and livestock. It is convenient to use the term **agricultural system** when talking about regional organization of farming systems. The regional basis opens additional avenues for analysis, including matters such as drainage and air and water pollution. At this level, economic and sociological studies may also include service roles (e.g. grain purchase, storage, and transport) of towns and villages within the region.

We can also evaluate agricultural systems at watershed, regional, national, or global scales with regard to impact on food security and ecosystem services that include water and air quality, and biodiversity. These effects result from aggregate contributions of individual fields and farms, which contribute in turn to outcomes at much larger geospatial scales. Crop ecology principles provide the basis for understanding and estimating these aggregate effects.

1.1.2   Important attributes

Productivity is perhaps the most important property of farming systems but there are others. Marten (1988) provided definitions used in this book for five main system properties:

Ecological attributes	Social attributes
Productivity	Equitability
Stability (variation, persistence)	Autonomy
Sustainability	

Productivity is explicitly defined by yield of useful product per unit land area. Expression per unit land area is multidimensional because various natural and human inputs, such as radiation, water, nutrients, and labor, also occur per unit land. Yield is therefore an integrative measure of efficiency relative to such inputs. Yields vary over years with weather and other causes. Marten (1988) used the term **stability** in reference to the degree of such variation; **sustainability** is then concerned with whether current production levels can be maintained over years at the same site. The simplicity and clarity of these definitions commend them for our purposes. Stability and sustainability are broad terms, however, and sometimes carry other definitions, so it is important to be explicit about the contexts in which they are used. Ecologists who work with natural plant communities give a different emphasis to stability, and **sustainable agriculture** is sometimes taken as a synonym for **organic agriculture**, implying (incorrectly) that it is sustainable, while other types of farming are not. Here we consider sustainability to be a variable attribute of all agriculture at all levels of organization (Fig. 1.1), and it can be evaluated in terms of system performance over time.

Table 1.1 Global typology of cultivated farming systems with relative areas and examples[a]

Farming system		Tropical and sub-tropical (62%)[b]			Temperate (38%)	
		Warm humid/sub-humid (26%)	Warm semi-arid/arid (12%)	Cool/cold (highland/montane) (24%)	Humid/sub-humid (22%)	Semi-arid/arid (16%)
Irrigated	(18%)	rice (e.g., East, Southeast Asia) rice–wheat (e.g., Pakistan, India, Nepal)	rice (e.g., Egypt, Peru)			cotton
Rainfed – high external input (crops, livestock, tree crops)	(82%)	rice–wheat (e.g., Pakistan, India, Nepal)		tea, coffee plantations (e.g., East Africa, Sri Lanka)	maize and soybean – Argentine pampas, US corn belt; small grains (wheat, barley, rapeseed, sunflower, oats) and mixed crop–livestock systems (e.g., West and North Central Europe)	
Rainfed – low external input (crops, livestock, tree crops)		staple crops in humid tropics (e.g., yam, cassava, banana in sub-Saharan Africa)	Mixed crop–livestock (e.g., Sahel, Australia)	Cereals and tubers (e.g., High Andes)	mixed crop–livestock systems (e.g., Europe)	wheat–fallow systems (e.g., Central Asia, Canada, United States, Australia)
Shifting cultivation	na	Amazon Basin, South East Asia, equatorial Africa				

[a] Adapted from Cassman and Wood (2005)
[b] Values in parenthesis represent percent of global total arable land area.

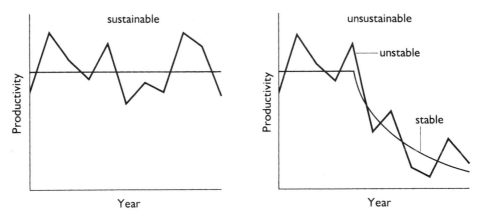

Fig. 1.2 Marten's (1988) views of the meaning of stability and sustainability of production.

Equitability (evenness of benefits within an agricultural system and between it and the larger society) and **autonomy** (the degree that agriculture is independent of larger society) are important in social and economic analyses. Subsistence farmers, for example, have a high degree of autonomy but contribute little to urban economies.

Productivity, stability, and sustainability are principal themes throughout this book. Relationships among these terms are illustrated in Fig. 1.2. We would add to this list of important attributes that agriculture must also be **sufficient** in supplying food for humans. Taken together, sufficiency and equitability determine food security and well-being at local, regional, and national levels. Given the continuing rapid increase in human population, agriculture must continue to expand in area, yield per unit area, or both.

Interactions between agriculture and society deserve comment before proceeding with ecological questions. Farming is generally an extension of the larger society in which it exists. Larger society provides markets for produce and sources for necessary inputs. In this context, most farming systems, particularly those of developed countries, have a low degree of autonomy. Within a country, farm and urban populations are usually exposed to the same legal and economic rules. If labor and capital are free to move between sectors, they compete for those means of production and come to have similar standards of living. That process usually leads to a reasonable degree of equity. Government regulation has important influences on equity, however. Price controls on inputs or outputs of farming, for example, can change the balance sharply.

Economists analyze the complex and diverse interactions involved in common units of currency. It is difficult (and in many cases impossible), however, to interpret economic findings or to extrapolate them to the future without also having an understanding of the biology, chemistry, and physics of production processes: of knowing, for example, about the finite, limiting relationships that exist between solar radiation, water use, and nutrient supply on one hand and agricultural production on the other.

In addition to essential flows of information, goods, and services that take place between urban and rural sectors, each sector also affects the other's environment. Air pollution originating in urban centers affects farm production; soil erosion as dust from

---

Box 1.1 Unifying themes for study and analysis of farming systems

The essential goal of farming systems is to provide food, feed, fiber, and fuel. Other goals include protecting the environment and ensuring (long-term) sustainability of the systems.

Productivity is determined by the structure and function of managed plant communities and access to resources that support their growth.

Efficient use of scarce resources is achieved through strategic decisions about cropping patterns and choice of crop and soil management practices.

Long-term sustainability of farm productivity depends on the maintenance of chemical, physical, and biological properties of soils.

Farming systems will continue to evolve in response to changes in markets, technology, and climate. Growth of population and income and scarcity of water will be major driving forces in the future.

---

fallow fields, and sediments and solutes in streams, affect the quality of life for all. Agriculture's need for goods and services stimulates growth of urban communities within food-producing regions with the result that urban sprawl may remove productive lands from agriculture. In fact, most cities were located in areas with good farmland because they were established before modern rail and air transport made it possible to move food long distances, and cities were provisioned by surrounding agriculture. Thus urban sprawl worldwide removes the most productive land while replacement can only occur on land of poorer quality for crop production. This too has implications for global food security and protection of natural resources as discussed later.

1.2   Unifying themes

Five themes listed in Box 1.1 unify the contents of this book. They relate contemporary issues in food supply and environmental management to attributes of farming systems as previously presented.

The essential goal of farming is to provide food, feed, fiber, and fuel. Other goals include protecting the environment and ensuring (long-term) sustainability.

Crop ecology has a tight focus on productivity because the essential role of farming is provisioning farmers and society with an adequate and stable food supply. The other attributes of agriculture differ in importance from farm to farm and from country to country. In contrast, the need for agriculture to be sufficient is universal although the form of sufficiency varies depending on type of agricultural system. In some developing countries, small-scale subsistence farming systems must meet the nutritional requirements of farm families directly because they do not have other sources of food, or cannot afford to purchase it from others. As such, subsistence farming must be highly risk adverse to ensure a minimum, adequate level of production. Farmers in developed regions have sufficient financial resources to buy food from both local and distant sources, allowing them to specialize in producing a few commodities in large quantity for regional and global markets.

Sufficiency of agriculture must also ensure global food security as the population increases from 6.8 billion in 2010 to a projected 9.2 billion by 2050 (United Nations Population Division). Prospects to meet future food demand can be evaluated from a crop ecology perspective based on rate of increase in crop yields (Section 4.4), amount and quality of available farm land (Box 7.2), and climate and water that support crop growth (Section 6.10). At issue is whether enough food can be produced without a large expansion of cropped area. Nearly all of the world's prime farmland, around 1500 Mha, is currently in production and most remaining land that might be used for agriculture lies under rain forests, wetlands, and grassland savanna that provide critical habitat for wild animals and flora. Thus, increased demand for food must come mostly from yield (productivity) increase on existing cropland.

Stability and sustainability play associative roles to productivity because both must be achieved at adequate levels of production. Stability of production depends upon weather conditions and management that minimize losses to competition from weeds, and damage by pests and diseases. Stability can be increased by irrigation to offset variation of water supply from rainfall (Chapter 14), fertilizer inputs to ensure adequate nutrient supply (Chapter 12), and by management practices, including biocides, to direct maximum possible proportions of primary production to crop and animal products. As farming moves into areas of low and variable rainfall, greater long-term production is most commonly achieved with less stability because strategies to increase productivity focus on making best use of good (higher rainfall) years and perhaps not cropping in years of lowest rainfall (Chapters 13 and 16). In many cases, perennial crops and livestock grazing systems, rather than staple food crops, are a more sustainable and stable use of marginal lands.

Major cropping systems Provisioning urban populations and global food security depend on relatively few major cropping systems located in regions with best farmland (Cassman & Wood 2005). The most productive systems presented in Table 1.1 include:

- irrigated lowland rice systems of south, east, and southeast Asia (24 Mha – Section 17.3).
- irrigated rice–wheat annual double-crop systems of the Indo-Gangetic plains in Pakistan, India, Nepal, and Bangladesh, and the central and south-central plains in China (17 Mha).
- rainfed wheat and small grains-based systems of north and central Europe (40 Mha).
- rainfed maize–soybean rotations in the temperate central prairies of North America and in parts of South America (85 Mha – Chapter 17).

These four high-yielding systems supply staple food grains for 2.7 billion, i.e. 40% of global population, but require only 12% of the total farmland used in the production of primary crops. Sustaining high productivity requires inputs to replace extractions and losses so discussion on sustainability revolves around security of supply of water, nutrients, and energy to sustain productivity. High-yield systems must be managed for efficient use of nutrients to avoid losses that can have a negative impact on water quality through eutrophication of ground and surface water resources.

Productivity is determined by structure and function of managed plant communities and access to resources that support their growth.

The primary process of plant growth and the basis for crop productivity is photosynthesis (Chapter 10). Solar energy allows capture of carbon dioxide (CO_2) that diffuses into foliage from the atmosphere with hydrogen (H) split from water (H_2O) acquired by roots. The resulting carbon skeletons (C–C–C) are used for synthesis of plant structures and for energy to sustain metabolism. Plants have evolved for efficient capture of solar energy and transport of CO_2 to sites of photosynthetic fixation within leaves. The evolutionary sequence from aquatic ancestors has, however, left plants with a significant challenge that limits productivity in most terrestrial environments. Absorption of CO_2 is inevitably associated with simultaneous loss of H_2O (transpiration) through open pores (stomates) on leaf surfaces to a drier atmosphere outside (Chapter 9). Plant metabolism operates in an aqueous medium within cells, which means growth requires that water lost by transpiration be replaced by uptake from soil via roots. The major mechanism for control of water loss and maintenance of internal water status is stomatal closure but that must always reduce CO_2 uptake and the efficiency with which intercepted radiation is utilized in photosynthesis.

So how have plants evolved to deal with this challenge? The most effective evolutionary solutions include development of alternative photosynthetic pathways, a wide range of leaf shapes, and metabolic processes that withstand internal water deficit and damage to photosynthetic pigments as discussed in Chapters 9 and 10. Productive crops are those that quickly establish foliage cover to intercept all incident solar energy and use it most efficiently to drive photosynthesis throughout the life cycle. This requires open stomatal pathways and adequate supplies of water to maintain internal plant water status. Water uptake from soil also provides the medium for transport of (essential) nutrients required for many metabolic pathways and constructed compounds. Of these nutrients, nitrogen (N) is especially important because amounts in plant biomass are exceeded only by carbon (C). Nitrogen also is a critical building block of proteins such as those that constitute the photosynthetic apparatus of leaves (Chapter 10). Legumes that can fix atmospheric nitrogen (N_2) by symbiotic association with certain bacteria provide most N in the natural biosphere. They are important components of farming systems for nutritive value as well as N fixation that contributes to productivity of non-leguminous crops (Chapter 8).

Optimizing efficiency of resource use in photosynthesis and yield formation concerns C assimilation to sugars while biomass formation results from subsequent metabolic steps required to synthesize structure and storage products, most frequently in seed (Chapter 11). Large increases in crop yield have been achieved by breeding improved cultivars (Chapter 4) and better crop and soil management practices (Chapters 12, 13, 14). In the 41-year period from 1966 to 2006, the yield of the three major cereal crops (rice, wheat, and corn) increased at a relatively constant annual rate of 52 kg ha^{-1}. Average yield is now about 3900 kg ha^{-1} and the relative rate of yield increase has fallen from 3.0 to 1.3% y^{-1}. Continually increasing investment in plant breeding has been required to obtain that result because gains are increasingly difficult to sustain as yields approach the biological limit. A challenge for students who seek a career in crop and soil science, or in farming itself, is to continue to raise yields while also protecting environmental quality through improved crop and soil management, a process called **ecological intensification** (Cassman 1999).

Defining yield The concepts of actual, attainable, and potential yields assist in assessments of farming systems and help to identify opportunities for improved productivity. For individual crop species, **actual yield** is that currently achieved (average) at any location while **attainable yield** is that possible by use of best technology with the most adapted cultivar. Record yields, those of the best farmers, and crop yields at research stations when grown with management practices to give highest possible yields, provide estimates of attainable yields. For individual farmers, the **exploitable yield gap** is the difference between the actual and the attainable yields for their environment. In practice, farmers do not necessarily aim to close this gap completely, however, because diminishing returns to inputs and management typically reduce economic returns as yields approach the ceiling. Rather, farmers seek an **optimum yield** that provides the best return to labor and investment. With competition for other uses for land, labor, and capital, and other regional and global sources of commodities, farmers learn that uneconomic practices are unsustainable in the long run. **Potential yield**, of a species, is the highest attainable yield achieved over years in its most favorable environment. Potential yields help define the genetic capability of crops and progress in breeding for greater yield but are not, by definition, available to all environments in which the crop is produced.

Animals are also an integral component of many farming systems. They transform forages and crop residues for human use and consumption and play a stabilizing/buffer role in many farming systems (Chapters 2 and 16). Crop–pasture rotations that sustain animals in farming areas play an important role in control of pests and diseases. With legumes included, pastures are an important source of N for crops in rotation with pastures because grazing animals remove only a small part of the nutrients they consume; the remainder is cycled in available form for pasture growth through urine and feces. Non-ruminant animals (hogs, chickens, etc.) rely for nutrition on grain products that can be used for human food. Greater proportions of meat in diets, therefore, increase demands on crop production (Chapter 2).

Efficient use of scarce resources is achieved through strategic decisions about cropping patterns and choice of crop and management practices.

Regions of good soil and favorable climate are able to produce crops more economically than other regions. Farmers in those regions concentrate on crops that provide greatest return to land, labor, and capital at an acceptable level of risk. In less-favored regions farmers typically choose lower value crops with less intensive and less costly systems of management. On a global basis, most farming is carried out in areas where productivity

Table 1.2 Some indices for analyses of efficiency of resource use by crops

Acronym	Name	Meaning	Units
R_i	Fractional interception of solar radiation	Proportion of incident radiation intercepted by a crop	$MJ\ MJ^{-1}$
RUE	Radiation-use efficiency	Biomass produced by unit intercepted radiation	$g\ MJ^{-1}$
HI	Harvest index	Ratio of grain mass/aboveground crop mass	$kg\ kg^{-1}$
DE	Digestible energy	Digestible energy content of plant food – different for ruminants and monogastrics	$MJ\ MJ^{-1}$
IE	Irrigation efficiency	Biomass production, or yield, per unit irrigation water applied	$kg\ ha^{-1}\ mm^{-1}$
WUE	Water-use efficiency	Biomass production, or yield, per unit evapotranspiration	$kg\ ha^{-1}\ mm^{-1}$
TE	Transpiration efficiency	Biomass production, or yield, per unit water transpired by a crop	$kg\ ha^{-1}\ mm^{-1}$
NUE	Nitrogen-use efficiency	Biomass production per unit of N available to the crop from indigenous soil resources and from applied fertilizer, manure, or compost	$kg\ kg^{-1}$
NIE	Nitrogen input efficiency	Biomass production per unit of applied N	$kg\ kg^{-1}$
NRE	Recovery efficiency from applied N	Amount of N in crop biomass that originated in applied N	$kg\ kg^{-1}$
NCE	Nitrogen conversion efficiency	Biomass production per unit N absorbed by crop that originated in applied N	$kg\ kg^{-1}$
LER	Land equivalent ratio	Yield of a mixed species crop (intercrop) relative to component single species crops	$kg\ kg^{-1}$
NER	Net energy ratio	Energy content of crop or crop product (biofuel) relative to energy used in crop production	$MJ\ MJ^{-1}$
CI	Carbon intensity	Greenhouse gas emissions (expressed in CO_2-equivalents) per unit energy in biofuel produced	$g\ CO_2e\ MJ^{-1}$
TE	Trophic efficiency	Ratio of production at one trophic level to food ingested from the adjacent higher level	$MJ\ MJ^{-1}$
GM	Gross margin	Financial return per unit area less variable costs of production	$\$\ ha^{-1}$

is restricted by unfavorable soils and/or weather. While crops can be chosen to minimize deleterious effects of extremes in temperature regimes, inadequate rainfall and poor rainfall distribution represent the greatest constraint to productivity, which can be corrected by access to irrigation in only a small part of total farmed area (Chapter 14). If the current relative rate of yield increase ($\sim$1.3% y^{-1}, Box 4.4) could be maintained at a compound rate to 2050, then production on current farmed area would rise by 80%, sufficient to feed the projected increase in population. Previous experience suggests this is unlikely, however, in turn highlighting a critical need for more efficient use of scarce resources, especially water and nutrients.

Measuring performance Various indices are used to measure efficiency of resource use in farming systems to compare effects of site, crop, cultivar, and management. Some indices associated with limiting resources in crop production systems are introduced in Table 1.2 to emphasize the scope of analyses in crop ecology. The term efficiency is strictly applied to unit-less comparisons (MJ MJ^{-1}, kg kg^{-1}, mm mm^{-1}, etc.) but, in this book, we apply the term more generally, consistent with most existing literature.

Indices find use individually but also in various combinations. For example, under optimal conditions of water and nutrients, crop productivity (crop growth rate, CGR, kg ha^{-1} d^{-1}) can be analyzed in terms of intercepted radiation and radiation-use efficiency according to:

$$CGR = (\text{incident radiation, MJ m}^{-2}\,d^{-1})\,(R_i)\,(RUE) \qquad (1.1)$$

Other combinations provide integrative measures of component efficiencies as in:

$$NIE = NRE \times NCE \qquad (1.2)$$

explaining that N input efficiency (NIE) is the product of N recovery efficiency (NRE) and N conversion efficiency (NCE), emphasizing that both NRE and NCE must be quantified to understand how crop and soil management practices affect the response to applied N.

The combination of high interception of radiation and adequate supply of N are recurring themes in this book. Both are required for high crop yields and N is a highly mobile element in cropping systems with potential to pollute adjacent ecosystems and water resources. Careful measurements and comprehensive analyses are required to establish sustainable management practices. Nitrogen fixation in cultivated croplands, including pastures and grazing land, is estimated at 50 to 70 Mt y^{-1} (Herridge *et al.* 2008) but high production by non-legume crops requires more N than legumes can provide. Consequently, N fertilizer produced industrially from atmospheric N_2 by the **Haber process**, provides the majority of N required to produce enough food for a large and expanding world population (Smil 2004). Current world use of N fertilizer is 100 Mt N y^{-1} with the greatest proportion used in developing countries: North America and Europe together use 29% versus 62% in Asia (FAOSTAT 2007). Sub-Saharan Africa is outstandingly the smallest user, 1.5%, despite its large population (800 million) and agricultural area.

Strategies and tactics Efficient use of scarce resources is achieved by enterprise selection (the **strategy**) and management practices (the **tactics**) that adjust the strategy to variations in weather, soil resources, and economic conditions. Given possibilities for modifying environment by tillage, drainage, fertilization, and other means, farmers seek practices that optimize yield, stability, and efficient use of scarce resources for individual fields. Commonly, however, what is grown is not the most productive, risk averse, or efficient, even when information and technology are available to achieve them. Subsistence farmers, for example, who consume all they produce, must pay attention to traditional foods of their culture and derive an adequate diet for a year-round supply of food. All farmers must view the whole farm as an integrated activity, considering factors such as labor and capital requirements of particular crops, need for rotation of crops, best use of marginal lands, and feed requirements for animals. Considerable variation also arises because farmers differ in personal objectives and initiative as well as in technological and business skills. Likewise, both farming and its commodity markets are subject to economic and political controls imposed by governments. Farmers in both developing and developed countries are increasingly dependent on work off-farm to supplement household income, leaving less time for optimal crop management. When decisions about what will be grown are influenced by such factors, economic and agricultural efficiencies are given less attention.

Market access plus climate and soil cause farming in most regions to focus on a limited number of crop species and types of pasture. In developed countries, attempts at achieving high diversity on individual farms commonly lead to mediocrity and failure because high diversity requires more information, management skill, and equipment than can be managed by individual farmers. California, with its great diversity of horticultural crops, might seem an exception. With irrigation and mild climate, California farmers can achieve high yields and high quality for a number of high-value vegetable, fruit, and nut crops. They supply over 50% of fresh and processed fruits and vegetables consumed in the USA despite large distances to principal markets. In this case, diversity of crops is regional rather than on individual farms because of risks associated with inelastic markets for fresh produce and the need for considerable labor and special machinery. In contrast, a high diversity of crops per farm is more common in densely populated regions of India and Japan where household gardens, horticultural production, field crops, and pastures become geographically intertwined on small farms. In those places, household needs, land scarcity, and advantage of an even distribution of labor demand over the year are important in favoring diversity.

Long-term sustainability of farming systems depends on the maintenance of chemical, physical, and biological properties of soils.

History documents a number of civilizations that have fallen from dominance because their agriculture could no longer provision large urban communities. Examples are, the Mesopotamia cultures that thrived between the Tigris and Euphrates Rivers (3500–300 BC) in what is now Iraq, and the Phoenicians who lived along the Mediterranean coast (2000–500 BC) in what is now Lebanon. Deterioration of soil quality to support crop production caused by salinization in Mesopotamia and erosion in Phoenicia caused the declines.

Soil properties that determine suitability to support crop growth, and water supply, are responsible for much of the variation in crop performance within and between fields, regions, and agro-ecological zones. Most plants grow best on deep, well drained soils of medium texture. Flat land or land with subdued topography is best suited to interventions of farming. Risk of soil loss to erosion is greatly reduced in level fields, and deep soil profiles provide a large capacity to hold water and nutrients to support crop growth and biological activity that helps maintain favorable physical and chemical soil conditions. All soils do not have such properties (Chapter 7). Soils of light texture (sands) hold little water or nutrients while heavy textured soils (clays) of higher fertility drain slowly with resulting problems of waterlogging. Shallow soils of any texture have limitations to water- and nutrient-holding capacity. Root growth and function can be reduced where soils are too acidic, alkaline, or saline.

Improving soils Soils can, however, be improved by farming. Additions of lime and macro- and micronutrients increase crop growth and accumulation of organic matter, which, in turn, can improve physical and chemical properties of soil (Chapter 12). Application of manure or compost provides nutrients and organic matter. Use of legume cover crops protects soils from erosion and contributes biologically fixed N for a subsequent non-legume crop. Tillage, which may involve plowing or disking operations, can have positive or negative effects on soil quality and crop production depending on the situation. Primary tillage distributes organic matter and nutrients more evenly within topsoil and can improve infiltration in some situations but more generally tends to reduce soil structure, especially if undertaken when soils are either too wet or too dry.

Development of herbicides has facilitated crop production with less tillage and erosion to the benefit of increased soil water storage, and reduced energy use. Production with little or no tillage, variously called **minimum tillage**, **zero tillage**, **conservation tillage**, has expanded greatly in recent years and is now widely practiced in wheat, maize, and soybean systems of North America, South America, and Australia. In the semi-arid US Great Plains, greater snow capture, less runoff and reduced evaporation from soil with no-till compared to disking allows intensification from a fallow–wheat system (one crop every two years) to two crops in three years in a fallow–wheat–sorghum or maize system.

Farming is inevitably an extractive activity with amounts of nutrients removed depending upon quantity and chemical composition of harvested products. Over time, extraction reduces soil fertility and can cause acidification, in turn reducing suitability to support future crop production. In contrast, grazing livestock return most nutrients they consume in feces but harvesting pasture or forage crops for hay or silage removes entire nutrient contents of aboveground biomass. Crops harvested for grain remove a greater proportion of nutrient content than the mass ratio (harvest index, HI) of yield to aboveground biomass because most nutrients are concentrated in grain. Some crops, e.g., sugarcane and biomass crops for energy (Chapter 15), are harvested completely, causing major extraction. Such extraction cannot continue indefinitely without replacement of essential nutrients as soil reserves become deficient. The amount of nutrient replacement required depends on the soil properties, nutrient requirements of crops grown, yield targets, and costs of nutrient inputs relative to value of the harvested output.

Evolution of farming methods Concern for availability, costs, and environmental impacts of commercial fertilizers leads to use of legume rotations and cover crops, manure, and compost to supply N and other nutrients. This is done by emphasizing potential cost savings and avoidance of environmental hazards that can result from inappropriate use of fertilizer. "Organic" production methods are even more prescriptive and exclude use of manufactured fertilizer, most pesticides and fungicides, and more recently transgenic crop cultivars. Legume–cereal rotations of the eighteenth century provide the model for these organic farming practices that are often now promoted without appreciation of their low productivity (Chapter 8). That individual crops can be grown to similar yields by organic methods as with "chemical" fertilizer is unquestionable. The problems are that additional time is required for rotations with legume cover crops, and that production of organic manures and compost require additional land, water, and nutrients. When that opportunity time and land are included in assessment of yields, organic systems are inevitably much less productive on a land area–time basis than those that use fertilizers.

Modern methods of **integrated farming** combine biological cycles for nutrient, weed, pest, and disease management with tactical use of fertilizers and other agrichemicals. Sustainable, efficient production depends on careful monitoring of soil conditions and requirements and use of water and nutrients. It also requires limiting losses of nutrients and applied chemicals to minimize negative impacts on quality of ground and surface waters, and emission of nitrous oxide (a potent greenhouse gas). Protocols of good agricultural practice (GAP) are applied and certification schemes are increasingly available for both product safety and ecological integrity of production systems (e.g., www.globalgap.org). These methods are essential to feed the human populations, while maintaining soil and water quality, and preserving as much land as possible out of agriculture for conservation of nature and water resources.

Farming systems will continue to evolve in response to changes in markets, technology, and climate. Growth of population and income, and scarcity of water, will be major driving forces in the future.

Historical analyses reveal that farming systems undergo continuous change in response to demographics, economic and environmental forces, and availability and adaptability of new technology (Chapter 16). Consumer demand shapes, and at the same time is shaped, by these forces that can be analysed in terms of food quantity and preference.

Markets   Rapidly increasing world population is a major force for change. It requires greater productivity of farmed land, which, in turn, motivates new, more intensive systems of production: fewer fallows, more crops per year, and more irrigated cropland.

Consumers now demand safe food and will pay premiums for quality, novelty, and, increasingly, for products grown by methods that are promoted as environmentally sound. To many consumers food safety relates to hazards of contamination by agrichemical (biocide) residues during production, and by microorganisms that cause disease (*E. coli*, salmonella, etc.). In reality, biological, chemical, and physical contaminants may occur throughout the production and market distribution process. Supermarkets, increasingly the major source of food products, and almost the exclusive source of imported products, seek protection from litigation for sale of unsafe food by adoption of monitoring and tracking systems from field to point of sale to consumers. Farmers seeking access to these markets require certification, which requires adherence to accepted production protocols and laboratory testing of products. Various "organic" labels rely on rules that prohibit the use of fertilizers and agrichemicals under the belief that omitting these types of inputs ensures food safety. Conventional farmers meet government safety standards by following other protocols that restrict type, quantity, and timing of agrichemical applications.

Preferred diets of consumers also change in response to availability of food types and consumer purchasing capacity. Consumption of livestock products, including meat and dairy, increases as incomes rise. Trends are similar across continents and cultures. Those have driven expansion of agricultural area and changes in cropping systems in major exporting countries such as the USA, Brazil, Argentina, and Australia. FAO production statistics (FAO 2008) reveal that Brazil has greatly increased soybean production to 60 Mt on 22 Mha and is now the second largest producer to the USA (81 Mt on 30 Mha). About 45% of Brazilian soybean production is exported. Most is destined to feed animals in Europe and China. Increasing affluence in developing countries also results in greater wine and beer consumption and more diverse diets that include a multitude of vegetables, fruits, and nuts, increasing the demand for those crops as well.

Energy During the first half of the twentieth century non-human, motive power on farms was supplied most by draft animals. Substantial areas of farmland were cropped for animal fodder. Since the 1930s, tractors have replaced draft animals on US farms releasing 40 Mha, previously used for forage crops, for grain and legume crop production. In the wheat belt of southern Australia, when the common rotation was fallow–wheat–oats, the fallow period stored water and mineralized N to increase the yield of wheat for human consumption and export, while oats supplemented pasture for working horses. Replacement of horses by tractors allowed large expansion in cropping area and the resultant increase in production continued with the use of fertilizers and legume rotations to replace fallow (Chapter 16). By the late twentieth century, there was concern over the extent of **support energy** required to mechanize agriculture (Chapter 15), seen by some as contributing to higher fuel costs, and by others as adding risk to human food supply by making it reliant on energy subsidies from external sources. More recently, with the rise in petroleum prices, agriculture is being asked to provide not only food, but also energy in the form of liquid biofuels for transport. Agriculture has always been a major producer of energy, and even now, well into the fossil fuel era, together with forestry still contributes 10% of world energy, mostly by direct combustion of biomass, and most importantly in developing countries. In fact, on a global basis biomass provides more energy than either hydroelectricity or nuclear power. Energetics and the energy values of crop components for food and fuel are discussed in Chapters 2 and 15.

Today, large-scale production of liquid biofuels from sugar, starch, and oil crops are competing with food production for land, water and nutrients (Chapter 15). The component issues bring together production and environmental issues that go the heart of crop ecology and require serious study. A current challenge is to analyze the potential for development of fully integrated systems for efficient food and energy production and reduced emission of greenhouse gases (Chapters 6 and 15).

Technology Farmers are the source of many innovations related to production, but ideas also flow freely from other sources. In modern times, major scientific breakthroughs from research conducted at universities, agricultural research institutions, and private companies contributed many landmark technologies including hybrid seed, semi-dwarf wheat and rice, conservation tillage, soil and plant testing to determine nutrient input requirements, and transgenic crop cultivars with resistance to certain insect pests and herbicides. Diffusion of new technology is sometimes very rapid (short-stature wheat and rice, herbicides, transgenic crops) and sometimes very slow (combine harvesters, zero tillage, integrated pest management). Two time-constants constrain changes on individual farms: the lifetime of machinery and tools (5 to 10 y) and the working lifetime of farmers as decision-makers (*c.* 25 y). These responses are randomized among farms, however, and so change within a region tends to be evolutionary with time (Chapters 16 and 17). For a given technology, the adoption pathway typically begins with a relatively small number of progressive farmers. Rate of adoption is largely determined by education required to implement the new practice and the economic return from it.

Increased use of inorganic fertilizers is perhaps the most significant technological change because renewable organic sources of nutrients are not sufficient to support food production adequate for the world's current population (Chapter 8). There is evidence that farmers in developed countries view fertilizer application as a risk-avoidance measure and, in some cases, tend to overfertilize as a result (Legg *et al.* 1989). Because the cost of fertilizer is tightly linked to the price of energy, expectations for continued increases in fossil fuel prices will motivate continued technological innovation to improve fertilizer use efficiency, with ensuing environmental benefits.

Transgenic cultivars (Chapter 4) are a recent innovation that has been adopted rapidly. They were used on over 123 Mha worldwide in 2008, 15 years after their first release in the mid-1990s. Two types account for nearly all of this area: (1) herbicide resistance to a highly effective, non-selective herbicide (glyphosphate), and (2) resistance to insect pests of the Order Lepidoptera from a class of δ-endotoxin proteins (called Cry toxins) that kill insects ingesting them (Chapter 4). Widespread adoption of transgenic maize and cotton cultivars has greatly reduced the use of biocides while the use of herbicide resistance in soybean and canola facilitates the adoption of conservation tillage. Conservation tillage systems increase soil moisture storage, which increases stability in rainfed cropping, and also reduce energy use, decrease soil erosion, and improve quality of topsoil (Chapter 12).

Recent technological advances in remote sensing of plant status and soil properties promise new, cost-effective approaches for site-specific management to accommodate within-field variability in soil properties (Box 12.1). These differences cause non-uniformity in crop nutrient status and irrigation requirements. Both real-time and long-term historical weather data, including solar radiation, are becoming increasingly accessible via the internet. Access to historical weather records allows use of crop simulation models to evaluate impacts on yield and yield stability from combinations of management practices (e.g. planting data, plant population, and maturity). Access to real-time weather data allows development of decision-support tools that provide farmers with information to guide tactical decisions about the timing and amount of N topdressings or irrigation (Chapters 13 and 14).

While all these innovations tend to increase stability (reduce yield variation), each can also serve to decrease it. Experience in developing countries seems to be that yield stability at first decreases with increasing use of fertilizer and modern cultivars and then increases as growers become more experienced and inputs are brought to optimal levels. Drainage for control of excess moisture in wet regions (Chapter 12) and for salinity control in dry regions (Chapters 13, 14, 16) has major beneficial effects in increasing both yield and stability. Other important changes have also come through plant breeding, irrigation techniques, and use of biocides. Both seed treatment with fungicides, and especially weed control with herbicides, have major benefits in stability. In addition to advances in attainable yield and harvest index from use of hybrids and semi-dwarf stature, plant breeding has contributed greater insect and disease resistance. With increasing farm and field size, bigger, more powerful machinery has contributed to improved quality, uniformity, and timeliness of tillage and harvest with major improvements in stability as a result. On the other hand, machinery has also permitted farming in more risky environments, leading to cropping systems that have inherently low stability, and cultivars that take advantage of fluctuations in resource supplies are less stable than unresponsive lines.

Climate It is very difficult to determine if a weather event is a short-term aberration – the 100-year drought or flood, for example – or a sign of a more permanent change. Most cropping practices are conservative in the sense that they accommodate significant year-to-year variation in weather while minimizing downside risks. If a weather event is just an aberration, farmers can stay with proven methods, but if the change is permanent or part of a longer term trend, then different crops and cultural practices may be needed.

Climate has changed in the past as demonstrated by large changes in sea level that have occurred over recent millennia as polar ice expanded and contracted. Natural fluctuations in Sun–Earth geometry and solar activity can change climates significantly, effects that operate slowly over hundreds and thousands of years. Agriculture has adapted to climate change in the past. Iceland and Greenland were successfully colonized one thousand years ago when the climate was more amenable to agriculture than at present. More recent records reveal reduced production and changes to cropping practice when Europe experienced a Little Ice Age in the 1500s.

The world is currently in an interglacial phase and, on the basis of historical evidence, a long-term cooling trend would be predicted. Now, however, there is concern that agriculture may soon face substantial limitations to agricultural productivity due to global warming and associated effects on rainfall patterns (IPCC-AR4 2007). The putative cause is rising atmospheric concentrations of greenhouse gases (GHG), principally CO_2, caused by combustion of fossil fuels and land-use change (Chapter 6). Models predict increases in temperature in the range of 1 to 2°C by 2050, variously distributed around the globe, and accompanied by a redistribution of rainfall. Such a scenario would require changes in optimal cropping systems for each region. For annual crops, simple adjustments in sowing date or cultivar may suffice to place a crop in a suitable environment. Some crops might be displaced geographically to higher altitudes or latitudes.

Prognoses of productivity are complicated and uncertainties in predictions of temperature and the fact that rising $[CO_2]$ will have a generally beneficial effect on crop growth (Chapter 10) and water-use efficiency (Chapter 9). Temperature effects, on the other hand, will be positive or negative depending upon existing temperature regime (cold or hot). Climate models also predict greater variability in temperature and rainfall due to **climate change**, which would decrease stability of production. How agriculture might respond to actual climate change depends very much on how rapidly the change occurs and its magnitude. Given uncertainties associated with climate change predictions, it is not surprising that these issues are now attracting considerable attention in the global agricultural research community.

1.3   Maintenance of agricultural systems

Agricultural systems depend on human effort for maintenance of their essential biological, soil, and human resources. Off-farm, maintenance takes form in provision of drainage and surface irrigation infrastructure, machinery, fertilizers, biocides, scientific and technological research, and education of farmers. On-farm, principal concerns are for maintenance of soil fertility and structure, weed control, protection from erosion, waterlogging in humid areas, and salinization in semi-arid regions.

Need for scientific effort is seen clearly with diseases and insects. Weaknesses in resistance of existing cultivars and continued evolution of diseases and insects necessitate major efforts in plant breeding just to maintain a status quo (Chapter 17). Similar problems occur when weed species develop tolerance to current herbicides. Competent diagnoses and correction are also needed in dealing with inevitable changes in soils and soil fertility. Much of the world's cropland is still new to farming, and nutrient deficiencies and imbalances will increasingly appear as original supplies are exhausted. Staying ahead of continuous change in pests and soil nutrient status and quality requires substantial research effort. This **maintenance research** also involves development of technologies to ensure that current crop and soil management practices avoid negative impact on water quality and wildlife.

Maintenance of human resources is perhaps less obvious. Skill levels of young farmers are developed mainly in traditional ways through apprenticeship with parents and others, but education, particularly basic literacy to facilitate continued learning, is essential (Ruttan 1982). Business affairs and agricultural technology are increasingly complex, such that advanced education is increasingly beneficial for success in farming.

A significant portion of benefits from research and education accrue to society generally, through lower cost and safer supplies of food, and through assurance of future supplies. Recognition of those relationships has led to government-sponsored programs including education, research, and extension in most countries. In developed nations, those commitments have weakened in the face of surplus food production capacity and declines in agrarian populations. There is now little support for maintenance research in many countries and the future impact of this trend remains to be seen. Real professional competence in breeding, agronomy, pathology, and entomology requires years of experience. Plant breeders, for example, must become familiar with a wide range of germplasm. Without continuing activity in these fields, society would lack the technological skills and human resources necessary for response to future food supply crises.

Government policy can affect viability of agriculture in other ways. A farmer's sense of land stewardship, no matter how strong, can be subverted by social rules, such as price controls that place returns below costs, subsidies that distort markets, or taxes that place farming at a disadvantage to other economic sectors. Special problems occur with maintenance activities such as terracing for erosion control. There, present worth of future benefits that accrue during a farmer's lifetime is generally less than costs. Continued farming without preventive practices may allow erosion at greater than tolerable rates, degrading land's production potential. Off-site consequences of erosion include silting of lakes, reservoirs, and streams that are common resources for all. Some governments share costs with farmers for such conservation projects in recognition of both farmers' responsibility for externalities and society's benefit from future productivity and protection of the commons.

Finally, issues with the "**commons**" also arise from off-farm flows of nutrients and agricultural chemicals, and with atmospheric pollution by wind-blown soil and smoke from burning agricultural residues. Some of these issues are examined in later chapters. Difficulties with the commons take a different form in pastoral cultures of Africa and Asia where animals owned by individuals graze common pastures. In these systems, there are no boundaries to delineate individual shares of the common resource and an individual's share depends upon the number of animals owned. Inevitably, commons are overgrazed to the detriment of all. Similarly, there are numerous cases where unconstrained aquifers, a common resource, are overappropriated by too many irrigation wells that cause water tables to fall, requiring all farmers to use more fuel to pump from greater depths and, in the extreme, to abandon irrigation. Regulations on the number of wells and amount of withdrawal are required to maintain such irrigated systems.

1.4   Review of key concepts

Production systems

- Agriculture can be viewed as a hierarchy of systems beginning with individual fields (cropping systems), extending to their integration into farms (farming systems) and farms into regions (agricultural systems). Biological, environmental, and social interactions and issues can be identified at each level in the hierarchy.
- Agricultural systems interact strongly with the larger society that consumes their products and in turn provides goods and services needed by farmers to produce food. Demand for products depends upon the size of population and its wealth. As a result, how and where farming is conducted are determined as much by economic issues of prices of inputs, including labor and land, as by productivity of crops and animals, and their impact on the environment.

- Six key attributes of farming systems are productivity, stability, sustainability, autonomy, equitability, and sufficiency. Metrics for each of these attributes allow evaluation of system performance over time.

Productivity, efficiency, and sustainability

- Productivity is the basic ecological attribute of agricultural systems. It determines efficiency of resource use and economic viability of farming. It also establishes **carrying capacity** for domestic animals and humans per unit area of agricultural land, in turn determining minimum areas that must be devoted to agriculture.
- Climate and soil determine what crops can be grown in a farming region but crops actually grown and intensity of production depend upon size of markets and upon availability of labor and other means of production.
- Time trends in productivity serve in defining stability (variation in yield) and sustainability (maintenance of yield over time). Sufficiency of supply, autonomy, and equitability determine food security and well-being at local, regional, and national levels.

Future challenges

- Agricultural systems are subject to continuing evolution in response to the development of knowledge, available technologies, and the amount and type of products that society requires. The future holds numerous uncertainties. Availability of energy and nutrients, water for irrigation, restrictions on use of biocides and antibiotics, and climate change are key factors, but increasing demand due to population and income growth will likely have greatest impact. Future stability of farming will depend greatly on these factors and on the ability of farmers and supporting services to respond.
- Continuing education of farmers and consumers, and support for competent research programs in plant breeding, entomology, plant pathology, soils and agronomy, are examples of services needed to maintain viable agriculture.

Terms to remember: autonomy; the commons; climate change; cropping, farming, and agricultural systems; ecological intensification; exploitable yield gap; equitability; greenhouse gases (GHG); non-uniformity principle; organic and integrated agriculture; productivity; stability; support energy; sustainability; yields, actual, attainable, optimum, and potential.

2 Trophic chains

Plants provide all energy for maintenance, growth, reproduction, and locomotion of every living organism on our planet. That energy, originating from the Sun, flows from plants through a web of herbivores, carnivores, and decomposers. This trophic chain – "who eats whom" – gradually returns carrier CO_2 molecules to the atmosphere. Fires, occurring naturally from lightening strikes, or provoked by human activities, are a more sudden, but chemically similar, release of solar energy accumulated by plants.

Humans and some other animals also use plant material (biomass) for construction but humans alone have combusted them under controlled conditions to provide heat for warmth, cooking, and both stationary power and traction. Once, animals were the only source of traction and, in the eighteenth century, consumed as much as one third of agricultural production. Biomass accumulated by plants during previous geological periods formed coal and oil (fossil fuels) that have driven the development of transportation, agriculture, and industry during recent centuries.

Agricultural systems have developed predominantly to provide food for humans in plant and animal products, but they also provide fiber and fuel. This chapter describes the chemical and energetic content of plant products and explains their relationship to nutritive value and carrying capacity of land for animals used in agriculture and for humans. Questions of energy use in agriculture and its potential to supply a greater proportion of society's demand for non-dietary energy, including the current focus on biofuel, are discussed further in Chapter 15.

2.1 Plant production

Growth of crops and pastures depends on the synthesis of organic compounds from CO_2 absorbed from the atmosphere and on inorganic nutrients and water absorbed from soil. The major component of biomass is carbon (C), which comprises 40 to 45% of plant dry matter. The major chemical compound of plant biomass is a carbohydrate, cellulose $[(CH_2O)_n]$, but many other more complex compounds are found in the structures of living cells.

The initial products are mainly simple sugars. These serve as chemical building blocks and, through respiration, provide energy needed for biosynthesis of new compounds and for maintenance of existing tissues. A simple scheme of the flow of energy and carbon in these processes is presented in Fig. 2.1. The format follows conventions used for

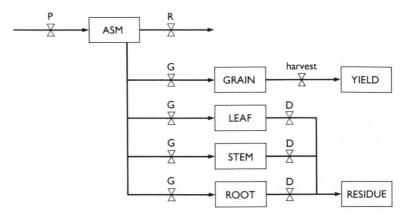

Fig. 2.1 Scheme for the path of carbon and energy in production of an annual grain crop. Gate symbols represent processes: **P** for photosynthetic production of new **AS**si**M**ilates in leaves; **R** for respiration related to maintenance, biosyntheses, and growth; **G** for growth of various plant parts; and **D** for the death of parts.

"state-variable models" (Forrester 1961) and is easily expanded with additional detail on control variables. The term **partitioning** describes the distribution of new assimilates to growth of various plant parts and to respiration. Total dry mass accumulated by a crop is **biomass**, and that portion useful to humans directly or indirectly as animal feed is **economic yield**, or just yield. The fraction yield/total plant mass is termed the **coefficient of economic yield**. It is difficult or impossible to measure the mass of crop roots in many cases, however, and **harvest index (HI)**, denoting useful fractions of aboveground biomass, is common in agronomic studies.

2.1.1 Composition of plant material

The usefulness of plant materials to humans and animals depends on their composition and on energy stored in the chemical bonds of their organic constituents. The cellular nature of plants dictates that a degree of **stoichiometry** must always exist among proportions of basic biochemical substances found in **structural** components of biomass. A rigid cellulosic wall encloses each cell and its dimensions are fixed once its growth ceases. Because growth involves an increase in size, mass, or number of cells, wall materials including cellulose, hemicellulose, pectin, and lignin normally constitute 70 to 80% of organic matter in vegetative parts of plants. The balance is composed mainly of proteins and nucleic acids found in the nucleus, various organelles, and cytoplasm, and of lipids that form membranes around those compartments.

Plant cells also contain varying amounts of **non-structural** material. Assimilates such as starch, sugars, and free amino acids not incorporated into walls or protoplasm are classified as non-structural materials. Temporary accumulations of non-structural material can usually be mobilized and used elsewhere in the plant. Such accumulations are sometimes referred to as "reserves" but one needs to be careful about teleological implications. Plants do not accumulate non-structural material with "intent"; rather,

Table 2.1 A scheme for analysis of the major biochemical classes in dry plant material. This approach borrows from the proximate scheme and from the neutral detergent fiber (NDF) scheme of Van Soest (1982)

Class	Procedure	Major components	Comment
Nitrogenous compounds	Kjeldahl N × 6.25	Proteins, amino acids, nucleic acids	6.25 converts to crude protein
Lipids	Extraction with petroleum ether	Fats, oils, waxes	
Wall material	Residue after boiling with neutral detergent	Cellulose, hemicellulose, lignin, wall protein	Wall protein; may deserve further analysis
Non-structural carbohydrates	Extraction with ethanol, followed by gas or liquid chromatography	Sugars (nitrogen free extract, NFE)	Treatment with amylase for starch analysis
Organic acids	As for non-structural carbohydrates	Organic acids	
Minerals	Ignition at 500 to 600°C	Mineral oxides	Actual minerals = c. 0.6 × ash
Remainder	By difference	Pectin, tannin	May deserve specific analysis

possession of the trait has conferred advantage in evolution. Storage of non-structural materials in fruits and specialized storage organs such as tubers is more permanent and those materials are generally available only to the next generation. In contrast to moderate concentrations of non-structural materials found in vegetative tissues, starch, oil, protein, or sugar can account for up to 75% of dry mass of storage organs and seeds.

The composition of biomass is constrained by the inclusion of structural components, yet it can vary over a considerable range depending on the types and amounts of various tissues involved, on cell size (which determines surface:volume ratio and thus amount of wall material), and on type and amount of non-structural material. In leaves, chloroplasts add a large amount of protein to the structural component; the principal carboxylase enzyme of photosynthesis accounts for about 25% of leaf protein in many plants. Xylem tissues on the other hand contain large amounts of specially differentiated vessel and tracheid elements with lignified wall thickenings. Those cells are dead and thus nearly free of nucleic acid, protein, and lipid. Vascular strands and sclerenchyma, another tissue in which lignified walls are common, contribute stiffness that aids the display of stems and leaves and hence the plant's ability to intercept light.

Methods of analysis The composition of biomass in terms of major biochemical classes is important information for ecological studies. The need of animal scientists for a better understanding of animal rations was a major force behind the development of simple methods for analysis of plant material. By 1900, agricultural chemists and animal feeders had established that energy (carbohydrates and lipids) and nitrogenous compounds (e.g., protein) were the main factors in animal nutrition, and **proximate analyses** involving solvent extractions had emerged as a standard method for distinguishing broad biochemical classes. Extensive tables of feed analyses constructed by that method (e.g., National Research Council 1982) are useful in ecological work.

To use such feed composition tables, one must know something about the methods of analysis. The steps are summarized in Table 2.1. The fraction soluble in petroleum ether (oils, fats, and waxes) is taken as **lipid**. Kjeldahl analysis is used to measure total nitrogen (N). Proteins and amino acids comprise most of that (nucleic acids in DNA are a very small component) and total N × 6.25 is taken as an estimate of **crude protein** (Table 2.1). The factor 6.25 comes from the observation that 0.16 ± 0.01 of the mass of plant proteins is N and $1/0.16 = 6.25$. Smaller, species-specific, factors are used for estimating protein contents of grains, and corrections for nitrate-N content are needed with some vegetative materials. In proximate analysis, residue remaining after extractions with acid and alkali is defined as wall material (**crude fiber**, CF). Treatment with acids and alkalis dissolves some wall material, however, and so the CF fraction lacks validity for some purposes. An alternative method (Van Soest 1982), in which wall material is the residue remaining after extraction with neutral detergent solution, gives better results. This wall fraction is called **neutral detergent fiber** (NDF). No wall material is dissolved and its contents of cellulose, hemicellulose, lignin, and protein are open to further analysis.

It is very difficult, even by highly refined methods of analysis, to account for all the components of biomass by any method of analysis. The proximate method takes the remainder after subtracting lipid, crude protein, CF or NDF, and ash from the beginning mass as **nitrogen-free extract** (NFE), i.e., as non-structural carbohydrate. With the CF method, dissolved wall material is included in the NFE. The scheme of analysis presented in Table 2.1 employs the NDF determination. In this case, the remainder comprises compounds such as pectin and tannin, due to incomplete extraction. Analyses of feeds are now commonly done with physical methods in which spectral distribution of reflected near-infrared radiation (NIR) provides information on moisture, starch, protein, lipid, and fiber contents.

Examples of composition Some examples of biomass composition are presented in Table 2.2. Large differences in content of wall material are evident between vegetative biomass (sorghum at anthesis, bluegrass, and alfalfa) and grains and soybean seed. The sharp increase of non-structural carbohydrate in sorghum with maturity is due to grain growth; note how whole-plant protein content is diluted to 7.4% (1.2% N) by increase in starch.

2.1.2   Energy storage in plant material

The substrates for photosynthesis, CO_2 and H_2O, are fully oxidized compounds. By contrast, organic constituents of plants are all reduced to some degree and will release their chemical bond energy as heat energy upon oxidation. **Gross heat content** (ΔH_c, heat of combustion) is determined easily by ignition of plant samples or chemical substances under an oxygen atmosphere in a "bomb calorimeter". Heat released is measured from the rise in temperature of a surrounding water bath. Engineers distinguish between "higher heating values", obtained when the water product of combustion is condensed to liquid, and lower values found when water remains as vapor. The higher values are used in this book and in biological studies in general because liquid water, not vapor, is the product in metabolism.

Heats of combustion (ΔH_c) indicate maximum amounts of energy that might be derived from materials in metabolism and thus serve as crude measures of their value as a food source. They also provide a measure of the cost for biosynthesis of the material.

Table 2.2 Approximate biochemical composition of several plant materials (% dry matter)

Class	Grain sorghum biomass (anthesis)	Grain sorghum biomass (mature)	Bluegrass forage (early bloom)	Alfalfa hay (midbloom)	Maize grain	Wheat grain	Soybean seed
Nitrogenous compounds[a]	12	7	17	18	11	15	43
Lipids[b]	3	3	4	3	4	2	19
Wall materials[c]	59	42	65	46	9	(11)	(10)
Cellulose	27	18	28	26	2	–	–
Hemicellulose	29	21	16	10	6	–	–
Lignin	3	3	4	9	1	–	–
Non-structural carbohydrates[d]	9	33	na	(15)	(74)	(68)	(13)
Sugars	9	3	–	–	(2)	(2)	(11)
Starch	0	30	–	–	(72)	(66)	(2)
Organic acids	2	1	na	na	–	–	–
Minerals[e]	6	4	4	5	1	1	3
Remainder	9	10	10	13	1	3	12
Total	100	100	100	100	100	100	100

Notes: [a] N $\times$ 6.25; [b] petroleum ether extract; [c] values in parentheses are "crude fiber", all others are by the neutral detergent method; [d] values in parentheses were gleaned from other sources; [e] ash $\times$ 0.6; na, not analyzed.
Sources: sorghum data from Lafitte and Loomis (1988b); other crops from National Research Council (1982).

Characteristic ΔH_c values for the major biochemical classes are: organic acids, 4 to 11 MJ kg^{-1}; carbohydrates, 15.6 to 17.5 MJ kg^{-1}; proteins, 22 to 25 MJ kg^{-1}; and lipids, 35 to 40 MJ kg^{-1}. Vegetative biomasses of most crop and pasture species, with their relatively small content of lipid and protein, average 17 to 17.5 MJ kg^{-1}. Woody materials contain more lipid (e.g., waxes) and their ΔH_c values are nearer 18 to 19 MJ kg^{-1}; oil seeds, such as soybean and sunflower, yield 21 to 25 MJ kg^{-1}. Only a portion of the heat content of protein can be released in metabolism, however, because amino groups (-NH$_2$) cannot be oxidized by higher plants and animals. Energy yields from metabolism of proteins for those organisms, about 17 MJ kg^{-1}, are similar to what they obtain from carbohydrates. Soil microorganisms that can metabolize reduced N perform the important function of recycling it to atmospheric N$_2$ (Chapter 8).

It is sometimes convenient to represent the costs of plant biosyntheses in terms of the amount of a simple sugar, glucose, consumed in the process (Section 11.2). For comparison with other substances, the heat of combustion of glucose at 20°C is 15.6 MJ kg^{-1}.

2.2   Trophic systems in agriculture

Humans and other animals differ in their capabilities in digestion. Animals are included in farming systems because they thrive on plant materials unsuited for human consumption while converting them to high-quality food for humans. Biological values of milk and

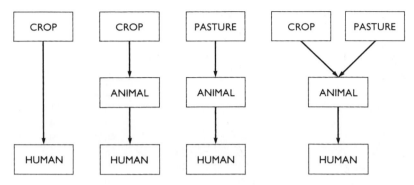

Fig. 2.2 The four basic trophic chains found in agriculture.

meat as human food are superior to those of most plant materials, and wool and hides are enormously useful. Further, some animals harvest their own feed by grazing, thus saving human labor. The biochemical composition of plant material and the capabilities of animal digestive systems are therefore pivotal determinants of farming systems.

Worldwide, domestic animals graze an area of land more than 2.5 times that given to crops (Box 7.2). In addition, the majority of production from arable lands goes directly to animals. The 6.8 billion humans that now inhabit Earth are accompanied by 1.7 billion large mammals (principally cattle), 3.1 billion smaller mammals (roughly equal numbers of sheep, swine, and goats), and 19 billion fowl (principally chickens) (FAOSTAT). The feed requirements of these animals are more than 2.5 times those of humans.

2.2.1   Trophic chains

Despite great diversity found among soils, climates, crops, and domestic animals, all farming systems can be classified within four basic **trophic chains** (Fig. 2.2). These food chains serve as maps of "who eats what". They begin with crops and follow some attribute such as energy, N, or dry mass through the animals that consume them. Agricultural systems, in contrast to most natural systems, have short trophic chains. In addition to food, agriculture also provides organic raw materials for industry. Fibers (e.g., cotton, hemp, flax, and wool), hides, and lipids (waxes and oils) are more obvious examples. Heavy emphasis given to soybean in world trade is based as much on industrial uses of its oil in alkyd resin paints, and more recently for biodiesel production, as it is on food and feed values of the oil and the high-protein oil-seed cake that remains after its extraction. A crop $\rightarrow$ industry path, then, parallels the trophic pathways of Fig. 2.2.

Trophic chains are transformed into complex webs when weeds, insects, birds, and soil flora and fauna are included (Fig. 2.3). Various organisms in the web can be grouped into functional classes according to their principal roles and level as producers or consumers. Autotrophic green plants of crops and pastures are termed **primary producers**, and their important characteristic is **net primary production** of biomass achieved through photosynthesis, associated metabolism, and nutrient uptake. Consumers of green plants

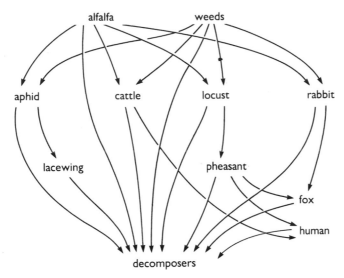

Fig. 2.3 The chain alfalfa → cattle → human expanded into a trophic web.

are considered **primary consumers**, or more simply, herbivores; carnivores dominate as **secondary consumers** of other animals. Microbial populations decompose waste and dead material from each of these levels, and so terminate trophic chains and webs. **Decomposers** perform the critical role of recycling chemical elements; without their activities C, H, N, O, and essential mineral elements would simply accumulate in dead material.

The real world is not quite as simple as this trophic classification suggests. Many species including humans are omnivores, consuming both plant and animal material, and an alternative scheme might arrange animals according to anatomy and capability of their digestive systems. Microbial populations participate in more than just a terminal role. In particular, they are important in the digestive tracts of animals (and insects) and thus could be considered primary consumers.

2.2.2   Biological efficiency in trophic chains

The production of organic materials is only the first step in bringing food to our tables in palatable forms. Losses of energy and carbon occur during plant growth (Fig. 2.1) and in subsequent stages of harvest and consumption, particularly where domestic animals are involved. One of our concerns is to minimize such losses.

Some rather broad principles that emerge from Fig. 2.3 serve as a foundation for farm management. The web makes the point that only a portion of production of one trophic level passes to the next level while the balance eventually passes to decomposers. In plant-to-herbivore transfers, for example, some plant material is inaccessible to grazers (e.g., roots), or indigestible by them (lignin). Additional losses occur at each level through respiration and senescence. Therefore, short trophic chains can pass a greater proportion of primary production to the last consumer than can long chains. A second

point is that crops and pastures usually include a number of plant species of little utility to our animals or to us. Such plants are weeds; they occupy "space" and utilize resources of light, nutrients, and water that would have supported additional crop production. Deficiencies of plant nutrients also result in lost production. Grazers such as rabbits, birds, and insects consume production that could have gone to domesticated animals, while parasites and imbalances in animal diets lessen production per unit feed consumed. Each of these problems represents a drain of organic material and associated nutrients from trophic flow to humans and they are therefore targets of management strategies that might reduce them.

Only rarely does wildlife consume enough to be a significant issue in agricultural production and most farmers view the presence of wildlife as a fringe benefit of farming. Predation of sheep by carnivores, attacks on ripening grain by birds, and devastation of pasture by rabbits (Australia) are examples where control measures may be required.

Measurements of the efficiency with which materials transfer from one trophic level to another (**trophic efficiency**) define how well a system is working. Trophic efficiency is calculated as net production by animals per unit feed input. This can be expressed in energy terms ($J\ J^{-1}$) or, for example, as mass of N in eggs per unit grain or grain N consumed. Trophic efficiencies vary with amounts of feedstuff supplied, how it is presented to animals, its biochemical composition, type and size of animal, ambient temperature, and other factors.

Theoretical charts of energy and material fluxes through populations of consumers become more complex when heat production from voluntary and involuntary work, digestion, and thermoregulation are included. Measurement of total heat production can be made in a whole-animal calorimeter, an elaborate and expensive apparatus, but the problem remains of identifying what proportions of heat come from each process. In addition, measuring the growth of cattle in other than live-mass terms is a difficult task, compared with measuring dry matter growth of plants with an oven and a balance. Live-mass measures are in fact subject to considerable variation because gut fill for ruminants can vary by 10 to 50% of empty-body mass (Agricultural Research Council Working Party 1980). Lactation efficiency (milk production/feed intake) also seems simple until associated changes in the animal's mass and composition are considered.

2.3   Animal and human nutrition

Issues to be considered here are the abilities of distinct digestive systems, digestibility of food and feed components, and the state of consuming organisms in terms of growth rate, and exercise, including work.

2.3.1   Digestive systems of animals

Through evolution, higher animals have arrived at diverse designs for gastrointestinal tracts that allow existence on various types of vegetation. All animals including humans can digest simple proteins, fats, and sugars, but they differ greatly in the digestion of complex carbohydrates such as starch and cellulose. Digestion of cellulose is important:

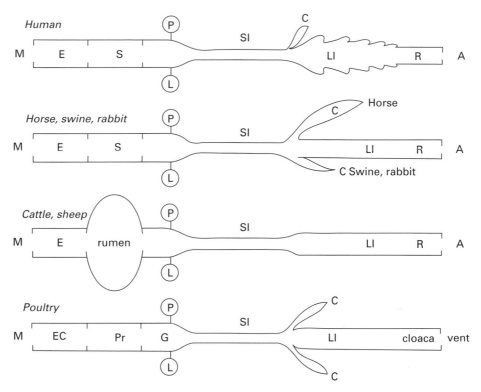

Fig. 2.4 Digestive tracts of humans and various domestic animals. Code: **M**, mouth; **E**, esophagus; **S**, stomach; **P** and **L**, pancreas and liver, respectively; **SI** and **LI**, small and large intestine, respectively; **R**, rectum; **A**, anus; **C**, cecum; **EC**, esophagus-crop; and **Pr**, proventriculus; **G**, gizzard.

first, because cellulose is the most abundant biochemical constituent of plants; and, second, because cellulosic wall materials enclose and protect cell contents. Acid and alkaline digestive secretions digest some wall material but serious attack on cellulose requires cellulase enzymes produced by microorganisms. Anaerobic bacteria that exist through fermentation have a principal role in cellulose digestion. Gastrointestinal tracts that provide a place and opportunity for bacterial fermentation of wall material gain access to the resulting products (bacterial protein and by-products such as volatile fatty acids) as well as the released contents of plant cells.

Digestion of plant material increases with time between intake and elimination ("passage time"). Passage time increases with length of digestive tract and thus with body size. Most animals depend on a large intestine (colon) and its appendages for fermentation (Fig. 2.4). The human colon, for example, is "sacculated", allowing an opportunity for some fermentation. Our body size and passage time are small, however, and we gain little from fermentation. Dietary fiber is beneficial to our digestive activity but, despite the sacculated colon, humans are dependent mainly upon readily digestible carbohydrates, fats, and proteins that can be released (by cooking or chewing) from the constraining influences of plant cell walls. In some animals, an appendage of the colon located near the junction of the small and large intestines, the **cecum**, increases capacity for fermentation. The human cecum is small. That of swine is larger but it does not entirely circumvent the need for relatively high-quality feed. In contrast, horses have a very large and well developed cecum. Coupled with a large body size and long passage time, this allows them to subsist solely on forages. In poultry, the gizzard grinds cellulosic fibers, increasing their surface area and subsequent rate of digestion in a small cecum. Rabbits overcome the passage-time problem by ejecting separate feces from cecum and rectum. Nutrient-rich cecal pellets are reingested for a second passage.

A multichamber stomach, the **rumen** (Fig. 2.4), serves as the principal site for fermentation in ruminants including cattle, sheep, goat, and buffalo. Cattle and sheep possess a complex, four-chamber rumen. This organ can retain fiber for continued attack while screening and passing digested materials to the intestine. Fermentation capacities of rumens are impressive. Most cellulosic materials except those intimately linked to lignin can be digested. Differences in digestion are evident from an examination of resulting fecal matter. Horse dung contains a considerable amount of undigested wall material and associated protein, whereas that of cattle is much lower in fiber and protein (Azevedo & Stout 1974). One consequence is that horse manure, because of its greater N content, has a greater value as fertilizer than that of cattle.

Differences in digestion cause animals to vary in their roles in agricultural systems and in their competitiveness with humans for foodstuffs. Fowl and swine, although competitive with humans because of their dependence on grain, are used in places where the ratio of humans to arable land is large and pasturage is scarce. Their utilization of excess grain gives them a role in buffering human food supplies. Populations of poultry and swine can be reduced by consumption during food shortages and their high reproductive rates allow rapid restoration when grain is abundant. Ruminants are less competitive with humans. They greatly extend human food supplies through use of grazing lands, forages, and coarse grains unsuited as sources of human food. Their reproductive rates are small, however, and their numbers cannot be adjusted quickly in response to variations in food supply.

2.3.2   Feeds and feeding

Despite difficulties in determining the fate of mass and energy after ingestion by animals, a number of important concepts emerge from feeding trials. These serve in calculating dietary requirements and expected yields of domestic animals.

Digestible nutrients Apparent digestibility of a feedstuff is determined by acclimating animals to it, collecting fecal mass produced, and subtracting it from feed mass. The difference is **total digestible nutrients** (TDN) in the feedstuff. Digestibility of attributes such as N and energy can be assessed from the same samples. Sufficient trials have been done with domestic animals that TDN can be calculated from proximate analyses as the sum of protein, lipid, and fiber, with the latter adjusted according to digestive abilities of each animal species.

Total digestible nutrients has been used as a basis for formulation of animal rations mainly in America, where tables of TDN values for most feedstuffs along with daily

TDN requirements of various sizes and kinds of animals were developed. Europeans have used an equivalency approach in which a feedstuff's merit was expressed relative to some standard feed such as starch, barley, or hay.

Box 2.1 Definitions used in analyses of the nutritional energetics of domestic animals

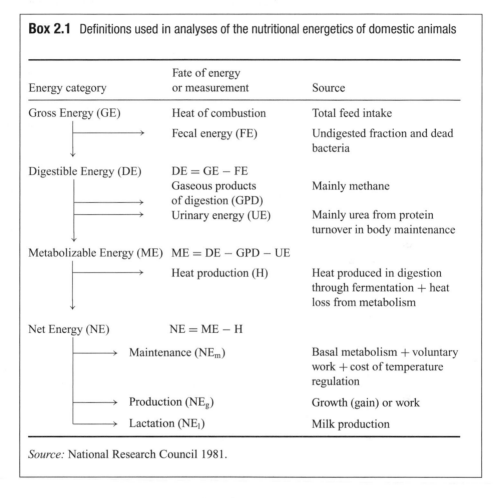

Energy category	Fate of energy or measurement	Source
Gross Energy (GE)	Heat of combustion	Total feed intake
	Fecal energy (FE)	Undigested fraction and dead bacteria
Digestible Energy (DE)	DE = GE − FE	
	Gaseous products of digestion (GPD)	Mainly methane
	Urinary energy (UE)	Mainly urea from protein turnover in body maintenance
Metabolizable Energy (ME)	ME = DE − GPD − UE	
	Heat production (H)	Heat produced in digestion through fermentation + heat loss from metabolism
Net Energy (NE)	NE = ME − H	
	Maintenance (NE$_m$)	Basal metabolism + voluntary work + cost of temperature regulation
	Production (NE$_g$)	Growth (gain) or work
	Lactation (NE$_l$)	Milk production

Source: National Research Council 1981.

Nutritional energetics Total digestible nutrients accounts for digestibility but it does not account for energy losses in digestion, thermoregulation, and later metabolism. Those costs are considered in a set of energy concepts summarized in Box 2.1. **Gross energy** (GE) content of a feedstuff is its ΔH_c, usually about 17 MJ kg^{-1}. Additional energy categories define the fate of energy within an animal. **Digestible energy** (DE), for example, is ΔH_c of TDN and is near 18.4 MJ kg^{-1} in most cases. Like TDN, DE is calculated from composition with an allowance for greater heat content of lipids. In ruminants, significant portions of digested energy are lost in urine, as heat during fermentation, and in gaseous products such as methane. In cold climates, digestive heat contributes to the maintenance of body temperature, but in hot climates animals must expend additional energy to cool themselves, and production efficiency is less in both cases. **Metabolizable energy** (ME) is that remaining from DE after accounting for gas production and urinary losses; **net energy** (NE) is the remainder after correcting ME for heat and urine losses.

Net energy is a measure of residual energy yield from a feed that can be used in maintenance, growth, or lactation. Maintenance (NE_m) is feed energy required to maintain body mass while fasting and resting and during voluntary work such as standing and eating, i.e., to support "basal metabolism". The basal metabolism of warm-blooded animals is roughly proportional to their surface:volume ratio and thus to $W^{0.75}$ ($W =$ body mass) and to the temperature difference with the environment. When $NE > NE_m$, the energy balance will support growth or lactation. Growth is measured as "gain" in live mass, including, for example, gain in wool produced by sheep. Calculations of NE_g (growth) and NE_1 (lactation) produced by a given feed must take account of metabolism required for biomass and milk production.

Both ME and NE have powerful applications in animal management. The British have developed equations for feed requirements of cattle and sheep from ME (Agricultural Research Council Working Party 1980); Americans generally employ NE for those purposes. The American system has been reduced to tables giving nutrient requirements of domestic animals (National Research Council 1984; and various handbooks) for use in ration formulation. Data from large numbers of carefully conducted feeding trials are summarized in regression equations used to calculate such tables.

The NE method is outlined in Box 2.2. As shown in that example, NE_g (2.4 MJ kg^{-1}) supplied by a feed is significantly smaller than its NE_m value (5.3 MJ kg^{-1}). That occurs because some feed mass is converted into structural material of growing animals rather than being respired. This effect is explained by the growth–yield concept introduced in Section 11.2. Production of milk also has a smaller cost than growth: NE_1 from alfalfa hay, for example, is 5.4 MJ kg^{-1}. Effectiveness of hay in milk production explains in part why milk is a cheaper source of protein for humans than beef meat and why hay (which costs less to produce than grain) is a common feed for dairy animals and for maintenance of breeding cattle. Low bulk density and slow digestion of **roughages** such as hay can, however, be limiting to daily intake in some cases. In contrast, feeds such as grains and soybean meal, with their greater bulk density and feeding value per kilogram, are termed **concentrates**.

Calculations presented in Box 2.2 illustrate how it is now possible to manipulate diets of confined animals in ways that make best use of available feedstuffs while optimizing animal performance and minimizing waste. The NE system is available in computer programs used by feeders to determine "least-cost" diets for given rates of production. Those programs compare current prices and net energy values of various feeds to arrive at a feed mixture and rate of feeding that provide best economic return. The main problems with NE systems lie with the expense of experiments needed to obtain NE values.

Influence of work Exercise requires food energy; animals in sparse or hilly pastures have smaller trophic efficiencies than those in level, lush pastures and in lots or barns. For sheep and cattle, the cost of level walking is 2 to 3 J kg^{-1} body mass m^{-1} traveled, and for climbing is 27 to 32 J kg^{-1} m^{-1} vertical lift (Agricultural Research Council Working Party 1980). The effects of exercise are readily apparent even in feedlots:

Box 2.2 An example of rations based on net energy

Example calculations of daily feed requirement for a half-grown, 250 kg steer with data drawn from NRC tables (National Research Council 1984) help in understanding NE concepts and differences in merits of feedstuffs. From tables, the animal's NE maintenance requirement is:

$$NE_m = 20.2 \text{ MJ d}^{-1}$$

Possible rates of growth are linked with feeding beyond the NE_m level:

$$NE_g \text{ for gain of: } 0.4 \text{ kg live mass d}^{-1} = 5.4 \text{ MJ d}^{-1}$$
$$0.8 \text{ kg d}^{-1} = 11.5 \text{ MJ d}^{-1}$$
$$1.2 \text{ kg d}^{-1} = 17.9 \text{ MJ d}^{-1}$$

Therefore total NE required for 0.8 kg gain per day is $20.2 + 11.5 = 31.7$ MJ d^{-1}. That could be met by various feeds. Net energy yields from alfalfa hay (National Research Council 1982; composition shown in Table 2.2) are $NE_m = 5.3$ MJ kg^{-1} and $NE_g = 2.4$ MJ kg^{-1}. Hay requirements for 0.8 kg gain d^{-1} therefore are:

for maintenance 20.2 MJ d^{-1}/5.3 MJ kg^{-1} = 3.8 kg d^{-1}
for gain 11.5 MJ d^{-1}/2.4 MJ kg^{-1} = 4.8 kg d^{-1}
$$\overline{\text{Total hay} = 8.6 \text{ kg d}^{-1}}$$

For comparison, NE yields from flaked maize grain (composition shown in Table 2.2) are $NE_m = 8.9$ MJ kg^{-1} and $NE_g = 6.1$ MJ kg^{-1}. Feed requirements with grain are only half those with alfalfa:

for maintenance 20.2 MJ d^{-1}/8.9 MJ kg^{-1} = 2.3 kg d^{-1}
for gain 11.5 MJ d^{-1}/6.1 MJ kg^{-1} = 1.9 kg d^{-1}
$$\overline{\text{Total hay} = 4.2 \text{ kg d}^{-1}}$$

The smaller amount of feed required with grain compared with hay explains why it is referred to as concentrate.

This example demonstrates the variable nature of trophic efficiency. At 17 MJ kg^{-1} feed in each case, gross energies for the two diets are 146 and 92 MJ for alfalfa and maize, respectively. Efficiency is limited by large loss of energy between GE and NE and by maintenance requirement. With 11.5 MJ retained in the animal body as gain, feeding efficiencies are $11.5/146 = 0.078$ for alfalfa and $11.5/92 = 0.125$ for maize grain. Trophic efficiency declines if intake is reduced because a larger fraction of NE is then expended on maintenance. From this example, it is clear that feeding grain is a more efficient practice for growing beef than feeding hay. Efficiencies with either feed are reduced sharply, however, when feeding requirements of parent animals are included in the budgets.

The steers in this example need 0.66 kg digestible protein d^{-1} (0.1 kg N d^{-1}) (National Research Council 1984). The alfalfa diet would supply 0.93 kg digestible protein (1.57 kg total protein), resulting in considerable waste to manure, whereas maize grain with only 0.43 kg digestible protein (0.46 kg total protein) requires supplementation with a protein source such as oil-seed meal.

trophic efficiencies are greater in level than in sloping lots. Exercise has a similar effect on food requirements of humans.

When animals are used for work, their energy requirements increase sharply. Brody's (1945) discussion of horses and other animals is instructive. It turns out that humans and horses are rather similar. Their maximum net efficiency (work accomplished/DE expended in muscle activity) increases asymptotically to about 0.25 as rate of work increases. Much work done by humans goes to lifting of the body, however, so effective work is considerably less than 0.25. In lifting a pencil from the floor, for example, we must also lift our upper body. By contrast, animals are used mainly for pulling or carrying, which are relatively efficient activities. Large (700 kg) workhorses can generate as much as 10 kW (13.3 HP) but for only a few seconds at a time. At that rate, a horse would need to consume 4.3 kg of good hay per hour to obtain the necessary 36 MJ of DE! (Horse rations are usually done on a digestible energy basis; good hay is about 45% digestible.) The maximum sustained effort of a horse for a ten-hour day is nearer 0.75 kW (1 HP) resulting in 27 MJ of work accomplished, corresponding to 40 km travel with a draft equal to about 10% of body weight (about 70 kg). Net efficiency is still near 0.25, but when maintenance requirement is added overall efficiency drops to 0.16. Total digestible energy requirement is then 174 MJ d^{-1} (66 MJ for maintenance + 108 MJ for work). That could be met with 21 kg of good hay per day, but that quantity is too bulky to be consumed and digested daily so a portion of diet is fed as concentrate. Efficiency of work accomplished to gross energy ingested is 7% [27 MJ work/(174 MJ DE/0.45 digestibility)].

Work animals are still used in many areas of the world. Cattle perform better on low-quality feedstuffs such as straw than do horses and mules but they also work at a slower rate. As a result, their maintenance overhead per unit work is greater and more human labor is required while animals perform a given task. In the USA, where horses and mules predominated over oxen in earlier times because of shortages of farm workers, nearly 40 Mha of farmland (20% of the total) was once given to their grazing and feed production. Conversion to tractor power released that land for general farming. The worldwide population of horses, mules, asses, and buffalo is nearly 280 million (FAOSTAT 2007). Assuming that they consume an average of 10 kg feed per day, the annual feed requirement is near 1 Gt of dry matter containing 15 to 20 Mt N.

2.3.3   Trophic efficiency in animal production

Efficiency of trophic transfers through an animal population is not calculated easily. In addition to the variable nature of animal production, problems arise from different bases of measurement used by plant and animal scientists. Crop ecologists emphasize dry-matter yield per unit area and, in some cases, heat of combustion (ΔH_c) or nutrient content. Given non-linear responses to feed quantity, animal scientists prefer to view animal intakes in derived concepts of DE, ME, and NE. The situation is further confused by variations in live mass, dressed carcass, and edible parts under different feeding systems, by measurement of output as fresh masses of gain, milk, egg, and wool yield per animal, and by maintenance costs of breeding animals. Readers are referred to

Table 2.3 Efficiencies of various animal production systems in conversion of feedstuffs to human-edible food

(a) Inputs of total digestible and human-edible (HE) energy in animal feed and return of human-edible energy in animal products. Efficiencies for return of human-edible energy are given relative to total and human-edible inputs. (b) Corresponding analysis for protein in the same amount of feed.

System	Product	Input Total	Input HE	Return HE	HE return/ Total in	HE return/ HE in
(a) Energy		(GJ)	(GJ)	(GJ)	(%)	(%)
Dairy	Milk, meat	83.5	19.0	19.3	23	101
Beef	Meat	86.0	7.82	4.47	5	57
Swine	Meat	6.16	2.46	1.43	23	58
Poultry	Meat	0.097	0.047	0.014	15	31
(b) Protein		(kg)	(kg)	(kg)	(%)	(%)
Dairy	Milk, meat	702	111.5	202.2	23	181
Beef	Meat	823	39.9	43.6	40	109
Swine	Meat	66	29.0	24.9	38	86
Poultry	Meat	1.27	0.48	0.36	31	75

Note: amounts of feed required for the breeding population are included along with that needed for the slaughter or milking animal.
Source: adapted from Bywater and Baldwin (1980).

Leitch and Godden (1953), Agricultural Research Council Working Party (1980), and Bywater and Baldwin (1980) for insights into data on "usual" trophic efficiency. The ARC publication provides the most complete information on body composition but fails to translate that to human-edible fractions.

Table 2.3 illustrates levels of trophic efficiency that can be achieved with good management. These calculations include feed requirements of entire herds in addition to cows milked or animals grown for slaughter. All have low efficiency in use of total feed compared to dairy, swine, and poultry because of greater overheads in breeding animals. Feed requirements of breeding cattle, however, are derived entirely from roughages unsuited for human consumption. In constructing this table, the authors considered grain portions of feed supply to be "human-edible" on the basis that crops such as wheat or potato could have been grown instead of feed grains. Trophic efficiency of the systems in converting these "human-edible" inputs (DE_h) to human-edible output declines as proportion of human-edible material in diets increases. All systems are quite efficient in returning human-edible output per unit human-edible input, however. Contrary to popular notions that grain feeding of cattle is inefficient, cattle have a clear advantage over swine and poultry, returning significantly more human-edible protein than they consume.

Comparisons of trophic efficiencies of animals fed entirely on human-edible food are now, at a time of increasing global food demand, often used to promote vegetarian diets. Grain to meat (mass) ratios differ between animal species and production systems, from about 2.2:1 for fish, to 3:1 for poultry and swine, and 8:1 for feedlot cattle while on grain rations. These ratios can be misleading because diets usually also include food not edible by humans. This is especially so for cattle, the major part of whose life cycle is spent on pasture or eating forage crops. In consequence, their life-cycle grain-conversion efficiency is about the same as for poultry and swine (CAST 1999). Likewise, feed conversion efficiency for commercial poultry, swine, and cattle production has increased steadily with improvements from animal breeding, health, and feed ration composition.

2.3.4   Human nutrition

Energy sources and protein are also main components of proper diets for humans. Early nutritional standards were drawn from surveys of eating habits of well-fed people rather than from exhaustive feeding trials. As a result, a high intake of protein (85–100 g cap^{-1} d^{-1}) was thought necessary and debates ensued about the undesirability of low-protein carbohydrate sources (grains, sugar, potato, and cassava were labeled "empty calories" by critics), and a need for protein sources with a high content of the amino acid lysine. These questions were settled in the 1970s with findings from N balance experiments. Daily requirement for an average healthy adult is now set at 0.6 g of high-quality (animal) protein kg^{-1} body mass. It was also found that any balanced diet, even those depending mainly upon grains and potato, is adequate in all essential amino acids providing enough is eaten to satisfy energy requirements (Payne 1978).

As with animals, required food intakes of humans vary with body mass, age, sex, activity, and other factors. Our knowledge of energy and protein requirements is now embodied in regression equations fit to average needs of various groups of people. In contrast to these daily "requirements", "safe levels" of intake and "**recommended dietary allowances**" (RDAs) are set at a somewhat higher level (e.g. the mean + 2 standard deviations) to include population extremes. The FAO/WHO committee thus uses 0.75 g protein kg^{-1} body weight as the *safe level* of daily intake whereas only 0.6 g is *required.* Readers are referred to consultative reports (FAO/WHO/UNU 1985; National Research Council 1989) for equations and detailed tables of requirements and allowances. Those publications also include information on other aspects of diet and food quality.

Table 2.4 presents a summary of current American RDAs for energy and protein intakes of adults. In contrast to the relative constancy of protein requirement, daily energy requirements can vary substantially for the same individual (range 1–7$\times$) as is explained in the footnote to the activity factor. Requirements and allowances for children are larger per unit body mass but smaller in total than for adults. Although these numbers should apply generally to Western nations, allowances are larger than for small people (Asians). People with high-fiber diets (as in India and Africa) need a somewhat higher intake of protein to offset reduced absorption of amino acids in the presence of fiber.

It is useful to have numbers representing energy and protein allowances for an "average" human. Table 2.4 serves to illustrate, after allowing for children and the large size of Americans, that 10.5 MJ (2500 kcal) of digestible energy and 50 g of digestible

Table 2.4 Recommended dietary allowances (RDAs) of energy and protein for median adults in the USA

Category	Age (y)	Median mass (kg)	REE[a] ×	Activity factor[b] =	Energy RDA	Protein RDA[c] (g d^{-1})
			Energy allowance (MJ d^{-1})			
Males	19–24	72	7.4	1.67	12.4	58
	25–50	79	7.5	1.60	12.0	63
	51+	77	6.4	1.50	9.6	63
Females	19–24	58	5.6	1.60	9.0	46
	25–50	63	5.8	1.55	9.0	50
	51+	65	5.4	1.50	8.0	50

Notes: [a] REE is resting energy expenditure, i.e. basal metabolism; [b] activity factors range from 1.0 for resting, 1.5 with very light activity, 2.5 with light work, 5.0 with moderate work, and 7.0 with heavy work, the factors given here are integrated daily values for the US population; [c] protein allowances are calculated as body mass × 0.8 g protein kg^{-1} body mass.
Source: adapted from National Research Council (1989).

protein (8 g N) would serve as reasonable estimates of adequate daily allowances. For comparison, FAO/WHO (1973) estimated that the average human needs 9.1 MJ and 30 g protein per day. The average American allowances translate to annual requirements of:

$$10.5 \text{ MJ energy d}^{-1} \times 365 \text{ d y}^{-1} = 3.8 \text{ GJ y}^{-1};$$

$$50 \text{ g protein d}^{-1} \times 365 \text{ d y}^{-1} = 18.2 \text{ kg protein (2.9 kg N) y}^{-1}.$$

With 17 MJ kg^{-1} as average energy yield from plant materials, 3.8 GJ could be obtained from 224 kg of digestible dry matter. Ingestion of this amount of even low-protein sources (9% protein) would satisfy protein and amino acid needs (0.09 × 224 kg dry matter = 20.2 kg protein or 3.2 kg N). It appears that the present world population of 6.8 billion people could be adequately fed for a year with 1.6 Gt of digestible grain or its equivalent containing at least 21 Mt of N.

2.4 Carrying capacity

Knowledge of food requirements allows us to calculate carrying capacity of particular farming systems, defined as the number of animals (or people) that can be supported by primary production from a given area of land. Carrying capacity is an old and important concept that relates to sufficiency of agriculture. It is apparent that carrying capacities depend upon performance of producer communities and thus on the environment and intensity of cultivation. Calculations of carrying capacity follow the Conservative Law based on dietary energy terms (GE, DE, or NE) or some component such as protein. The approach is illustrated with DE as follows:

$$K = (k_c k_d \text{GE} P)/(\text{DE}_d \text{ or DE}_y) \tag{2.1}$$

where:

K = carrying capacity (animals ha^{-1}),
k_c = fraction consumed,
k_d = fraction digested,
GE = gross energy content of the feed (MJ kg^{-1}),
P = net production (kg biomass ha^{-1} y^{-1}), and
DE$_d$ and DE$_y$ = daily and yearly energy requirements (MJ cap^{-1}).

Box 2.3   Carrying capacity of a grazing system

Ryegrass is an important pasture species in the cool climates of Europe, New Zealand, and elsewhere. With intensive management, annual production of 15 000 kg forage ha^{-1} is common. Taking 17 MJ kg^{-1} as the gross energy content of forage, a total of 255 GJ ha^{-1} is available for grazing. With $k_c = 0.7$ and $k_d = 0.7$, 125 GJ of that energy would be obtained as DE by ruminants. The annual DE requirement for a Holstein–Friesian milking cow is near 64 GJ with an additional 19 GJ needed to support replacement heifers (when one third of cows are replaced each year) (Bywater & Baldwin 1980). Total DE$_y$ per cow unit (1.5 cow and 0.5 heifer) is thus 83 GJ; and $125/83 = 1.5$ cow units could be supported per hectare. An output of 9525 kg milk ha^{-1} y^{-1} containing 28 GJ in the form of fat and protein would be obtained. This corresponds to 11% of total energy in forage and 22% of the grazed, digestible, part. Using average energy allowance developed earlier (in Table 2.4), and ignoring losses that occur in transportation and distribution, each hectare would support a population of seven humans (28 GJ in milk/3.8 GJ cap^{-1} y^{-1} = 7.4).

   The high protein content of milk is perhaps as significant for human nutrition as the energy. The 9525 kg of milk produced would contain about 314 kg protein or enough to sustain 17 people (314 kg protein/18.2 kg protein cap^{-1} y^{-1} = 17.2).

   Carrying capacity is illustrated in Box 2.3 with production from an intensively managed pasture. In that example, forage is indigestible by humans but, after conversion by dairy cattle, it can feed seven humans. A wheat crop with a similar 15 000 kg ha^{-1} of aboveground production and a harvest index of 0.4 would supply 6000 kg of human-edible grain. Only about 80% could be consumed because some grain is needed as seed and there are losses in storage and processing. Digestibility of wholewheat grain for humans is about 0.85 (compared with 0.96 for refined flour). An energy content of 17 MJ kg^{-1} would again apply. Our average annual dietary allowance for energy could be met by production of 329 kg grain [3.8 GJ cap^{-1} y^{-1}/(0.8 consumed × 0.85 digested × 17 MJ kg^{-1} grain) = 329 kg cap^{-1} y^{-1}]. One hectare of high-yield wheat would therefore support the energy requirements of 18 humans (6000/329 = 18.2) or twice what is accomplished by dairying.

   The [N] of wheat grain ranges over 2 to 3% of dry matter depending on cultivar and environment. The composition of wheat protein is sufficiently different from the average that 5.9 rather than 6.25 is used as the factor for conversion to protein from [N]. On that basis, the wheat crop would supply humans with at least 453 kg crude protein ha^{-1} y^{-1} (6000 kg grain × 0.8 consumed × 0.8 digested × 0.02 N × 5.9 kg protein/kg N). That would be sufficient for 25 humans (453 kg protein/18.2 kg protein cap^{-1} y^{-1} = 24.9).

A simpler approach serves in rough assessments of national and world food budgets. Grains dominate in world food supplies (*c.* 60% of both energy and protein) and protein and energy contents of other foods can be expressed in equivalent amounts of grain. The wheat example illustrated needs to allow for losses in harvest, storage, distribution, and food preparation chain, as well as to cover variations in production, seed supplies, and diet diversity. One measure of a desirable level of "original" (on-farm) production is the **standard nutritional unit (SNU)**:

$$1 \text{ SNU} = 23 \text{ MJ cap}^{-1}\text{day}^{-1} \times 365 \text{ days} = 8.4 \text{ GJ cap}^{-1}\text{y}^{-1} \qquad (2.2)$$

This amount of gross energy is contained in 500 kg grain. A convenient feature of this number is that each tonne of a system's production expressed in grain equivalents can be equated with generous annual nutrition of two humans.

The SNU is large enough (more than twice necessary human intake of digestible energy) to accommodate significant diversity in diets since portions of farmland can be given to production of fruits, vegetables, and animal products rather than to grain. Year-to-year variations in yields can be accommodated through grain storage and increased animal production in abundant years for consumption in poor years. Animal populations can be increased or decreased according to feed supply but this strategy is not viable with humans. Two key points made in this chapter are that domestic animals insulate us from variations in primary production and do so with reasonable efficiency.

2.4.1 The yield–area relationship

The production of a farm, region, or country is defined mathematically as the area harvested times yield per unit area. This is such an obvious relationship that its significance is often overlooked in discussions of carrying capacity and input–output relationships of agricultural systems. There are two basic reasons why yields may be small. One is that climate, through shortness of season, limited rainfall, or other factors, may be restrictive. The other is that farming is not intense and actual yields are small relative to attainable yields. Limits arise from factors such as poor management of nutrient supply or weeds, i.e. from things that are preventable through human effort and technology. Intensity thus depends as much on increased inputs of knowledge as it does on material resources. The important point is that the area required to achieve a given production is inversely related to the yield (Fig. 2.5). With small yields, the area required is much larger than with a more productive agriculture. This is a critical non-linear component of calculations of carrying capacity. Options for increasing food supply are therefore simple: increase yields and/or the area under cultivation.

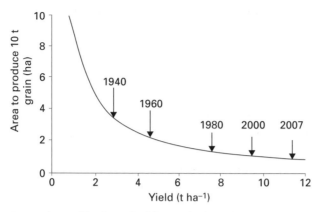

Fig. 2.5 Area of land required for production of 10 t grain as a function of yield per hectare. The arrows identify progression of US three-year mean maize yield, 1940 to 2007.

2.5   Review of key concepts

Primary production

- Plants obtain energy for growth from sunlight in photosynthesis and build biomass using carbon dioxide from the air and inorganic nutrients from soil. All consumers and decomposers in the biosphere ultimately rely on plant biomass to provide energy and nutrients.
- Plant species vary in their chemical composition of carbohydrates, proteins, lipids, and vitamins that determine suitability as food for consumers. Humans have selected agricultural species on the basis of suitability for nutritional, medicinal, and industrial purposes.

Trophic chains

- Are webs of "who eats whom (or what)?" They start with plants (producers) and proceed through successive levels of consumers and decomposers. Energy and nutrients are lost at each step. A major distinction among consumers occurs between monogastric animals, including humans, that require simple carbohydrates (sugars and starches) for energy, and ruminants that can digest cellulose, the major component of plant biomass.
- Cropping systems are managed with rotations, nutrients, seedbed preparation, and weed, insect, and disease control so that the greatest possible proportion of primary production is diverted from competing consumers and decomposers to yield.
- Animal nutritionists recognize feed materials in terms of digestibility to supply net energy, first for maintenance and then for work, lactation, or weight gain. Differences in those traits help explain the roles played by different livestock species in agricultural systems.

- Domestic animals are major consumers of plant production. Their efficiency in trophic transfers to humans varies with type of digestive system and with the amount and quality of feed they consume.

Carrying capacity

- Farming plays the central role in providing food and fiber for humans. Level of production determines carrying capacity for domestic animals and humans per unit area of agricultural land.
- The standard nutritional unit (SNU) of 500 kg grain is a measure of primary (on farm) production required to feed one human for a year. It is large enough to accommodate significant diversity in diet, i.e. the proportion of farmland given to vegetables and fruits, to provide seed for the following crop and allowance for waste; 1 t ha^{-1} supports two people while 10 t ha^{-1} supports 20.

Terms to remember: biomass, structural and non-structural; carrying capacity; cecum; concentrates; consumers, primary and secondary; crude protein; decomposers; economic yield; energy, digestible (DE), gross (GE), metabolizable (ME), net (NE); fiber, crude and neutral detergent; gross heat content; harvest index (HI); lipid; net primary production; nitrogen-free extract; partitioning; primary producers; proximate analysis; roughage; rumen; standard nutritional unit (SNU); stoichiometry; trophic, chain and efficiency.

3   Community concepts

Crop communities can be described in simple terms. Species and cultivar define genetic content while density, spacing pattern, plant size, and stage of development define structure. The type of community is termed a **monoculture** when only one crop species is grown in a field at a time; the terms **polyculture** and **mixed cropping** apply to communities with two or more cohabiting crop species. Other definitions and terms exist but these are the traditional ones employed by agronomists. Most arable farming involves rotations of monocultures over time whereas pastures are mostly polycultures.

Critical issues in crop ecology, which we will examine in detail in this chapter, include the impact of community structure on resource capture and yield of crop production systems and interactions between component plants.

3.1   Community change

Concepts of community structure evolved from complementary work by agronomists who study managed communities and by botanists concerned with natural communities. In agriculture, small differences in production are important and agronomists study how production rate, competition, limiting factors, and genetic expression influence behavior of simple communities. Botanists, faced with highly diverse, natural systems give greater attention to community species composition in relation to adaptive traits and evolution. A background in plant ecology is useful for agriculturalists, and vice versa. Natural communities are subject to continuing change as different species of plants invade a site and displace earlier occupants. This process is termed **succession** and the sequence followed, the **sere**. Annual species capable of aggressive occupation of space are usually pioneer invaders of unvegetated sites. If soil organic matter is scarce, the ability to exist on scant supplies of available N or to obtain N through symbiosis are important traits for pioneers. With time, soil organic matter accumulates from the deposition of plant residues, improving soil water-holding capacity and nutrient stocks to support greater plant density. This in turn makes individual plants compete more intensely for resources. Where rainfall and nutrients are not strongly limiting, height growth becomes a critical factor for competitive success. Taller perennial shrub and tree species invade, shading out pioneer herbaceous species. Persistence of individual species within a community becomes increasingly dependent on the ability of replacement seedlings to develop under shade of their maturing parents. Communities eventually approach equilibrium with their environment and a more stable condition, termed **climax** community, is reached.

Succession in similar climates around the globe leads to similar vegetation; hardwood and boreal forests and tundra are found in comparable latitudinal sequences in Europe and Asia. The dry mid-latitudes of Asia support grasslands similar to those of the North American plains. Control by climate is so strong that when vegetation is destroyed, as by fire or agriculture, recovery generally follows the same sequence towards the same climax. Climate and vegetation together control soil formation (Chapter 7) and areas where agriculture is now practiced correspond in important ways with original vegetation. The rich, sub-humid grasslands of Ukraine, Argentina, and the US Corn Belt, for example, have all become major centers of agriculture and produce a similar set of crop species.

The widespread occurrence of vegetation with similar general appearance and with the same or similar species led early botanists to ideas about communities as "super organisms" with relatively fixed complements of participating species. Gleason (1926) challenged that proposition with evidence that community structure was more a result of fortuitous immigration of plants. It is now clear that cohabitation involves a large measure of randomness in accord with Gleason's "individualistic" view: communities are composed, from available species, of those able to persist at the site. Survival within a community, however, requires adherence to social rules imposed by competitors. This occurs because all plants compete for the same resources. If one species adjusts osmotically and takes the soil towards dryness, all would-be cohabitors must escape that condition or have a similar ability. If some species grow tall or are aggressive in uptake of N, only those able to accommodate to that new environment will persist. The explanation, then, for similar endpoints to succession, is that a limited range of physiological tactics is possible for dominants in each climate and soil type.

Because cohabiting species experience similar selection pressures, they frequently come to similar evolutionary solutions for survival under a given set of conditions ("convergent evolution"). Dry Mediterranean climates, whether in Chile, Australia (Chapter 16), or California, lead to grasslands generally composed of annual grasses and forbs, and perennial species found in those communities usually have some form of summer dormancy. With greater rainfall, chaparral vegetation composed of a floristically highly diverse group of shrubs predominates. Those diverse shrubs, however, are sometimes remarkably similar in physiological and morphological traits that lend adaptation to life with severe summer drought.

3.2   Biomass accumulation

In this section we will examine the nature and analysis of crop growth and its relation to interception of light, the principal determinant of crop growth rate.

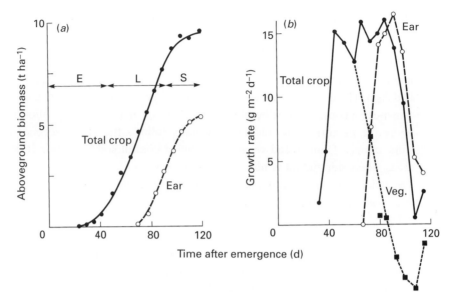

Fig. 3.1 (*a*) Accumulation of total biomass and ears over time by an average crop of maize. E stands for "exponential phase", L for "linear phase", and S for "senescent phase". (*b*) Growth rates of total crop, ears, and vegetative parts of the maize crop. Grain growth exceeds crop growth late in the season as materials are transferred from vegetative parts. (Data from Bair 1942.)

3.2.1   Biomass dynamics

Primary production by plant communities results in accumulation of biomass with time. Within a season, accumulation generally follows a sigmoidal (S-shaped) curve. Trees continue to accumulate biomass (as dead xylem) for many years. In that case, the sigmoidal pattern for each year is superimposed with a larger one for the entire life of the community. The simplest pattern of biomass (*B*) accumulation per unit area occurs with communities of annual plants as shown in Fig. 3.1*a*. The slope of the curve is **crop growth rate** (CGR), defined as rate of change of *B* with time, $\Delta B / \Delta t$, with units of kg ha^{-1} d^{-1} or g m^{-2} d^{-1} (Fig. 3.1*b*). Such curves are established by sampling crops periodically. Crop growth rate is then calculated as increase in biomass, ΔB, between two dates divided by Δt. More accurate values are obtained by taking the first derivative (dB/dt, i.e., value of $\Delta B / \Delta t$ as $\Delta t \rightarrow 0$) of an equation fitted to the data. The procedure is explained further in Box 3.1.

It is easy to sample aboveground and tuberous underground portions of annual crops but proper sampling of roots is difficult. Most data on *B* and CGR reflect only above-ground material. When soil resources are not limiting, roots constitute only about 10% of peak crop biomass. Where water or nutrients are limiting, however, roots may be a much larger fraction and errors from neglecting root biomass become significant.

Sigmoidal biomass accumulation curves can be divided into an early "**exponential phase**", a "**linear phase**" when growth proceeds at a largely constant rate during mid-season, and a final "**senescent phase**". These periods are indicated in Fig. 3.1*a*. Growth

---

Box 3.1 Equations to describe and analyze patterns of crop growth

During the exponential phase, absolute growth rate, dB/dt, is proportional to B (and through B to leaf area):

$$dB/dt = \mu B \tag{3.1}$$

where μ is the **relative growth rate** (RGR). After rearrangement:

$$\mu = 1/B\,dB/dt \tag{3.2}$$

Units for B cancel so units for μ (RGR) reduce to t^{-1}. The integrated form of Eq. 3.1 defines B at any time t during exponential growth that starts at time 0 with initial biomass of B_0:

$$B = B_0 e^{\mu t} \tag{3.3}$$

Sigmoidal shapes can be applied to the entire growth curve by various equations. The simplest is the logistic function that introduces a limiting term $(B_{max} - B)$ to gradually reduce the initial relative growth rate, μ, to zero as $B \rightarrow B_{max}$. Crop growth rate (CGR or dB/dt) is then $(B_{max} - B)\mu B$, while the integrated form is most easily expressed as:

$$B = B_{max}/(1 + ae^{-ct}) \tag{3.4}$$

where maximum CGR $= B_{max}\,c/4$ occurs at time $= (\ln a)/c$ when $B = B_{max}/2$. This equation is useful to compare growth curves of individual crops although it suffers the limitation of symmetry. The more complex Richard's Equation (Hunt 1978) has an extra parameter that allows asymmetric sigmoidal curves. One of its solutions is the logistic function.

---

rates of seedlings are limited by leaf area and light interception and the exponential phase results from positive feedback of expanding leaf area on growth rate. As leaf area increases, light interception and photosynthesis increase, and so CGR, which includes leaf growth, increases. That continues until foliage covers land area completely and the canopy intercepts all incident light. The term **cover** denotes the fraction of land area obscured by leaves when viewed from above; it can be measured experimentally from photographs, by noting the frequency with which a "point quadrat" (a thin needle) encounters leaf or ground when repeatedly projected vertically into the community, or indirectly with light meters placed on the soil surface.

The equations in Box 3.1 express growth rate per unit (i.e., "specific" or "relative" to) existing biomass. The parameter μ assumes its largest value during the seedling phase but declines rapidly as plants increase in size, some parts cease growing, and the canopy closes. During exponential growth, μ relates closely to leaf area and photosynthetic activity. Values of μ for full-cover crops are essentially meaningless, however, because dB/dt is then largely independent of B and leaf area.

With complete cover, crops enter the linear period during which light interception and photosynthesis are maximal. Crop growth rate then varies mainly with changes in solar radiation. With determinate annuals such as maize (Fig. 3.1*b*) and sunflower, the declining phase of crop growth is due to maturation and senescence. These are synchronized phenomena in each cultivar of annual crops; all members of a field planted to a given cultivar obey the same phenological rules (Chapter 5) and die at the same time. Crop growth rate also declines as radiation and temperature decrease towards the end of summer and also when standing crop biomass becomes so great that the requirement for maintenance respiration (Section 11.1) usurps most supply of photosynthate. Thus, sigmoidal patterns of biomass accumulation are also seen over individual seasons in pastures and sugarbeet crops even though no senescence or death of plants occurs.

Seasonal production is greatest when full cover is achieved early in the growing season and is then maintained as long as weather is favorable. Perennial plants have an advantage over annuals in that the lag of the exponential phase is bypassed: cover is established very quickly by the growth of new leaves supported by reserves from perennial roots and stems. A price is paid, however, through costs of producing and maintaining perennial parts such as underground rhizomes of pasture grasses. Other than tree and vine crops and forage species, there is little agricultural use of long-lived species. Some perennials such as cotton are in fact grown as annuals and in some cases storage organs of perennial and biennial species (such as tubers of potato and tuberous roots of beet and carrot) form the basis for annual crops.

3.2.2   Leaf area and light interception

Because CGR depends on photosynthetic activities of leaves, relationships exist between CGR and the area of leaves in the foliage. The term **leaf-area index** (LAI or *L*) describes the sum of areas of all leaves (only one surface is counted) per unit area of ground. Areas of individual leaves can be measured with light-cell planimeters in which leaves are passed over an array of light cells. Leaf areas may also be established from regressions of the product of leaf length (*l*) and width (*w*), i.e., as area $= alw$ where *a* is the regression coefficient.

Due to mutual shading among leaves, light interception by foliage is a diminishing-returns function of LAI. Japanese workers were among the first to develop techniques for examining light penetration into foliage canopies in relation to leaf area. By comparing leaf area and light interception in successive horizontal strata beginning at the top of canopies, they found that light flux at any level (*I*) could be related to leaf area above it and light flux incident on top of the canopy (I_0) with a simple equation known as the Bouguer–Lambert Law (Monsi & Saeki 1953):

$$I = I_0 \exp\left(-k_L L\right) \text{ or } \ln\left(I / I_0\right) = -k_L L \tag{3.5}$$

where k_L, the **attenuation coefficient**, relates to the fraction of light intercepted per unit LAI (*L*). The equation has the same form as the familiar Beer's Law for light absorption by homogeneous media but canopies are not homogeneous, hence the Bouguer–Lambert name. The diminishing returns relationship is illustrated in Fig. 3.2*a* for clover and grass canopies. When these curves are plotted as $\ln(I/I_0)$ versus LAI, straight lines of

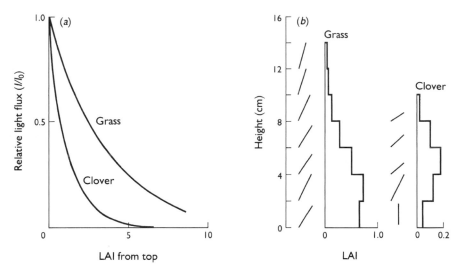

Fig. 3.2 (*a*) Attenuation of sunlight in stands of clover and ryegrass as a function of leaf-area index penetrated (after Stern & Donald 1962). (*b*) Typical patterns of leaf display by clover and ryegrass. Clover leaves tend to horizontal display (noted by the angled line) and to cluster some distance above the soil surface. Ryegrass leaves are more erect and the distribution with height is pyramidal (after Warren Wilson 1959).

slope – k_L are usually obtained. The k_L values of grass and clover differ because of differences in canopy architecture. Clover, with a tendency towards horizontal leaf display (Fig. 3.2*b*), is very effective in light interception. It has a large k_L whereas grass, with vertical leaves, has a small k_L. With k_L equal to 0.7 (clover), near 90% interception is achieved with LAI = 3. With k_L equal to 0.5 (grass), 90% interception requires LAI = 4. The k_L also varies with solar altitude and azimuth and thus with the time of day and year. These issues of light interception by foliage canopies are explored further in Section 10.4.

Light interception is particularly useful as an index of canopy development. In contrast to measurements of LAI and k_L, interception is measured easily from readings with appropriate light sensors exposed horizontally above vegetation and at ground level [R_i, fractional interception = $(1 - I/I_o)$]. Interception (like k_L) varies with leaf display and solar angle, and observations taken near the solar noon correlate best with cover and CGR. The daily average obtained from continuous measurements is needed for some purposes.

Examples of relationships between CGR and LAI and between CGR and fractional interception $(1 - I/I_0)$ are presented in Fig. 3.3. The plateau response of CGR with increasing LAI is explained by the diminishing relationship between light interception and LAI. Linear relationships between CGR and light interception, first reported by Shibles and Weber (1965), are now known to occur with most crops (Monteith 1977). The slope of that relationship, using absolute rather than fractional interception, embodies the non-linear relationship between CGR and LAI and defines **radiation-use efficiency** (RUE) in units of g dry matter MJ^{-1} sunlight intercepted.

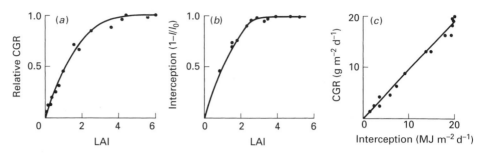

Fig. 3.3 (*a*) Growth rates of soybean crops as a function of total leaf area. The data come from three crops with different spatial arrangements of the plants and different maximum growth rates and so are presented relative to the maximum CGR of each community. (*b*) Fractional light interception as a function of leaf-area index for the same soybean crops. (*c*) Absolute crop growth rates of soybean crops as a function of total light interception estimated from meteorological data. (Adapted from Shibles and Weber 1965.)

3.2.3   Growth analysis

Crop growth rate, leaf-area index, and light interception (or cover) define the main features of a community that determine primary production. Crop growth rate, relative growth rate (RGR), and leaf-area index are concepts and terms developed by early English crop ecologists. They also developed mathematical methodologies for estimating these and other variables from periodic sampling of crops to measure leaf area and total aboveground biomass on a dry weight basis. That approach is sometimes termed growth analysis (Radford 1967; Hunt 1978). Compared to the usefulness of CGR and LAI, we find little meaning or use in net assimilation rate (NAR; crop growth rate per unit leaf area) and leaf area duration (LAD; sum of LAI × time). Leaf area duration recognizes that duration of LAI is important to seasonal production but study of Fig. 3.3*a* reveals that a non-linear relationship rather than a simple sum must be used because CGR is not a linear function of LAI. Leaf-area index greater than that needed for full cover increases LAD but has no effect on CGR and seasonal production because light interception is not increased. In contrast, linear relationships between CGR and interception offer a sound foundation for linear relationships between B and summations of interception over time (Monteith 1977).

3.2.4   Productive structure

Leaves are key elements in light interception and crop productivity but leaves cannot exist alone. They must be displayed appropriately on stems and petioles and must be supplied with water and nutrients. The **productive structure** of a crop thus includes stems and roots as well as leaves. Definitions from growth analysis for **specific leaf area** (SLA; leaf area/g dry mass leaf) and its inverse, **specific leaf mass** (SLM), are useful here. As an example, SLM for sugarbeet is near 400 kg leaf mass ha^{-1} leaf area. A roughly equal amount of dry matter is required for supporting stem and petioles, and each kg of leaves needs about 0.15 kg of fine roots to supply water and nutrients.

Therefore, about 3000 kg dry matter ha^{-1} is required to achieve a full-cover display of LAI $= 3$.

Annual plants follow characteristic partitioning patterns in which early growth is given to productive structure while growth of fruit and vegetative storage organs is delayed until later in the season. That was illustrated in Fig. 3.1b by patterns of dry matter accumulation in maize. Nutrient uptake occurs mainly during vegetative growth when large amounts of nutrients, especially N, are needed to create photosynthesis apparatus in leaves. Later, as grain fills and leaves die, that N may be transferred to grain. In most annual crops, as much as 70% of whole-crop N is found in grain or seed at maturity. That pattern accounts for changes in composition of sorghum biomass presented previously in Table 2.2 and for generally small concentration of N in crop residues.

Factors that limit growth or photosynthesis rate, such as temperature, water, and nutrient supplies, also limit leaf expansion and development of cover. These constraints can be manipulated to some extent through choice of cultivar and with tillage practices that influence soil temperature (Sections 7.6 and 12.5.4). Time to full cover is less for dense plantings than for wide-spaced plantings and for large-seeded plants (large initial leaf area) than for small-seeded ones. Those practices have the effect of increasing B_0 in Eq. 3.3. Shortening the period of incomplete cover also is important in reducing opportunities for development of weed competition.

Aboveground productive structures form the useable yield of forage grasses and legumes but in contrast to reproductive organs, such as grain or tubers, only small amounts of biomass can accumulate in fine stems and leaves. As a result, CGR for forage plants declines after a brief period at full cover, owing to maturity or to "feedback inhibition" of photosynthesis (Section 10.2.7). The solution is to graze or mow forage crops periodically to allow production of new stems and leaves. Another problem with forages is that older stems generally have greater lignin and lower N content, which causes digestibility to decline with age. For these reasons, sums of repeated harvests of pastures or forage crops during a growing season are generally much larger and more nutritious as feed than biomass obtained in single, end-of-season harvests. Regrowth after each harvest follows lag and linear phases as leaf area expands.

3.2.5   Production rates

The main portion of seasonal yield accumulates during "linear growth" and it is useful at this point to consider the magnitudes of production rates during that period.

Potential production rate Photosynthesis is driven by energy absorbed in light quanta. The amount of sunlight incident on a crop therefore sets an upper limit to its production. Assuming that nutrients and water are abundant, potential gross (before respiration) and net (after respiration) productivities of crops with complete cover can be estimated at ambient atmospheric CO_2 concentration. To do that, we need to borrow information about spectral characteristics of solar radiation from Section 6.4 and photosynthetic ability of leaves from Section 10.2. Steps in the estimation are presented in Box 3.2 for a full-cover canopy of leaves during linear growth. Each MJ of incident solar

---

Box 3.2 Stepwise calculation of potential productivity of a crop surface per MJ solar radiation and per day

Incoming solar radiation	1 MJ m^{-2}
Incoming photon flux density[a]	2.06 mol m^{-2}
Loss by reflection (8%)	-0.16 mol m^{-2}
Loss by inactive adsorption (10%)	-0.21 mol m^{-2}
Quanta for photosynthesis	1.69 mol m^{-2}
[CH$_2$O] produced at quantum yield of 0.10	0.169 mol m^{-2}
and as glucose	5.07 g m^{-2}
Average loss to respiration (33%)	1.67 g m^{-2}
Net production per MJ[b]	3.40 g m^{-2}
Daily net production at 20 MJ m^{-2}	68 g m^{-2} day^{-1}

Notes: [a] based on the spectral composition of average sunlight in the photosynthetically active range, 0.4 to 0.7 μm; [b] with glucose (30 g mol^{-1}) as the product, 5.3% of the energy in total radiation and 12% of that in the photosynthetically active range is conserved in biomass. *Source:* adapted from Loomis and Williams (1963).

---

radiation contains about 2 mol quanta of **photosynthetically active radiation (PAR)** in the active spectral region (0.4 to 0.7 μm wavelengths). Some radiation intercepted by the crop is lost by reflection and through absorption by non-photosynthetic structures, leaving 1.69 mol quanta available for photosynthesis. Maximum observed quantum yield of photosynthesis is near 0.10 mol CO_2 assimilated per mol quanta reaching the chloroplasts. Gross photosynthesis is therefore 0.169 mol (5.07 g) carbohydrate MJ^{-1} radiation. If a third of that production is then used in respiration, net increase in dry matter for each MJ sunlight is 3.4 g MJ^{-1}. That ratio of organic production to radiation defines potential radiation-use efficiency. With a moderate daily total radiation of 20 MJ m^{-2} d^{-1} (40 mol quanta m^{-2} d^{-1}), 68 g carbohydrate m^{-2} d^{-1} would be produced.

The amount of chemical bond energy stored in dry matter can be determined from heat energy released during combustion to CO_2. The heat of combustion of carbohydrate (glucose) is near 15.6 kJ g^{-1}, therefore the potential energy efficiency of net photosynthesis is 3.4 g MJ^{-1} × 15.6 kJ g^{-1} = 0.053 in use of solar radiation. Because PAR amounts to only about 45% of total energy in sunlight, potential energy efficiency in that waveband is near 0.12. By contrast, the best artificial solar cells now approach 0.4 efficiency.

Record rates of production It is instructive to compare these estimates of potential production with record performances of crops under conditions of adequate water and nutrients and without limitations from insects, disease, and weed pressure (optimal growth conditions). Considerable care is needed in assessing such records because many published estimates do not meet essential standards of sampling or attention to cultural conditions (Loomis & Gerakis 1975; Monteith 1978). Data on daily productivity of crops presented in Table 3.1 were selected as valid measurements against which fair

Table 3.1 Some high daily crop growth rates (CGR), radiation-use, and photosynthetically active radiation (PAR) conversion efficiencies for various cultivated crops and forages according to photosynthesis types (PS)

Crop	PS type	CGR $(g\,m^{-2}\,d^{-1})$	Irradiance[a] $(MJ\,m^{-2}\,d^{-1})$	Radiation use $(g\,MJ^{-1})$	PAR conversion $(\%)$[b]	Location
Sugarbeet	C3	31	12.3	2.52	9.5	United Kingdom
Potato	C3	23	(16.7)	1.38	5.4	The Netherlands
Maize	C4	29	(18.8)	1.54	6.1	New Zealand
Maize	C4	52	30.8	1.69	6.4	California, USA
Maize	C4	52	20.9	2.49	9.8	New York, USA
Sudangrass	C4	51	28.9	1.87	6.7	California, USA
Millet	C4	54	21.3	2.53	9.5	Australia
Napiergrass	C4	39	(16.7)	2.34	9.3	El Salvador

Notes: [a] daily total solar irradiance, estimated values in parentheses; [b] energy content of organic production/ energy in PAR; $MJ\,MJ^{-1} \times 100$.
Source: adapted from Cooper (1970).

comparison of potential productivity can be made. Largest daily crop growth rates shown there belong to a group of crop plants of the family Gramineae distinguished by C4 photosynthesis and a high content of carbohydrates. C4 plants are capable of very rapid rates of photosynthesis. No record observation of dry matter production rates by C4 crops ($30-54$ g m^{-2} d^{-1}; $300-540$ kg ha^{-1} d^{-1}) exceed our estimated potential daily net production (68 g m^{-2} d^{-1}) despite the fact that some of those crops experienced radiation levels as great as 30 MJ m^{-2} d^{-1}. Broad-leafed species with C3 photosynthesis (characterized by smaller rates of photosynthesis than in C4s) and those that accumulate compounds such as oil and protein had smaller daily rates of production ($20-30$ gm^{-2} d^{-1}; $200-300$ kg ha^{-1} d^{-1}). C3 and C4 systems of photosynthesis are examined in detail in Section 10.1.2.

Annual yields of 50 t ha^{-1} or greater are achieved by real crops but only over long growth periods. Length of season and sensitivity of C4 plants to low temperature (high latitude) are revealed in Fig. 3.4a as major restrictions of annual production. Best C4 crops outyield best C3s at low latitudes where they receive abundant radiation, but above latitude $40°$, C3 crops are more productive. Differences in short-term productivity also persist into seasonal yields. Mean growth rates of the best crops in Fig. 3.4b are 22 ± 4 and 13 ± 2 g m^{-2} d^{-1} (mean $\pm$ SE) for C4 and C3 crops, respectively. Most crops, including those with special adaptations to water or temperature stress, and those with large protein and oil content, will be less productive.

3.3   Responses to crowding in monocultures

Crowded plants interact strongly with each other in a number of ways. Controlling variables include the number of species involved, plant density (ρ, number of plants

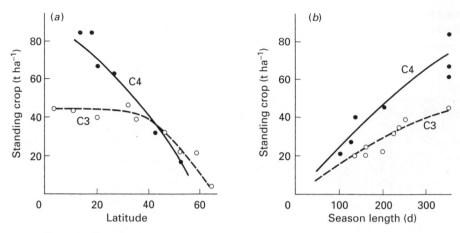

Fig. 3.4 (*a*) Latitudinal distribution of annual aboveground production by record crops (data sources are given in Loomis and Gerakis 1975). (*b*) Dependence of those record yields on length of growing season (after Monteith 1978).

per unit area), and plant size. Here we look at density effects and spacing control in monocultures, but first we discuss the concepts of competition.

3.3.1   Concepts of competition

Seedlings are small and have little or no interaction with neighbors but as plants grow, they overlap to an increasing extent in both aerial and soil environments. Average space per plant is $1/\rho$ and area of sunlight per plant at full cover declines with increasing density. As a result, growth rates and morphologies of individual plants differ dramatically with density.

Crowded plants clearly interfere with each other for light. Evaporative demand leading to transpiration per plant is less for dense than for sparse stands but the area for receipt of rainfall and CO_2, and the soil volume supplying stored moisture and nutrients, are also less. The terms **interference** and **competition** are used interchangeably in reference to crowding phenomena. Interference occurs when the supply of resources at a site is less than the collective ability of closely spaced plants to use them. Interference may also occur because chemical substances released from a plant or its residues affect the growth of neighbors. That type of interference is termed **allelopathy** (Putman & Tang 1986). Although relatively common in natural systems, allelopathy, particularly self-allelopathy, is unusual in crop plants. Agriculturally relevant examples occur mainly with pasture species. It may be that allelopathic compounds are expensive in synthesis (Section 11.2), or tend to be toxic to humans and animals, and have therefore been lost under selection by humans.

Regardless of whether resources are limiting, or whether allelopathy is involved, interference among plants is largely **indirect** (passive) through changes in the surrounding environment. For example, as one plant depletes soil N and water, a second plant then finds less of those resources available to it. Interference is easily demonstrated when

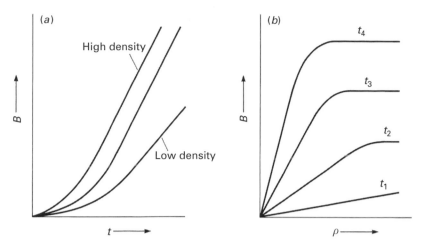

Fig. 3.5 (*a*) Time course of biomass (*B*) accumulation beginning at emergence by crop communities with different plant densities. (*b*) Biomass accumulation as a function of plant density (ρ) at different times after emergence.

plants grown at a low density achieve a larger size than those at greater density, or when crowded plants respond with expansive growth after thinning or receiving fertilizer, which alleviate resource constraints.

In contrast, crowding responses of animals may be either indirect (one animal reduces forage supply below the amount needed for a second animal) or **direct** (active) as occurs when two animals contest physically for territory. Although the phrase "competition for resources" is acceptable in crop ecology, its use should not imply direct physical contention. The passive nature of plant competition led to the use of the term interference (Harper 1977).

3.3.2   Density effects

The general nature of crowding responses in monocultures (**intraspecific competition**) is illustrated in Fig. 3.5*a*, where the usual exponential pattern of biomass accumulation is modified sharply by changes in density. In particular, the "lag" associated with the exponential phase is shortened as density increases: ten apical meristems m^{-2} can provide new leaves and increase light interception and growth rate much more rapidly than one. As a result, CGR (initial slopes of curves in Fig. 3.5*b*) increases with increasing density until cover is complete. Crop growth rate at full cover, however, is affected little by density. The denser crop has significantly greater biomass early in the season but the relative magnitude of that advantage declines as the season progresses. The curve for low density in Fig. 3.5*a* fails to reach full cover and is fated to a small final yield.

During the early stages of crop growth, seedling plants grow without interference and all plants have similar mass ($b = B/\rho$). Thereafter mass per plant is affected according to the intensity and duration of interference and b varies as a reciprocal of density. Curves of this shape suggest analyses on a logarithmic or reciprocal basis that tend to

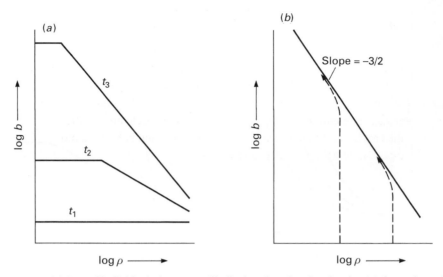

Fig. 3.6 (*a*) Log of individual plant mass (*b*) displayed against log density (ρ) for various times after emergence in a Kira plot. (*b*) Kira plot of the self-thinning phenomenon. The dotted tracks depict the course of log *b* up to and then along the $-3/2$ limit.

straighten the relationship (Fig. 3.6*a*) (Kira *et al.* 1953). As plants increase in size, onset of interference is evident as a break from constant *b*. In this case, plants grew within the space allotted them at planting, adjusted their growth accordingly, and none died.

Seeding densities chosen for field and vegetable crops result in growth responses illustrated in Fig. 3.6*a*. The aims are to have mature plants sufficiently crowded to use all resources efficiently, yet not so crowded that some plants die or are unproductive. Neither should density so weaken stems that plants fall down (lodge) due to high fruit load or extreme weather conditions. Each plant then produces less than it would with unlimited space but production from the community is maximized. Plants of large final size must be sown with wide spacing; as a consequence, achievement of full cover is delayed. As noted above, wide spacing allows opportunity for invasion by weeds. That problem can be partially offset with large seeds, which enter exponential growth with large initial leaf area giving greater growth rates to achieve earlier cover.

3.3.3 Self-thinning

Some variation in space and mass per plant is usual even with precision planting, and quite large variations occur with broadcast (random) seedings. When crowded, the smallest plants with least resources become further disadvantaged and die. Resultant open space allows expanded growth of neighbors. In this case, competition leads to winners and losers.

Where mortality occurs, Kira plots of log (mean *b*) against log ρ show the interesting relationship illustrated in Fig. 3.6*b*. Masses of individual plants in closed stands eventually reach a limit defined by $b = a\,\rho^{-3/2}$, where *a* is a proportionality factor. That leads to

---

Box 3.3 Plant density and spacing control interference in crop communities

Precise control of plant density as well as uniform spacing are easily achieved with modern planters using large, uniformly sized, seed. Small seed can be coated with clay or other materials to form large, uniform size. Precision planters allow control not only of average space per plant but also of their spatial arrangement. Foliage of a sugarbeet plant, for example, is displayed in a circle around the axis; for that shape, the hexagon ("French orchard") pattern illustrated below is a more efficient solution than square or rectangular arrangements.

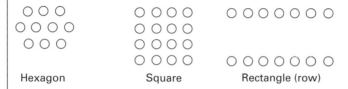

Hexagon   Square   Rectangle (row)

In hexagons, each plant encounters six equidistant neighbors and circular foliage packs efficiently. Yield advantage gained by that arrangement is small, however, and it is not used with crops. It seems that most plants have sufficient plasticity to adjust to whatever space they are provided above or below ground. Row patterns reduce the number of planter and harvester passes needed and fertilizer can be distributed efficiently in bands (Section 12.4). More time is required for canopy closure but weeds can be controlled by cultivation or directed herbicide sprays. Row culture can also serve as a means to conserve and ration a limiting supply of soil moisture (Section 13.7).

---

the linear relation, $\log b = \log a - 1.5 \log \rho$. Dotted lines in Fig. 3.6b depict time-course trajectories for two stands, differing in initial density, towards and then along the $3/2$ limit line. Communities apparently reach a ceiling in the amount of biomass that can be supported per unit area; individual plant masses can then increase further only as space is created by death of neighbors. This relationship has been observed with a wide range of plants, including annual crops and trees, and is termed the **self-thinning rule**. One explanation is that as plants increase in size they need larger areas of sunlight to supply enough carbohydrate for maintenance respiration. The $3/2$ term results from a correlation between plant mass (proportional to volume occupied, length^3) and sunlit area (length^2).

Self-thinning is of great importance in natural systems and in forestry. In agriculture, we attempt to avoid it by precise control of crop density and spacing, as illustrated in Box 3.3, because dead plants tie up nutrients and carbon that would have benefited survivors. In addition, large plant densities create conditions that are conducive to disease and lodging of grain crops. Self-thinning is common, however, during early stages of growth with heavy seedings of forage legumes and grasses.

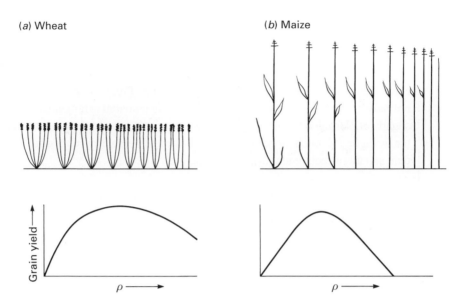

Fig. 3.7 (*a*) Tillering of wheat varies over a broad range of plant densities. That results in a wide range of optimum densities for grain production. (*b*) Tillering is suppressed in maize and plasticity rests mainly with variations in ear number and grain number per ear. The optimum plant density occurs over a narrow range.

3.3.4 Plasticity

Optimum density for a crop is influenced by many factors including resource levels and length of growing season. Potential plant size and degree of **morphological plasticity** for successfully occupying varying amounts of space are also important. Branching (including tillering of grasses) is a basic feature of plasticity. Tillering allows small grain crops to reach similar numbers of heads and final yield per unit area over a wide range of densities (Fig. 3.7*a*). By contrast, maize can be precision-planted to final stand and modern cultivars have been selected to tiller little, or not at all. As a result, that crop has a relatively narrow range of optimum densities (Fig. 3.7*b*).

Many crops are sown with drills. That gives reasonable control of row width and seed per meter of row, but spacing between plants within rows is variable and a degree of morphological plasticity is required. Broadcast sowing is much less precise. In that practice, seeding rates are denoted in kg of seed ha^{-1} and seeds are broadcast on the surface and then incorporated with soil by harrowing. Seed distribution is random and each plant encounters variable amounts of space between itself and neighbors. A high degree of morphological plasticity is needed and *b* is variable. Given such plasticity, a wide area can be occupied by aggressive tillering; with close spacing, tillering is suppressed by crowding and self-thinning is common.

3.4 Competition in polycultures

Main features of **interspecific** competition in crop stands derive from differing requirements of individual species for resources and their corresponding abilities to gain access

Fig. 3.8 Pure and mixed stands of hypothetical species X and Y. The understory niche in a community of Y has very limited light resources for the growth of X; when the crops are mixed, little production can be expected from X.

to them in their shared environment. Here we discuss effects of physiological and morphological differences between plants but begin with a discussion of the niche concept and its application to competition between plants.

3.4.1   Niche concept

Competition in mixed stands is introduced by considering competition between unlike animals. If animals eat different things and do not prey on each other, there is very little interference between species. Such animals are considered to occupy and exploit different **niches** (places) in the community. Definitions of niche vary. The issue can be kept relatively simple by saying that in crop ecology, we define niche as the physical space and associated resources utilized by a plant.

Some experiments with animals indicate that intensity of competition between two species is greatest when their niches are similar. Those studies support the view that species with overlapping niches may compete so intensely that they cannot continue to coexist within a community. That hypothesis, known as Gause's exclusion hypothesis, is of interest in evolution and animal ecology.

Whether Gause's hypothesis applies to plants as well as animals remains unestablished. The theory proposes that interference will be more severe among closely similar plants (i.e., monocultures) than in polycultures among plants differing in growth habit and resource requirements. Opportunities for use of different space or different resources (**niche differentiation**) are much less with plants than with animals. Plants are not mobile, as are most animals, and all plants require the same resources of light, carbon dioxide, water, and nutrients. Except for legumes, which are capable of using atmospheric as well as soil N, there is near-perfect overlap among plant species in their resource requirements.

The example in Fig. 3.8 illustrates two plant species cohabiting in time, but differing in size. In pure stands of X and Y, interference may become intense; indeed it must be if the crops are to achieve maximum yield or efficiency of resource use, but it is equal among all plants, and all individuals perform well. Competition is intense but it does not affect survival or reproductive success, i.e., competition in monocultures is not "important" in an evolutionary sense (Weldon & Slauson 1986). When X is interplanted with Y having quite different size, the outcome is very different. Placed in the understory, X is

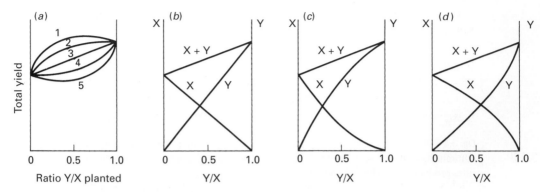

Fig. 3.9 (*a*) Possible outcomes of total yield of binary communities created by replacement of species X with species Y. (*b*) Total yield and component contributions with neutral interference. X and Y respond to each other as they do to themselves. (*c*) Y wins. (d) X wins.

suppressed strongly by shading and perhaps also by interference for nutrients and water unless Y is widely spaced. If X is unable to reduce nutrient or water supply to a level limiting to Y, large Y plants may respond almost as if they were competing only against themselves. In this case, competition is severe and "important" for X and "unimportant" for Y. Unlikeness has led to winners and losers. That, of course, is exactly what we hope will happen when a crop encounters weeds.

The most limiting resource in a competitive situation is easily diagnosed in extreme cases, for example, where nutrients and water are abundant, competition for light may be inferred. Interactions occur, however, where more than one resource is in limited supply (Donald 1958). Such situations are not so easily diagnosed. A species that has advantage in light capture will have a greater supply of substrates for root growth. As a result of a larger root system, it may capture more of the limiting soil resources, enhancing leaf growth and light capture. Competing species are then distressed in their access to soil resources as well as to light. Conversely, a primary advantage in root display or nutrient uptake can have secondary effects, for example through greater leaf display. In these situations, competition for one factor induces increased competition for other factors. Effects can be additive and therefore stronger than expected from primary competition alone.

3.4.2   Replacement experiments

Competitive interactions among unlike plants can be evaluated in **replacement experiments** with treatments consisting of a series of communities with varying proportions of two cohabiting species (de Wit 1960). Figure 3.9*a* depicts the range of possible total yields (biomass or economic) that might be obtained as a pure stand of species X is progressively diluted by species Y. Over-yielding (line 1) indicates that the polyculture exploited the environment more thoroughly than either crop in monoculture. If both species perform better in the mixture than in pure stand, they **complement** each other, which means that the two crops must have some form of niche differentiation.

Conversely, a total yield less than either pure stand (line 5) indicates possible mutual **antagonism**. The straight line (3) indicates the possibility of a neutral interaction.

Separate yields of component species must be examined in each of these cases for proper interpretation of interactions involved. That is done in Fig. 3.9*b–d* for the case where total yield suggested a neutral interaction. Component analysis reveals that interference may indeed be neutral (Fig. 3.9*b*), or that Y, which is superior in pure stand, wins (Fig. 3.9*c*), or that X, the poorer yielding in pure stand, wins (Fig. 3.9*d*).

3.4.3 Land equivalent ratio

Yields of mixtures may also be compared on a relative basis. If each component of a 50:50 mixture yielded 0.6 times as much as in a pure stand, then 1 ha of polyculture would provide the same total yield as 1.2 ha equally allocated to X and Y as sole crops. The ratio 1.2/1 ha = 1.2 is termed the **land equivalent ratio** (LER). More generally, LER can be calculated as:

$$LER = \sum_{i=1}^{n} Ym_i / Yp_i \qquad (3.6)$$

where, Ym_i, and Yp_i, represent the yields of the *i*th species in the mixture and pure stand respectively. Land equivalent ratio is numerically the same as **relative yield total** (RYT) of polycultures introduced by de Wit and van den Bergh (1965). That term is now not widely used; LER is preferred for its implication of optimization of unit area of land.

An LER >1 indicates that the mixture is superior in yield to the sum of the pure stands, i.e., that complementation exists. Serious mistakes about management practices can be made from experiments on mixed cropping unless pure stands and mixtures are both formulated independently in optimal ways (i.e., with spacing and ratios of plant numbers that provide the best performance of each community). Most reports on the performance of polycultures involve errors of this type and the reported LER values are then meaningless. In addition, LER is affected strongly by environment and management level: values from the same mixture may be <1 or >1 depending upon plant density, fertility, water supply, or soil pH.

3.4.4 Complementation in mixtures

Intensity of competition and its impact on productivity of competing plants depends on the degree of niche differentiation. Complementation in use of resources results from niche differentiation. In polycultures, complementation can occur in three ways through use of differences in:

- space, especially with regard to root systems;
- time during the growing season as a result of different phenological development rates and maturities; and
- sensitivity to uncertain stress factors such as drought, temperature extremes, or disease and insect pests.

Each approach deserves discussion.

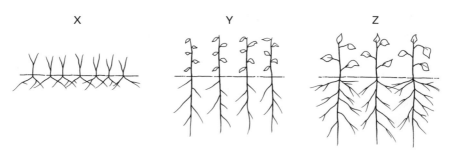

Fig. 3.10 Shallow-rooted species X and deep-rooted species Y could complement each other in a binary mixture but species Z, with both shallow and deep roots, could fully exploit the soil profile in monoculture. Species X and Y can be viewed as phenotypic defects that might be corrected by selection for type Z plants in a breeding program.

Different space　There is little opportunity for complementary use of aerial resources because CO_2 and light can be fully intercepted by any canopy at full cover. Full cover is just as easily achieved with monocultures as it is with polycultures. In addition, lower leaves of sun plants adapt to low light levels within canopies as well as a "shade" plant might. With less than full cover, as is frequently the case when soil resources are limiting, it makes little difference what canopy is displayed because all leaves are then in bright sun. Then, final yield is determined largely by limiting soil resources.

Complementation through the use of different space is sometimes possible when soil resources are limiting and species have marked differences in rooting habit (Fig. 3.10). With a mixture of species X and Y, soil resources not accessed by shallow-rooted X could be captured by deep-rooted Y. The two species might complement each other and yield in mixture could exceed that of pure stands of either species (i.e., LER > 1). In this case, spatially non-uniform treatment of land through mixed cropping increases the use of scarce soil resources.

The possibilities for complementation are greatest where soil resources, either water or nutrients, are limited (Spitters 1980; Ofori & Stern 1987). Surface layers of soils generally have the greatest supply of nutrients (Chapter 7) and can be accessed by most species, whereas water resources are distributed throughout the profile. As a result, niche differentiation for water resources is more likely than for nutrients in mixtures not involving legumes. Supplied with adequate nutrients and water, however, each species easily acquires all its needs from its usual rooting zone and the basis for complementation is reduced (LER declines). When the limiting resource is non-mobile (P or stored moisture), advantage goes to the species that, through rapid or directed root growth, reaches it first. Where the limiting resource is mobile (nitrate ions or a downward flux of soil moisture), advantage passes to the species with roots best displayed for interception.

A plant breeder could view existence of complementation (LER > 1) of X and Y at low fertility as an indication of genetic defect. For the example in Fig. 3.10, versions of X and Y might be selected that, like species Z depicted there, would each be capable of exploring the entire soil profile. With limited soil resources, yields of both pure stands would be improved by more expansive rooting but the advantage previously shown by the polyculture would disappear because a mixture of improved cultivars would find no more

Table 3.2 Estimated LER for yield and nitrogen for one cycle of rotated maize and bean monocultures, each grown on 1 ha, and two crops of maize–bean polyculture per year in Kenya, which require an equivalent production area

Yields[a]	Monoculture	Polyculture
Maize	1200 kg grain ha^{-1}	800 kg grain ha^{-1}
Bean	800	600
Yield/cycle	2000	2800

$LER_Y = 800/1200 + 600/800 = 1.42$

Estimate of N harvested per crop cycle with 1.5% N in maize grain and 3% in bean[b]

	Monoculture	Polyculture
Maize	$1200 \times 0.015 = 18$ kg N ha^{-1}	$800 \times 0.015 \times 2 = 24$ kg N ha^{-1}
Bean	$800 \times 0.03 = 24$ kg N ha^{-1}	$600 \times 0.03 \times 2 = 36$ kg N ha^{-1}
Total N harvested	42 kg N ha^{-1}	60 kg N ha^{-1}

$LER_N = 12/18 + 18/24 = 1.42$

Sources: [a] yield information came from local farmers; [b] N content of maize grain and bean seed are from National Research Council (1982).

nutrients or water than are accessed by either alone. Alternatively, instead of altering X and Y through plant breeding, they could simply be replaced with a monoculture of Z with the desired rooting habit.

Interactions between legumes and non-legumes can be viewed as involving the use of different space because non-legumes depend on soil N whereas legumes can obtain much of their N from air (Chapter 8). Maize–bean polycultures, common on infertile soils in tropical regions, are a principal example with crops. A situation with LER > 1 is illustrated in Table 3.2 with an approximate N budget for a low-N site in Kenya. Maize plants at a density of 10 000 plants ha^{-1} (1 m^2 area per plant, which is 7.5 times that used in the US Corn Belt where maize is planted at much higher density (Chapter 17), obtain enough N to produce one ear of grain. At that low density, abundant light and water remain for an understory bean crop.

The small yield of the Kenyan maize monoculture in Table 3.2 is explained by N deficiency; the poor yield of the bean monoculture indicates problems with disease, other nutrients, or acid soil. For comparison, maize adequately supplied with N easily yields 5 to 10 t grain ha^{-1} in Kenya. When the acid soil is neutralized with lime, and fertilized with phosphorus, 2 to 3 t ha^{-1} can be expected for beans. Correction of soil problems leads to vigorous growth by taller growing maize and LER would likely drop below 1.

Different time Long growing seasons create opportunities for niche differentiation in time. When a single species is unable to utilize the full season, growth duration may be extended by adding other species. The most common examples are pasture communities constructed by sowing mixtures of cool and warm-season grasses and legumes. That not only extends the growing season but also may provide a more even and prolonged supply of green forage than could be obtained with monocultures. Land equivalent ratio values seldom exceed 1 (Trenbath 1974), however, so the principal advantages seem to come from improved animal nutrition (high-protein legume forage balances more productive, cellulosic grasses), N fixation, and from a more even distribution of forage.

In the tropics, a slower growing legume such as pigeon pea is often planted as an alternate-row polyculture with a cereal crop like sorghum that matures much earlier. The more rapidly growing sorghum with a fibrous root system exploits upper soil layers for nutrients and water during the first months of the rainy season and then matures, leaving the remainder of the rainy season and deep residual soil moisture to provision the slower growing and deeply tap-rooted pigeon pea. This case involves both space and time differentiation.

Different sensitivities to uncertain stresses Occurrence and severity of drought, temperature extremes, disease and insect pests vary considerably from year to year and are difficult to predict. Where farmers do not have access to irrigation or pest control options such as biocides, polycultures of species with differential tolerance to these stresses provide a degree of risk management. When a stress event occurs, growth of sensitive species decreases, reducing competition for resources, and allowing more tolerant species to exploit them. This can result in LER > 1. In years without stress, however, polycultures may underperform relative to monocultures. Thus, this type of niche differentiation is most useful in harsh environments where the probability of stress is high.

3.4.5   Advantages and limitations of polycultures

Any advantage from polycultures compared with monocultures depends on achieving LER > 1. As we have seen, that is most likely where soil resources are limiting, and rooting habits differ, or one species is a legume. Considerable attention has also been given to whether LER > 1 can also result from fewer disease or insect problems than occur in monocultures. The basis for this is that some insects and diseases can be limited in their dispersal and spread when susceptible host plants are interspersed with resistant plants. With insects, it is also possible that one species may serve as a trap for insects, reducing infestations of the other, or that it may serve as a breeding place for predators. Evidence of these possibilities is equivocal. Examples of both increased and reduced insects and disease have been observed depending on the type of polyculture and they have not been well correlated with effects on yield.

In general, a larger number of hosts in a polyculture also encourages a greater diversity of pests and diseases. That problem can be particularly severe with soil-borne organisms. Since host plants are present in every crop, sanitizing effects obtained with rotated monocultures are lost. For example, maize–bean polycultures are reasonably successful in the tropics because neither species is faced with severe attack by soil-borne organisms. The system would not work in the American Corn Belt, however, where corn rootworm (*Diabrotica longicornis*) builds up quickly when maize is present in the same field every year.

A summary of polyculture variants is presented in Box 3.4. Polycultures are not widely employed with arable crops in developed countries for two reasons in addition to loss of

---

Box 3.4 Polyculture variants

Agricultural polycultures with cohabitation in space and time involve varying degrees of niche differentiation and competition.

Intercropping Planting one crop within open areas of another.

Some variants may actually have little or no competitive interaction. During early years of a new orchard, for example, spaces between young trees can be planted with an annual crop.

Overlap systems Two or more species overlap for portions of their life cycles.

These are sometimes used to extend growth duration in mild climates. That can be done by interplanting a maturing monoculture with a second monoculture. Interplanting maize with beans in some tropical regions is an example.

Companion cropping A special case of overlap cropping.

Most common cases involve seeding a small grain as a companion crop with forages. The grain crop serves to reduce soil erosion in the seedling forage stand while smothering weeds and providing an economic return as the forage crop becomes established. The forage crop also suffers shading, however, and little or no hay is harvested in the first year. Selective herbicides are now available for weed control and forage crops can be established without a companion crop. The advantage of herbicide, in addition to smaller cost, is that significant production of hay and N are obtained from legumes in the first year.

Relay cropping Two or more monocultures are grown in sequence in the same year.

This is sometimes defined as a polyculture technique but is actually just a form of crop rotation. Little interspecific competition occurs. Relay cropping is found in vegetable gardening, with rice production in Asia, with wheat–soybean (winter–summer) double cropping in the USA, and in many other forms.

---

rotation benefits. First, fertility levels are generally high and LER > 1 is rare and, second, improved performance, if it occurs, generally does not compensate for difficulties encountered in balancing competition, planting and harvesting, and optimizing management (e.g., weed control and supplies of nutrients and water). Optimal management of two or more crops in a polyculture in terms of planting date, water and nutrient supplies, and other factors is more complex or even impossible compared with monocultures. Possibilities for mechanical harvests and for controlling weeds with herbicides also are greatly restricted. Management of pasture polycultures (e.g., grass–legume mixtures) faces several of these issues and must also ensure the persistence of desired species over time. With an adequate supply of soil N for grasses, legumes languish; with N deficiency, grasses perform poorly while legumes and weeds (unpalatable or toxic species) increase. In some systems, rotation of pastures through a cropping phase allows them to be reconstructed periodically with suitable species. In Europe, the legume component has in many cases been abandoned as a source of N for intensive pastures (see

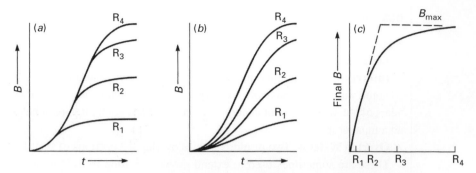

Fig. 3.11 (*a*) Time course of biomass accumulation by crops with different initial supplies of a limiting resource ($R_1 \rightarrow R_4$). Growth slows as the resource is exhausted. (*b*) Time course of biomass accumulation with varying degree of limitation by a resource whose supply is never adequate. (*c*) Total biomass ultimately produced by the crops in (*a*) or (*b*) as a function of resource level. The broken lines that indicate the initial slope and asymptotic yield, B_{max}, the two parameters of the simplified, linear Liebig–Blackman response.

Table 8.3). Pastures there are now managed as pure stands of grass supplied with fertilizer or manure N. Management of extensive grasslands found in semi-arid regions is limited to tactical deployment of grazing activity and stocking rates to favor dominance by desired species. Efforts are sometimes made to introduce legumes and to supply them with needed soil amendments (e.g., P, S, or Mo); that approach has generally not been profitable.

3.5   Community response to limiting factors

Yield is a community property that reflects the collective ability of crowded plants to capture light and access soil resources. Community response to additions of a limiting resource is a critical determinant of productivity and, in high-yield systems, of environmental impact also.

3.5.1   Single-factor responses

Returning to sigmoidal growth as seen in an annual monoculture (Fig. 3.1), consider what may happen when supply of a resource such as N or water is less than adequate. Two patterns of behavior over time are possible. In Fig. 3.11*a*, the resource is exhausted after a varying period of normal growth with adequate supply; thereafter, growth is restricted. This pattern is common with N and water resources. The second pattern is illustrated in Fig. 3.11*b*. In this case, the resource is always limiting as occurs, for example, when mineralization of N from soil organic matter proceeds at a rate slower than plants can use it. An unfavorable soil pH or toxicity (aluminum for example) has a similar effect.

Response functions of biomass yield (B) versus resource level (R) can be obtained for the examples in Fig. 3.11*a,b* using B values observed late in the growing season after

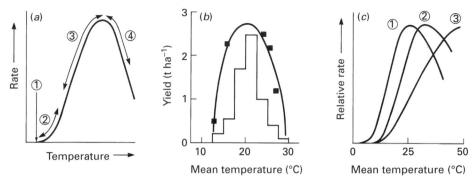

Fig. 3.12 (*a*) Shape of the complex response function observed with most plants for growth rate or development rate as a function of temperature. The numbered portions are explained in the text. (*b*) Curve of cardinal temperatures for bean production defined from observed yields and mean growing-season temperatures at five locations (filled squares) in Latin America. The histogram describes the present distribution of bean yields in Latin America. (Redrawn from Laing *et al.* 1983.) (*c*) Generalized growth and development responses for (1) temperate grasses and legumes, (2) tropical legumes, and (3) tropical grasses. (Adapted from Fitzpatrick and Nix 1970.)

treatment differences are fully developed. Both of these examples produce a plateau response function (Fig. 3.11*c*). The same general curve is found when growth rate is graphed against resource level. Plateau curves can be fitted empirically by several expressions including the negative exponential:

$$B = B_{max}(1 - e^{-cR}) \tag{3.7}$$

where B_{max} is the asymptotic yield, unlimited by resource level (R), and c is a constant. The slope (dB/dR) defines the **sensitivity** of yield to changing levels of R:

$$dB/dR = c(B_{max} - B) \tag{3.8}$$

It is sometimes convenient to simplify the curve to two intersecting straight lines (Fig. 3.11*c*), one reflecting initial slope of response (equal to $B_{max}c$ at $R = 0$), and the asymptote (B_{max}) (Cerrato & Blackmer 1990). That concept was first applied by J. Liebig to soil nutrient responses and by G. E. Blackman to photosynthesis rate with increasing light. The terms **Leibig** and **Blackman** response are still in common use.

The number of response functions encountered in crop ecology is relatively small. Examples of plateau responses (Fig. 3.3*a*) and exponential decay (Fig. 3.2*a* and Eq. 3.3) have already been presented. Observed dependence of growth and development rates on temperature in controlled environments (Fig. 3.12*a*) has a different shape. Whereas a plateau response is explained reasonably well by initial slope and asymptote, responses to temperature apparently integrate several processes. A set of hypotheses can be proposed to account for the numbered regions in Fig. 3.12*a*: the threshold (1) could involve activation energy for a limiting enzyme; an increase in frequency of collisions between enzymes and substrates would account for region (2); enzyme saturation or substrate limitation would slow the increase (3); and injury to critical enzymes could explain region (4).

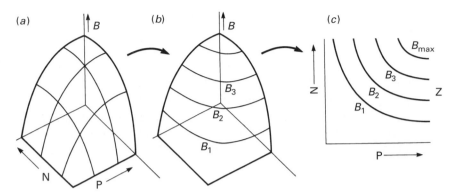

Fig. 3.13 (*a*) Response surface of yield to variation in levels of N and P constructed from intersecting plateau functions. (*b*) The same surface shown with elevation (yield) contours. (*c*) The contour lines transferred as isoquants to the two-dimensional, N–P plane.

Fine detail of growth response to temperature is difficult to see under field conditions but **cardinal temperatures** corresponding to minimum, optimum range, and maximum temperatures for growth can usually be drawn from field data. The example given in Fig. 3.12*b* explains the geographical distribution of bean production in Latin America. Optimal responses are also seen with factors such as soil pH. As with temperature, species may differ significantly in threshold and maximum and in general breadth of response. Such response functions give quantitative definition to the term **tolerance**. The wide range between minimum (12.2°C) and maximum (29.1°C) temperatures for bean in Fig. 3.12*b* gives it a "broad" (in contrast to "narrow") tolerance to temperature. Fitzpatrick and Nix (1970) make the useful point that crop plants can be classified into three broad groups characterized by the growth responses to temperature as illustrated in Fig. 3.12*c*.

3.5.2   Multiple-factor interactions

The curvilinear nature of many single-factor responses offers the possibility that two or more factors can simultaneously limit crop performance. Consider a simple example with plateau responses to N and P as nutrients. Through construction of a three-dimensional diagram (Fig. 3.13*a*; visualize this in the corner of a room), interaction of two factors results in a response surface that can be shown with contour lines of equal yield (Fig. 3.13*b*). Reduced to a two-dimensional drawing of yield isoquants (Fig. 3.13*c*), it depicts yield obtained with any combination of the two factors. An important point from this figure is that field situations can always be analyzed as single-factor responses. The response to N alone is qualitatively similar at all levels of phosphorus; the principal differences are that slope of response and maximum yield differ with P level.

Graphical presentation of yield responses to more than two simultaneously limiting factors is more difficult. Fortunately, Liebig's Law, that yield is determined by the most limiting factor, generally works better than more elaborate equations. This can be explained with an example involving nutrient limitations. With N supply sufficient for

$0.5B_{max}$, amounts of other nutrients such as phosphorus needed to support that yield are also much less than are needed for B_{max} and therefore less likely to limit growth.

Solutions are also more difficult when independent factors limit sequentially. That is illustrated in crop simulation models that advance time on a daily or weekly basis using numerical methods. Multiple limiting factors then must be included with situation-specific response functions in multiplicative fashion. With a shorter, hourly time advance and internal feedback control, selection of just one most limiting factor for each hour provides realistic interaction over time in crop models using pure response functions (Ng & Loomis 1984; Denison & Loomis 1989). The most-limiting factor then changes hourly as temperature, water, or substrate become limiting in turn [an example is presented in Fig. 11.6; see also Loomis *et al.* (1990)].

The real world of farming is manageable by simpler approaches providing one is alert to what factors limit crop performance. With N and P both limiting, for example, additions of P in excess of that needed by a single crop are possible because P is not readily leached from soils. The situation for N is more complicated. While good farmers know their land and rotations well enough to define the amount of N required to bring their crops towards maximum yield with Liebig's Law, application methods and amounts may result in significant losses that can have negative environmental consequences. Indeed, optimizing yields and minimizing N losses represents a formidable challenge in high-yield agriculture (Chapters 8 and 12).

In general, most multifactor issues in farming can be simplified into a series of simple single-factor decisions: which species, then which cultivar, then what spacing and planting date, then amount of fertilizer and how and when to apply it, and so on.

3.6   Review of key concepts

Communities

- Crops and pastures are grown as communities of closely spaced, interacting plants. Field crops are grown mostly as single species but many polyculture communities are used, especially in pastures but also in subsistence tropical agriculture.
- Sunlight, rainfall, and attainable yield are intrinsic properties of an individual production field.
- Farmers manage plant communities, not individual plants, to produce yield.

Biomass dynamics

- Crops accumulate biomass in a sigmoid pattern. Early vegetative growth is proportional to leaf-area index (LAI) so crop growth rate (CGR, accumulation of biomass) increases exponentially with time while the specific growth rate (RGR) remains constant. Later when crops achieve full cover, RGR falls and CGR remains constant until senescence commences.
- Examination of seasonal growth curves reveals two principles of community productivity. Seasonal production is greatest when (i) the duration of growth extends over

the available growing season (whether set by temperature or a limiting resource, e.g., water, nutrients) and (ii) duration of full cover is maximized.

- Interception of sunlight determines production rate. Potential production rate of full-cover crops corresponds to about 3 g biomass MJ^{-1} solar radiation, and maximum efficiency of energy storage in chemical bonds of biomass is near 5% of solar radiation. Record crops approach those rates of production.

- Growth responds in predictable ways to variation in levels of limiting factors although response functions to temperature, pH, and essential nutrients differ among crop species. Those differences define adaptation of each species to the environment.

Competition

- As plants increase in size, adjacent plants interfere (compete) with each other for limiting resources of light, water, nutrients, and CO_2. After canopies close to full ground cover (LAI 3–4), CGR of well-nourished crops are limited by solar irradiance. As a consequence, CGR is a plateau function of LAI.

- Crowding within crops restrains each plant to less than its potential growth rate and final size. In monocultures, interference between plants is determined by density and spacing pattern. If competition is equal, all plants survive and contribute equally to community yield. Precise spacing is not always possible, however, and varying degrees of morphological plasticity are needed. Planting densities employed with crops are designed to reach full cover quickly but with a level of interference that optimizes output of economic yield.

- Niche differentiation in time or space reduces inter-specific competition for resources such that polycultures may exploit an environment more thoroughly than a mono-culture. When that occurs, total yield can exceed combined yields of the component monocultures, i.e., land equivalent ratio, LER > 1. This outcome is most common with cereal–legume combinations grown at low N fertility.

Terms to remember: allelopathy; attenuation coefficient; cardinal temperatures; climax; competition, intra- and inter-specific; cover; crop growth rate (CGR); cropping patterns, companion, intercrop, mixed, monoculture, polyculture, relay; growth phases, sigmoidal, exponential, linear and senescent; interference; land equivalent ratio (LER), relative yield total (RYT); leaf area index (LAI); Liebig and Blackman responses; morphological plasticity; niche concept and differentiation; photosynthetically active radiation (PAR); productive structure; radiation-use efficiency (RUE); relative growth rate (RGR); self-thinning rule; sere; specific leaf area (SLA) and mass (SLM); succession.

4 Genetic resources

Crop plants carry information acquired during their evolution and breeding that determines their performance in agricultural fields. That information is held in the genetic material of living plants and is subject to change through mutation and through recombination into new patterns, and can be lost. Proper management of this resource involves knowing the capabilities of germplasms, maintaining and improving their genetic constitution, and employing them advantageously in farming.

The terminology relating to genetic resources is in a state of flux. Here we use **germplasm** to denote the totality of genes and genetic combinations found in a species, or a major portion of it. A **genetic population** (sometimes "line" or "strain") describes a smaller group of individuals (plants or seed) that share common ancestry and genes, and thus common traits. This use of the term population differs from that of population ecologists, who use it to denote closely related individuals that cohabit in time and space.

4.1 Genetic diversity in agriculture

Genetic diversity in farming can be defined at several levels. Here we deal with diversity of species and cultivars.

4.1.1 Species diversity

Thousands of plant species have been cultivated at some time or place, and several hundred are currently employed as crops, yet most crop production is derived from only a small number. Data presented in Table 4.1 account for most global production from arable land and permanent crops. Production shown there equals about 450 kg dry matter (c. 8.7 GJ) for each present world inhabitant, compared with 500 kg and 8.4 GJ of the SNU (Section 2.4). More than 65% of that production is grain, of which large portions are fed to animals and reach us indirectly in animal products. Additional contributions to the food supply are derived through other sources of animal feed (residues, forages grown on arable lands, and permanent grazing lands) but good data about those sources are not available. About 16% of energy and about 34% of protein in our diets comes from livestock products, including meat, fish, and dairy products. Small additional amounts come from home gardens but there is scant information about those sources.

Table 4.1 World areas and production of edible dry matter (DM) and crude protein by principal crops in 2007 derived from the FAOSTAT website. Cereal and pulse production values have been adjusted for an assumed 0.86 DM fraction

Crop	Area (Mha)	Production (Mt dry matter)	Mt (protein)
Wheat	214	521	78
Maize	158	681	75
Rice (paddy)	156	568	51
Sorghum and millet	82	83	9
Barley	55	114	15
Other (oat, rye, etc.)	31	55	7
Total cereals	**696**	**2022**	**235**
Potato	19	62	6
Cassava	19	43	1
Sweet potato, yam, etc.	15	35	2
Total roots and tubers	**53**	**140**	**9**
Sugarcane	23	159	0
Sugarbeet	5	35	0
Total sugar crops	**28**	**194**[a]	**0**
Total pulses[b]	**72**	**48**	**12**
Soybean	90	190	82
Cotton (seed)	33	64	12
Rapeseed	31	44	8
Peanut (in shell)	23	32	8
Sunflower	21	24	3
Other (sesame, mustard, etc.)	10	7	1
Total oilseeds	**208**	**361**	**114**
Vegetables and melons[c]	53	182	na
Fruits including berries[c]	54	111	na
Tree nuts[c]	9	6	na
Total principal crops[d]	**1173**	**3064**	**369**

Notes: [a] sugar production; [b] includes dry beans, pea, chickpea, broad bean, lentils, pigeon pea; [c] vegetables, melons, fruits, and berries assumed to have 20% dry matter; nuts in shell assumed to have 50% edible dry matter; [d] total area of land under crop is 1563 Mha, including 152 Mha in permanent crop including all tree and vine crops (fruits, nuts, berries, grapes, rubber, cocoa, coffee, and tea) (see Box 7.2 for more detail of land use in agriculture).

Four species from the family *Gramineae* – wheat, rice, maize, and barley – account for more than half of the total production shown in Table 4.1. Even discounting amounts fed to animals, cereals are clearly the major sources of energy and protein in human diets. About 20 additional crop species, including pulses, oilseeds, tuber and sugar crops, round out our food supply. The number of agricultural species expands considerably only when forage, pasture, and horticultural species are included.

Tremendous diversity is available in horticultural species (see *Bailey's Hortus*; Staff of the L. H. Bailey Hortorium 1976). Fruits and vegetables are important sources of vitamins, minerals, and roughage, and they add flavor and diversity to our diets, but their contribution to food supply is small. Less than 8% of world plant production comes from vegetables, melons, fruits, berries, and nuts; while potato alone accounts for more than 30% of that.

Limited species diversity for staple food crops is not a recent phenomenon; rather, it seems always to have been that way. Archeological evidence indicates that domestication was a gradual process and that a very small number of crops satisfied basic food needs within each region where agriculture developed (Smith 1990). Recent movement of species around the globe has resulted in much greater diversity of crops for each region than was available to earlier cultures. In the USA, for example, maize, sorghum, wheat, oat, barley, rice, beans, potato, sugarcane, and sugarbeet all serve as carbohydrate sources whereas native Americans cultivated mainly maize, beans, and squash.

There are several reasons why cereals continue as dominant crops: they satisfy basic dietary needs for protein and energy; as annual generalists they fit well in a broad range of environments; and the dry grain has high energy density and is easily stored and transported. To be useful, a plant must contribute protein and energy, or supply a material such as cellulose that can be converted by animals into a basic food. Cereals, with high digestibility and 8 to 14% protein content and 60 to 80% carbohydrate, supply protein and energy as well as or better than other groups of plants.

Relatively high resistance of cereals to disease and insect pests is perhaps unexpected until we recall that resistance is the common trait while susceptibility is rare. We have yet to observe wheat succumb to measles, for example (or to encounter humans infected by the rust fungus of small grains). Insects and disease are in fact serious problems in agriculture but in any one region, cereal crops face few such hazards, allowing focused efforts for relief through breeding and management.

Considerable research has been given to new crops, and to resurrecting old ones, with little success. Grain amaranth, for example, was once a staple crop in the Americas. Amaranths offer similar protein and carbohydrate contents to cereals but they suffer disadvantages in their small seed (with a large proportion of indigestible seed coat) and small harvest index. Those factors perhaps contributed to their displacement from Meso-America in prehistoric times by maize. The high lysine content in amaranth seed proteins has been cited as justification for efforts to improve their potential as crops, but cereal grains are now nutritionally adequate in that regard. One difficulty for "new crops" such as grain amaranths is that a large investment in research is needed just to evaluate whether their performance could be competitive with existing crops.

Fodder beet is a crop that was advanced to new roles through plant breeding. French scientists developed it as a sugar crop for temperate zones after the French navy's defeat by the British at Trafalgar in 1805 deprived France of West Indian sugar. Although successful, sugarbeet lacks a large germplasm bank and requires considerable "maintenance" research to sustain adequate resistance to diseases and insect pests. More recently, the roles of canola, soybean, sunflower, rapeseed, narrow-leafed lupin (Australia), and safflower as oil crops, and triticale (derived from hybrids of wheat and rye), have been significantly advanced by plant breeding within the past hundred years. Despite the small number of successful new crops, however, these examples justify continuing efforts with other species.

4.1.2   Agricultural cultivars

During domestication, and more recently through breeding, genetic diversity of crops has been maintained in populations with relatively narrow genetic bases. In earlier times, when farmers simply saved the best and brightest seed from their own fields, distinct local populations emerged through natural selection to local environments and farming practices. Coupled with the removal of undesirable plant types and/or positive selection of desirable types, such populations differentiated regionally into **landraces**. In modern times, landraces have gradually been displaced by carefully bred, uniform populations possessing distinctive traits. These are given names as **cultivars**. The term cultivar (cultivated variety) is preferred to **variety**, which applies in botanical taxonomy to subgroups of species; botanical varieties are given Latinized names whereas cultivars are not.

Propagation method, mating system of the species, and tactical choices about the type of matings to allow determine the genetic structure of a cultivar. **Clonal lines**, for example, are propagated vegetatively without sexual reproduction and the opportunities that it offers for genetic exchanges. As a result, clonally propagated cultivars of crops such as potato, asparagus, bermuda grass, and sugarcane generally consist of a single genotype. The same is true for most tree and vine crops and many ornamental species. An advantage of clonal propagation is that superior genotypes can be propagated indefinitely without change. By contrast, hybridization, segregation, and recombination can occur with each generation of seed production and considerable attention must be given to controlling changes in genetic content to maintain desired genetic traits.

Pollination types Species that can reproduce with pollen from a nearby plant are normally **outbreeding** and even selected populations are highly heterogeneous unless the pollen source and seed parent are controlled by isolation. **Self-pollinated species** (reproduction with pollen from the same plant), in contrast, such as barley, cotton, rice, and wheat are largely **inbred** and thus homozygous at most gene loci (but not completely, since outbreeding rates of 1 to 2% are common). The mating systems of several important crops and tactics employed for cultivar development and propagation are summarized in Table 4.2.

4.2   Change in genetic structure

Genetic structure of a cultivar, landrace, or genetic population is defined by particular genes, their relative frequency, and by their combination with other genes. It may change during reproduction through natural or induced mutation of genes, and through loss of genes, introduction of genes from other populations, and through recombination of genes into new patterns. **Selection** is the key process affecting the relative frequency of genes. Plant breeders artificially select for new populations composed of progeny from individual plants found to possess certain visual traits (**phenotypes**). Artificial

Table 4.2 Reproduction and planting methods, and types of cultivars used with several important crops

HYB, hybrid; IB, inbreeding; IBL, inbred line; IBOB, inbreeder with some outbreeding; OB, outbreeding; OPP, open-pollinated; PP, polyploids; SYN, synthetic[a].

Crop	Reproduction	Propagation	Cultivar types
Wheat	IB	seed	IBL
Rice	IB	seed	IBL, HYB
Maize	OB	seed	HYB, OPP
Barley	IB	seed	IBL
Sorghum	IB OB	seed	HYB, OPP, IBL
Oat	IB	seed	IBL
Potato	OB	clone	HYB[b], IBL
Cassava	OB	clone	HYB[b], IBL
Sugarcane	OB	clone	HYB[b]
Sugarbeet	OB	seed	HYB, PP
Dry bean	IB	seed	IBL
Soybean	IB	seed	IBL
Peanut	IB	seed	IBL
Cotton	IB OB	seed	IBL, HYB
Sunflower	OB	seed	HYB, OPP
Alfalfa	OB	seed	SYN[a], OPP
Clover	OB	seed	SYN[a], OPP
Ryegrass	OB	seed	OPP

Note: [a] synthetic cultivars are composed in various ways, see Section 4.3; [b] these species flower weakly.
Source: adapted in part from Simmonds (1979).

selection is fundamental to all efforts at crop improvement. Selection also occurs naturally in seed-propagated plants in response to pressures imposed by environments of agricultural fields.

4.2.1 Artificial selection

In this method, a "screen" (test or standard) is established that identifies phenotypes possessing a particular trait. Individual plants that meet that criterion are then accepted or rejected. The aim is to isolate favorable or unfavorable **genotypes** from the population. Screens might be very simple: for example, plant height, flowering date, or response when inoculated with disease. Devising screens for improvement of integrative traits such as partitioning pattern, drought resistance, or efficiency in use of N is much more complex.

Success in selection depends on many factors including whether desired traits are controlled by a single dominant or recessive allele, or are polygenic. Polygenic traits are influenced by many gene loci, each with partial control, or an indirect influence, over the trait. The mating system of the species is also important. With highly homozygous, inbreeding species, selection of individual plants can be effective with both single-gene and polygenic traits if desired traits can be easily identified in the offspring. In contrast,

Box 4.1 Unintended consequences of selecting for a specific trait

A danger to be avoided in selection, the inadvertent loss of desired traits, can be illustrated with selection for virus resistance in sugarbeet. The US sugarbeet industry depended during its early years on seed imported from Germany. In the 1920s, curly top virus transmitted by leaf hoppers from surrounding desert vegetation reduced sucrose yields by half in the western states and the industry nearly collapsed. The problem was solved by collecting plants that did not show disease symptoms in farmers' fields. These were grown for seed at a USDA station near Aberdeen, Idaho. The screen for true resistance (or tolerance) involved plantings during warm summers when seedling plants would be exposed to massive flights of viruliferous hoppers from adjacent desert vegetation. The program was highly successful in the development of resistant cultivars. In yield trials, these gave superior performance in the presence of the disease.

Many years later, however, when the selected populations were compared with the original German lines under disease-free conditions in controlled environments (see figure), it was found that the important ability to grow rapidly in the cool weather of spring had been lost through selection during summer (Ulrich 1961). In the field, loss of cold-climate performance had been offset by greater benefit of disease resistance. Yields and production efficiencies would have been even greater, however, if cool-climate ability had not been lost.

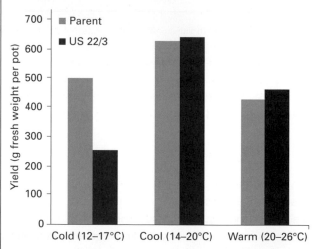

Effect of temperature on yield of sugarbeet cultivar selected for disease resistance in a warm climate compared with a parental cultivar that had not undergone selection.

repeated cycles of mass selection of many individuals may be necessary to capture a desired trait from a highly heterozygous population. In all cases of screening, selection for one trait must not come at the expense of other desired traits. An example of this is presented in Box 4.1.

The recent development of detailed chromosome maps of nearly all major crops, and the ability to determine the presence or absence of particular genes or DNA segments, allow selection for genotype, in addition to phenotype, through **marker-assisted selection (MAS)**. **Molecular markers** are pieces of DNA that can include a gene, part of a gene, or are located close to a gene that controls a desired trait. The presence of the marker is highly correlated with the presence of the desired trait. Non-destructive tests that extract DNA from a small piece of leaf or harvested seed test for the presence or absence of the desired marker. Seeds from plants with the marker are used in subsequent crosses. Such non-destructive selection methods reduce the number of selection cycles and time required to produce new cultivars.

Marker-assisted selection has been more successful for improving **qualitative traits** under the control of one or two major genes, such as soybean nematode resistance or augmenting concentration of a specific seed protein, than for complex **quantitative traits** such as yield or drought resistance. This is because the size of segregating populations increases exponentially with the number of genes involved, and the probability of incorporating unfavorable alleles linked to markers increases. Despite greater effort and costs, some seed companies are using MAS methods to increase frequency of markers associated with yielding ability in breeding lines. These methods identify markers called **quantitative trait loci** (QTLs) that are associated with a polygenic trait and use statistical analysis to quantify impacts of specific QTLs on the desired trait. Because many genes contribute to expression of quantitative traits, the impact from single QTLs is relatively small. Gilmartin and Bowler (2002) and the Plant and Soil Sciences eLibrary (http://plantandsoil.unl.edn/croptechnology/2005/pages/index.jsp) offer more information about marker-assisted selection.

Successful selection is only a part of plant breeding. Captured genes must then be placed into combinations that result in agronomically useful phenotypes. This is usually done by making specific crosses (F_1 generation) and then testing for desirable recombinants in segregating F_2 and later generations (see Section 4.3).

4.2.2   Natural selection and genetic shifts

Heterogeneous crop populations are subject to natural selection. That occurs when genetic combinations differ in reproductive success. Phenotypes that are well suited to particular environments achieve greater reproductive rates than poorly suited phenotypes, and their genes increase in frequency in the population over generations. This process is termed **genetic shift**. Successful genotypes generally have greater "fitness" to local climate and crop management practices. Adaptation to factors such as disease, soil pH, aluminum or heavy metal toxicities, and nutrient deficiencies also occurs. And climatic adaptation, for example, may involve changes in phenological response (Chapter 5) or in tolerances to extreme temperatures. Gradual divergence between original and selected populations is observed in most cases. With intense selection, as occurs when non-dormant alfalfa is transferred to a region with severe winters, genetic shifts can be rapid and dramatic. Subterranean clover has undergone rapid shifts of this sort during its movement around Australia (Gladstones 1967). The same phenomenon is seen with

Table 4.3 Genetic shift in scorings for spring growth habit of "Balboa" rye

Seeds from various sources were planted in Tennessee (origin of the cultivar). An erect habit, due to culm elongation and indicating earliness in flowering, was scored as 1; scores ranged to 10 for prostrate growth and lateness in flowering.

Seed source	Iowa	Kansas	Illinois	Tennessee
Erectness score	10.0	7.4	5.6	1.9

Source: data from Hoskinson and Qualset (1967).

ecotype formation in natural populations. Ecotypes are subpopulations of wild species with sets of common traits lending fitness to local environments.

Data presented in Table 4.3 reveal genetic shift in a rye cultivar grown as a winter cover crop and forage in the Mississippi basin. The original cultivar was released in Tennessee but was adopted widely in other states. Seed production was done in each locale. After 10 to 20 y of reproduction in locations with more severe winters, the comparison presented here was made in Tennessee (mild winter) with seed from several sources. It reveals a strong shift in growth habit related to flowering, which involved changes in vernalization requirement (requirement of a cold period for flowering; see Section 5.2) through natural selection at locations with more severe winters.

Natural selection is for *individual* fitness rather than for performance of the entire *community*. This is a major problem for agriculturalists. It was explained in Chapter 3 that vigorous individuals (e.g., tall, rank plants) may acquire above-average shares of local resources and thus achieve a larger reproductive rate than their suppressed neighbors. Such selection for individual competitive ability represents a reversion towards "wild-type", causing a decline in harvest index and reproductive yield per unit area. Reversion is avoided in agricultural populations by careful management and roguing of seed fields. In Kansas and Nebraska, shattercane, a weedy variant of grain sorghum, is a reversion towards wild-type. It is difficult to control once established because it is resistant to herbicides and cultural practices used on grain sorghum. In many irrigated rice-growing areas worldwide, wild-type red rice with phenological development, stature, and grain size similar to those of domestic rice presents a similar problem.

Natural selection is also seen in loss of genes relating to insect and disease resistance in the absence of those pests, and in increase in frequency of resistance genes in their presence. Selection pressures imposed by farming practices on insect pests, diseases, and weeds are also very strong, allowing those organisms to adapt quickly to new circumstances. Weed populations, for example, commonly shift towards greater tolerance of herbicides and cultivation practices employed in a farming system. The shift can be rapid when a herbicide is highly effective and widely adopted by farmers, because together these place large selective pressure on few tolerant survivors. Once weeds become tolerant, use of other herbicides with different modes of action and changes in cultural practices, including tillage and crop rotation, must be employed for control. Of special interest is management to safeguard value of herbicide- and insect-resistant **transgenic** crops as discussed in Box 4.2 (see also Section 4.3.4).

Box 4.2 Development of herbicide and insect resistance associated with transgenic crop cultivars

Glyphosate is a strongly effective, non-selective herbicide to which transgenic culti-vars of various crops are resistant. Most of almost 100 Mha of soybean worldwide are now sown to glyphosate-resistant cultivars. The consequence is that a large number of weed species have evolved tolerance as seen in the rate of application required for adequate control, often increasing to uneconomic levels. Herbicide tolerance develops by inadvertent section of tolerant individuals among weed populations and is aided when herbicide is applied unevenly or at non-killing dosage. In the USA, widespread use of glyphosate-resistant crop cultivars in maize–soybean and cotton-based cropping systems has promoted a shift from predominance of perennial grass and broadleaf weed species to primarily annual broadleaf weeds (Johnson *et al.* 2009). To date confirmed glyphosate-resistance has evolved in six broadleaf and three grass species (www.weedscience.org/In.asp).

Another group of transgenic crops, including cotton and maize, have resistance to insect attack by stem borers and ear worms. In this case, refuge areas provide a source of non-resistant alleles in the pest population, ensuring that individuals with one Bt resistant allele will mate with others without resistance to produce heterozy-gous, susceptible offspring. The result is smaller probability of indirect selection for resistance. Reproductive mobility of pests governs the type and size of refuge areas required. Stem borer and ear worm (Family *Lepidoptera*) are highly mobile by long-distance flight. For these pests, refugia of around 20% of total crop area are required. In contrast, root worm (Family *Coleoptera*) is less mobile and requires smaller refugia. Other methods to delay the build-up of resistance to Bt transgenic cultivars include: (1) deployment of Bt cultivars expressing different toxins, either as multilines in the same field or as "pyramided" or "stacked" transgenes in the same cultivar, and (2) tissue-specific expression of Bt toxin in organs most at risk of damage and only during the most sensitive growth phase (Bates *et al.* 2005).

In contrast, refugia are not a useful defense against build-up of glyphosate tol-erance because most weeds are self- or cross-pollinating so pollen movement is limited. Instead, cultural practices are necessary: crop rotation and use of herbicides with different modes of action. More timely glyphosate application at proper dose rates and even coverage of all weeds are advised at times when weeds are most vulnerable.

The use of integrated pest management (IPM) practices that include cultural and biological controls on target pests, in tandem with transgenic crops, also helps avoid build-up of pest resistance. Rotation of crops and their cultivars, residue management, use of parasites and predator organisms, and balanced nutrition are among the most effective practices.

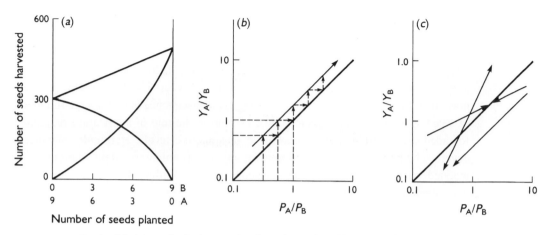

Fig. 4.1 (*a*) A hypothetical example of numbers of seed harvested from genotypes A and B, grown in a replacement series, as a function of numbers of seed planted. The figure is similar to that presented in Fig. 2.9*d*. (*b*) Data from (*a*) displayed in a log–log plot of the ratio of harvested seed (Y_A/Y_B) as a function of the ratio of seed planted (P_A/P_B). The 1:1 line corresponds to the cases where no selection for A or B occurs over generations. In the example from (*a*), however, A increases in frequency in each planting cycle (the dotted line) because the $\alpha_{AB} = 1.2$. The course of selection is seen in the solid line. (*c*) Stable (converging to an equilibrium value $\alpha = 1$) and unstable patterns of selection.

4.2.1   Relative reproduction rate

Natural selection occurs because individual plants that do well in particular environments have high reproduction rates and contribute more progeny to subsequent generations. The example presented in Fig. 4.1 with simple populations of two genotypes helps in understanding the relationships between reproductive rate and natural selection. Figure 4.1*a* depicts seed yields of two genotypes of an inbreeding species grown together in a replacement series (Section 3.4.2). The reproductive rate of each genotype can be calculated directly as the ratio of seeds harvested to seeds planted. The ratio of those rates for the two components is termed the **relative reproduction rate** (α). For component A relative to component B:

$$\alpha_{AB} = (Y_A/P_A)/(Y_B/P_B) = (Y_A P_B)/(Y_B P_A) \tag{4.1}$$

where Y and P represent, respectively, the numbers of seed harvested and planted. When seeds harvested from one generation are used as planting stock for the next, and $\alpha_{AB} > 1$, frequency of A in the mixture increases over generations. That corresponds to a left-to-right progression in planting ratio in Fig. 4.1*a*. If α_{AB} were < 1, A would decline in frequency over generations. Eq. 4.1 assumes a useful linear character after conversion to logarithmic form and rearranging (de Wit 1960):

$$\log a_{AB} = \log(Y_A/Y_B) - \log(P_A/P_B) \tag{4.2}$$

Plots of log (harvested ratio) versus log (planted ratio) (Fig. 4.1*b*) make it easier to see how a mixture changes over generations. With $\alpha_{AB} > 1$, the mixture is unstable and

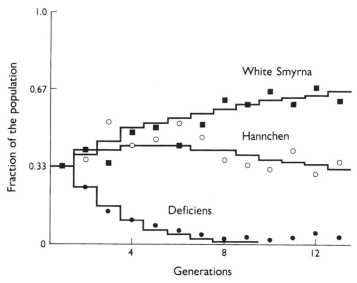

Fig. 4.2 Changes over generations in the frequency of three barley cultivars grown in a blend at Aberdeen, Idaho. The data points are from Harlan and Martini's (1938) experiment; the lines are de Wit's (1960) predictions based on relative reproductive rates observed in the first year. (Redrawn from de Wit 1960.)

proceeds towards pure A; the reverse is true with $a_{AB} < 1$. In some situations, α_{AB} changes as proportions of A and B in the mixture change. Such mixtures may be either stable or unstable (Fig. 4.1c).

This method of analysis is applicable to any mixture, whether of genotypes within a particular population or of different species. It can also be used to follow the course of marker genes in successive generations of outbreeding species. The power of the approach is seen in Fig. 4.2 where the 25-y course of an unstable cultivar mixture of barley is explained by "α" values measured in the first year. This example involves a rather stable environment (Idaho, USA), however, and rather large phenotypic differences in an inbreeding species. Most genetic changes, particularly those in outbreeding species, are not so easily seen or predicted.

4.3   Cultivar development

Continuous effort to develop improved cultivars is an important feature in all agricultural systems. Superior, predictable performance in the field is a principal objective in plant breeding. Adaptations to local environments, including tolerance of normal stresses, appropriate pest and disease resistance, yield performance, and suitable quality for intended uses, are attributes that determine superiority. There are many different approaches to the development of new cultivars, dictated in part by mating habits and genetic structure of the species. Here we examine outbreeding and inbreeding species, multilines, and the role of biotechnology in cultivar development. Narrowing of genetic structure so that all individuals display desired phenotypic traits is involved in most cases.

4.3.1   Cultivars of outbreeding species

Selection and isolation are the principal means to limit diversity in outbreeding open-pollinated species. Isolation serves to protect such cultivars from foreign pollen. New cultivars are created by hybridization followed by repeated mass selection. A major difficulty with outbreeding species is that plants are generally highly heterozygous and selection efficiency (i.e., heritability of a selected trait) may be poor. That problem is circumvented easily when selected phenotypes are propagated as clones (e.g., sugarcane), but other approaches are required for seed-propagated species. Two methods employed with open-pollinated crops are illustrated with alfalfa and maize.

Alfalfa possesses self-incompatibility systems such that only foreign pollen is accepted by stigmas, and all seeds are hybrid. Progeny are diverse through gamete sorting during meiosis in seed and pollen parents, and because pollinating bees visit a diversity of pollen sources. When bees are uncontrolled, pollen sources may include all local alfalfa fields. Alternatively, alfalfa plants from specific superior sources (e.g., clonal lines) can be brought together for open pollination in isolated fields. Bulked seed from such fields, although also highly diverse genetically, represents a more uniform result that is released as a **synthetic cultivar**. Synthetics can be reconstituted at regular intervals from their original parental lines, in this case clones. Without reconstitution, synthetics eventually become ordinary open-pollinated cultivars. Most advanced cultivars of alfalfa and various clovers used in forage production are synthetics, or began as such.

Maize, in contrast to alfalfa and clover, is grown for grain, and uniformity in reproduction is important. Maize plants are wind pollinated and self-compatible, accepting foreign pollen as well as their own. Prior to 1930, cultivars of maize were all open-pollinated. Some were landraces while others, as a result of isolation and repeated selection, formed distinctive cultivars. Mass and individual selection techniques, however, were frustratingly inefficient. The problem of how to further narrow open-pollinated populations to desirable genotypes was solved through inbreeding, which fixes genes into homozygous combinations, followed by production of **specific hybrids**. Crossing two highly homozygous inbred lines leads to genetically and phenotypically uniform (and heterozygous) F_1 hybrid seed that can be used directly as planting stock or as parents in additional crosses for more complex cultivars (and cheaper seed sources) such as "double crosses" ($F_1 \times F_1$).

In the USA, hybrid maize quickly displaced open-pollinated cultivars: the transition was complete in the state of Iowa in only 10 y. While "hybrid vigor" and greater attainable yield are often credited for the rapid adoption, the principal reasons were that new hybrids solved several long-standing problems through resistance to fungal stalk rots and greater uniformity. Once widely adopted, the greater attainable yield of hybrid maize was a major contributor to the steady increase in actual farm yields since World War II.

Seed production for F_1 maize hybrids is done in fields planted with alternate strips of male and female parent inbreds. In earlier times, contamination by pollen from tassels of female parents (maize is "dioecious" with separate inflorescences – male [tassel] and female [ear]) was avoided by removing tassels from female plants before they shed pollen. F_1 hybrids are preferred as planting stock because the cost of seed production has been reduced through use of vigorous inbreds and male-sterile female parents. Genetic segregation in F_2 and later generations results in less uniformity and lower yield, which motivates most farmers to buy newly constituted, hybrid seed for each crop.

Heterosis (hybrid vigor) Unlike cotton and some other crops, inbred lines of maize are not used in commercial production because they are generally smaller and less vigorous than their open-pollinated parents owing to **inbreeding depression**. That occurs because maize germplasms carry many inferior recessive genes. These appear as deleterious double recessives in inbred lines and make those lines weaker plants than their open-pollinated parents. In hybrids and open-pollinated populations, inferior genes are masked by dominant alleles, or have additive effects with unlike alleles. A key step in hybrid development is the search for superior *combining ability* in specific crosses between inbred lines. A favorable combination of genes in F_1 offspring restores plant vigor to the same or greater level than in the open-pollinated parents, a phenomenon known as **heterosis**, or hybrid vigor.

The basis for hybrid vigor probably involves gene interactions as well as genetic dominance and additivity. In addition, inferior genes are exposed to elimination through selection among inbred lines. Significant advances in vigor of maize inbreds have been achieved in that way but they are not yet equal in productivity to hybrids. Because inbred lines still tend to be small in stature, proper comparison of inbreds and hybrids requires that inbreds be grown at a higher density than hybrids. Few such trials have been conducted properly.

Specific hybrids offer a rapid and efficient way to generate superior cultivars of some open-pollinated species. In addition, inbreds, and their specific combinations, represent an information system with proprietary opportunities for commercial plant breeding. Hybrid efforts continue with many crops, mainly by private breeders. Hybrid breeding is particularly suited to clonally propagated crops such as sugarcane because they retain their genetic structure over vegetative generations.

4.3.2 Cultivars of inbreeding species

Inbreeding is common in annual plants whereas outbreeding is more prevalent in perennials. Selection from an inbreeding population can lead directly to a true-reproducing, **pure-line cultivar**. Breeding, then, mainly involves transfer of genes between pure lines by making specific crosses. After crossing, new pure lines are developed from the segregating populations through repeated selection. In **back-cross breeding**, involving repeated crossing to the desirable parent, selection is made for recombinants having the desired parent's phenotype coupled with the transferred trait. With a relatively constant phenotype, farmers need make only small adjustments in production practices when using the new cultivar.

Pure lines are employed widely in agriculture. Continuing gene mutation and small amounts of out-crossing ensure some heterogeneity in pure-line cultivars, which allows slow genetic drift in response to selection pressures imposed by management. In some cases, heterozygotes have an advantage in competition and thus a greater reproductive rate, which helps perpetuate diversity. Such heterogeneity is usually not a problem in production, and farmers can use their own crops as seed for several generations.

4.3.3   Multilines and blends

There are instances where genetic diversity is deliberately included in agricultural communities in the form of population-based cultivars or physical mixtures of genotypes or cultivars. The example in Fig. 4.2 resulted from a simple physical mixture, or **blend**, of several existing cultivars. More sophisticated blends composed of cultivars or lines chosen for similarity and compatibility in mixture find occasional use. Blends of closely related isolines of one cultivar, differing only in disease-resistance genes, for example, are termed **multiline** cultivars. These are useful with a disease such as leaf rust, which has many strains. The disease may occur in the multiline each year but spread slowly because most plants offer resistance.

Fields are seldom uniform throughout their area in features such as soil type, fertility, drainage, and slope. It would seem that a genetically diverse cultivar with an array of genotypes differing in adaptation might be able to exploit such spatial variability. Achievement of superior yield through genotype selection over microsites is unlikely with annual crops, however, and at present no means exists for matching genotype and microsite at planting. Fortunately, most genotypes possess considerable physiological and morphological plasticity to spatial variability. In addition, large gains in yield and stability are achieved through changes in farming practices (e.g., improved tillage, drainage, and fertility) and by choice of field boundaries that reduce or restrict spatial variation (Section 12.1).

Use of genetically diverse cultivars works better with perennials (e.g., pastures) because there is then sufficient time for selection of the best genotype at each microsite. It also may be useful in Mediterranean climatic regions where pastures are dominated by annual species (Rossiter & Collins 1988). Those pastures reseed themselves each year from their own production. When pastures are established with diverse seed sources, best adapted genotypes come to dominate the stands, including at the microsite level.

A theory relating to the use of genetic diversity in multilines revolves around possibilities for slowing secondary spread of diseases from epicenters within a field by presenting pathogens with mosaic patterns of plants varying in resistance genes. The theory is supported by observations that disease problems seem to decrease with increased diversity of a cultivar (clone < inbred < outbred). Similar evidence has been developed with epidemiological models (Kampmeijer & Zadoks 1977). The approach appears well suited to air-borne pathogens, such as black stem rust of wheat, having numerous races (pathotypes). Van Der Plank's (1963) terminology for **horizontal resistance** (HR; general resistance) and **vertical resistance** (VR; specific resistance) is useful in this case. Host resistance to stem rust is centered on major genes at single loci. A particular VR gene provides resistance to only a few of the hundreds of races of the pathogen. A wheat

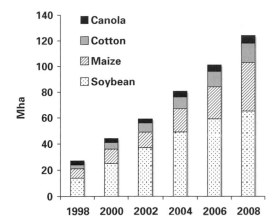

Fig. 4.3 Global area of transgenic crops 1996 to 2008. (Adapted from James 2008).

cultivar can thus be completely resistant to some pathogen races but not to all. In the absence of a single source of general resistance (HR) to all races of a disease, the usual tactic is to release new cultivars carrying the major gene for specific resistance (VR) to currently dominant pathogen races. Other races will then increase in frequency over years, however, and unless we again change to a new wheat cultivar carrying appropriate VR, a disease epidemic and drastic yield decline may occur.

While theoretical justification for use of multilines is sound, they are not widely used in practice because grain purchasers and consumers want uniform grain quality and farmers desire uniformity in crop management requirements and maturity. Instead, to avoid "boom and bust" cycles in crop production, pathogen races must be monitored to allow breeders time to identify new sources of resistance and incorporate them into new cultivars, and time for farm advisors to advise farmers about these changes. In fact, much public-sector investment in crop improvement has focused on maintenance breeding to stay ahead of rapidly evolving diseases and insect pests of wheat and rice. Some argue that so much effort has gone towards maintenance breeding that funding has been inadequate to increase attainable yields, which may have contributed to yield stagnation in some countries (see Section 4.4).

4.3.4   Biotechnology and transgenic cultivars

New methods for gene transfer based on a variety of molecular techniques are now used routinely to supplement traditional methods of plant breeding. These biotechnology methods use recombinant DNA to insert a gene from one plant, microbe, or animal species into the genome of a crop cultivar or line. This approach provides a powerful tool to introduce new genetic traits not found in crop germplasm. The important contributions so far have been from the development of insect- and herbicide-resistant cultivars. Since release of the first commercial transgenic cultivars in the mid-1990s, adoption has been rapid in both the developed and developing countries (Fig. 4.3). In 2008, of

125 Mha planted to transgenic cultivars (i.e., 8% of global cropland), 45% was sown in developing countries (James 2008).

One form of insect resistance was derived from *Cry* toxin proteins that destroy the digestive systems of *Lepidoptera* pests of corn (European stem borer, ear worm, and root worm) and cotton (boll worm). Death comes quickly when these pests consume plant tissue containing the protein. Genes that produce *Cry* proteins are obtained from a common soil bacterium, *Bacillus thuringensis* (Bt). The bacterium itself has long been cultured for use as an insecticide spray, and regulations governing organic agriculture allow use of Bt insecticides. Because *Cry* proteins are highly specific to target insect pests and are not toxic to humans, or wildlife, widespread use of Bt maize and cotton cultivars has enabled a large reduction in the use of highly toxic insecticides worldwide.

Transfer of a bacterial gene was also used to create crop cultivars resistant to glyphosate, which is a highly effective, broad spectrum herbicide of relatively low mammalian toxicity. Use of glyphosate-resistant transgenic crop cultivars provides much greater flexibility in timing herbicide application in relation to other field operations, improving efficiency in large-scale, mechanized agriculture. Glyphosate-resistant cultivars are also useful in no-till systems where weed control is difficult due to large amounts of crop residue that can shelter emerging weed seedlings from herbicide contact and reduce the activity of conventional herbicides.

Other transgenic crop breakthroughs have been touted, such as golden rice with an augmented concentration of iron and zinc to fight widespread micronutrient deficiencies in developing countries, and substantial improvement in quantitative traits such as yield, drought resistance, and N-use efficiency. Actual proof of concept and deployment in farmers' fields, however, has been slow. In the case of nutritional traits, such as golden rice, the delay is due to the development of biosafety standards and testing procedures, and some questions about whether yellow-colored rice will be accepted by consumers. In the case of quantitative traits, there is scant scientific evidence to substantiate claims of improvement.

Despite these delays, expanded use of transgenic crops is expected. Likewise, biotechnology tools such as MAS are widely used to improve efficiency of conventional breeding for qualitative traits controlled by one or two major genes. In contrast, the impact on qualitative trait improvement is likely to be limited in the foreseeable future as argued by a number of plant physiologists and ecologists (see Box 4.3).

4.3.5   Seed production

Advances in plant breeding have made farmers' choices of genetic material increasingly important. Modern cultivars are usually much superior to earlier cultivars and landraces, and they also may differ greatly from each other in adaptation to local conditions and in yield quality. This leads to the problem of how farmers can be well informed in their choice of cultivars. A cultivar's traits can be defined through field trials and farmer experience, and that knowledge is generally available from local seed merchants and extension agencies. Advanced cultivars of crops such as maize, soybean, and cotton are

---

Box 4.3 Biotechnology in the context of natural selection

In a paper entitled "Darwinian agriculture: when can humans find solutions beyond the reach of natural selection?" Denison *et al.* (2003) focused on the challenge to increase the attainable yield of major food crops. They postulated that tradeoff-free traits under simple control of one or two genes have already been optimized by evolution. Their argument is that simple traits, which increase competitiveness of a single plant competing for resources against neighboring plants of the same or different species, have already been fully exploited by natural selection. Indeed, evolution has proven to be a very good genetic engineer! Even relatively complex traits like C4 photosynthesis (Section 10.1), which involve perhaps ten or more genes, have evolved independently from C3 progenitors in 19 species. Therefore, complex quantitative traits such as yield, drought resistance, and N efficiency that provide competitive advantages to single plants in competition with other plants are not likely to be improved by up- or down-regulation of one or two genes, which is the current capability of biotechnology methods. Likewise, it is unlikely that single genes transferred from microbes or animals could have much impact on these quantitative traits.

In contrast, traits that confer disadvantages to individual plants but that give yield advantages to a community of uniform plants in a crop stand are those most likely to be helped by genetic improvement. The shorter stature of semi-dwarf wheat and rice that gave rise to the green revolution is an example of such a trait because shortness is a disadvantage to competitive ability of individual plants. Hence, Denison *et al.* conclude that biotechnology has greatest potential for significant impact on traits that confer yield advantages to a community of similar plants. Such traits include disease and insect resistance in pure stands of crop plants. Biotechnology may also help accelerate breeding for traits that improve performance in the face of rapid changes in soil fertility, pest pressure, and climate changes that occur too quickly to allow for natural selection (Denison 2007).

---

produced and sold by private companies under brand names. For inbreeding crops such as wheat and rice, public-sector state, province, and national agricultural research institutions produce and sell seed. In addition, most nations now have specific requirements surrounding the naming and release of new cultivars and production and the sale of agricultural seed. The rules generally provide that seed produced and processed with independent inspection may be sold with a "certification" of genetic purity, germination percentage, and contamination by weed seed. These requirements are demanding, so most high-quality seed is now produced by specialist growers.

Many farmers still save their own seed, particularly of inbreeding species, which remain reasonably true over several generations (small grains and grain legumes). They return to certified sources when superior cultivars are released or when their own stocks deteriorate. Uncertified seed, with or without a cultivar designation, is also used in many countries. Landrace "cultivars" still predominate in most areas where farmers are isolated from roads, markets, and infrastructure.

Special circumstances for the production of seed arise when regions where a crop is grown are climatically unsuited for seed production. Sugarbeet is a biennial, flowering in the second year, after vernalization during a period of low temperature during winter. The danger exists that plantings for sugar production made in early spring (or fall in areas with mild winters) will flower prematurely and adversely affect sugar yield. In that case, "bolting-resistant" cultivars having a strong vernalization requirement are employed to avoid premature flowering. Seed production is conducted in other localities where winters are sufficient for strong vernalization yet mild enough for the plants to survive. Similar problems occur with alfalfa production in temperate zones. Strong winter hardiness is required but seed yields in those regions are usually much less than can be obtained in milder climates. Seed yields of alfalfa amount to perhaps 100 kg ha^{-1} in Wisconsin compared with the 800 kg ha^{-1}, of better quality, that is obtained with irrigation in California's Central Valley. As a result, much of the USA's supply of alfalfa seed comes from California. Seed fields of winter-hardy cultivars are inspected and certified to minimize genetic drift towards non-dormant, non-hardy types that occurs in California's mild winters.

4.4 Genetic advance and maintenance of diversity

Agricultural productivity in developed countries has increased several-fold during the past 150 y. It would be useful to know how much of that advance is due to genetic improvement and how much to improvements in management, including changes in nutrition and planting density. These are not easy questions to answer. The simplest approach is to compare present cultivars side by side, at the same time and place, with the earlier cultivars that they displaced (Austin *et al.* 1980; Duvick & Cassman 1999).

4.4.1 Genetic advance

Unfortunately, comparison of historical cultivars is biased towards the performance of recent releases because these cultivars have been selected against current atmospheric [CO_2] and soil properties, and current strains of insect and disease organisms. Older cultivars were selected under different environmental conditions. Such bias is seen in a comparison of rice cultivars released by the International Rice Research Institute (IRRI) in the Philippines (Peng *et al.* 1999). In 1996, 12 widely used rice cultivars released between 1966 and 1995 were grown in a replicated field study at two locations with good management to give the highest possible yields. Results showed a linear yield increase of 75 kg ha^{-1} y^{-1}. The oldest cultivar, IR8, gave the smallest yield of 7.2 t ha^{-1}; the most recent cultivar yielded 9.6 t ha^{-1}. But IR8 was widely used in agronomic studies at the IRRI research farm in the late 1960s when yields of 9 to

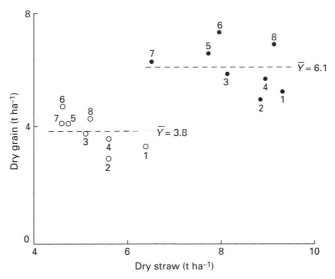

Fig. 4.4 Grain yields of eight English wheat cultivars released between 1908 and 1978 plotted against their straw yields. Cultivar no. 1 is the earliest release, no. 8 the most recent. Open circles, Paternoster Field, 38 kg N applied ha^{-1}, mean yield = 3.8 t ha^{-1}, range 1.7 t ha^{-1}; filled circles, Camp Field, 104 kg N applied ha^{-1}, mean yield = 6.1 t ha^{-1}, range 2.3 t ha^{-1}. (Data from Austin *et al.* 1980.)

10 t ha^{-1} were often achieved, and reported in a number of publications. Therefore, instead of genetic advance in attainable yield, it appears that this historical cultivar comparison tracks the impact of maintenance breeding. As the yield of older cultivars declined in response to changes in soil, atmosphere, and disease and insect pest populations, breeders selected new cultivars to perform in the face of these evolving conditions. Although similar studies with optimal management to achieve yields that approach attainable yield levels have not been conducted with maize, lack of increase in contest-winning yields under irrigated conditions in the USA since 1984 suggests that attainable yield has remained unchanged (Cassman *et al.* 2003).

Surprisingly, comparisons of historical cultivars reveal little or no genetic advance in the photosynthetic abilities of leaves (Section 10.2). Dense canopies of new and old cultivars generally have very similar abilities in biomass production. The genetic component of yield advances has come mainly through advances in harvest index through more conserving patterns of partitioning. Less stem growth, for example, allows more assimilate for reproductive activities. Benefits from improved resistance to lodging are also important. Not so obvious in those experiments are significant improvements in insect and disease resistance. All of these traits should contribute to more efficient use of scarce resources of nutrients, water, land, and labor, and they should provide greater safety and yield stability.

Progress with English wheat during the twentieth century is illustrated in Fig. 4.4 at two levels of N supply. As has been found in other studies with wheat, most genetic gain

Table 4.4 Contributions of various changes in production practices to net increase (4.28 t ha^{-1}) in maize yield between 1930 and 1979 in Minnesota

Sources of increase	Magnitude (%)	Sources of decrease	Magnitude (%)
Genetic gains	+59	Less manure	−15
Fertilizer N	+47	Less organic matter	−13
Plant density	+25	Erosion	−8
Herbicides	+23	Insects	−8
Machinery	+21	Rotations	−6
		Other factors	−25
Total	+175		−75

Source: adapted from Cardwell (1982).

came from improvements in harvest index. Total aboveground biomass produced by these cultivars varied over only a small range (about 10%) but cultivars released before 1950 had HI near 0.35 whereas later releases approached 0.45 to 0.50 and had smaller straw yields. Plant height was reduced sharply, accounting for less biomass partitioned to straw.

This example provides a good test of the role of nutrition in yield improvement. The genetic advances were made during the same period when farmers were increasing the use of N fertilizers. Many claims of genetic advance ignore this point and credit all yield increases to breeding. In this case, mean yield increase due to an additional 66 kg N ha^{-1} was 2.3 t grain ha^{-1} (35 kg grain kg^{-1} N). This is the same amount as can be ascribed to the genetic yield advance demonstrated by the range of yields within the +N field. Austin *et al.* (1980) found that these wheat crops accumulated considerably more N than was applied as fertilizer (total uptake was 60 to 75 kg N ha^{-1} in the −N field and 161 to 215 kg N ha^{-1} in the +N field) indicating that wheat is an effective scavenger of N released from soil organic matter. The later releases not only had greater N uptake, but they also partitioned a larger fraction (nearly 80%) of that to grain. "N-use efficiency" can be said to have improved.

Cardwell (1982) took a different approach in his analysis of contributions of various changes in farming practice to maize yield increases in Minnesota (USA), from 1930 to 1979 (Table 4.4). During that 49-y period, the state average yield increased by 4.28 t ha^{-1} from 2.01 to 6.29 t ha^{-1}. By regressing the time-course of average yield against information on farming practices and cultivars, he was able to assign 59% of yield gain to cultivars, 47% to fertilizer, and 25% to increased plant density and closer rows. Herbicides and machinery contributed 23% and 21%, respectively. Those gains total more than 100% but they were offset during the same period by less input of organic N, continuing soil erosion, and other factors.

Cardwell's study demonstrates the difficulty of distinguishing between genetic and management components of gain, since 2.53 ha^{-1} (59% of 4.28) of genetic gain could never have been achieved without at least 60 kg ha^{-1} additional N. Positive contributions from drainage and from P and K fertilizers were not included in the analysis and it seems that Cardwell may have overestimated genetic gains by perhaps the -25% shown for "other factors". Tollenaar (1989) was able to link genetic advance during 1959 to 1988 with other aspects of technology. His field experiments in Ontario, Canada, were conducted in a way that allowed each cultivar to be grown at its optimum density. Machine harvestability (machine-harvested yield/hand-harvested yield) improved dramatically with year of introduction, owing to advances in lodging resistance even as optimal density increased.

The disturbing point that emerges from these studies is that easy genetic gains, such as advances in harvest index, may have been achieved already with most crops. We use the word "easy" but the advances were not necessarily obvious or easy for plant breeders to accomplish. Some breakthrough in photosynthetic or metabolic efficiency (both unlikely) may be needed for further quantum jumps of similar magnitude. Other possibilities do exist, however, for designing plants that are better suited to community conditions. We examine some of these issues in Chapters 5 and 11 while questions relating to more efficient use of N are explored in Chapter 8.

4.4.2   Germplasm collections

Most crops are supported by germplasm banks that can be searched for alternative sources of disease and insect resistance and other traits. The size of the genetic base available differs greatly between species. Those for cereals are generally very large, whereas those for fruit and vegetable crops are generally small. The principal genetic bases of clonally propagated species such as sugarcane and asparagus, and most tree and vine crops, can be traced to a few individual plants. Seed propagation allows greater opportunity for continuing generation of new diversity through mutation and recombination. Even there, however, modern breeding may rest on few truly excellent lines. Concern about whether our genetic bases are broad enough to provide resistance to new diseases, or traits suitable for unknown, future changes in soil and climate and new technologies justifies significant efforts to expand and conserve genetic materials. As landraces disappear, it is increasingly important to have programs for preservation of the diversity resident in older strains, special lines, and their wild relatives.

Plant breeders generally maintain their own small germplasm collections centered around entries useful in their breeding programs. Sharing information and seed with other plant breeders was the tradition in that profession for many years, but such sharing has decreased markedly with the present emphasis on intellectual property and commercialization of useful genes and transgenic organisms. Fortunately, 11 international crop research centers have assembled large germplasm collections of all the major food crops. An International Treaty on Genetic Resources for Food and Agriculture mandates open access to these resources for farmers, crop researchers, and breeders (http://www.planttreaty.org/). Collections for some inbreeding species include more than

60 000 entries. Outbreeding species, by contrast, can generally be managed in fewer populations. Certain elite lines may be kept as clones or, as is the case for maize, as inbred lines.

One problem of germplasm collections is that seeds have finite lifetimes in storage and seed-propagated plants are subject to mutation and selection with every generation. Debate continues on how best to deal with the problem of collection size and genetic change. One solution is to treat inbreeders in the same way as outbreeders by compositing diversity in heterogeneous populations. Collections can be stored as seed for extended periods before reproduction, and changes in gene frequency can be controlled to some extent through reproduction in different environments. Maintenance of diversity is greatly simplified then, but searching composites for desired traits can be much more difficult than searching through a collection of pure lines because phenotypic control becomes dispersed by genetic exchange. Clonally propagated materials present special problems because they cannot be stored as seed. Some can be maintained in plantings indefinitely (asparagus and bermuda grass) or for very long periods (fruit trees) with little care. Others, such as sugarcane and potato, are burdensome because they require frequent propagation. Protocols for cryopreservation of vegetative material and seed at low temperature are improving rapidly but remain costly. DNA banks are also being explored although DNA must also be preserved at very low temperatures.

Genetic collections are of little value unless plant breeders have time to explore and utilize that diversity. Maintenance of genetic resources has meaning only as new cultivars are developed from them through breeding. One of today's ironies is that although more attention is being given to the preservation of germplasm and to the creation of new diversity through biotechnology, research administrators, particularly in universities and government agencies, are increasingly reluctant to support conventional plant breeding. It is strange, and potentially dangerous to sustainability of agriculture, to see one of our most successful forms of technology viewed as too mundane. The main effect of biotechnology may be to increase the diversity of germplasm collections.

4.4.3   Present and future yield advance

On a global basis there has been a steady gain in crop yields since the mid-1960s (see Box 4.4). But the yields of some crops appear to be approaching a plateau in several countries. Rice yields in Japan and Korea have changed little since the early 1980s. Plateauing yields, despite high prices paid to rice farmers in these two countries, lead to the hypothesis that average farm yields in a region or country begin to plateau as yields approach 80% of the climate-adjusted attainable yield as estimated by simulation (Cassman 1999). Stagnation occurs because achieving attainable yield requires perfect management of all crop and soil factors that influence plant growth and development throughout the growth cycle. Although a few superior farmers may come close to this state, it is neither profitable nor feasible for all farmers to do so. Therefore, average farm

---

Box 4.4 Linear or exponential rates of yield gain

The figure below, using data from FAOSTAT, presents yield trends of the three major cereals since 1966. The first modern cultivars of rice and wheat were released in the mid-1960s to initiate the **green revolution**. The fitted regressions reveal that yields have been increasing at a linear rate. That aside, linear rates of gain mean that relative rates (%) steadily decrease as average yields increase. Thus, average maize yield was 2.26 t ha^{-1} in 1966 and the linear rate of annual yield gain was 2.8% of that amount. By 2006, however, at the same rate of absolute gain, annual relative increase had fallen to 1.3%. The relative rates of gain for wheat and rice have also fallen to about 1.3% of 2006 yield levels.

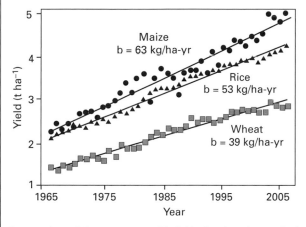

Progression of the average world yield of maize, rice, and wheat from 1965 to 2005 (data from FAOSTAT).

Despite these recorded linear rates of yield gain, economic studies of future food security often use exponential rates of increase in estimating future food supply and demand. For example, Rosegrant *et al.* (2002) estimate a 1.3% annual rate of increase in the global demand for rice, wheat, and maize from 1995 to 2025. Without a continuing increase in the rate of yield gain or a large expansion of crop area, annual production of all three cereals will fall below the 1.3% needed to meet future food demand. The 40-year time series presented here seems appropriate for estimating long-term yield trends because science and technology applied to genetic and agronomic improvement of crop yields have been relatively constant during this period.

---

yields level off well below attainable yield. It is noteworthy that since the publication of this hypothesis in 1999, rice yields in China and wheat yields in Mexico's Yaqui Valley are also indicating signs of plateauing at about 80% of climate-adjusted attainable yield (Cassman *et al.* 2003).

4.5   Review of key concepts

Genetic diversity in agriculture

- Thousands of species have been cultivated at some time but a small number of crops, cereals, and tubers now provide most human dietary energy, supported by a wide range of pulses, fruits, vegetables, and animal products. Four grass species, wheat, rice, maize and barley, account for more than half of crop production. This is not a new phenomenon. Since development of agriculture, humans have always relied on few species for staple food and cereals have significant advantages. They are relatively resistant to disease and insect pests, and provide grain, which is highly digestible with a reasonable protein content, and is easily stored.

- Large diversity exists in extensive collections of living plants and seeds of a wide range of crop species and their ancestors. Most diversity is found in species that reproduce sexually. The range of diversity in many vegetatively propagated crops is relatively small.

Agricultural cultivars

- Cultivars used in agriculture have been selected from naturally occurring populations, bred by cross-pollination between populations, and more recently by direct gene transfer using biotechnological methods. The objective is to achieve high yield and quality with uniform behavior to facilitate crop management. Access to broad genetic diversity is vital for the development of new cultivars.

- Uniform expression of yield and resistance traits by a cultivar generally requires a relatively narrow genetic composition. Most plant breeders strive towards a high degree of genetic purity in their commercial cultivars to provide predictability in response to climate and management. In that case, genetic diversity exists in cropping systems rather than individual crops.

- On the other hand, heterogeneity is not necessarily damaging to performance and many outbreeding species are bred as heterogeneous populations (e.g., synthetics). In the absence of effective horizontal (HR) or vertical resistance (VR) to air-borne disease, multilines, through their potential for limiting expression of disease, offer an alternative to cultivar substitution, although complete resistance is obviously preferred. Marker-assisted selection provides a powerful tool to improve selection efficiency for multiple sources of disease resistance and for other traits under control of multiple genes.

Crop improvement

- Breeding programs have two parts. First, breeding to maintain the utility of current cultivars, especially maintaining resistance to diseases and insects. Second, breeding to improve yield and quality. Because disease and insect pests continually evolve to overcome plant defenses, exploration of germplasm banks for new sources of resistance and "maintenance" breeding require significant commitments. A balance is required to maintain current productivity while seeking increases for the future.

- New gene technologies offer great advantage to breeding programs but they still require extensive field testing and evaluation by traditional techniques. The use of transgenic cultivars is expanding rapidly at the same time that many consumers remain unconvinced of their safety.

- Spectacular yield gains achieved after World War II are now showing signs of deceleration as farm yields approach current attainable yield levels in some countries. It is uncertain that attainable yield can be increased further without some basic change to plant metabolism (e.g., photosynthetic capacity). While genetic advance in primary productivity remains elusive, important opportunities remain for advances in matching genotype to specific environments and cropping systems to optimize the use of light, water, and nutrients.

- Genetic populations are subject to change. Natural selection is a strong force even within inbreeding populations with both positive and negative effects on cultivar performance. Adaptation to prevailing climate, disease, and management is beneficial to performance but resistance to diseases and insects not currently present may be lost and positive selection for crop yield and quality may not occur without explicit efforts.

- Natural selection operates on individual plants, through the reproductive success of individuals within communities. Where reproductive success is increased through competitive ability, a greater proportion of community resources may be expended on vegetative growth to the detriment of economic yield.

Terms to remember: breeding, inbreeding, outbreeding, back-cross breeding; cultivar, ecotype; germplasm; genetic shift; genotype; heterosis (hybrid vigor); inbreeding depression; landrace; lines, pure, synthetic, clonal; marker-assisted selection (MAS); molecular markers; multilines and blends; pollination, self and open; population; phenotype; quantitative trait loci (QTL); relative reproduction rate; specific hybrids; transgenics; resistance, horizontal (HR) and vertical (VR); green revolution; selection; traits, qualitative and quantitative.

5   Development

In the improvement and management of crops in which fruit and seed comprise economic yield, particular attention must be paid to the timing and extent of reproductive development. Successful cultivars are able to complete reproduction within available growing seasons, avoid stresses at vulnerable stages and, for maximum yield, balance available time and resources between vegetative and reproductive growth.

Anatomical and physiological bases of organ formation are important background for crop ecology, requiring integration of information from genetics, plant physiology, morphology, and development. Apical meristems of shoot and root have the capacity for unlimited growth and produce the elongating body of the plant. Shoot meristems progress with periodic production of new leaves at stem nodes separated from each other by internodes. Intercalary meristems in internodes also contribute to shoot elongation while new apices in leaf axils (axillary meristems) provide branching. Lateral meristems, mainly vascular cambia, increase girth. Meristems convert from production of leaves to flowers in a direct response to environmental signals of temperature and daylength, indirectly to environment through assimilate supply, and in some cases with age.

This chapter deals with coordination and timing of initiation, growth, and longevity of vegetative and reproductive parts, and also with seed germination. It provides the necessary background to cultivar improvement (Chapter 4) and crop management (Chapters 13, 14, 16, and 17).

5.1   Developmental time

Through evolution, plants have arrived at a variety of timing mechanisms that control development. These improve the chance that germination and reproduction occur during favorable growth conditions. Analysis of crop development is helped by the recognition of distinctive developmental events, termed **phenostages**, such as "emergence", "flower initiation", and "first flower" that signal changes in pattern of development. Rate of advance within intervening **phenophases** is **developmental rate**, and study of progress of crop development in relation to environmental conditions is termed **phenology**.

5.1.1   Developmental stages

Numerical phenological scales, that provide quantitative description of development, have been devised for many crops. The Feekes (Large 1954) and Haun scales

Table 5.1 The decimal scale for phenological description of cereals

A second digit is used to divide each phenophase in up to ten further steps.

Value	Phenostage
0	germination
1	seedling growth
2	tillering
3	stem elongation
4	booting
5	inflorescence emergence
6	anthesis
7	milky grain
8	doughy grain
9	ripe

Source: from Zadoks *et al.* (1974); see also Tottmann *et al.* (1979).

(Haun 1973) are still widely used for wheat and related cereals. The decimal scale, summarized in Table 5.1, is an improvement on those earlier scales. Most phenostages are recognized by observation, but some (the most important is flower initiation) can be detected only by dissection and microscopic examination. Such detailed observations of development are needed if crop response is to be understood. Illustrated descriptions of meristem differentiation (e.g., Moncur 1981; Kirby & Appleyard 1984) are helpful in focusing on importance of "hidden" stages of development, such as the sensitivity of reproductive yield to environmental stresses during early flower development.

Individual plants in a crop community vary in their development rate because of differences in genotypes and microenvironments, so it is usual to record occurrence of a phenostage when 50% of individuals in a population have achieved it. In some crops, synchronized development permits critical management operations such as application of herbicides or fertilizer to be made at an optimal time for most individuals. However, tight synchrony usually occurs only with cultivars having a narrow genetic base, such as F_1 hybrids of maize and homozygous pure lines of self-pollinated crops such as pea and small grains.

5.1.2   Developmental rate

Plants proceed through stages from germination to maturity at rates that vary with environment. Active meristems produce organs faster or slower depending upon the supply of assimilate and general chemical activity that is positively linked with temperature. Thus it is necessary to distinguish "physiological" time of an event or age of an organ from the corresponding chronological time or age.

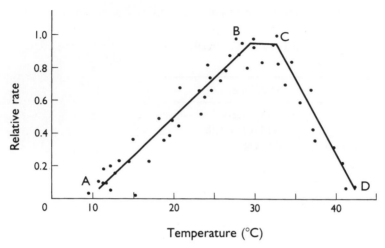

Fig. 5.1 Rate of growth and of development to flowering in maize in response to temperature (after Coehlo & Dale 1980; see also Warrington & Kanemasu 1983).

The concept of physiological time is evident in appearance of new leaves by a stem meristem. The interval between **appearance** of successive leaves is the **phyllocron** and number of leaves is the **phyllocron index**. Rate of leaf appearance is the reciprocal of the phyllocron. In most plants, rate of appearance of organs lags behind their initiation rate so that primordia accumulate within apical buds. Measurements with a range of wheat cultivars, for example, reveal that up to flower initiation, when the last leaf has been initiated, the number of leaves initiated (P) on the apex is related to emerged leaf number (L) by the simple relationship $P = 2L + 4$ (Jamieson *et al.* 1998). The inference is that four primordia are formed before the first leaf emerges and then primordia are formed at twice the rate that leaves emerge. Such relationships allow estimates of timing of flower initiation from observations of leaf appearance, but only in retrospect once the final leaf has emerged.

The influence of temperature on physiological processes is non-linear (Fig. 3.12) but workable linear approximations can be established for individual regions of response. This is illustrated in Fig. 5.1 for growth response of maize to temperature. The rate of organ initiation, which is the central response of development, is closely related to growth by a common response of cell division to temperature. Therefore, the same form of optimal response also describes the rate of crop development. Three temperature ranges are evident in this response: in range AB, growth and developmental rates increase positively with temperature; range BC is optimal; while in CD, temperature is supra-optimal and both growth and developmental rates decrease as temperature increases. This non-linear response of development to a wide temperature range is often overlooked when applying simple phenological indices to analysis of crop development (Section 5.3).

Cardinal temperatures (A, B, C, and D) differ for species and cultivars and for individual phenophases but the form of response has general application to all species. Individual response functions can be derived from field data provided a sufficient environmental range is included. Otherwise, they must be established through experiments in controlled environments.

Rates of aging towards senescence, especially by leaves, may depart from the pattern shown in range CD of Fig. 5.1. It seems that leaf senescence is accelerated by decreasing water status, and mobilization of nitrogen, as well as by high temperature.

5.1.3   Determinate and indeterminate crops

Flowering is perhaps the single most important phenostage in crop development because it signals change to growth of fruit and seed essential for economic yield of most crops. A useful operational distinction can be made between crops depending upon timing of flowering relative to vegetative growth. In **determinate** crops, including most cereals and sunflower, flowering occurs over a period of a few days following termination of vegetative growth. Typically, vegetative growth ceases because flowering involves conversion of apical meristems of shoots to reproductive structures.

In **indeterminate** crops, including cotton and many grain legumes such as soybean, flowering overlaps with vegetative growth and can be prolonged for weeks or even months. This occurs because flowering progresses from axillary meristems while the apical meristem continues to produce new leaves and new axillary positions. Under some conditions, reproductive growth in indeterminates so monopolizes assimilate supply that apical activity ceases. Such crops behave as facultative determinates with a single flush of fruit. Cotton behaves in this way but, given a sufficiently long season, it may resume vegetative growth after a "cut out", resulting in a second flush of flowering and yield.

The distinction between determinate and indeterminate crops has great significance to crop adaptation and management. Indeterminate cultivars of fruit and vegetables such as tomato and melon (see also Fig. 11.13) are frequently chosen for home gardens and fresh markets because a single planting provides produce daily over extended periods. That is a disadvantage with processing crops, however, for which repeated harvests are expensive and some fruit become too ripe. An alternative strategy, common with crops processed to frozen or canned products, is to employ staggered plantings of determinate cultivars.

Progressive ripening is disadvantageous to mechanical harvesting so facultatively determinate cultivars, with a flush of flowering and fruiting over a short period, have been selected for processing tomato and lima bean to allow harvest in a single pass by machine. Greater yields are obtained with those determinate cultivars than with repeated harvests of indeterminates. Indeterminate field crops grown for dry seed or grain must be selected against losses from shattering of first-maturing fruit. Pods of most wild-type *Phaseolus* beans, for example, spring open at maturity, scattering their seed. Pods of domesticated types remain closed, and mature, dry pods accumulate on plants as flowering progresses.

Determinate and indeterminate flowering habits have different significance in stressful environments. Prolonged flowering of indeterminate crops compensates for flower loss or seed abortion caused by transient stresses of high or low temperature or water shortage. In contrast, determinate crops are vulnerable to isolated periods of stress that occur during flowering. In wheat, for example, a single stress from frost or heat at flowering can abort most flowers, so that grain yield is negligible. Determinate crops do, however, have a yield advantage under resource-limiting conditions provided they avoid such transient stresses.

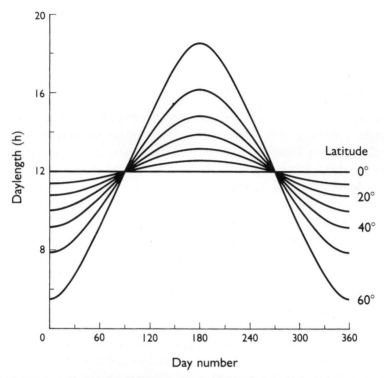

Fig. 5.2 Length of day (sunrise to sunset) in northern latitudes according to day of year. The pattern is shifted by 182.5 d for the southern hemisphere. The equations for calculation are presented in Box 6.1.

The sharp transition in growth pattern provides a basis for an **"optimum switchover"** from vegetative to reproductive growth (Paltridge & Denholm 1974). Harvest index (HI) increases during seed filling because there is no further vegetative growth. An optimum balance is achieved, and HI and yield are greatest, when assimilation of the shoot equals yield capacity of seed (see also Section 13.4). In indeterminate crops, each additional flowering unit comprises a leaf and stem portion that subtends it. As a result, vegetative material increases as yield increases. Harvest index remains low and partition of growth between vegetative and reproductive parts is not optimal (Paltridge *et al.* 1984).

5.2   Developmental switches

Shoot meristems of some species switch from initiation of leaves to flowers at a certain developmental stage as determined by temperature. In others, the switch occurs as a specific response to daylength (photoperiodism) and/or low temperature (vernalization). Daylength and low temperature operate as inductive switches that change the type of organ initiated and are important factors in crop adaptation. Daylength is precisely and invariably related to latitude and day of year (Fig. 5.2) and it is not surprising that plants have evolved adaptive responses to it. Outside the tropics, daylength is repeated twice annually, each followed by progressively more distinct seasons in the move to higher latitudes. In consequence, combinations of **photoperiodic** and **vernalization** responses have proven particularly successful in selecting appropriate seasons for growth and reproduction.

5.2.1 Photoperiodic control of flowering

The flowering response of plants to daylength, called **photoperiodism**, is one of a number of photomorphogenic responses of plants that also include initiation of tubers and root thickening in some crops.

Some plants **(day-neutral plants, DNP)** are insensitive to daylength and for them time to flowering is controlled by temperature. Most plants, however, respond to various combinations of changing photoperiod. These can be divided into two broad groups: those that flower in response to lengthening days are termed **long-day plants (LDP)**, while **short-day plants (SDP)** flower in response to shortening days. In a few plants, such as Italian ryegrass, a single photoperiodic cycle can induce an irreversible shift to reproductive behavior. More commonly, several consecutive cycles are required.

The distinction between SDP and LDP lies in their responses to shortening or lengthening days (lengthening or shortening nights) and not to some absolute daylength. Some LDPs flower at daylengths well below 12 h while some SDPs respond at longer daylengths. Thus a cultivar of chickpea (LDP) flowers only when daylength exceeds 8 h, whereas one of soybean (SDP) flowers only with daylengths less than 14 h (Roberts & Summerfield 1987).

Recognition of daylength depends on the ability of phytochrome pigment systems in leaves to measure the duration of night (rather than the duration of day) through the ratio of red and far-red light received. The importance of dark periods is evident because interruptions of a few minutes, even less in some species, are just as effective in promoting flowering of LDPs as is increasing daylength. Manipulation of photoperiod with artificial lighting is useful in commercial horticulture and to some extent in plant breeding.

Photoperiodic responses may be either **facultative** or **obligate** as illustrated in Fig. 5.3 by linearized responses of developmental rate (DVR) to daylength (DVR = reciprocal of the time [d] to complete the phase). In plants with facultative responses, flowering is promoted by changing daylength but proceeds at all daylengths. Other plants have obligate photoperiodic responses because there are daylengths above (for SDP) or below (for LDP) which they will not flower. Most plants respond to daylength over a finite range, but in a few species the responsive range is narrow enough that it can be expressed practically by a single value. The former response has been described as quantitative and the latter as qualitative. With very short days, responses of all plants are more complex and DVR usually declines. In such cases it is difficult to separate inductive effects of short days from an associated effect of small supply of assimilate for growth.

Definition of cultivar response is required for breeding and selection and for predicting crop development in the field. Linear approximations of response functions are frequently adequate for that purpose. That is done by establishing **maximum** and **minimum DVRs** occurring, respectively, at **critical** and **threshold daylengths**. Then, **photoperiodic**

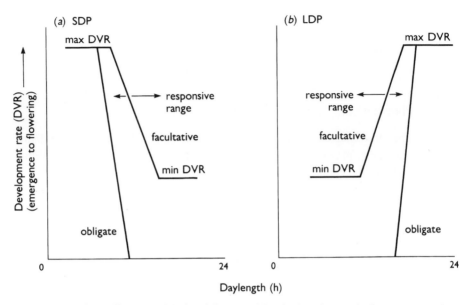

Fig. 5.3 Approximate linear models for obligate and facultative photoperiodic responses of (*a*) short-day (SDP) and (*b*) long-day plants (LDP).

sensitivity is the slope of DVR versus daylength within these limits (the responsive range), as illustrated in Fig. 5.3. For plants with an obligate response, minimum DVR is zero, but is finite for facultative plants. In species of high photoperiodic sensitivity, threshold and critical photoperiods coincide and the responsive range approaches zero.

Photoperiodic responses in a range of agricultural species are presented in Fig. 5.4. They correspond with the generalized linear models of Fig. 5.3 and also illustrate how responses of closely related plants may differ in range and sensitivity. Those differences offer the potential for manipulation in breeding programs. A classic example of such manipulation, breeding and latitudinal distribution of soybean, is presented in Box 5.1 and discussed again from a genetic perspective in Section 5.5.

As a generalization, plants from tropical regions are either DNP or SDP without vernalization responses and those from higher latitudes are DNP or LDP, often with a vernalization response. There are notable exceptions to those rules, including strawberry clover and soybean that are SDP from mid-latitudes and the pasture legume stylo, an LDP from the tropics (Roberts & Summerfield 1987). Sunflower is a species of temperate origin that now has SDP, LDP, and DNP cultivars (Goyne & Schneiter 1987). Such a wide range of responses is more common in species that are naturally widely distributed, as is seen in photoperiodic and vernalization responses of ryegrasses from Europe and North Africa. In those, and other species, such variation is readily manipulated by selection and breeding to suit specific production requirements of many regions. Moreover, manipulation of genetic variation in photoperiodic responses become increasingly important in a world of changing climate as discussed later in Box 5.3.

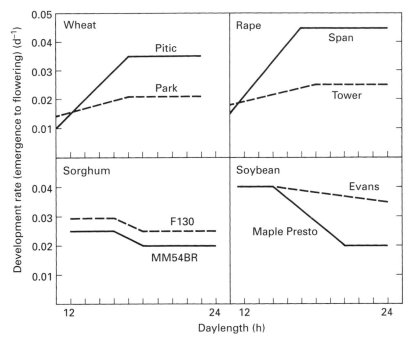

Fig. 5.4 Photoperiodic responses of wheat and rape (LDP) and sorghum and soybean (SDP) cultivars (after Major 1980).

Box 5.1 Photoperiodic responses in soybean

Soybean, in which the mysteries of photoperiodism were first seen by Garner and Allard (1920), provides an elaborate example of the use of photoperiodic response to match cultivar to environment. Twelve phenological "maturity" groups serve to identify zones of cultivar adaptation in Canada, the USA, and Central America. Groups OO, O, and I are adapted to the longer days and cooler temperatures of Canada and northern USA (latitude 40–50°), groups II to X are adapted progressively further south to latitude 30°. Two recent groups, IX and X, extend adaptation to southern Florida and further to latitude 10° in Central America.

A reference cultivar chosen from each group serves as a maturity standard against which other cultivars are rated. Differences between photoperiodic responses of groups are small as is difference in daylength between contiguous zones. For example, for a May 20 planting the difference is less than 1 h between 30° and 40°N, that covers the range for groups II to X. Nevertheless, cultivars grown outside their zone of adaptation flower too soon if planted to the south and too late if planted to the north. As short-day plants, northern cultivars (groups O, OO, I) have narrower geographic adaptation than those adapted to shorter daylengths of the south (V, VI, VII). North American cultivars flower just 30 days after planting under the short days of tropical regions.

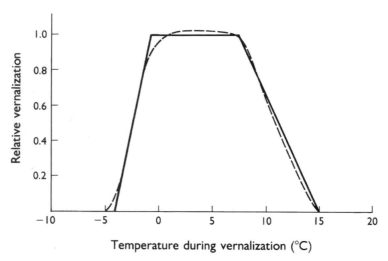

Fig. 5.5 Vernalization response of flowering in winter cereals (based on data for "Petkus" rye from Salisbury 1963; see also Weir *et al.* 1984).

5.2.2　Vernalization

In some species, germination of seed, initiation of flowering, or bud break require or are hastened by prolonged exposure to temperatures down to −4°C. The process involved in these varied responses is called **vernalization**. This section concentrates on flowering responses, a subsequent section deals with seed dormancy. Many winter annuals, biennials, and perennials require vernalization for flowering. As with responses to daylength, vernalization responses may also be **absolute** or **facultative**.

Winter cereals, for example, follow the form of response presented in Fig. 5.5 for "Petkus" rye. During exposure to low temperatures, a stimulus is recognized and "accumulated" by shoot apices. Exposure to high temperatures (>30°C for wheat) reverses the process. That is called **devernalization**. Differences in vernalization response are determined by allelic variation in three vernalization loci (VRN1, VRN2, VRN3) (Pugsley 1982; Trevaskis *et al.* 2007; Distelfeld *et al.* 2009). When activated, VRN3 encodes a protein that acts as a long-distance "flowering" signal (the long-sought flowering hormone – florigen?). It is produced in leaves and moves to apices where it switches meristematic activity from production of leaf to flower primordia. The flowering process is thus initiated and subsequent progress depends on photoperiodic response, mediated by another locus, PPD1, and promotive effects of warm temperature.

5.2.3　Interaction of daylength and vernalization

Wide ranges of vernalization requirements are found in commercial cultivars of oat, barley, rye, and especially of wheat. Few genes are involved and are now readily manipulated in breeding programs. Consequently, wheat cultivars (LDPs) can be classified into three phenological groups according to environmental controls over development as summarized in Table 5.2.

Table 5.2 A summary of thermal (T), photothermal (PT), and vernalization (V) responses of wheat cultivar types

All responses are facultative except for the vernalization response of winter wheat.

| Type | Phenophase | | | |
	S–E	E–I	I–A	A–M
Winter	T	PT, V	PT	T
Intermediate	T	PT, V	PT	T
Spring	T	PT	PT	T

Note: S–E, sowing to emergence; E–I, emergence to flower initiation; I–A, initiation to anthesis; A–M, anthesis to maturity.

Winter wheats have an absolute requirement for vernalization before they can develop beyond the vegetative phase. These cultivars are used in continental climates with moderate winters, such as Kansas and Ohio (USA), England, and Ukraine. Crops, there, experience prolonged cold following sowing in autumn but the winter climate is not so intense that they do not survive.

Spring wheats have no vernalization requirement and develop reproductively in response to increasing temperature and photoperiod. Such cultivars are sown in spring in regions where winters are too severe for the survival of wheat (e.g., Canada and North Dakota, USA) but they are also sown in autumn in milder climates (e.g., southern Australia, Argentina, Chile, South Africa, and Pacific coastal regions of North America). The distinction between winter and spring wheats is thus a phenological one based on a requirement for vernalization rather than the season when they are usually sown.

Intermediate wheats exist between these two extremes. They display a wide range of facultative vernalization responses in which exposure to cold enhances but is not essential for reproductive development.

An example of interaction between temperature and daylength is presented in Fig. 5.6 with leaf production by "Pitic 62" wheat over a range of daylengths at constant temperature. This cultivar is LDP with a facultative vernalization response. These plants had previously received varying degrees of vernalization by subjecting imbibed seed to temperatures in the range 1 to 3°C for six weeks. Vernalization promoted flower induction especially at daylengths less than 12 h. The number of leaves produced by vegetative apices varied from 6 to 14. Leaf initiation ceased when apices switched to production of flowers. Apices produced most leaves without vernalization, especially under short days. These results emphasize the role of developmental responses in canopy formation.

Intermediate wheats find many applications. Incorporation of some "winterness" ensures that crops sown under warm temperatures in autumn following early rains in Mediterranean environments (the "break of season") do not develop prematurely in early spring. In this way flowering is delayed until after late destructive frosts. Also, as with

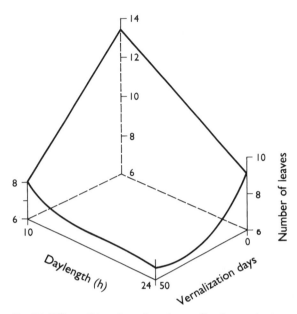

Fig. 5.6 Effect of duration of seed vernalization on leaf number in wheat grown at constant temperature over a range of daylengths (after Levi & Peterson 1972).

winter wheats, they can be safely used for winter grazing because flower initiation and stem elongation are delayed, thereby protecting reproductive organs.

Summary of interactions Table 5.3 records interactions between photoperiod and vernalization in control of flowering in a number of agricultural species. More elaborate classifications of these interactions have been made. Salisbury (1981), for example, distinguishes 25 groups.

5.3   Quantifying phenological response

The complexity of phenological development can be quantified most easily with mathematical models. This section presents a model structured to allow easy incorporation of functional relationships between developmental rate and controlling environmental factors. It is then used to explain the utility of two well-established phenological indices, **thermal** and **photothermal units**, often used to provide simple analyses under more restricted conditions. The two units are shown to be subsets of the general model.

5.3.1   A model of developmental rate

This model is adapted from Robertson (1973). Chronological progression through a phenophase depends upon accumulation of daily development (DVR, d^{-1}). The phase

Table 5.3 Photoperiodic and vernalization responses of some agricultural species

	SDP	DNP	LDP
Obligate photoperiodic response	soybean	soybean	oat
	rice	cotton	annual ryegrass
	dry bean	potato	canary grass
	maize	rice	red clover
	coffee	sunflower	timothy grass
		tobacco	spinach
			radish
Facultative photoperiodic response	soybean		cabbage
	cotton		spring barley
	sugarcane		spring wheat
	rice		spring rye
	potato		potato
	sunflower		sunflower
			red clover
Positive vernalization requirement	onion	onion	winter oat
		carrot	winter barley
		broadbean	perennial ryegrass
			winter wheat
			sugarbeet

Source: after Vince-Prue (1975) and other sources.

is complete after d days when $\Sigma\text{DVR} = 1$. DVR is set as a multiplicative function of responses to temperature $[f_1(T)]$, vernalization $[f_2(V)]$, and daylength $[f_3(L)]$:

$$\text{DVR} = f_1(T)f_2(V)f_3(L) \tag{5.1}$$

This form allows easy, but effective, representation of the controls that vernalization and photoperiodism exert on development. Function $f_1(T)$ holds units of DVR (d^{-1}), while $f_2(V)$ and $f_3(L)$ are scalar functions (range 0–1) that modify expression of $f_1(T)$. Species lacking vernalization or photoperiodic responses are represented by setting $f_2(V)$ or $f_3(L) = 1$. These three functions are linear or non-linear depending upon particular phenophase–cultivar combinations as established by experiment. Observations made in controlled environments are valuable because they extend combinations of temperature and daylength available from field sites.

The model presented as Eq. 5.1 has been applied to the analysis of phenological development of sunflower as described in Box 5.2 and is incorporated in winter wheat models (Weir *et al.* 1984; Ritchie & Otter 1985; Porter 1993). In the latter case, a vernalization response is included in the responsive phenophase E–FI (Table 5.2). In a recent analysis, Jamieson *et al.* (2007) compare the nature and performance of various models of phenological development for predictions of leaf number and duration to flowering in intermediate wheat.

Box 5.2 Control of phenological development in two cultivars with different photoperiodic responses

Hammer *et al.* (1982) used the model of Eq. 5.1 to define the phenological development of "Sunfola 68–2" and "Hysun 30" sunflowers. Observations made in a phytotron indicated functional forms for $f_1(T)$ and $f_3(L)$ during the phenophase "emergence to head-visible" as follows.

$$DVR = f_1(T)\, f_2(L)$$

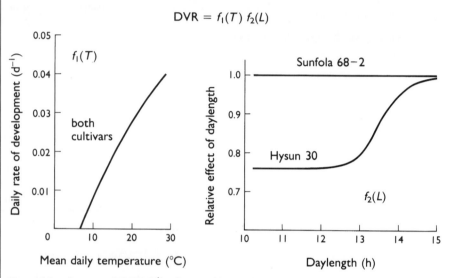

Rate of development (DVR, d^{-1}) of two cultivars of sunflower from emergence to head visible in response to temperature (T) and daylength (L) in the form of the general phenological model (Eq. 5.1) (after Hammer *et al.* 1982).

Data from time-of-sowing experiments in a number of locations were then used to estimate parameter values of a quadratic response to temperature and a switch function used to portray influence of photoperiod. Vernalization plays no role in development through any phenophase of sunflower so $f_2(V) = 1$. Both cultivars had similar temperature responses but whereas "Sunfola 68–2" was day-neutral, "Hysun 30" responded as a facultative LDP for daylengths from 12.5 to 15 h. The model explained 96% of variation in independent observations of heading date over a wide range of temperature and daylength.

5.3.2   Thermal units

Dependence of developmental rate on temperature was recognized by the French naturalist Reaumer in 1735 when he invented a linear temperature model (and a thermometer scale!) for analysis of development. Thermal models have since been used widely in the analysis of crop development and in applications such as scheduling planting dates for vegetable crops.

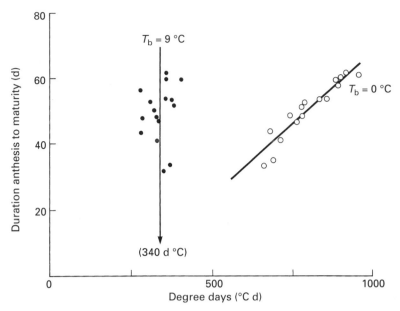

Fig. 5.7 Thermal analysis of the duration of the phenophase anthesis to maturity in four cultivars of wheat. For each cultivar, thermal units calculated to a base temperature (T_b) of 9°C explain the duration of all times of sowing. Calculations to $T_b = 0$°C do not (after Weir *et al.* 1984).

In its simplest form, thermal analysis relates DVR directly to mean daily temperature $[T_d = (T_{max} + T_{min})/2]$ in excess of a base temperature (T_b), i.e., as:

$$\text{DVR} = f_1(T) = (T_d - T_b)/TU \qquad (5.2)$$

If $T_d < T_b$, no thermal time is accumulated.

Daily accumulation $[\Sigma(T_d - T_b)$; units are "degree-days" (deg d)] serves as an estimate of physiological time expressing development through a phenophase rather than chronological time. Summation of thermal time corresponding to completion of a phenophase (i.e., in *d* days when $\Sigma\text{DVR} = 1$) defines the **thermal unit (TU)** for that phenophase.

Comparison of the method of calculating thermal time with response of DVR to temperature (Fig. 5.1) reveals that it describes the linear approximation, line AB, to the rising part of the curve. Eq. 5.2 is successful only when T_d remains within that linear range. Some modifications have been made to improve the scope of TU. Thus the Corn Heat Unit calculates daily accumulation of thermal time to a maximum T_d of 30°C, acknowledging that DVR does not increase beyond that temperature (line BC). Gilmore and Rogers (1958), also for maize, included a reverse linear function in their calculations of thermal time (line CD) to account for decreasing DVR for $T_d > 30$°C.

Duration of phenophase A–M in wheat is determined by temperature and is relatively constant for various cultivars. Thermal analysis presented in Fig. 5.7 reveals that TU calculated against $T_b = 9$°C provides a common description of thermal duration of

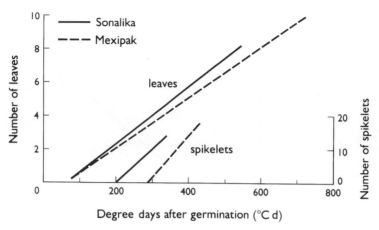

Fig. 5.8 Thermal dependence of leaf and spikelet formation in two cultivars of wheat (adapted from Stapper 1984).

that phenophase (340 deg d) for four cultivars. Calculation against $T_b = 0°C$ does not.

Thermal analysis may also be used to describe developmental advance within phenophases. Figure 5.8 presents field data on the appearance of leaves and spikelets of two spring wheat cultivars. The data illustrate that rate of organ appearance was linearly related to temperature in that experiment. Spikelets were produced faster than leaves and the two cultivars had different DVR for both organs.

Caveats on the use of thermal unit Use of daily mean temperature (T_d) introduces a confusing empiricism into the determination of TU. The TU for a particular phenostage of a cultivar can be different at locations with the same mean temperature but differing in diurnal amplitude. Differences are particularly evident in comparisons between arid (large amplitude) and humid (small amplitude) regions. Explanation is found in the response presented in Fig. 5.1. A wide amplitude around optimum temperature promotes a smaller daily advance than at optimum temperature because crops spend many hours at temperatures that limit development. As a result, most published TU values are useful only in environments with similar diurnal amplitude to those where they were established. The TU can be made more generally applicable, however, by accumulating it from hourly temperature, which if necessary can be estimated by sinusoidal interpolation from daily maximum and minimum values. Suitable equations for interpolation of diurnal temperature are included in various crop simulation models, e.g., that of Denison and Loomis (1989).

Finally it is important to re-emphasize that success of TU does not demonstrate that development, unlike growth, has a linear response to temperature. The basic calculation remains a linear approximation to one part of a non-linear response to temperature (Fig. 5.1). Modifications referred to previously in the calculation of thermal units for maize (Gilmore & Rogers 1958), and use of hourly rather than daily time steps, seek to account for that non-linearity of response. Linear approximations can be avoided by using the general model of phenological development (Eq. 5.1).

Table 5.4 Thermal (TU) and photothermal units (PTU) for development of the spring wheat cultivar Olympic

Phenophase	Unit	Value	T_b (°C)	L_b (h)
Sowing–emergence	TU	78 deg d	3	–
Sowing–stem extension	TU	315 deg d	4	–
Stem extension–booting	PTU	6600 deg d h	4	0
Sowing–anthesis	PTU	6846 deg d h	2	6
Anthesis–maturity	TU	416 deg d	8	–

Source: adapted from O'Leary *et al.* (1981).

5.3.3 Photothermal units

A second important phenological index, the **photothermal unit (PTU)**, finds utility in the definition of phenological development of LDPs. In this model, DVR is made proportional to the product of mean daily temperature (T_d) and daylength (L_d), above base values, T_b and L_b, respectively:

$$DVR = f_1(T)f_3(L) = (T_d - T_b)(L_d - L_b)/PTU \qquad (5.3)$$

If $T_d \leq T_b$ or $L_d \leq L_b$, no photothermal time accumulates.

As with TU, a cultivar is characterized by PTU (units, deg d h) required for completion of a phenophase. L_b is the extrapolation of the photoperiodically responsive phase to DVR = 0 (Fig. 5.3). For obligate LDPs, L_b will approximate threshold photoperiod, but for facultative LDPs it is merely a mathematical extrapolation. For species with facultative responses, PTU analyses will be most successful when daylength remains above the threshold for sensitivity. Photothermal units work best for all species when daylength does not exceed the critical photoperiod.

The PTU, like the TU, is established empirically by iterative regression from field and/or laboratory data. It leads to descriptions of phenological development of cultivars, such as presented in Table 5.4 for the spring wheat "Olympic".

Photothermal unit analysis has been used formally to describe only promotive effects of increasing temperature and daylength on the development of LDPs. But the general model (Eq. 5.1) is readily adapted to short-day plants also. For example, Kiniry *et al.* (1983) investigated the interaction between temperature and daylength on the development of various maize cultivars. The response of one of them, "TX60", from emergence to tassel initiation is presented in Fig. 5.9. It demonstrates, for this SDP, that development can be described by a constant TU (290 deg d > 8°C) for optimal daylengths between 10 and 12.5 h. Above 12.5 h, development slows linearly to 17.5 h, the maximum daylength studied. At 17.5 h, the TU for tassel initiation is 450 deg d (>8°C).

This developmental response of "TX60" can be expressed in the form of the general model (Eq. 5.1) by a pair of equations, namely:

$$DVR = f_1(T) = (T_d - 8)/290; (10 < L_d \leq 12.5) \qquad (5.4a)$$

$$DVR = f_1(T)f_3(L_d)$$
$$= [(T_d - 8)/290][(1 - 0.074(L_d - 12.5)]; (12.5 < L_d < 17.5) \qquad (5.4b)$$

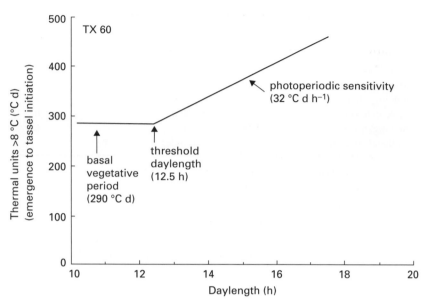

Fig. 5.9 The dynamic nature of temperature and photoperiod response of the phenophase emergence to tassel initiation in maize. Below the threshold daylength of 12.5 h, the thermal duration (basal vegetative period) is 290 deg d ($>8°C$). The photoperiodic sensitivity between 12.5 and 17.5 h daylength is 32 deg d h^{-1} ($>8°C$) (adapted from Kiniry *et al.* 1983).

5.4   Seed germination and dormancy

Seeds are key organs of propagation and dispersal. They preserve the genetic resources of species and also serve to disperse genetic diversity that is generated during sexual reproduction. In nature, this requires that seeds do not germinate prematurely under transiently favorable conditions, and further, that germination is synchronized with conditions most favorable for subsequent growth and reproduction. An understanding of the range and action of environmental factors and physiological conditions that determine the germination of seed are important to the successful storage of seed and the establishment of field crops. It is also important to management of species composition of crops and pastures by attention to soil seed banks that accumulate under a given cropping system.

This section considers the physiologic and anatomic conditions that render seed dormant and environmental factors that control the establishment of germinable seed. It also discusses soil seed banks and their management for community productivity and persistence.

5.4.1   Storage

Dry seeds have low metabolic rates and consequently can survive in that state for considerable periods on their generally small reserves. Longevity is increased at low temperature and oxygen concentration that further depress metabolic activity. For that reason, farmers avoid high-temperature locations for storage of seed; and scientists, concerned with long-term maintenance of germplasm, carefully dry seed for storage at low temperature. The national seed collections that have been assembled in a number of countries store seed to around 5% moisture and $-15°C$. Germinability is tested every five years or so; most species can be maintained in that way for 15 to 20 y before there is a need to regrow the collection in the field.

5.4.2   Germination and establishment

Seed germination, defined by emergence of the radicle from the seed coat (testa), can be studied in the laboratory or field. In contrast, **seedling establishment** relates only to field performance, occurring when seedlings have grown sufficiently to emerge from the soil and establish functional shoot and root systems.

Major environmental factors that determine germination are moisture, temperature, and aeration. The first step, **imbibition** of water by seed, is a physical process dependent upon the colloidal properties of seed and permeability of seed coats. For seed with permeable coats, the rate of imbibition is determined by water content of the medium and the degree of contact with it. For that reason, imbibition of seed is less predictable in fields than in laboratories. Contact between seed and soil is critical to the continued uptake of water and depends upon relative sizes of seed and soil particles and, for recently sown seed, subsequent repacking of soil.

Imbibition of water initiates metabolic activity leading to mobilization of reserves, growth of the embryo, and ultimately germination of seed. Rate of embryo and seedling development is then determined by prevailing temperature and supply of oxygen for respiration. Consequently, initial stages of imbibition can proceed with seed immersed in water, but germinating seed or growing seedlings cannot persist under such conditions.

Seedling establishment requires continued development after germination. Provided there is sufficient moisture to maintain hydration and oxygen for respiration, rate of establishment depends largely on temperature. Development during that time continues to depend upon the mobilization of reserves to meet growth requirements of root and shoot systems. The size of seed reserves thus sets limits to possible duration and hence to a major practical issue also, maximum depth in soil from which seed can successfully establish. Complex environmental responses of some seed to light, $[O_2]$ and $[CO_2]$, and alternating temperatures have evolved as adaptations that restrict germination at locations where establishment is improbable. However, most responses of that type have been bred out of crop plants, so when moisture and aeration are adequate, rate of germination and ultimate establishment depend largely upon temperature.

The rate of germination of individual species provided with adequate water and aeration shows the same form of optimal response to temperature presented in Fig. 5.1. Provided temperature range is not extreme, laboratory measurements of time to 50% germination can usually be explained by such linear approximations. The response of seedling establishment to temperature is more complex due to interactions between seed depth and soil physical properties, such as impedance, but in general seeds that germinate rapidly will establish most rapidly and successfully also. In fact, rapid and uniform seedling establishment is one of the most important factors governing subsequent crop growth and productivity. But physical conditions favorable for germination

Table 5.5 Thermal units (TU, deg d) calculated above base temperature (T_b) for seedling establishment of various agricultural species

Species	TU	T_b	R^2 (%)
Wheat	78	2.6	46
Barley	79	2.6	39
Oat	91	2.2	32
Maize	61	9.8	91
Sorghum	48	10.6	96
Pearl millet	40	11.8	97
Field pea	110	1.4	10
Soybean	71	9.9	87
Peanut	76	13.3	99
Navy bean	52	10.6	86
Rapeseed	79	2.6	45
Safflower	70	7.4	68
Sunflower	67	7.9	73
Linseed	89	1.9	37
Buckwheat	37	11.1	90
Amaranthus	32	11.7	86

Source: adapted from Angus *et al.* (1981).

and establishment typically exist for short rather than long periods, and rapid establishment gives less chance of seed and seedling loss by fungal infection or predation. Seed planted too deep may exhaust reserves regardless of rate of germination.

Table 5.5 presents an analysis of crop establishment, i.e., duration of sowing to emergence, for monthly sowings of a number of species over two years. It applies thermal unit (TU) analysis (Section 5.3). Base temperatures (T_b) were derived by iterative regression to find TUs with least variation over sowing dates. In most cases, degree of fit is good as quantified by R^2 values, which estimate the proportion of observed variation explained by the regression. Failures indicate either that TU is not an adequate index of response to temperature for those species, perhaps because temperatures varied beyond those for which a linear approximation is appropriate, or that factors other than temperature influenced germination.

5.4.3   Dormancy

Dormancy is detected when seed will not germinate under conditions of moisture and temperature known to be suitable for a particular species. It may arise in either seed coat or embryo. Many crop plants have been selected during domestication to remove such impediments to ready germination but seeds of most species exhibit some degree of dormancy, varying from year to year in response to environmental conditions during seed-set. In contrast, most weed and many pasture species, particularly self-regenerating annuals, rely on environmental cueing for successful establishment and some form of distributed germination for persistence in the soil **seed bank**. They achieve this by a variety of dormancy mechanisms.

Hard seeds have seed coats that restrict uptake of water or exchange of gases. Low $[O_2]$ or high $[CO_2]$ can prevent or delay germination but low permeability to water is the more common limitation. This morphological condition is general in important families such as *Leguminosae*, *Chenopodiaceae*, and *Malvaceae*. Hard seeds are isolated from their external environment and germinate only as seed coats break down by action of soil solution, temperature and moisture fluctuations, and decomposition by soil micro-organisms. Hard seed with no other impediment will germinate following treatment to break seed-coat impermeability. This can be achieved physically or chemically in various ways but most easily, a technique used routinely by seed growers with clover seed, is "scarification" (abrasion). Seeds of many clover cultivars have high germinability only after such treatment.

Immature embryos Some seeds are unable to germinate when formed because embryos are immature. There are two forms of this condition. Embryos of some species complete development only following imbibition of water. These combined processes of maturation and germination take between several days to weeks in various species. In others, embryos are unable to germinate at maturity but can develop further ("after-ripen") in dry storage without imbibition. This is an important characteristic for all seed crops. Germination in the head ("shot grain") can be a persistent problem in wheat and other small grains exposed to rainfall after maturity. An after-ripening requirement has been deliberately bred into cultivars for regions where such problems occur.

Embryo dormancy Completely developed embryos may be incapable of germination because of true physiological dormancy or presence of a chemical inhibitor. Wild oat has an inhibitor in the seed coat while in other species it resides in the embryo itself. Seeds of some species have this form of dormancy at formation; in others, it develops later **(secondary dormancy)**. These forms of dormancy are induced and "broken" by a seemingly bewildering array of responses to environmental signals including light, temperature, vernalization, nitrate concentration, and various alternating patterns of these factors.

Dormant seeds are undesirable for sowing crops because farmers rely on full and uniform germination to achieve a crop stand of appropriate density and spacing (Chapter 3). Seed-bed preparation and planting are done carefully and at the "right" time. Pastures present a different situation. There, seed-bed preparation and placement are usually less precise so that some proportion of dormant (usually hard) seed helps overcome losses due to variable conditions during establishment.

5.4.4   Seed banks

Seed in surface litter and soil comprises a seed bank. Seeds are added intermittently but lost gradually by germination, death due to exhaustion of reserves, decay, consumption, and physical transport by wind and water. Because of dormancy, seed of each species exists in a range of age classes and states of germinability. A large part of the germplasm of a species may reside in the seed bank; for this reason, input–output dynamics (Fig. 5.10) of seed banks are a focal point in studies of weed ecology and in persistence and productivity of pastures (Roberts 1981).

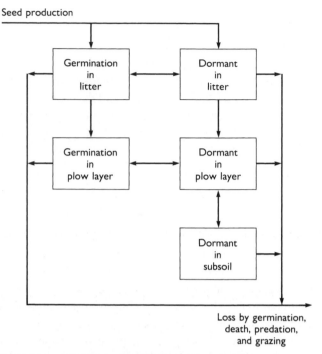

Seed production

Fig. 5.10 A compartment model of the soil seed bank emphasizing location and dormancy. Seed passes through soil layers by natural movement and predators, but mainly by tillage. Losses occur by consumption, death, and germination. Seed may gain and lose dormancy in response to environmental factors and internal physiological activity. Although single compartments are shown, seeds of individual species have different responses and dormant seed may include a range of dormancy states.

Not surprisingly, seed banks commonly have thousands of seeds m^{-2}. Most species rely upon annual additions to maintain their place in plant communities. Under rapidly changing vegetation, species composition of seed banks may be quite unlike those of existing plant communities. Seeds of some weed species are extremely persistent, with examples of arable weeds persisting in soil for as long as 40 y.

Seeds are formed in a range of positions above and below ground, in various seasons, and with varying degrees of dormancy. Seeds of some species exist in a hazardous surface litter where special adaptations, such as mucilaginous seed coats, are required to provide adequate contact with water for germination. Others, such as those of subterranean clover, are produced in the safer environment of the surface soil, and some have special adaptations, such as hygroscopic awns of wild oat and many grasses, that assist entry into soil. Other seeds rely on ingestion or gathering by animals and insects, or more simply upon the presence of surface cracks, to enter the soil.

Soil microenvironment attenuates rapidly from the surface (Chapter 7). Water penetrates most effectively, temperature less so (there is little diurnal variation below 0.5 m), and light does not penetrate at all. It is the vertical distribution of these factors and of

$[O_2]$ and $[CO_2]$ that determine induction or breaking of dormancy and germination. The depth at which seeds germinate and their level of reserves then determine the probability of successful emergence and establishment.

Managing seed banks Under cropping, disposition of seed banks is dominated by tillage, which rearranges seed, burying some and uncovering others. Tillage also incorporates surface litter into the soil, changing energy balance and hence temperature and moisture regimes of surface soil. Seed buried to depth of the plow layer (15–20 cm) experiences a subdued diurnal and seasonal temperature range, low $[O_2]$, and high $[CO_2]$. Those conditions are conducive to induction or maintenance of dormancy. However, redistribution of seed by tillage also exposes previously buried seed to fluctuating environments near the surface that encourage germination. On balance, tillage is an effective way to reduce weed-seed load of soil, even though seed burial contributes to persistence.

The recent widespread adoption of zero tillage has changed the dynamics of seed banks (Chauhan *et al.* 2006). Most seeds are now found in the surface layer so that weed emergence is more rapid. Species with seed adapted to survival, germination, and establishment in that environment become the dominant weeds. Seed buried at depth by fauna remains undisturbed, however, and seedlings do not establish, even if germination commences.

In pastures, seed banks are closely associated with the surface soil. Regeneration of annual pastures requires special attention. Seed production is managed through variations in timing and intensity of grazing, mowing, herbicides, fertilizer, and fire. Sufficient seed must be retained to establish a competitive pasture of desired composition each year and to accommodate losses that arise when conditions suitable for germination do not persist sufficiently long for seedling establishment.

In annual pastures, continuing production of some hard seed aids regeneration in subsequent years by accommodating false breaks that lead to death of a cohort of germinating seed. A high degree of dormancy ensures persistence of pasture species but there are circumstances in which they thwart efforts at pasture improvement through introduction of new species or cultivars. Beale (1974) measured the persistence of subterranean clover seed in self-regenerating annual pastures on Kangaroo Island, South Australia. Attempts to renovate pastures with more productive cultivars of lower estrogen content (estrogen causes infertility in sheep) were commonly unsuccessful. A survey of 10 pastures (age 9–19 y) established a mean seed bank of this species of 243 kg ha^{-1} (range 36–2260 kg ha^{-1}). New cultivars established but were soon lost to competition and pastures reverted to dominance by the original cultivars. Observations on pastures where seeding was prevented revealed that 20% of seed remained viable after three years, leaving considerable reserves for continued re-establishment of old cultivars. Interestingly, only 11% of seed-bank decline was due to germination. Other losses, particularly consumption by mice, were dominant; sheep were excluded from these plots.

Fire is commonly used in the management of crops and pastures. It removes litter, controls disease, modifies the microenvironment of soil surfaces, and has major effects on the size and disposition of seed banks. Fire destroys most seeds in litter but conditions others, particularly hard seed in surface soil, for germination. Species composition of many natural systems is determined by fire that exposes mineral soil to provide a seed bed. In such communities there is a wide range of adaptation of seed to survive fire and to germinate after it, including loss of dormancy due to heat treatment and mechanisms that release seed protected from heat in protective fruit.

5.5   Crop improvement

Complex developmental responses of crop plants derive from their long evolutionary history before and since domestication. Under natural selection, adaptation is generally restricted to local environments. Wide geographic distribution of individual species is achieved by evolution of phenological populations each adapted to distinct local conditions. An important question in crop ecology is what developmental responses (wide or restrictive) are consistent with the managed conditions we might now create in agriculture?

Some crops are generalists, finding roles in a wide range of environments. Typically, that is achieved with growth durations less than the available season so that planting and harvest can be fit to a suitable portion of it, or by having a range of cultivars with different phenological responses. Identification of phenological patterns has been an important aspect of crop improvement. This has been achieved by screening wide ranges of germplasm in time-of-sowing experiments at a range of sites to identify best cultivar–time-of-sowing combinations. Simulation models now make such efforts more efficient by helping identify the most promising phenological pattern for a given location or region and the most representative testing sites.

Interesting contrasts are evident in various solutions reached by plant breeders for controlling development. Many crops, the prime example being soybean, are bred with attention to carefully tuned photoperiodic responses for local environments (Box 5.1). A common thrust with others such as wheat has been to simplify developmental response through suppression or elimination of photoperiodic controls resulting in wide geographic adaptation of improved cultivars. In those cases, temperature remains the principal control over developmental rate (Fig. 5.1). The task is often straightforward because relatively few genes are involved (Section 5.2).

Simulation models are playing an increasing role in crop management and improvement. Phenological submodels, as introduced in Section 5.3, are the best performing parts of such models. Values of phenological parameters are established from observations of development of crops in multisowing date, multilocation experiments. In *post-ante* analyses, flowering dates of many crops are readily reproduced within one or two days using such cultivar-specific relationships, recorded temperature data, and progression of photoperiod calculated from latitude. Precision of these phenological submodels has been important in crop improvement because experience reveals that the first and easiest adaptation of crops to new environments is achieved by adjusting phenological development. A case study for adapting to potential climate change is given in Box 5.3.

Box 5.3 Simulation of optimal development for spring wheat at two locations in southern Australia under present and changed climate conditions

The matter of optimal development and switch-over from vegetative to reproductive growth to maximize yield of crops in water-short environments is discussed in Section 5.1 and further in Section 13.4. Briefly, crops will achieve maximum yield when they use water to flowering to build a yield potential (grain number) that can be met by water remaining for growth during grain filling. Any other combination of water use will result in smaller yield. In wheat, partitioning of biomass between vegetative growth and grain is determined by phenological development.

Wang and Connor (1996) used a wheat simulation model to study yield variability at two locations (Mildura and Wagga Wagga) in southern Australia. The locations are distinct in climatic conditions and yield and response. The average annual temperature at Mildura is 1.5°C greater than at Wagga Wagga (17 vs. 15.5°C) but annual rainfall is almost half (293 vs. 570 mm). The latitudes (34°11′ vs. 35°06′ S) are similar so that annual distribution of daylengths is comparable (Fig. 5.2). Mean yields at Mildura are about half of those at Wagga Wagga (1.13 vs. 2.01 t ha^{-1}).

The model was used to study yield responses to variations in two phenophase parameters, P_1 (TU, deg d from sowing to stem extension) and P_2 (PTU, deg d h from stem extension to anthesis), over 100 years at both locations under present and changed environmental conditions. Changed conditions were increases in mean annual temperature of 3°C and [CO_2] from 350 to 460 ppm, as predicted by global circulation models. The optimization sought mean yield that exceeded a threshold minimum of 1 t ha^{-1} with least inter-annual variation. The results of the simulations are summarized below.

Scenario	P_1 (deg d)		P_2 (deg d h)		Mean yield		cv	
	M	WW	M	WW	M	WW	M	WW
Present	240	240	7600	8000	1.04	2.07	0.16	0.21
+1.5°C 460 ppm	220	240	7800	8400	1.08	2.12	0.12	0.20
+3°C 460 ppm	240	360	8600	8600	1.12	2.03	0.12	0.29

Mean yield (t ha^{-1}) and its coefficient of variation (cv) for optimal combinations of P_1 and P_2 for present conditions and two future climate change scenarios at Mildura (M) and Wagga Wagga (WW).

The model predicts optimal combinations under present climate for Mildura ($P_1 = 240$ deg d, $P_2 = 7600$ deg d h) and Wagga Wagga ($P_1 = 240$ deg d, $P_2 = 8000$ deg d h). These values are comparable with cultivar Olympic (Table 5.4) that was widely grown during that period. Yield variation (cv) is smaller than has been recorded in practice, especially at Mildura, because the model assumes optimum sowing time and agronomic practices.

Studies such as that presented in Box 5.3 are valuable because they identify the phenological characteristics of cultivars best adapted to given climatic conditions. In this case, the analysis reveals that yields can be maintained for changed climate without greater variability by breeding longer season cultivars (greater P_1 and P_2), especially at Wagga Wagga. That range of phenological parameters is found in current Australian spring wheats and greater values of P_2 are available in winter cultivars.

5.5.1   Gene-based models of phenological development

The most recent developments in phenological analysis are attempts to relate cultivar parameters of crop models to their genetic constitution. The intention is to assist evaluation of new gene combinations for crop productivity and for adaption to new and changing environments. This approach seeks equations based on the genetic constitution of individual cultivars to replace current phenological parameters (TU, PTU).

In progress so far, linear regressions, for example of the form:

$$PTU = a + b_1 \text{ allele } 1 + b_2 \text{ allele } 2 \ldots \ldots \ldots b_n \text{ allele } n \qquad (5.5)$$

have been successfully fitted for six alleles ($E_{1-5,7}$, $e_{1-5,7}$) that control flowering in soybean (Messina *et al.* 2006) and to one photoperiodic (Ppd, ppd) and three vernalization alleles (Vrn_{1-3}, vrn_{1-3}) that control flowering in wheat (White *et al.* 2008). In these equations, dominant and recessive alleles take values of 1 or 0, respectively. The soybean study involved a set of 14 near-isogenic lines (NIL) in which dominant alleles lengthen time to flowering and maturity in response to increasing daylength. The wheat study selected 29 cultivars (spring and winter types) from the International Winter Wheat Performance Nursery that evaluates cultivars in many locations worldwide.

To test the approach, new equations were used as alternative expressions to standard cultivar-specific phenological parameters in either CROPGRO-Soybean (Boote *et al.* 1998) or CSM-Cropsim-CERES-Wheat Version 4.02.0 (Jones *et al.* 2003) models. Given good statistical fit of allele-based equations (R^2 generally > 0.8), it is not surprising that comparable predictions of crop phenological development and yield were achieved by either method. In the case of soybean, an independent test of methodology was made comparing simulations with yield data for seven commercial cultivars grown at eight locations in the State of Illinois (USA) over a five-year period. For this, the seven cultivars were "genotyped" using molecular markers so that allele-based equations of phenological development could be established by comparison with relationships established for the original 14 NILs. The authors report that correlation between predictions from parameter-based and gene-based models were weaker than expected. R^2 values (forced through the origin) for comparison of days to maturity and yield, by the two models, were 0.75 and 0.54, respectively. From the perspective of phenological analysis, the comparison is limited by small differences between cultivars. The variation in crop maturity for these cultivars was small, ranging from day of year 242 to 258.

The hope is that further development will make it possible to "genotype" cultivars more simply by molecular marker analysis than to establish parameters by current methods of multisowing, multilocation trials. Another goal of this work is to extend the range of alleles included in crop simulation models beyond those for phenology, and from that to design synthetic cultivars for improved environmental adaptation and greater yield.

It is likely that progress will be slow for a number of reasons, despite current enthusiasm for genomic research and cataloguing genomes of crop species. First, the method also requires extensive field evaluation of phenological responses to extend data beyond the few species/cultivars so far studied comprehensively. Second, the caution expressed earlier (Section 5.3) on limited portability of thermal units established with daily mean temperature will further complicate the objective to establish widely applicable allele-based equations for individual cultivars. It is possible that existing parameters will need to be re-calculated and new values established using diurnal temperature data. Third, while phenological development is a good place to start this work, success is required there before the approach can be safely extended to the many other gene combinations that determine crop productivity. Vernalization provides a good example. The mechanisms remain ambiguous (Section 5.2) and so genetic characterization and physiological experiments need to move along together so that genetic equations can represent the actual gene-based processes. For the moment, analysis of impact of Ppd and Vrn alleles in wheat (White *et al.* 2008) would more usefully concentrate on prediction of FI before being extended to crop yield. The many remaining processes of yield formation and realization in the model (Jones *et al.* 2003) use parameters that do not have a direct link to controlling genes, as do phenological responses.

Finally, it is not clear how phenological parameters of synthetic genotypes can be constructed from established responses since equations are specific to individual combinations of many genes. Meantime, standard phenological parameters that have provided guidelines for optimal development patterns during recent decades retain utility. This is shown in an example of optimal adaptation of wheat development to changing climate presented in Box 5.3.

5.6   Review of key concepts

Morphology

- Development is determined by initiation, differentiation, and growth of new organs that change plants' morphology during their life cycles. Rate of organ production depends upon temperature and supply of assimilate and essential nutrients.
- In determinate species, flowering signals the end of vegetative growth; in indeterminate species, vegetative and reproductive growth proceed together.

Flowering

- Stem meristems switch from formation of leaves to flowers in response to daylength (photoperiod) and temperature. There are many combinations of these responses, e.g., long-day, short-day, and day-neutral plants, each with or without a previous cold requirement (vernalization). These responses allow plants to adapt successfully to a wide range of climates.

- In agriculture, flowering responses are manipulated to provide cultivars best suited to specific environments, or they can be suppressed to provide cultivars of wide adaptation. A critical consideration is that sexual reproduction avoids periods when environmental stresses are most likely to occur and so reduce yield regardless of growing conditions at other times.

Yield

- Crop productivity relies upon continual provision of new organs as sites for utilization or storage of assimilate. It also requires a balance between production and activity of vegetative (leaves and stems) and reproductive organs (seed and fruit).
- Successful crops complete reproductive development within available growing seasons. In environments of variable growing season, this involves developmental responses that allow complete use of both long and short seasons for maximum yield.
- Yield advance has been achieved more by adjusting phenological development to environment than by changing intrinsic growth capacity of plants.
- Work with crop simulation models has encouraged quantitative analyses of phenological development but much work remains to define phenological response of agriculturally important cultivars. Quantitative and genetic analyses will play an increasingly important role in the design of cultivar–management combinations, particularly in variably stressful environments where optimum combinations require assessment of crop performance over more years and locations than traditional field experimentation can provide.

Terms to remember: daylength, minimum, maximum, and critical; cardinal temperatures; determinate and indeterminate crops; developmental rate; devernalization; imbibition; optimum switchover; phenophase; phenostage; photoperiodism; phyllocron; plants, day-neutral, short-day and long-day; thermal and photothermal units; responses, facultative and obligate; seed, bank, dormancy (embryo and secondary), germination and hard; seedling establishment; vernalization; wheat, winter, spring, and intermediate.

Part II

Physical and chemical environments

Mammals have thermoregulatory abilities and mobility to avoid unfavorable features of current weather. In contrast, plants are rooted in place and must accept that rates of metabolic processes are determined by ambient conditions they cannot avoid. Crop ecology gives special emphasis to environment as the main determinant of what will grow, how rapidly, and for how long.

Crop communities extend a strong influence over their local microenvironment. Nearly all cropping practices are directed towards, or have the effect of, modifying chemical or physical aspects of that environment. The next two chapters deal specifically with these issues, beginning in Chapter 6 with the aerial environment, giving emphasis to solar energy as the driving force for productivity, and continuing in Chapter 7 with properties of soils.

6 Aerial environment

Electromagnetic radiation is a central feature of crop environments; its energy determines soil and air temperatures, wind movements, evaporation, and photosynthesis. This chapter examines radiation sources and their roles in the macro- and microclimates of crops. When objects absorb radiation their temperature increases. That heat energy may remain in the object or it may be radiated as new long-wave radiation, transferred to another object, or dissipated in evaporation of water. All of these subjects are covered here. We begin with a review of several physical laws important in radiative transfers of energy among plants, soil, and atmosphere, as well as from the Sun.

6.1 Radiation concepts

Two types of electromagnetic radiation, distinguished by their sources and spectral distributions, are important in crop environments. **Solar radiation** from a very hot thermal radiator, the Sun, is termed short-wave radiation (SW) because most energy is received in relatively short wavelengths, 0.3 to 3 μm. In contrast, **thermal** radiation from objects on our planet, including soils, plants, and atmosphere occurs at longer wavelengths because these radiating bodies are at much lower temperatures. Such long-wave radiation (LW) is found mainly between 5 and 100 μm.

6.1.1 Thermal radiation

All objects with a temperature greater than 0 K are sources of a continuous spectrum of electromagnetic radiation that, because of its source, is termed **thermal radiation.** Intensity and spectral distribution of thermal radiation may be compared with those from a reference "black body". Black body is a physicists' term for a perfect emitter of thermal radiation, represented by a hollow sphere pierced by a pinhole. The term black body is used because such spheres are also perfect absorbers of radiation: light entering the hole has essentially no chance of escaping by reflection.

Intensity and spectral distribution of radiation from black bodies varies with temperature as illustrated in Fig. 6.1. Maximum wavelengths of the curves (λ_{max}) in Fig. 6.1 are predicted by **Wien's displacement law**:

$$\lambda_{max} = 2897/T \ (\mu m) \tag{6.1}$$

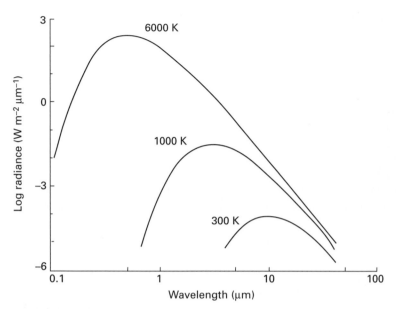

Fig. 6.1 Spectral distribution of black body radiation as calculated with Planck's distribution for bodies at 300, 1000, and 6000 degrees Kelvin (K).

where T is absolute temperature (K). Total energy emitted over the entire spectrum (E) can be calculated with the **Stefan–Boltzmann law**:

$$E = \varepsilon \sigma T^4 \ (\text{W m}^{-2}) \tag{6.2}$$

where ε is **emissivity** (effectiveness as a radiator relative to the black-body standard) with values between 0 and 1, σ is the Stefan–Boltzmann constant (5.67×10^{-8} W m^{-2} K^{-4}), and T is the absolute temperature (K). Although the distributions shown in Fig. 6.1 extend beyond infrared regions to very long wavelengths in the radio spectrum, most energy is emitted around λ_{max}. Energy flux calculated with the Stefan–Boltzmann law is **radiance** per unit area of source (sometimes also referred to as intensity) whereas **irradiance** is the amount of energy received per unit area some distance from the source. **Flux density** is also used to describe the strength of a beam of radiation in either energy or quantum terms.

Most Earth objects, including crops and soils, are near 300 K. At this temperature, λ_{max} is 9.6 μm, i.e., at a long wavelength in the infrared region (LW). At this temperature, emissivities of crops and soils are all near 0.95, indicating that they are nearly as effective as black bodies in emission. By contrast, the surface temperature of the Sun is near 6000 K and λ_{max} is near 0.5 μm. This is a short wavelength in the visible portion of the spectrum. Radiation from the Sun and the Earth thus differs markedly in spectral properties (Table 6.1).

Thermopile pyranometers of Kipp (Europe) and Eppley (USA) designs are standard instruments for measurements of SW irradiance. Because they depend on differential heating of a black absorber and a white reflector, thermopiles are equally sensitive over

Table 6.1 Characteristics of thermal, black-body radiation from objects at various temperatures

Temperature (K)	λ_{max} (μm)	Range for 95% of energy (μm)	Emission with $\varepsilon = 1.0$ (W m^2)
0	–	–	0
300	9.6	3–100	459
1000	2.9	–	56.7×10^3
6000	0.48	0.3–3	73.5×10^6

a wide range of wavelength. The receivers are protected from convection by domes of special glass. Photoelectric cells constructed of layered semiconductors (e.g., silicon) provide an electric current proportional to the number of quanta (or "photons", if visible to the eye) received rather than to energy. Silicon photocells are sensitive only to particular wavebands. They can be calibrated to measure SW irradiance, however, providing the spectral properties of the radiation remain constant. Fitted with appropriate filters, silicon cells are used to measure the number of quanta of **photosynthetically active radiation** (PAR, 0.4 to 0.7 μm). Human eyes are sensitive to this same waveband so we also refer to it as "visible" radiation. Usual units for PAR are mol m^{-2} s^{-1}; 1 MJ of typical SW irradiance carries about 2 mol PAR.

6.1.2 Two more radiation laws

Kirchhoff's law explains that effectiveness of a body in absorbing radiation of a given wavelength is the same as its emissivity at that wavelength. Crops and soils vary considerably in their absorption and reflectance of SW radiation but they all have LW emissivities near 0.95; i.e., they also absorb nearly all the LW radiation they receive. Long-wave radiation absorptivity and emissivity of air also obey Kirchhoff's law but they occur in complex spectral bands and are not generalized easily.

   The cosine law relates irradiance received by a surface to its angular display to the source of radiation. A surface, such as a leaf, displayed normal to the Sun's rays (at a right angle to them) receives maximum radiation per unit area. If the surface is inclined to the beam, energy is distributed over a larger area and irradiance is less. Irradiance of that larger surface (I) can be calculated with the **cosine law**:

$$I = I_0 \cos \theta \qquad (6.3)$$

where I_0 is irradiance normal to the rays and θ is angle of incidence formed between the rays and the normal to the inclined surface (Fig. 6.2). The law can also be expressed in terms of surface inclination angle: $I = I_0 \sin (90 - \theta)$. When the surface is horizontal, these two angles, θ and $90 - \theta$ are solar zenith angle and solar altitude, respectively (see Box 6.1). More generally, when the beam of radiation lies outside the plane of the surface's normal, two angles are needed to define the three-dimensional geometry of

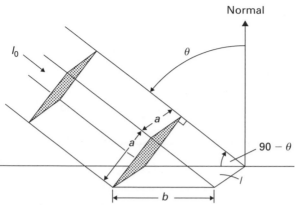

Conservation Law:

$$Iab = I_0\, a^2;$$
$$I = I_0\, a/b = I_0 \cos \theta.$$

Fig. 6.2 Derivation of the cosine law beginning with a beam of radiation with flux density of I_0 normal to the beam. The energy in this beam (I_0 times the area a^2) is distributed over a larger surface, also of width a but with length b, having an angle of incidence θ to the beam. Because the energy is distributed over a larger area ab, the irradiance I is less than I_0. From the conservation law, $I_0 a^2$ must equal I_{ab}.

irradiance, one for incidence (θ) and one for azimuth (ϕ), which is solar direction taken from due south: then $I = I_0 \cos \theta \cos \phi$.

6.2   The SW source

Our Sun is a rather ordinary celestial body, one among an estimated 10^{16} thermonuclear stars of various sizes and ages in the Universe. The Sun is of moderate size and is composed almost entirely of fluid gases. It probably formed about 4.6×10^9 years ago from dust and gas dispersed at the "beginning" of the Universe approximately 15×10^9 years ago. What is remarkable and exciting to us is the vastness of space and energy relations involved. The Sun's diameter is 540 000 km; its mass, 2.1×10^{27} t, comprises 99.9% of all the mass in our solar system. That mass generates enormous gravitational force, leading to low-grade thermonuclear reactions in the solar core where hydrogen and neutrons fuse to helium. During that process, mass is lost and energy is released as radiation: $E = mc^2$ (energy emitted = mass loss × velocity of light^2). The Sun emits energy at the rate of 4×10^{26} W; corresponding to loss of mass of 4.6×10^6 t s^{-1}. Even at that enormous rate, less than one-billionth of the Sun's mass will be converted to energy before it collapses to a cooling white dwarf about 5×10^9 y from now.

The outer surface of the Sun, the photosphere, is the principal source of sunlight. Through telescopes, it appears as a granulated surface composed of very large, gaseous, convection cells, broken periodically by long-lasting eruptive storms called sunspots, and by short-lived flares. The 6000 K spectrum of radiation from the photosphere is

---

Box 6.1 Location and movement of the Sun

Solar position relative to an Earth observer can be calculated from latitude (L in degrees), day of year [D in degrees $= 360$(day number/365)], and **declination** angle (d). In practical terms, declination is the latitude on Earth where the Sun appears directly overhead at solar noon. As indicated in Fig. 6.4, d varies annually between $+23.5°$ on June 22 (Sun is directly overhead at 23.5° N) and $-23.5°$ on December 22 (overhead at 23.5° S).

Values of d for particular days can be computed with sufficient accuracy from:

$$d = -23.5 \cos (D + 9.863) \tag{6.4}$$

and with this, solar altitude (α) at any hour of the day from

$$\sin \alpha = \cos L \cos d \cos h + \sin L \sin d \tag{6.5}$$

with azimuth (ϕ) from due south from

$$\sin \phi = \cos d \sin h / \cos \alpha \tag{6.6}$$

The Sun reaches its highest **altitude** at local solar noon [$\alpha_{noon} = 90 - (L - d)$] when azimuth is 180° for someone at a northern latitude (0° in the southern hemisphere). It is useful to remember that Earth turns 15° each hour (360°/24 h) and to know that the **sunset hour angle**, h_o, may be found from:

$$\cos h_o = - \tan L \tan d \tag{6.7}$$

The sunrise angle is $-h_0$. **Daylength**, then, is $2h_0$ in degrees and $0.1333h_0$ in hours. Distributions of daylength with latitude and day of year are presented in Fig. 5.2. A twilight period lit only by diffuse skylight precedes sunrise and follows sunset but is not included in $0.1333h_0$. Length of the twilight period increases with latitude.

---

modified slightly by atomic absorption of solar gases, and by radiation from coronal streamers that extend millions of kilometers into space. As a result, solar temperature calculated with the Stefan–Boltzmann law (5760 K) differs slightly from that obtained using measured peak wavelength and Wein's law (6170 K).

6.3 Sun–Earth geometry

Radiation from the Sun diverges in all directions but our distance from it is so great (1.5×10^8 km) that we generally consider that radiation arrives in parallel beams from a point source. In fact, angular difference between light from opposite edges of the Sun's photosphere is 0.2° to an Earth observer (the angle is larger when coronal streamers are included). This means that solar shadows are not really sharp but have a diffuse boundary (penumbra) between light and dark as is illustrated in Fig. 6.3. As a result, canopies composed of small leaves distributed over a large vertical distance (e.g., in a

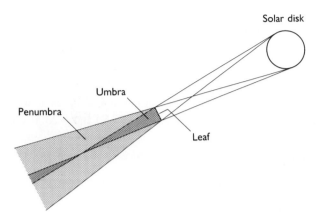

Fig. 6.3 Diffusion of a leaf's shadow due to penumbral effects caused by the finite size of the Sun. A full shadow from direct beam radiation occurs only within the umbra. The distance d, beyond which shadows are fully diffuse, equals approximately 100 times leaf width. (Not drawn to scale.)

pine forest) cause unintercepted light to appear as diffuse shade rather than as shadows and sunflecks.

Solar energy received by the Earth can be measured with radiation sensors displayed outside our atmosphere and normal to the Sun's rays. The amount varies slightly with sunspot activity, which affects X-ray, ultraviolet, and radio portions of the spectrum. A larger variation of several percent occurs during each year because the Earth follows a slightly elliptical orbit around the Sun and the Sun–Earth distance varies accordingly. For practical purposes 1360 W m^{-2}, the radiant energy received at mean Sun–Earth distance, is taken as the **solar constant**.

Average solar radiation received by the Earth is affected by its rotation and curvature. It can be calculated by spreading sunlight intercepted, taken to be equal to the projected area of the Earth, over its total area. Earth has a diameter of 6365 km and thus a projected area (πr^2) of 3.18 $\times$ 10^{13} m^2 and a total surface area ($4\pi r^2$) of 1.27 $\times$ 10^{14} m^2. Mean input of energy to the Earth's atmosphere is thus 1360($\pi r^2/4\pi r^2$) = 340 W m^{-2}; and the annual receipt of energy is 1360 W m^{-2} $\times$ 1 J s^{-1} W^{-1} $\times$ 3.15 $\times$ 10^7 s y^{-1} $\times$ 3.18 $\times$ 10^{13} m^2 = 1.36 $\times$ 10^{24} J y^{-1}. The cosine effect causes an unequal distribution of this energy between equatorial and polar regions, and the Earth's rotation causes it to be unequally distributed between day and night. The resulting circulation and mixing of air and ocean currents between regions are the driving force behind weather systems.

As is depicted in Fig. 6.4, unequal distribution of radiation over the Earth's surface is further modified by inclination of Earth's axis of rotation at about 23.5° to the plane of its orbit. During the annual circuit of the Sun, inclination of the axis combined with progress in orbit leads to variations in daylength and to alternation of the seasons as Earth's poles are alternately inclined towards and away from the Sun. At one **solstice**, near December 22, the north pole is inclined away from the Sun. In the northern hemisphere, days are short and it is midwinter; in the southern hemisphere, days are long and summer prevails. The situation is reversed six months later near June 22 at the other solstice. **Equinoxes**

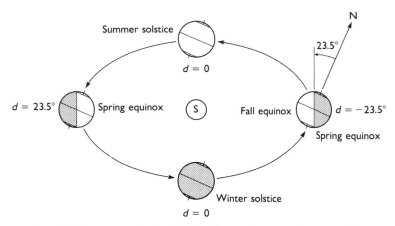

Fig. 6.4 Sun–Earth geometry during the Earth's annual progression around the Sun. Solstices and equinoxes are labeled according to seasons in the northern hemisphere; the declination angle, d, is indicated for each position.

(near March 21 and September 24) mark times of the year when the Earth's axis lies in a plane normal to the Sun's rays. Both poles are irradiated then and daylength is 12 h at all latitudes. The equinox between winter and summer is referred to as the vernal or spring equinox; the autumnal or fall equinox comes between summer and winter. Dates of these events vary slightly year to year because the Earth completes a revolution around the Sun in 365.256 days and our Gregorian calendar adjusts for the odd 0.256 d with an extra day in leap years. The Nautical Almanac is a useful reference for exact times of events for any given year.

Earth's orbital ellipse is not constant but varies with periods of about 97 000 y. Sun–Earth geometry also changes over time, owing to cyclic variations with a period of 41 000 y in the tilt of the Earth's axis of rotation (presently 23.5°). These matters vex calendar makers and high priests who must have things on proper dates. They are a source of excitement to geologists, however, who find in them explanations for past climates, including glacial cycles, and as a basis for predicting future climates. Fortunately, these changes occur too slowly to influence agricultural ecology within a human lifetime.

Knowledge of solar position relative to an Earth observer (or crop) is useful for many purposes, including crop ecology, landscape planning, building design, and navigation. The equations for calculation are presented in Box 6.1.

6.4 SW penetration of the atmosphere

Solar radiation is strongly modified during passage through the atmosphere, owing to absorption in ultraviolet and blue regions by ozone (O_3), in the visible by O_2, and in the infrared by CO_2, water vapor, and other gases. Absorption spectra for various atmospheric gases, illustrated in Fig. 6.5, have enormous influence on the radiation balance of the Earth. Absorption by gases in the infrared region, beyond 1 μm wavelength, occurs

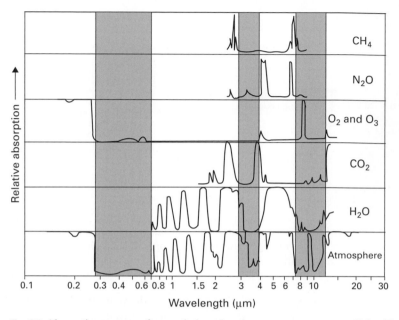

Fig. 6.5 Absorption spectra of normal air and various component gases. (After Fleagle & Businger 1980.) Note the PAR band in the atmospheric spectrum at 0.3 to 0.7 μm and how the width of the atmospheric "windows" to infrared at 3 to 4 and 8 to 12 μm are limited by the absorption bands for CO_2.

in "bands". Each band is composed of many individual lines relating to vibrational and rotational frequencies of gas molecules. CO_2 and water vapor have the strongest absorption bands.

Small molecules and dust also scatter light out of the direct beam. Scattered light may be rescattered several times before leaving the atmosphere or reaching Earth as **diffuse skylight** from all portions of a luminous sky. Blue light is scattered more strongly than are longer wavelengths, so skylight appears blue to our eyes.

Absorption and scattering of solar radiation in the atmosphere are proportional to the amount of air through which radiation passes. The mass of air between an observer and the Sun is much greater when it is near the horizon than when directly overhead. "Air mass" (m) is defined as 1.0 when the Sun is directly overhead; $m = 1/\cos \theta$ (where θ is the zenith solar angle), therefore a solar altitude of 30° ($\theta = 60°$) corresponds to an air mass of 2 (see Fig. 6.2). At that altitude, and with clear skies, irradiance on a horizontal surface is reduced by 50% by geometry of incidence ($\cos 60° = 0.5$) and by an additional 15 to 20% through absorption and scattering.

Spectral distributions of solar irradiance at the Earth's surface with air mass 2 and air mass 6 (solar altitude of 10°) are presented in Fig. 6.6. In Fig. 6.6*a*, irradiance is plotted against wavelength, λ. Photosynthetically active portions of the spectrum with air mass 6 are much more strongly depleted of short wavelengths (0.4 to 0.5 μm) than of wavelengths between 0.5 and 0.7 μm. This effect is known as the "red shift". Absorption bands in infrared regions (0.7 to 2.5 μm) match those of water shown in Fig. 6.5. In Fig. 6.6*b*, irradiance is plotted versus frequency, *v* (*v* is the velocity of light,

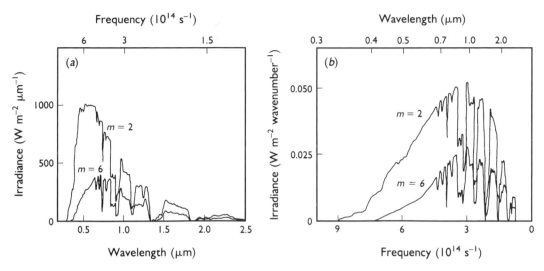

Fig. 6.6 (*a*) Wavelength distribution of short-wave irradiance of the Earth's surface for air masses (*m*) of 2 and 6. (*b*) The same spectra as in (a) plotted as a function of frequency. (Adapted from Gates 1980.)

3×10^8 m s^{-1}, divided by λ). Areas under the curves in Fig. 6.6*b* are proportional to the number of quanta in the radiation because energy per mole quanta (E_q) is proportional to frequency *v*:

$$E_q = nhv \ (\text{J mol}^{-1}), \tag{6.8}$$

where *n* is Avogadro's number and *h* is **Planck's constant** (6.63×10^{-34} J s).

Distributions of direct and diffuse irradiances and their sum, **global** irradiance, as functions of solar altitude are illustrated in Fig. 6.7*a*. These measurements were made under clear skies near Melbourne (38° S) at various times of the year. Skylight contributed more than 30% to global radiation with low solar altitudes but it was less than 10% with high angles.

Total irradiance decreases and diffuse radiation increases as atmospheric turbidity increases with greater content of water vapor and/or dust. Absorption and reflection of radiation by clouds also have important effects on irradiance at ground level. With overcast skies, radiation is completely diffuse and much less reaches the Earth's surface. Fifty percent cloudiness seems to be an average condition in mid-latitudes during summer, therefore only about half the hours approach a clear-sky condition. As a result, average daily insolation in humid regions is commonly 20 to 30% less than in arid and semi-arid regions. Latitudinal distribution of daily total global irradiance with clear skies is presented in Fig. 6.7*b* as a function of time during the year. This figure is based on 70% transmission through the atmosphere (humid conditions): maximum values there are near 24 MJ m^{-2} d^{-1} compared with more than 29 MJ m^{-2} d^{-1} received in midsummer at Davis, California (38° 30′ N), and other semi-arid locations. The fraction of total SW flux received as PAR (0.4 to 0.7 μm) varies from about 0.44 in semi-arid regions with clear skies and small air mass to about 0.50 where a greater proportion is received as diffuse skylight.

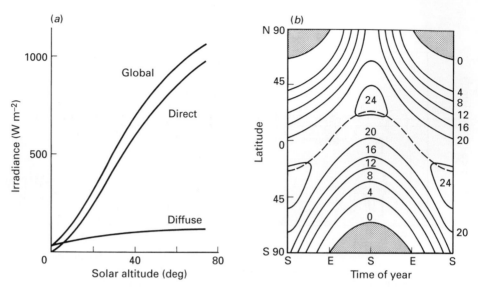

Fig. 6.7 (*a*) Hourly global, direct, and diffuse radiation observed with clear skies over the course of a year near Melbourne, Australia. (Calculated with regressions given by Paltridge & Platt 1976.). (*b*) Latitudinal distribution of expected daily total of short-wave radiation over the course of the year. S marks the time of the winter and summer solstices, and E the equinox. The units are MJ m^{-2} d^{-1}. Sun's zenith position at solar noon (broken line) is also shown. (Adapted from Neiburger *et al.* 1982.)

6.5   Radiation balance

Mean temperature of Earth varies little year to year because average receipt of solar radiation calculated earlier as 340 W m^{-2} is balanced by a similar LW flux to space. Examination of the fate of SW radiation and how it is balanced through LW radiation to space illustrates several important principles. The principles apply to Earth as a whole but, as will be developed later, to individual crops also.

A summary of average SW and LW fluxes is presented in Fig. 6.8. The magnitude of "greenhouse effects" (explained further in Box 6.2) is indicated in the figure by the large downward flux of LW radiation from the atmosphere. A **radiation balance** based on conservation of energy can be constructed at any level of the atmosphere.

At the top of the atmosphere:

$$SW_{incident} - SW_{reflect} - LW_{emit} = 0 \qquad (6.9a)$$

With values from Fig. 6.8:

$$340 - 105 - 235 = 0 \text{ W } m^{-2} \qquad (6.9b)$$

There is a good reason why incoming and outgoing radiant fluxes balance at the surface of the atmosphere; if they failed to balance, the Earth's temperature would simply increase or decrease until they did.

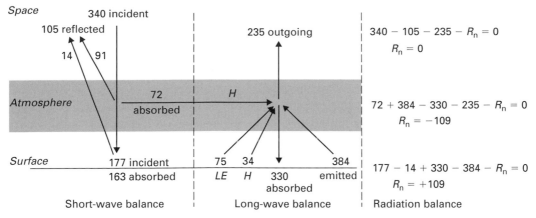

Fig. 6.8 Global average short-wave and long-wave radiation fluxes and balances. The units are W m^{-2}. The radiation balances are constructed following Eq. 6.9 (LW reflectance at the surface is ignored, i.e., LW absorbance is taken as 1.0). (Adapted from Fröhlich & London 1985.)

Box 6.2 The greenhouse effect

The atmosphere is relatively transparent to SW radiation but it is almost opaque to LW radiation from the Earth owing to strong absorption by water vapor, CO_2, N_2O, CH_4, and other "greenhouse gases". λ_{max} for terrestrial radiation is near 9.6 μm, however, and an **"atmospheric window"** to LW radiation exists at 8 to 14 μm (Fig. 6.5). This window is an important route for LW loss to space. Radiation emitted at shorter or longer λ (i.e., 4–8 and beyond 14 μm) is completely absorbed by a few meters of air.

When air is warmed by absorption of SW and LW radiation (and also by condensation of water in clouds and by contact with warm objects) its emission of LW increases. This emission of LW by the atmosphere is directionally random and Earth's surface receives a portion as a large downward flux. That flux amounts to 15 to 20 MJ m^{-2} d^{-1} in mid-latitudes and is a major factor in maintaining a warm Earth. Earth's surface is usually warmer than the atmosphere, however, and is the source of an even stronger upward flux of LW (as much as 30–40 MJ m^{-2} d^{-1} in mid-latitudes). Because very little LW comes from outer space (its temperature is near 0 K), a gradual, net "upwelling" of LW radiation to outer space occurs.

Window glass behaves in an analogous way to the atmosphere. It is also relatively transparent to SW radiation and relatively opaque to infrared. That, plus entrapment of warm air, are the principles underlying the use of glass houses for culture of crops in cold climates. By analogy, the roles of atmospheric CO_2, water vapor, other gases, and clouds in maintaining a warm Earth are referred to as the greenhouse effect.

At Earth's surface, additional terms must be included in the balance. One is the downward flux of LW from the warm sky:

$$\text{SW}_{\text{incident}} - \text{SW}_{\text{reflect}} + \text{LW}_{\text{incident}} - \text{LW}_{\text{reflect}} - \text{LW}_{\text{emit}} \neq 0. \qquad (6.10\text{a})$$

As is indicated by $\neq$, radiation terms seldom balance to 0 at the surface. That is because radiation absorbed or emitted by objects at the surface may be exchanged with other fluxes of energy, including evaporation or condensation of water and heating or cooling of crops and soils. For the example in Fig. 6.8, average SW flux incident to the surface is 177 W m^{-2} (340−91 reflected by the atmosphere; −72 absorbed by the atmosphere). Reflection from the surface is 14 W m^{-2}, so 163 W m^{-2} is absorbed there. Earth's absorptivity for LW is near 0.95 so reflection of LW from the surface can be ignored in this example. Due to the "**greenhouse effect**" (Box 6.2), the sum of short- and long-wave fluxes absorbed at the Earth's surface ($163 + 330 = 493$ W m^{-2}) is much greater than SW entering the atmosphere. It is also greater by 109 W m^{-2} than LW emitted (384 W m^{-2}):

$$163 + 330 - 384 = 109 \text{ W m}^{-2} \qquad (6.10\text{b})$$

This difference in radiant fluxes, termed **net radiation** (R_n), represents exchanges with other forms of energy. Inclusion of R_n as an additional term in the radiation balance brings the result to zero:

$$(SW_{\text{in}} - SW_{\text{out}}) + (LW_{\text{in}} - LW_{\text{out}}) - R_n = 0. \qquad (6.11)$$

R_n at the Earth's surface (109 W m^{-2}) is partitioned (Fig. 6.8) between evaporating water (75 W m^{-2}) and heating of air (34 W m^{-2}). By convention, R_n is positive when the absorbing surface or plane dissipates radiant energy to other forms (temperature increase, evaporation) and negative when LW flux is increased by loss of energy from other forms (cooling, condensation). In Fig. 6.8, $R_n = +109$ W m^{-2} at the surface is seen to be balanced by $R_n = -109$ W m^{-2} within the atmosphere where air cools and water condenses to clouds. The released energy is further dissipated in LW fluxes.

Similar imbalances in radiant fluxes occur at all levels above ground within crops and in the air above them. The magnitude of imbalance varies with time of day, atmospheric conditions, and crop status. In practice, R_n can be measured with net radiometers, which are simple, flat-disk, differential thermopiles shielded by polyethylene domes transparent to both SW and LW. One surface of the sensor is exposed horizontally to downwelling SW and LW while the other receives upwelling fluxes. Hot and cold thermocouple junctions are attached to upper and lower surfaces, respectively, so difference in heating is proportional to net radiant flux. Alternatively, the net LW portion of Eq. 6.11 can be estimated with empirical formulae such as those evaluated by Jensen (1974). Net SW of a crop can be estimated from measured incoming SW if we know the **albedo** (the term for reflectance from a complex surface). Monteith and Unsworth (1990) report that crops generally have albedo in the range 0.18 to 0.25, so SW$_{\text{absorbed}} \approx 0.8SW_{\text{incident}}$. Albedos of wet and dry soils and snow may differ considerably from 0.2, however. The small value used as a world average in Fig. 6.8 (14 reflected/177 incident $= 0.08$) is due to the low reflectivity of oceans.

SW radiation changes dramatically between day and night whereas LW fluxes change only moderately as can be seen by introducing day and night temperatures into the Stefan–Boltzmann relation. Consequently, diurnal and seasonal patterns of R_n are coupled rather closely with receipt of SW radiation. In tropical regions, daily and annual R_n totals are strongly positive whereas in polar regions they are strongly negative. At mid-latitudes, daily R_n is positive in summer and negative in winter.

6.6   Energy balance

Net radiation is an important concept because it accounts for energy exchanges by crop communities that influence soil and air temperatures and rates of production and water loss. The objective of this section is to construct an **energy balance** for R_n that identifies how energy is partitioned within the crop environment.

6.6.1   Components of R_n

Three forms of energy are involved in R_n exchanges. The most obvious are **sensible heat fluxes** related to increases or decreases in temperatures of air (H) and soil, plants, or water (G). Such thermal energy is termed sensible because it can be "sensed" with a thermometer. Also involved are **latent-heat fluxes** related to changes in state of water. Ice↔liquid water entails input or release of **latent heat of fusion** of water (F; 334 J g^{-1} H$_2$O), and liquid water↔vapor (evaporation and condensation) involves **latent heat of vaporization** (L; 2442 J g^{-1} at 25°C). Large amounts of energy are exchanged in these processes but the temperature of water does not change and energy cannot be "sensed" with a thermometer; it therefore is considered hidden or "latent". Radiant energy also is converted to **chemical-bond energy** in photosynthesis (P) and then released during respiration (R) as sensible heat or thermal radiation.

Sensible and latent heat contents of air are transported by convection. Through mixing into higher altitudes and transport to polar regions, excess energy of lower latitudes is eventually lost to space as LW radiation. Clouds, with their very large exchanges of latent heat, are an especially complex component of these processes. Ocean currents also transport enormous quantities of sensible heat between tropical and polar regions.

6.6.2   Conservation of energy

The Conservation law can be used to construct an **energy balance** for R_n exchanges at a crop surface in terms of energy forms defined above as:

$$R_n - H - G - LE - P - R = 0. \tag{6.12}$$

Latent heat of vaporization (L) is multiplied by amount of water evaporated or condensed (E) to calculate the energy flux involved. Each term in the energy balance deserves attention: H and G are important determinants of ambient temperatures in the crop environment; LE is a central issue in water relations; and P and R are determinants of primary production and carbon cycling. Our convention in signs considers a flux to one of these components positive (+) as it gains energy and negative (−) as it loses energy.

Exchanges occur among terms in the energy balance as well as between terms and radiation. A soil surface, for example, can be warmed (+G) through contact with warm air (and the air is cooled, −H), or through condensation of water (−LE). Conversely, soil may be cooled (−G) by transfer of heat energy to air (+H) or by evaporation of soil water (+LE). Similar exchanges among energy terms occur on plant surfaces. It is important to note that while SW radiation is the original source of energy (geothermal fluxes are very small), R_n terms can be quite divorced from it in time. Heat energy lost from soil during winter (−G) probably represents energy gained during the previous summer (+LE). Similarly, energy consumed in melting glacial ice today (+LE) is restoring that lost many thousands of years ago during freezing (−LE). Enormous amounts of water on the Earth's surface and very large amounts of energy involved in phase changes combine to give water a strong buffering role in global temperatures.

In crops, the G, P, and R terms are generally small (c. 1 to 5% of R_n). Discussion of G will be delayed to Chapter 7; its magnitude is limited by the small conductivity of soils and because soils are protected from radiant exchanges by crop cover. The terms P and R are considered in Chapters 10 and 11, respectively. Over an annual cycle, $G \approx 0$ (no net change in soil temperature) and P approximately equals R because production of organic material is balanced by decay. That means that LE and H are the main exchanges between crop surfaces and atmosphere and, for climatological analyses, the energy balance can be simplified to:

$$R_n - H - LE = 0 \tag{6.13}$$

LE dominates in wet systems and H in dry ones. Global averages for LE and H are included in Fig. 6.8; in that example, Eq. 6.13 is satisfied by $109 - 34 - 75 = 0$ W m^{-2}. Examples of diurnal patterns of R_n, LE, and air temperature are presented in Fig. 6.9a. Peak R_n fluxes in this figure are near 700 W m^{-2} and daily average is 195 W m^{-2}. Close coupling in time between incident solar radiation, R_n, and LE is demonstrated here.

6.6.3   Sensible heat flux to air

Air is a poor absorber of SW radiation and is heated and cooled mainly by contact with soil and plant surfaces, and by absorption and emission of LW radiation. Changes in heat content of air are determined by its specific heat (c_a), density (ρ_a), and change in temperature, ΔT_a (K):

$$H = \rho_a c_a \, \Delta T_a \tag{6.14}$$

(An analogous equation for soil heat flux is $G = \rho_s c_s \, \Delta T_s$.) The specific heat of air (c_a) is only 1 kJ kg^{-1} deg^{-1} at 27°C and 100 kPa where density is 1.17 kg m^{-3} (1.2 kJ m^{-3} deg^{-1} is a convenient value to remember for $\rho_a c_a$ as the volumetric heat capacity of air). The heat capacity of air, therefore, is only about 0.0003 that of liquid water (4.184 MJ m^{-3} deg^{-1}). For air to carry large quantities of heat as it does, very large volumes of air and large temperature changes are necessary.

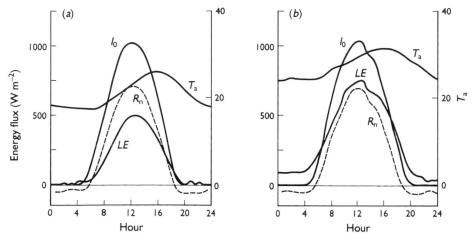

Fig. 6.9 (*a*) Hourly distribution of solar radiation, R_n, and latent heat exchanges (*LE*) of a ryegrass crop on a clear day in late spring at Davis, California, USA. Air temperature (T_a) is also shown. The daily totals of SW radiation (I_o) and R_n were near 31 MJ m^{-2} and 19 MJ m^{-2}, respectively; winds were light all day. (*b*) The same crop two days later with similar conditions except that winds were strong. In this case, $LE > R_n$. (Adapted from Pruitt 1964.)

Air warmed (or cooled) through contact with crop and soil surfaces is moved away from surfaces by convective action of prevailing winds and by buoyancy of warm air (or denseness of cold air). Warmed air is less dense than surrounding cool air and it rises away from a surface as a bubble even in the absence of wind. The reverse occurs as air is cooled. The basis for density changes is found in the Ideal Gas Law:

$$PV = nRT_a \text{ (L kPa)} \tag{6.15}$$

where P is air pressure in kPa (100 kPa = 0.987 mean atmospheric pressure), V is its volume in liters, n is moles of gas (mean molecular mass of air is 29 g mol^{-1}), R is the universal gas constant (8.314 L kPa mol^{-1} K^{-1}), and T_a is air temperature (K). At sea level and 300 K (27°C), 1 mole of gas occupies 24.9 liters, so average density is 1.17 g L^{-1} (29/24.9). As T_a increases, V increases because P is fixed by the general mass of air. As a result, n/V decreases, and air becomes buoyant relative to cooler air. Rising bubbles of light air are replaced by downward fluxes of cooler and thus denser air. Buoyant movements of air are observed most easily in up-slope and down-slope winds common to mountainous areas, in desert mirages, and in rough air encountered by small airplanes.

6.6.4 Latent heat exchanges

Within crops, the principal latent heat exchanges involve evaporation of water from leaves (transpiration) and soil, and condensation (dew formation). Collective evaporation (or condensation) of water by both leaves and soil is termed **evapotranspiration** (ET). As long as surfaces remain wet (wet soil or freely transpiring leaves), evapotranspiration

> **Box 6.3** Daily evapotranspiration (ET) of a crop
>
> Six millimeters of water is a typical daily total ET at mid-latitudes during summer. That would require an R_n flux near $+14.7$ MJ m^{-2} d^{-1} (14.7 MJ m^{-2} d^{-1}/2442 kJ g^{-1} H$_2$O $\rightarrow$ 6020 g H$_2$O vapor m^{-2} d^{-1}). Six millimeters of water seems like a small amount but the latent heat involved in its evaporation (or condensation and precipitation) equals the energy content of about 15 t dynamite per ha! Latent heat exchange through condensation in a typical rainstorm can equal many megatonnes of dynamite.
>
> Six millimeters of water occupies a large volume as vapor. Using the gas law [Eq. 6.15 with $P = 100$ kPa (1 atm) and 18 g mol^{-1} as molecular mass of water], 8060 L of pure water vapor, filling a column 8.1 m high, would accumulate over each m^2 land. Unless the vapor mixed quickly with a large volume of air, we would have considerable difficulty breathing! This does not occur, however, due to dilution through mixing with bulk atmosphere. As also for sensible heat flux to air, mixing occurs through wind action and buoyancy. Buoyant transport occurs because moist air is less dense than dry air (molecular mass of water is 18 g mol^{-1} compared to 29 g mol^{-1} for air).

dominates over sensible heat flux. In temperate climates, latent heat exchanges involved in freezing and thawing of water are also important since the length of growing season is reduced according to the time required for thawing of soils in spring.

Latent heat of vaporization can engage very large amounts of energy. With $R_n = 500$ W m^{-2} of ground surface and 2442 J g^{-1} as latent heat of vaporization at 20°C, R_n could (potentially) be dissipated by evaporation of 500 W m^{-2}/2442 J g$^{-1} = 0.205$ g H$_2$O m^{-2} s^{-1} or 737 g m^{-2} h^{-1}. Given 10^4 cm^2 m^{-2} and density of water as 1 g cm^{-3}, a depth of water equal to 0.074 cm would evaporate per hour. Energy balance terms are commonly expressed in equivalent mm of evaporation. In this case, evaporation rate is 0.74 mm h^{-1}. A calculation of energetics of daily evaporation is presented in Box 6.3.

Evaporation of water into air also requires substantial mixing because a limit, termed **saturation vapor pressure** (e^*), exists to the amount of water vapor that can be held in air. This e^* varies with temperature as depicted in Fig. 6.10a. Values of e^* as a function of temperature can be found in handbooks of physics and chemistry or calculated from an empirical equation as:

$$e^* = 0.611 \exp[(17.27T)/(T + 237.3)] \text{ (kPa)} \tag{6.16}$$

where T is in °C. The value of e^* at 20°C is 2.34 kPa. Returning to the example with 6 mm evaporation d^{-1} at 20°C (Box 6.3), mixing into dry air must extend to at least $(8.1 \times 1.013)/0.0234 = 350$ m. Even greater mixing is involved in the real world because air is never completely dry and ambient vapor pressure (e_a) seldom approaches saturation.

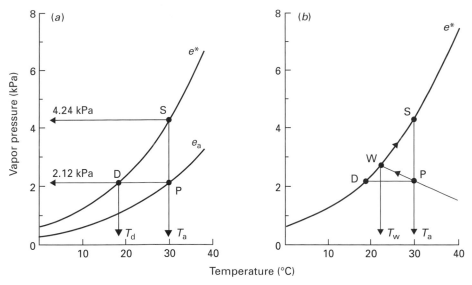

Fig. 6.10 (*a*) Saturation vapor pressure (e^*) and actual vapor pressure (e_a, 0.5 saturation) in air as a function of temperature calculated from Eq. 6.16. S is the saturation point for the air at P; D is the corresponding dew point. (*b*) Wet-bulb diagram for air at P following Eq. 6.17. The slope of line P–W is $-\gamma$.

6.6.5   Vapor pressure and evaporation

Evaporation from a wet surface depends not only on the rate of energy input but also on the gradient in vapor pressure to the surrounding air, the **vapor pressure deficit** (vpd $= e^* - e_a = \Delta e$). As a result, evaporation into unsaturated air differs from that into saturated air. It is convenient to consider these points now before returning to the question of how vapor is mixed with air. Normal air is usually less than saturated with water vapor and is found to the right of the e^* curve in Fig. 6.10a. Air at point P, for example, is at 30°C and $e_a = 2.12$ kPa water vapor pressure (15.1 g m^{-3}). The value of e^* at that temperature (point S) is 4.24 kPa. The air therefore is only 50% saturated, i.e., $e_a/e^* = 0.5$ and is said to be at 50% "relative humidity". If air at point P is cooled without change in water content, it reaches saturation at point D, corresponding to its **dew point temperature**, T_d, of 18.4°C. Vapor status of air can therefore be characterized by T_d as well as by e_a.

    Evaporation into unsaturated air "Adiabatic" exchanges can take place between H and LE within a volume of air. In adiabatic processes, total energy content in the volume remains constant ($-H = +LE$, and *vice versa*, independent of R_n). When evaporation occurs, T_a and thus air's heat content declines; with condensation, T_a rises. In such closed systems, evaporative cooling occurs along the line from P in Fig. 6.10b towards saturation at point W and ceases when air is saturated ($e_a = e^*$).

    Similar cooling takes place when a thermometer bulb is enclosed in moist cloth and exposed to rapidly moving air as in a "sling psychrometer". When air in the cloth reaches saturation at point W, the thermometer registers **wet bulb temperature**, T_w. This serves as one way to measure actual vapor pressure of water in air, e_a. At any point along the line PW:

$$e_a = e_{Tw}^* - \gamma(T_a - T_w) \text{ (kPa)} \tag{6.17}$$

where γ, the slope of line PW in Fig. 6.10b, is the psychrometric parameter. γ varies with atmospheric pressure and with heat capacity of air but remains close to 0.066 kPa deg^{-1} between 20 and 30°C. Psychrometers now are generally constructed with electrical humidity sensors (rather than wet and dry bulbs).

Progress forward and backward along line PW in Fig. 6.10b describes adiabatic evaporation and condensation in unsaturated air. Both adiabatic and non-adiabatic processes occur under field conditions. A crop system becomes open (non-adiabatic) when air within the canopy is exchanged with outside air. A warm, dry wind, for example, can supply energy for evaporation of water from leaves and from soil; LE can exceed R_n then because additional energy comes from cooling air. In Chapter 9, this concept is introduced into equations that estimate water loss by crops into unsaturated air.

Evaporation into saturated air Inside a leaf, air is saturated and **equilibrium evaporation** takes place. If e^* within the leaf equals e_a, no evaporation occurs because no gradient $e^* - e_a$ for vapor movement exists. In that case, R_n goes to H and to heating the leaf ($+G$). As leaf temperature (T_1) increases, e^* in interior spaces increases, a vapor pressure gradient is restored, and evaporation (LE) continues. Beginning at point W in Fig. 6.10b with saturated air in equilibrium with free water, an input of radiant energy causes evaporation to proceed along the e^* curve. For equilibrium evaporation, the ratio H/LE is given by γ/s where s is the slope of the e^* curve at T_1. Because $R_n = LE + H$, equations for LE and H can be written as:

$$LE = R_n s/(s + \gamma) \text{ (Wm}^{-2} \text{ or mm)} \tag{6.18a}$$
$$H = R_n \gamma/(s + \gamma) \text{ (Wm}^{-2} \text{ or mm)} \tag{6.18b}$$

As T_1 increases, a greater proportion of R_n goes to LE because $s/(s + \gamma)$ increases: it is 0.50 at 7°C and 0.75 at 27°C.

An important concluding point is that evaporation can cool unsaturated air and contribute heat energy to evaporation. In saturated air, by contrast, air temperature must rise if it is to hold more water vapor, thus robbing heat energy from evaporation.

6.7 Turbulent transport

Our calculation of the volume of water vapor produced through evaporation (Box 6.3) demonstrated the need for means by which vapor is mixed with very large volumes of air. Diffusion processes are simply much too slow and such mixing can occur only with convection. Buoyancy, involving rising bubbles of light warm or wet air countered by a downward flux of dense cold or dry air, is one type of convection. Another type occurs with turbulence created by the interaction between a lateral wind and vegetation.

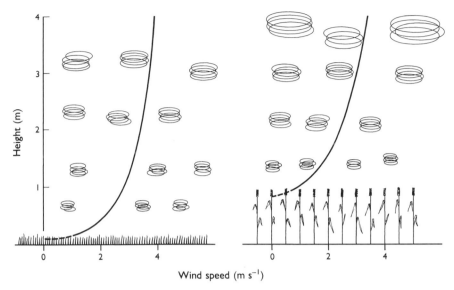

Fig. 6.11 Wind profiles and eddy size over short-grass and small-grain crops. The profiles are fit to Eq. 6.19.

Together, buoyancy and wind turbulence can involve a great depth of atmosphere in mixing. The eddies of air thus created carry all properties of air (temperature, water vapor, CO_2, and O_2) characteristic of their source (crop or atmosphere) and these properties are mixed (transported) between higher and lower altitudes.

Wind profiles illustrated in Fig. 6.11 serve as a starting point for discussion. The surface exerts a "drag" on air, due to friction, so wind speed (u) near the ground is less than at greater heights; at the surface it is essentially zero. Smooth surfaces create only small eddies while rough surfaces create larger ones. Coarse vegetation such as forests are quite rough to wind while closely grazed pastures are smooth. Topographical features such as hills and valleys, as well as isolated trees and hedgerows, add additional dimensions of roughness. Eddies near surfaces tend to be small and have rapid circulation but size increases and circulation intensity declines with height. The effect of wind, then, is to create a mixing zone of turbulent air above a crop.

Mean wind speed increases logarithmically with height for several meters above a **zero plane** near the top of the crop. Wind speed (u_z) at height z above that plane is:

$$u_z = a \ln(z - d) - \ln(z_0) \ (\mathrm{m\,s^{-1}}) \qquad (6.19)$$

where a is a proportionality constant, z is the height above soil surface, d is **displacement** of zero windspeed above soil and is a further correction for **roughness**. The value of d is normally between 0.6 and 0.8 times crop height. The layer of air that obeys the logarithmic relationship constitutes a "crop boundary layer". Because wind speed is proportional to $\ln(z - d)$, measurement of u at any two heights within this layer above the crop establishes most of the profile.

A **transport coefficient**, k, calculated from the wind profile, describes its effectiveness in transport of sensible heat, water vapor, and CO_2 between crop and atmosphere. Lessened wind speed near (and within) the crop demonstrates that friction is extracting kinetic energy from moving air. Height gradients of horizontally imparted kinetic energy ($0.5\rho_a u^2$ where ρ_a is density of air), or more simply, momentum ($M = \rho_a u$), are calculated directly from measurements of horizontal wind speed and air density. Existence of a concentration gradient (in this case a gradient of momentum) in a fluid system subject to mixing and diffusion results in transport between regions of unequal concentration. The **general transport equation** describes net flux (F) of that property between regions and can be used to establish the transport coefficient for momentum, k_m.

For momentum transfer in turbulent air:

$$F_m = k_m(M_1 - M_2)/\Delta z \tag{6.20}$$

where ($M_1 - M_2$) is the momentum gradient over a vertical distance Δz. The equation is commonly presented as a simple differential equation:

$$F_m = k_m \rho_a(du/dz) \tag{6.21}$$

where du/dz is the gradient of wind speed with height. Values of k_m derived from such measurements define effectiveness of wind in vertical transport of, in this case, the property of momentum; $1/k_m$ is then resistance to momentum transport and decreases with increasing wind speed. The transport parameter k is given various names depending upon the flux involved and units employed. It is also known as the "eddy diffusivity coefficient" and when the gradient is expressed in concentration units (i.e., mass m^{-3} m^{-1}), k has dimensions of m s^{-1}. The physics of turbulent transport in fluid media is complex; references such as Monteith and Unsworth (1990) and Rosenberg $et\ al.$ (1983) can be consulted for more detail on theory and applications.

An important point about atmospheric transport is that eddies of air that transport momentum also transport sensible heat, water vapor, and CO_2 between crops and a deep layer of the atmosphere. Analogous transport equations can be written for those fluxes:

$$F_H = k_H \rho_a c_a(dT_a/dz) \tag{6.22a}$$

$$F_w = k_w(de_a/dz) \tag{6.22b}$$

$$F_c = k_c(dC/dz) \tag{6.22c}$$

where ρ_a and c_a are respectively density and specific heat of air, T_a is air temperature, e_a is concentration of water vapor, and C is concentration of CO_2. Subscripts H, w, and c refer to sensible heat, water vapor, and CO_2. Values of all k parameters are roughly equal (the "**similarity principle**" or Reynolds analogy), differing slightly owing to buoyancy and to molecular diffusion within eddies. In principle, k_m can be obtained from wind speed, and the same value used for k_c and k_w. Evaporative and photosynthetic fluxes of a crop can then be calculated from measurements of CO_2 and H_2O concentration profiles. In practice, however, wind movements are often too variable and confounded with buoyancy for accurate measurements.

An alternative, and increasingly applied analysis, of transport of water vapor and CO_2 to and away from crop surfaces is made possible by measurements of the vertical velocity component of eddies themselves and concentrations of the gases they carry (Heilman *et al.* 1994, 1996). The changes are rapid and small and so measurements that are made at one height within the boundary layer must be frequent (e.g., 5 Hz) and precise. This is now possible for wind (sonic anemometer), CO_2 (open path infrared gas analyzer), and water vapor (Krypton hygrometer). These instruments require frequent attention to condition and calibration and so are suited to diurnal measurements and not, as yet, to long-term unattended operation.

The method of analysis gives rise to the name – **"eddy covariance"** technique. It turns out that "eddy" flux (F) can be calculated as the covariance between instantaneous deviation of vertical wind speed (w′) from its mean value and instantaneous deviation in gas concentration (s′, here representing water vapor or CO_2) from its mean value multiplied by mean air density (ρ_a), i.e., as:

$$F = \rho_a \overline{w' s'} \tag{6.23}$$

Instruments and data processing software are available from various commercial companies. Eddy flux measurements of CO_2 above crops have been very useful to estimate net C flux and identify if C is being sequestered or lost to the soil-crop system (e.g., Verma *et al.* 2005).

Scintillometry is a recent development in measurement of turbulence strength. Sensor–receiver pairs, separated from 250 m to 8 km apart, measure loss of energy from a near infrared (880 nm) beam to characterize activity of turbulent eddies. Sensible heat flux (H) can be derived when combined with standard meteorological measurements of wind speed, temperature, and humidity (Anandakumar 1999; Meijninger & de Bruin 2000; Cain *et al.* 2001). Additional measurement of R_n and G then allow calculation of ET at scales appropriate to field crops and also satellite measurements of land use. Combined with open path CO_2 analyzers, the technique has the further potential to measure crop or vegetation CO_2 exchange.

6.7.1 CO_2 transport

A calculation of CO_2 depletion from air due to crop growth is presented in Box 6.4. This is similar to our earlier calculation for water vapor and helps in understanding the significance of turbulent transport and the great depth of air involved in supplying a crop with carbon. Actual patterns of CO_2 depletion over a maize field during calm and windy days are illustrated in Fig. 6.12.

The necessary mixing height in the CO_2 example (550 m) is greater than the minimum height we calculated earlier for evapotranspiration. Measurements from airplanes reveal that the turbulent mixing zone (or "crop boundary layer") is frequently as much as 1500 m thick. Thickness of the layer and rates of transport within it increase with wind

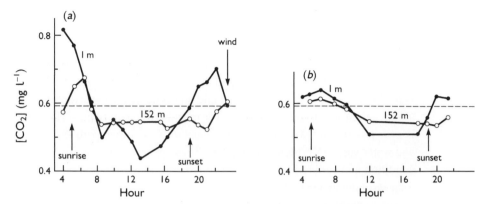

Fig. 6.12 Diurnal changes in CO_2 concentrations in the air 1 m above ground within an Iowa, USA maize field and at 152 m height. (After Chapman *et al.* 1954.) (*a*) A still day. (*b*) On a day with light winds of 4 to 10 m s^{-1}.

Box 6.4 The growing crop as a sink for CO_2

A crop with a moderate growth rate of 30 g biomass m^{-2} d^{-1}, and a carbon content of 0.44, has a net uptake of 13.2 g C or 48.4 g CO_2 m^{-2} d^{-1}. (The molar mass of CO_2 is 44 g.) To support this growth and also make up for nighttime respiratory losses of 12%, the crop must acquire at least 55 g CO_2 m^{-2} d^{-1}. The present concentration of CO_2 in our atmosphere is near 0.67 g m^{-3} or about 380 ppm, v/v. To acquire 55 g CO_2 the crop must completely exhaust CO_2 from 82 m of air (55 g m^{-2}/0.67 gm^{-3} = 82 m) during daylight hours. Some CO_2 comes from within the crop by plant and soil respiration but [CO_2] is seldom depleted by more than 1/6 (63 ppm), so mixing height must be at least $82 \times 6 = 492$ m.

---

speed. It is clear that turbulent transport could be limiting to crop performance. With little wind, CO_2 supply could become limiting to photosynthesis while water vapor concentration and air temperature within the crop would increase. Vertical transport would then be governed by buoyancy processes.

6.8   Advection

Crops grown in arid regions sometimes encounter environments that cause LE to be greater than is predicted from energy supplied in R_n. An example is presented in Fig. 6.9*b* with irrigated ryegrass, where LE > R_n throughout daylight hours. As an indication of evaporative demand imposed by wind, e_a at 1600 h was 1.3 kPa compared with e^* near 6.5 kPa ($T_a = 38°C$). Data from irrigated sudan grass (Table 6.2) illustrate a similar condition. In these cases, additional energy for evaporation of water came from surrounding unirrigated landscapes where, because of lack of water, $R_n{\rightarrow}H$. As warm, dry air moved across these crops, energy was extracted and air cooled

Table 6.2 Energy balance of irrigated sudan grass on a summer day in Tempe, Arizona, USA

Advected sensible heat flux from air supports a rate of LE greater than would occur with R_n alone; u is wind speed and β is the Bowen ratio.

Hour	R_n	LE (km^{-2} h^{-1})	H	G	u (m s^{-1})	β
0–2	−100	150	−113	−138	1.3	−0.75
2–4	−100	150	−126	−126	1.0	−0.83
4–6	−88	126	−75	−138	0.9	−0.71
6–8	163	439	−351	75	1.0	−0.87
8–10	1260	1330	−176	100	0.9	−0.13
10–12	1160	2310	−276	113	3.2	−0.12
12–14	2450	2640	−264	75	3.6	−0.10
14–16	1920	2350	−364	−63	3.5	−0.16
16–18	715	1730	−904	−113	4.6	−0.54
18–20	−151	715	−703	−163	5.2	−1.00
20–22	−151	326	−326	−151	2.8	−0.98
22–24	−163	251	−289	−126	1.9	−1.15
Daily totals:						
	15770	25020	−7950	−1300	(kJ m^{-2} d^{-1})	
	6.44	10.2	−3.25	−0.53	(mm)	

Source: data adapted from Penman *et al.* (1967).

Box 6.5 Bowen ratio and measurement of crop ET

The ratio H/LE, known as the **Bowen ratio** (β), finds use in estimating ET by partitioning the major components of energy balance (Eq. 6.12), simplified by omitting P and R and rearranging to:

$$ET = LE = (R_n - G)/(1 + H/LE) = (R_n - G)/(1 + \beta) \qquad (6.24)$$

As a separate step, β can be estimated from gradients of temperature and water vapor at two heights, typically 0.5 and 1.5 m above a crop as follows:

$$\beta = [k_H \rho_a c_a (\Delta T_a/dz)]/[k_w(\Delta e_a/dz)] \qquad (6.25)$$

where e_a is vapor pressure and other terms as defined earlier. The equation is further simplified by assuming $k_H = k_w$, so that they cancel each other.

Outputs from sensors for ΔT_a and Δe_a can be combined with those for R_n and G and processed automatically as a single instrument as in the original EPER, the Energy Partition Evaporation Recorder (McIlroy 1971). Now, integrated systems available from commercial instrument makers can operate unattended for periods up to one week.

(H→LE); as a result, the Bowen ratio (β = H/LE) (see Box 6.5) was always negative (Table 6.2). Lateral transfers of energy of this type are termed **advection**. Without a wind for a continuing supply of warm air, leaves would cool by evaporation, e^* within leaves would decline, and LE would be restricted. When prevailing winds are either drier or warmer than air within crops, moving air serves as a continuing source of additional energy for evaporation. Advection also occurs with vertical transfers of air.

Advection contributes to "border effects" that cause plants at margins of fields to grow differently from those in the center. At a border between short and tall plants, tall plants are exposed to greater evaporative demand through side-lighting and advection. Other factors such as reduced competition for aerial and soil resources are also involved in border effects.

Most microclimate transition is complete within a few meters from crop edges, but several hundred meters may be needed for a truly steady-state wind profile and pattern of microclimate factors. This is not achieved within fields of common size in most farming systems. The distance from a change in vegetation is called the wind **fetch**. Deserts have large wind fetches over dry vegetation. In humid regions, mosaics of pastures, crops, and woodlands are similarly supplied with moisture and transitions between fields are not as dramatic as in dry regions.

6.9 Microclimate

Radiation and energy budgets and turbulent transport are all involved in strong control that crops extend over their own microclimates. Microclimates are defined by vertical distributions of air properties within crops and are driven by R_n exchanges on leaf and soil surfaces. With water present, water vapor and cool temperatures dominate these profiles; without water, temperatures increase.

6.9.1 Microclimate profiles

Idealized day and night profiles of various microclimate variables within a wet crop are illustrated in Fig. 6.13. Flux of R_n measured at any level within a canopy is the balance of upward and downward radiant fluxes across the plane. A positive value (daytime) means that radiant energy is going to various terms of the energy balance. In the full-cover crop shown in Fig. 6.13, exposed leaves account for most R_n exchange. During daytime, the exposed-leaf plane is also the region where air is being cooled by contact with transpiring leaves. During nighttime, sky is a poor source of LW radiation compared with a crop that has warmed during the day, so R_n is then negative. Leaves within the canopy are surrounded by warm objects whereas uppermost leaves are exposed to the sky. As was the case during day, exposed leaves are the principal surfaces involved in radiation loss from the canopy. This effect is seen in Fig. 6.13 as a sharp decrease in R_n flux and a lower T_a at the level of upper leaves.

Transport in Fig. 6.13 is from regions with high levels of a factor ("sources") to regions with low levels of the factor ("sinks"). The water vapor profile, for example, demonstrates that the entire crop including soil is a source for water vapor while the drier atmosphere is the sink. During daytime, exposed leaves are the principal sinks for CO_2 while soil (where CO_2 is generated by "soil respiration") and the atmosphere are sources of CO_2.

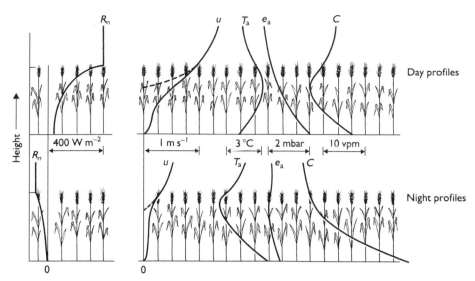

Fig. 6.13 Idealized profiles of R_n, wind speed (u), T_a, e_a, and CO_2 concentration (C) in a small-grain crop during day and night. Arrows indicate higher to lower concentration. (Adapted from Monteith & Unsworth 1990.)

Energy sources and sinks exist at all levels within crop canopies through repetitive interception and emission of LW radiation and exchanges among energy balance terms. Air heated at the soil surface, for example, may be intercepted by leaves; leaves warm and the energy then disperses further as *LE* or *LW* fluxes. This is an important factor in cropping strategies with limited moisture. Transpiration demand can be lowered by reducing leaf surface area but remaining leaves are then exposed to greater LW flux from soil, and surrounding air is both warmer and drier. These factors work to increase transpiration per unit leaf area, somewhat offsetting the intended savings of water (see Chapters 9 and 13).

6.9.2 Dew and frost formation

At night, exposed leaves are colder than other parts of canopies because they emit more energy than they receive. This explains why frost and dew accumulate on exposed leaves rather than on leaves lower within canopies. Dew formation is illustrated in Fig. 6.13 where upper leaves are sinks for water vapor during night. With nighttime $R_n = -50$ W m^{-2} (LW$_{out}$ = 350 and LW$_{in}$ = 300 W m^{-2}), the maximum rate of dew formation would be -0.020 g H$_2$O m^{-2} s^{-1} (-50 W m^2/2480 J g^{-1}, the latent heat of vaporization near 10°C). This corresponds to 72 g m^{-2} h^{-1} or a depth of about 0.07 mm h^{-1}, LE is given a negative sign here because latent heat content of air within the crop is declining. With leaf temperature at 0°C, the same rate of cooling would freeze a larger amount of water because its latent heat of fusion is only 335 J g^{-1}. Thus $-50/335 = -0.149$ g ice m^{-2} s^{-1}.

Moisture supply also limits dew formation. Without wind, air near a leaf can be depleted quickly of vapor. Resupply by diffusion is very slow and dew formation is greater when light winds provide turbulent transport. By contrast, freezing damage is greatest on still nights because temperatures can drop rapidly once a supply of moisture for freezing is exhausted.

Strong winds prevent both dew and frost formation because leaf cooling is prevented by advection of warm air from above the crop. Wind machines employed for frost protection in orchards work on that principle. Other methods for frost protection include heaters, which create vertical mixing by buoyancy as well as line-of-sight radiation heating, and sprinkler irrigation. Protection with sprinklers depends on the latent heat released through freezing of irrigation water; water in plant tissues is protected from freezing by presence of solutes, which lower its freezing point by 1 to 3 degrees below 0°C.

6.10   Climate and weather

Climate and weather both derive from continuous motion of the atmosphere that is driven by spatially unequal energy exchanges as the Earth orbits the Sun. Climate of each region is an average condition of meteorological variables (sunshine, temperature, humidity, wind, etc.) during the annual seasonal cycle. Weather is the short-term, day-to-day or week-to-week, variation from that average condition. Climate has importance in determining what type of agriculture is possible in given regions (e.g., perennial crops or pastures, summer or winter annual crops, or irrigation, etc.). Weather, in contrast, is the determinant of tactical management required to adjust individual farming enterprises to inevitable inter-annual variation. For example, winter crops are a suitable option for Mediterranean climates (Chapters 13 and 16) but their performance depends upon adjusting choice of crops and cultivar, time and density of sowing, and fertilizer strategy to variable rainfall conditions from year to year.

We start this section with a discussion of global circulation leading to agroclimatic analysis and the importance of topography and finish with a discussion of climate change and anthropogenic (human induced) global warming (AGW).

6.10.1   Global circulation

Atmospheric motion takes form in large circular eddies around centers of high and low barometric pressure. Warm air rising in the tropics spreads away from the equator in both directions. As it is cooled at high altitude by LW losses, it sinks to the surface near 30° latitude, forming subtropical high-pressure cells. The Earth's rotation imparts momentum to the air, which causes these high-pressure cells to rotate in the horizontal plane (clockwise in the northern hemisphere and counter clockwise in the southern hemisphere). Descending high-pressure cells are linked at intermediate latitudes with low-pressure cells of ascending air (with a rotation opposite to that of adjacent highs). Strong west-to-east (westerly) movement of the cells occurs at 30 to 70° latitude in both hemispheres; easterlies dominate at the equator and poles.

At intermediate latitudes, frontal systems develop where cold and warm air masses are forced together. The continuing progression of highs and lows with warm and cold air results in daily weather. These patterns are strongly modified by oceans and continents. R_n exchanges on land are limited by supply of water and small heat conductivity of land masses. During summer, continents tend to be dominated by $+H$ and ascending low-pressure cells and during winter by $-H$ and descending highs. Mountain ranges have a strong effect on weather, particularly on distribution of rainfall. As moisture-laden surface winds are lifted over mountains, they cool adiabatically to dew point, clouds form, and rain falls. This "orographic" effect contrasts to frontal and thunderstorm rainfall in areas without mountains. One consequence is that downwind of mountains air has less moisture, so cloudiness and rainfall are less. That effect contributes to the semi-arid and desert nature of many parts of the world. In western North America, for example, much of the moisture content of prevailing westerlies is deposited on the Pacific slopes of the Cascade and Sierra Nevada mountains and in the Rocky Mountains while eastern slopes receive much less rainfall. Ocean currents also have a marked effect. The Gulf Stream of the North Atlantic ocean brings warm water and moderate climates to high latitudes in Europe. Similar currents from the Philippines moderate the climates of Japan and Alaska. Coasts of California, Chile, Morocco, and southwest Africa, by contrast, are influenced by cold currents. There, equilibrium vapor pressures are low and coastal climates are dry.

In recent years, a number of general circulation models (GCMs; very large computer models) have been constructed from physical principles to simulate weather and climate. These models, atmosphere only (AGCMs) and atmosphere–ocean versions (AOGCMs), work at scales around 300 km and are suited for calculation of radiation and energy balances in zones of simple orography. Recently, the problem of relevance to topographically more complex regions has been addressed by the development of regional climate models (RCMs) that are nested within GCMs and work at local scales of 50 km or less. The models were initially developed for short-term weather forecasts but current emphasis is on long-term projections, especially the potential for accelerated global warming resulting from "greenhouse gases" (GHG) released by human activity (see Sections 6.5 and 6.10.4).

6.10.2 Topographical influences

Hills and mountains influence weather and microclimate in ways other than through orographic rainfall. Temperature declines with altitude, for example, due to the adiabatic expansion of air. The barometric pressure at 5500 m is approximately 50 kPa or only half that at sea level (100 kPa). As a result, rising air expands adiabatically and from the gas law, Eq. 6.15, its temperature declines. The normal "**lapse rate**" of T_a for an unsaturated atmosphere is $-1°C$ per 100 m rise in altitude. Water is condensed from saturated atmospheres on cooling, and the lapse rate then is only about $-0.6°C$ per 100 m owing to the release of latent heat in cloud formation. Temperature lapse has an effect on climate similar to increasing latitude so that in the tropics one can progress from a wet tropical climate at sea level to a polar one and observe their characteristic vegetations, simply by climbing a high mountain. Lapse rates are also involved in

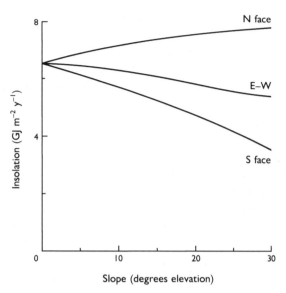

Fig. 6.14 Effect of slope and aspect on annual insolation at 35° S latitude. (Adapted from Jacobs 1955.)

desertifying effects of orographic rainfall. Moist air rising on one side of a mountain range cools at about 0.6°C per 100 m but after losing its moisture in rainfall it warms at a rate near 1°C per 100 m during its descent on the opposite side. The result is that downwind air is not only drier but also hotter.

Sites at similar altitude but with different slope and aspect have different climates. Sun-facing slopes receive greater insolation than reverse slopes (from the cosine law) and thus are warmer and have greater potential evapotranspiration. That is seen in Fig. 6.14 where annual insolation on a 10°, north-facing slope in the southern hemisphere is nearly 30% greater than for the reverse slope. Sharp differences in natural vegetation and in farming are common between north and south slopes. The effect is particularly evident in France and Germany where vineyards and orchards are generally placed on south-facing slopes. Temperature differences also occur between valley bottoms and surrounding hillsides. During night, when R_n is negative, air is cooled by contact with surfaces and dense, cold, air drains from hillsides into valleys. As a result, crops grown in valley bottoms have greater risk of damage from frost than those on higher slopes. Accumulation of cold air near the ground, whether due to advective drainage or to LW cooling of the surface, causes an **inversion** of the normal lapse rate. At night, inversions eliminate buoyancy as a component of turbulent transport.

6.10.3   Agro-climatic analyses

Modern methods of data capture and analysis have revolutionized descriptions and analyses of suitability of land for agriculture and other uses pioneered by Trumble (1939) and Trewartha and Horn (1980). The starting point remains short-term analysis to match average temperature and rainfall to requirements of individual crops, but that is now extended by consideration of seasonal patterns in water availability determined by characteristics of terrain and water retention of soils. The framework for analysis is provided by **geographic information systems** (GIS) that provide spatially referenced layers of data to describe climate, topography, and soil type. Analyses can then be made using models of water balance, crop development, crop growth, and yield (Chapters 5, 10, 13) to answer questions regarding land availability, crop suitability, crop productivity, and for example, requirements for irrigation and fertilizers. Output is in tables or maps for areas of similar response. A major advantage of this approach is the facility to include additional layers of geo-referenced data as they become useful or available, to continually improve the models and extend the breadth of possible applications.

An important example is the agro-ecological zone (AEZ) approach developed by the FAO over the last two decades (Fischer *et al.* 2002). The most recent version (AEZWIN) is available for general use as FAO (2007). The FAO/UNESCO Digital Soil Map of the World (DSWM) provides the land surface database comprising *c.* 2.2 million grid-cells at five arc-minute intervals of latitude/longitude. Climate databases include historical data, outputs from weather generators, and from global circulation models of possible future climate scenarios. Other digital global databases of topography, soils, terrain, and land cover provide the necessary components to supply water, energy, and nutrients for plant growth and productivity. The result is a global coverage for AEZ assessments of agricultural potentials in developed and developing countries to quantify impacts on land productivity of optional management strategies under historical climate variability as well as potential future climate change. The scale of these databases is, however, too coarse to be useful for tactical management of individual fields (Chapters 12, 13, and 16).

6.10.4 Climate change

Future climates are likely to be different from those we now experience, for two reasons. One is that climates are always subject to variation and change, owing to systematic variations in Sun–Earth geometry ("Milankovich theory"), solar activity, continental drift, and long time-constants associated with ocean–atmosphere equilibria for temperature, sea level, CO_2, and other gases (Lamb 1977). Over very long time scales, continental drift not only changes latitudinal position of land masses but also sea level and avenues for ocean circulation.

Historical context The natural range of the Earth's temperature is quite large. During recent geological times, global mean temperature has been as much as 10 to 15°C warmer than the present 15°C. During the past million years, however, it has generally been much cooler than 15°C; 18 000 y ago at the peak of last Ice Age, for example, global mean temperature was about 5°C cooler than now and the northern hemisphere was perhaps 15°C cooler (Schneider *et al.* 1990). Accumulation of ice during that period caused sea level to fall nearly 120 m below present. More recently, the world experienced a Medieval Warm Period during the tenth to twelfth centuries, when Greenland was settled, followed by a Little Ice Age during the seventeenth to ninteenth centuries when freezing of the river Thames in London, England was a common occurrence.

On shorter time scales, some of the most dramatic changes occur when explosive volcanism introduces large amounts of dust and sulfuric acid into the atmosphere. These agents lead to a reverse greenhouse effect by reducing penetrance of SW. The explosion of Mt Tambora in Indonesia in 1815, for example, lowered temperatures in the northern hemisphere by 1°C in 1816. In New England (USA), temperatures were 3°C below normal in a "year without a summer" (Bryson & Murray 1977). Crop failures were widespread and commodity prices increased dramatically. One current theory is that ice ages may be triggered by volcano eruptions.

Anthropogenic global warming (AGW) Terrestrial data reveal an increase in mean global terrestrial temperature of 0.74°C from 1900 to 2005, with evidence of a faster rate of increase during the last 50 years. The question that now exercises science is if warming is being accelerated by human activity. The concern is that current trends in atmospheric concentrations of GHGs, particularly continuing increase in [CO_2], will so affect transparency of the atmospheric window (Fig. 6.5) as to increase atmospheric and surface temperatures. Reconstruction of past [CO_2] from gas trapped in ice cores and from ratios of carbon isotopes in tree rings indicates values below 300 ppm (v/v) during 100 000 y prior to 1850. Continuous measurements made during 1960 and 2008 at the Mauno Loa Observatory in Hawaii reveal a slightly accelerating (non-linear) increase from 310 to 380 ppm. If this trend continues, atmospheric [CO_2] is projected to reach around 550 ppm during the present century.

Atmospheric [CO_2] is normally stabilized through equilibrium with oceans where it is sequestered as HCO_3^- and CO_3^{2-}. Time-constants for atmosphere–ocean exchanges are too large (1000s of years), however, to accommodate the present CO_2 surge. Only a shallow (about 600 m) mixing-layer of oceans is involved in annual cycles of temperature and CO_2. Major movements of CO_2 into deep ocean storage occur only at high latitudes where dissolved CO_2 is carried to depth by descending cold water. Dissolved CO_2 is eventually released through up-welling water in mid-latitudes. Capacity for transport of CO_2 into deep waters is small such that oceans accommodate large changes in biospheric carbon cycling very slowly. Solubility of methane (CH_4) in cold waters and ice can be as great as 0.5 mol CH_4 mol^{-1} water so deep salt-water pools of CH_4 represent an enormous reservoir of another potent GHG.

Recent increases in [CO_2] result mostly from the combustion of fossil fuel but also, in part, from the disturbance of biospheric carbon pools in soils and forest biomass (Trabalka & Reichle 1986). Methane from anaerobic microbial metabolism in wet lands and digestive systems of animals and insects, nitrous oxide (N_2O) released from soils, and chlorofluorocarbons from refrigerants and aerosol products, also affect transparency of the window. Increased [CO_2] would narrow the effective width of the window through greater absorption at 10 to 15 μm; increased [CH_4] would have an even larger effect near 7 to 8 μm. Methane is about 15 and N_2O about 280 times more absorptive on a molar basis than CO_2. The challenge is substantial for accurate projections of cloudiness, solar radiation, and e_a as atmospheric GHG concentrations increase. Water vapor concentrations are highly variable and have large direct effects on radiation balances and climate through variable cloudiness, but water vapor is also the major GHG (Fig. 6.5).

Future for agriculture Predicting the effects of increases in [GHG] with GCMs is difficult and the cause of much present debate in scientific circles. A comparison of sensitivity of these models reveals a wide range, 1.5 to 11.5°C, of increase in average global temperature to a doubling of [CO_2] (Stainforth *et al.* 2005; Kiehl 2007). In practice, the range of GCM–RCM combinations for individual regions is large with substantial variation among projections. In a Europe-wide comparison of 11 RCMs, Fronzek and Carter (2007) conclude that variation is too great to allow confident choice for impact analysis.

A change of several degrees is very large in comparison to those of the past century but small in comparison with Earth temperature variations during past geologic time. Not all scientists accept that world temperatures have risen to the extent reported, and will continue to increase or that GHG are the major cause of increases that have occurred in recent decades (IPPC-AR4 2007). While global temperature has increased since the end of the Little Ice Age around 1850, they do not accept that the trend continues or that the major driving force is increasing atmospheric [CO_2]. In the first place, terrestrial temperature data are notoriously difficult to analyze for long-term trends due to changes in instrumentation, modification to measurement sites, and discontinuance of many over recent decades. Heating of urban sites is a special problem and how to remove the effect is uncertain. Further, oceans occupy 70% of global area and their temperatures are not included in the analyses. Alternative analyses (Idso & Singer 2009) stress that terrestrial temperature has not increased since 1998 and satellite measurements of the temperature of the lower troposphere reveal a decrease since 1998. Current climate models, they argue, grossly overestimate the warming attributable to rising [CO_2], which is a minor GHG compared to water vapor. To the extent that GHG cause global warming, mitigation must include controlling water vapor, which accounts for 90% of radiative forcing. Finally from a crop-ecology perspective, they point out that higher [CO_2], rather than being an atmospheric pollutant, is positive to agricultural productivity (Section 10.2.2). Warnings of major imminent catastrophes due to global warming that appear regularly in the press and popular scientific literature should be viewed with caution.

Warming of the Earth by GHG would be accompanied by changes in rainfall but there is even less agreement on the direction and magnitude of those changes and no projections yet about the degree of variability. Restructuring our use of fossil energy to reduce releases of CO_2 would require many years of effort and a large portion of world wealth to accomplish. In 1997, many industrialized nations signed the Kyoto Protocol to reduce GHG emissions at 1990 levels. Now ten years later, as negotiations for a second agreement are stalled, most of these countries produce more emissions than in 1997 and are struggling to identify achievable targets. Energy policy is strongly on the international agenda, however, and agriculture will be asked to use less energy and contribute more (bio)energy to a broader mix of alternative (non-fossil fuel) sources (Section 15.5). The current direct contribution of GHG from agriculture of around 6 Gt y^{-1} CO_2 equivalent (10% of world emissions) is dominated by N_2O from soils

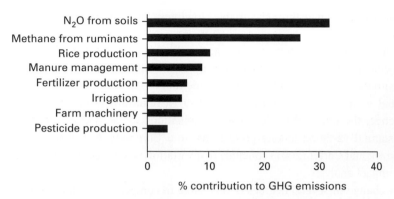

Fig. 6.15 The relative composition of direct greenhouse gas emissions from agriculture. The annual total of 6.5 Gt CO_2 equivalent is about 10% of total world emissions. Emissions of N_2O from soils include applied N fertilizer and manure, as well as from N transformation of indigenous soil N. (Adapted from Bellarby *et al.* 2008.)

and methane from ruminants and rice production, as displayed in Fig. 6.15. But to that should be added approximately equal emissions from land clearing.

Agriculture may, of course, be strongly affected if climates do change. The greatest changes will be in winter temperatures at high latitudes where such trends have occurred during the past century (Schneider *et al.* 1990). If that is the case, snow pack in mountainous regions and thus supplies of irrigation water in many arid regions could be strongly affected. Although predicted warming during the growing season in tropical and temperate regions is less than present inter-annual variations of temperature, agriculture there could be impacted on significantly. But there is significant potential for agriculture to adapt to changing climates, to maintain and in some places to increase productivity (Box 5.3). The next century will prove a very challenging period for agriculture that will require increasingly detailed attention to all levels of understanding of crop ecology, and potential manipulation from genes to production systems, issues that are covered widely in other chapters.

6.11   Key concepts

Radiation

- Sun emits radiation at wavelengths $< 3\,\mu m$ (peak $0.7\,\mu m$), from its surface at 6000 K calculated by Wein's displacement law. At mean Earth–Sun distance, radiation strength outside the atmosphere (solar constant) is 1360 W m^{-2}. Earth's climate is determined by atmospheric effects on penetration of this short-wave (SW) flux to the surface. Earth is a sphere and spins on a tilted axis in an elliptical orbit so receipt of SW radiation varies diurnally, and seasonally with latitude, creating strong vertical and horizontal motions (wind) in the atmosphere.
- SW radiation is reflected by clouds, absorbed by atmosphere, and is incompletely adsorbed by terrestrial objects (reflectivity, i.e., albedo is *c.* 10%). Terrestrial objects are heated and radiate energy at long wavelength (LW) $>5\,\mu m$ with a maximum around

10 μm for objects at 300 K. Long wavelength radiation emitted by terrestrial objects is absorbed almost completely by other terrestrial objects (black bodies) and strongly by water vapor and greenhouse gases (H_2O, CO_2, CH_4, and N_2O) in the atmosphere.

- Radiation balance of a crop has SW and LW components, each with a downward and upward flux. Downward SW flux is diurnally variable depending on latitude and day of year. It can be large during daytime and is zero at night. Upward SW flux, due only to reflection (c. 10%), is small. LW fluxes depend on temperature (K) of radiating bodies. The atmosphere is cooler than Earth so downward LW is small relative to upward LW. Compared with SW fluxes, LW fluxes are relatively constant diurnally.

Energy balance

- Net radiation (R_n), the downward versus upward balance of SW and LW fluxes, is partitioned to sensible heat (H) and latent heat (LE) and chemical-bond energy through photosynthesis and respiration. R_n exchanges within a crop canopy lead to distinctive microclimates depending upon whether the system is wet or dry. In crops well supplied with water, R_n is expended mainly in evapotranspiration (LE ≫ H) with the result that large amounts of water are used during crop production.
- In dry systems, R_n goes to H (H ≫ LE) and there can be little crop growth.
- Lateral and vertical transfers of H (advection) to crops sometimes support evapotranspiration rates greater than those predicted by radiation balances alone.
- Enormous quantities of CO_2 and water vapor are exchanged between crops and the atmosphere. Turbulent transport and buoyancy engage a great depth of air in those exchanges. The same eddies of air that transport those gases also transport H.

Climate and weather

- Regions of the Earth have different patterns of rainfall and temperature due to geographical position and topography relative to the Sun and to atmospheric and oceanic circulation. The latter are driven by large, latitudinal differences in radiation balance.
- Resulting variations in temperature and rainfall constitute our weather. Averaged over years, the general patterns define climates. Climate determines suitability of optional agricultural production systems; crop management seeks to reduce risk of low productivity caused by weather variation.
- Climates are subject to change due to natural variations in Sun–Earth geometry and other factors. The possibility exists that climates could change significantly during the next century, even if human-induced GHG plays only a minor role. If that is so, it presents a new challenge for farmers and all scientific disciplines involved in the support of agriculture and natural resource management.

Terms to remember: advection; albedo; balance, radiation and energy; black body; Bowen ratio; crop boundary layer; eddy turbulence; emissivity; equilibrium evaporation; evapotranspiration; fetch; flux density; geographic information system (GIS); greenhouse effect; inversion; irradiance; lapse rate; radiation, diffuse, direct, global, long-wave, net, photosynthetically active (PAR), short-wave, solar, thermal, visible; laws, cosine, Kirchoff, Stefan–Boltzmann, and Wien; solar, altitude, constant, equinox, solstice and zenith angle; heat, latent (fusion and vaporization) and sensible; temperature, air, dew point and wet bulb; vapor pressure, saturation and deficit; transport coefficient.

7 Soil resources

Soils are formed *in situ* over long periods under the influences of climate and vegetation during which they develop characteristic vertical profiles. Inorganic materials are the major component of soils. These include partially weathered parent materials, secondary minerals, and dissolved salts. Other components are air, water, organic matter in various stages of decay (with the most reduced form called humus), and living organisms including plant roots. Typical agricultural soils have a bulk density (dry mass per unit volume) near $1.3 \, \text{g cm}^{-3}$ [$1300 \, \text{kg m}^{-3}$ or $13 \times 10^6 \, \text{kg (m depth)}^{-1} \text{ha}^{-1}$]. Organic matter ranges by mass from 1 to 5% in mineral soils, and can be 80% or more in peaty soils. In typical mineral soils, water content at drained capacity accounts for 0.1 to 0.4 times the soil volume but some organic and volcanic soils hold much more. Understanding the physical, chemical, and biological properties of soils as media for plant growth provides insight into plant adaptations to soil conditions and crop management practices to overcome soil-related constraints.

7.1 Soil chemistry

We begin with a review of several concepts important to the study of soils and crops. Familiarity with these concepts is fundamental to crop ecology. Soil composition is dominated by an abundance of insoluble compounds of aluminum, silicon, and calcium, and soil chemistry centers on interactions between those solids and the water phase, called the soil solution.

7.1.1 Solutions

Many ions in soil solutions are in equilibrium with sparingly soluble minerals and with ion-exchange complexes discussed later. Concentrations range between 1 μM and 1 mM for various ion species in well-drained soils. By contrast, soils in semi-arid regions may contain 50 mM or more of soluble Na^+ and Cl^-, leading to a saline condition.

Except in very dilute solutions, ionic concentration is not the best measure of chemical potential. Ability of an ion to enter into reactions is limited by association with water or other ions and its true **activity**, a, is *less* than its **molar concentration**, M. The relation between activity and concentration is summarized in an activity coefficient, γ, range 0 to 1, so that $a = \gamma M$. Activity is denoted with () while [] are used for concentration.

Because γ is usually near 1 in dilute solutions, concentration suffices in approximate calculations.

Many reactions in soil solutions depend upon mass action. A dynamic equilibrium is reached in which forward and backward reactions occur at the same rate. In the reaction $A + B \leftrightarrow C + D$, the **equilibrium constant**, K_b, is defined from activities of reactants and products according to:

$$K_b = (C)(D)/(A)(B) \tag{7.1}$$

For dissociation of $KNO_3 \leftrightarrow K^+ + NO_3^-$ in water, as an example:

$$K_b = \left(K^+\right)\left(NO_3^-\right)/(KNO_3) \tag{7.2}$$

Where solutes are in equilibrium with a sparingly soluble solid, activity of the solid is taken as 1 and the equilibrium constant is termed the **solubility product**, K_s. Chemists have measured K_s values for mineral solids common in soil, and these can be found in standard references. For example, solution of $CaCO_3$ (calcite) $\leftrightarrow Ca^{2+} + CO_3^{2-}$ in water:

$$K_s = \left(Ca^{2+}\right)\left(CO_3^{2-}\right)/1 = 4 \times 10^{-9} \text{ at } 25°C \tag{7.3}$$

7.1.2   Oxidation–reduction reactions

Oxidation involves loss of an election by an atom thereby increasing its valence; whereas reduction occurs when an atom gains an electron. Oxidation and reduction are always coupled, with one partner serving as the electron donor and the other as the acceptor. All elements can exist at several oxidation–reduction levels. Oxidized ferric iron (Fe^{3+}) and reduced ferrous iron (Fe^{2+}) are examples. In soils, reduced C, N, and S from organic compounds are the principal electron donors.

In aerated soils, organic compounds in dead plant material are oxidized in microbial metabolism with oxygen as the electron acceptor ($O_2 \rightarrow 2\,O^{2-}$) leading to the formation of water. When O_2 is not available, a "reducing environment" is created in which oxidized forms of iron (Fe^{3+}), manganese (Mn^{4+}), nitrogen (NO_3^- and NO_2^-), and sulfur (SO_4^{2-}) serve as electron acceptors for bacteria. Iron and manganese are also subject to reduction at low pH without biological catalysis. Reduced forms of all these elements can be toxic to plants if present at high concentration.

Each half of an oxidation–reduction couple represents one-half of an electrolytic cell, i.e., an electrode. Electrical potential (voltage) of a half cell serves as a measure of its tendency to donate or accept electrons. In solution under standard conditions, the half reaction $\frac{1}{2}O_2 + 2e^- + 2H^+ \leftrightarrow H_2O$ at equilibrium generates 1.23 V. $Fe^{3+} + e^- \leftrightarrow Fe^{2+}$, by comparison, generates 0.77 V. When the two reactions are coupled, ferrous iron (Fe^{2+}) is oxidized to the ferric (Fe^{3+}) form and oxygen is reduced to form water. Individual electrical potentials cannot be isolated in soil systems but average values can be measured for whole soils. Redox potentials of 400 to 600 mV are found in well-aerated soils; values of 0 to 200 mV may occur with anaerobic conditions.

7.1.3   Equilibrium thermodynamics

Whether a chemical reaction will occur, its direction, and endpoint at equilibrium depend on activities of products and reactants, temperature, and tightness of bonding in products. Thermodynamics expresses these factors in energy terms to predict final endpoints of chemical and physical processes from initial states.

Key thermodynamic concepts involve changes in **enthalpy** (ΔH), **entropy** ($T\Delta S$), and **free energy** (ΔG) that occur between initial and final equilibrium states of a reaction system. Enthalpy, ΔH, is the heat energy released or absorbed in reaction; we met it earlier as ΔH_C from combustion of organic materials (Section 2.1). The **first law of thermodynamics** holds that total energy of an isolated system is constant. If heat is absorbed in a reaction ($+\Delta H$), temperature of the surroundings must decline. Entropy (S) is a measure of randomness in a system. Degree of randomness is seen in crystalline solids $<$ liquids $<$ gases, and in tendencies of dyes to diffuse throughout solutions. Chemical and physical reactions are driven strongly by a universal tendency for components to become more disordered. That entropy always tends to increase is the **second law of thermodynamics**. The product $T\Delta S$, where T is temperature in Kelvin scale, expresses entropy change in energy units.

Free energy change represents the amount of work that might be done by a system from changes in enthalpy and entropy. It is defined by the relation:

$$\Delta G = \Delta H + T\Delta S \tag{7.4}$$

The great importance of ΔG is that it tells us the direction and magnitude that a reaction will take. If ΔG is negative [e.g., if the reaction releases heat energy ($-\Delta H$) and/or randomness increases ($+\Delta S$)], the reaction may occur spontaneously and continue until ΔG equals 0 at equilibrium. A positive ΔG indicates that the reverse reaction may occur. Changes in concentrations of reactants or products, or in temperature (an exchange of heat between the system and its surroundings), may change $+\Delta G$ situations to $-\Delta G$ and allow a reaction to proceed. Negative ΔG does not ensure reaction, however. For example, ΔG equals -16.6 kJ mol^{-1} N for the reaction $N_2 + 3H_2 \leftrightarrow 2NH_3$ but the reaction must be catalyzed in order to bring N_2 to a reactive state. In soils, microorganisms and enzymes catalyze many processes, including NH_3 formation.

Obtaining quantitative values of ΔG for a particular reaction can be complex. Fortunately, ΔGs for formation of many compounds from their elements, and for principal biochemical processes, are tabulated in handbooks. These are denoted $\Delta G°$ for "standard conditions" (reactants and products at 1 molal, 25°C, and 1 atm) and can be adjusted to actual concentrations and temperature of a particular system as follows:

$$\Delta G = \Delta G° + 2.3RT \log Q \tag{7.5}$$

where R is the gas constant, T is temperature (K), and Q is the quotient of product of reactants to product of products at their actual activities. Redox potential is related to ΔG by the relation:

$$\Delta G = nF\varepsilon° + 2.3RT \log Q \tag{7.6}$$

Table 7.1 An international classification of particle sizes in soil texture

Texture class	Size range (mm)
Gravel	>2
Coarse sand	2 to 0.2
Fine sand	0.2 to 0.02
Silt	0.02 to 0.002
Clay	<0.002

Note: the US Department of Agriculture classifies five grades of sand covering the range 2 to 0.05 mm, silts between 0.05 and 0.002 mm, and divides clays into coarse (0.002 to 0.0002 mm) and fine (<0.0002 mm) categories.

where n is the number of moles of electrons involved, F is Faraday's constant, and ε° is the redox potential for the standard state.

7.2   Soil formation

Here we explain soil formation processes because they inform about soil properties and suitability for agriculture. Early ideas about soil genesis began with observations in the 1800s by V. V. Dokuchaev that humus content of blackland soils in Russia varied with rainfall, and by E. W. Hilgard that soils from humid southeastern and semi-arid western portions of the USA differed greatly in type and amount of secondary clays. In both cases moisture regime exerts a strong influence on soil formation through effects on chemical processes, vegetation productivity and type. Hans Jenny later incorporated these ideas into a comprehensive framework in *Factors of Soil Formation* (Jenny 1941) that identified climate, vegetation, parent material, topography, and time as the determinants of soil formation.

7.2.1   Parent material and secondary minerals

Development of soils begins with a **parent material** consisting, usually, of fragmented bits of one or more **primary minerals** (principal ones are augite, feldspar, hornblende, mica, olivine, and quartz). Those minerals originate as igneous, sedimentary, or metamorphic rock under influences of heat and pressure and are fragmented over time to the size of gravel, sand, silt, and clay (Table 7.1). Fragmenting may occur *in situ* or the materials may have been transported and deposited as wind-blown loess, stream alluvium, or glacial till. The relative proportions of various particle size classes found in soils determine their **texture** (Table 7.1).

Primary minerals of soils consist of crystalline lattices of oxygen ions (O^{2-}) arranged in tetrahedra and octahedra (Fig. 7.1). Oxygens are held together by metal ions, principally Si^{4+} and Al^{3+}, which have small size and large charge density. Primary minerals are distinguished by having silica (SiO_2) and alumina (Al_2O_3) arranged in ribbons or

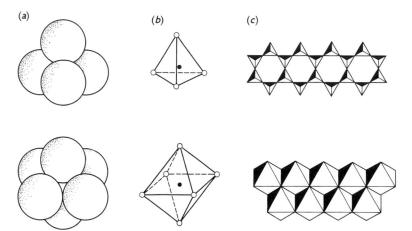

Fig. 7.1 Structural aspects of the alumino-silicate minerals found in soils (adapted from Jenny 1980). (*a*) Basic tetrahedron (four-sided) and octahedron (eight-sided) units formed of either four or six O^{2-} ions surrounding and concealing an Al^{3+} or Si^{4+} ion. (*b*) Schematic views of the tetrahedron and octahedron. The open and filled circles represent oxygen and aluminum or silicon ions, respectively. (*c*) *Top*: Arrangement of tetrahedra into ribbons or plates; each shaded triangle represents a tetrahedron. *Bottom*: Plate formation from octrahedra; one surface of each octahedron has been shaded.

plates bonded together or substituted with varying proportions of other metals such as Ca^{2+}, Mg^{2+}, Fe^{3+}, Na^+, and K^+. Quartz is pure SiO_4 in tetrahedra (its formula is SiO_2 because oxygens are shared by adjacent tetrahedra). Augite, hornblende, and olivine are also tetrahedral silicates with Fe^{2+} and Ca^{2+} and/or Mg^{2+} acting to hold the units together while mica and feldspar are alumino-silicates. Micas are composed of flaky layers of aluminum and silicon oxides generally arranged in tetrahedra; Mg^{2+} and Fe^{2+} substitute for Al and Si within the tetrahedra and K^+ is involved in binding layers together. Feldspar is composed of aluminum and silicon oxides in a fixed 1:3 ratio; K^+, Na^+, and Ca^{2+} are found within the lattices.

In the presence of water, primary minerals undergo a slow process of **weathering** through a variety of chemical transformations. Hydrogen ions from water ($H_2O \leftrightarrow H^+ + OH^-$), for example, may displace base cations from silicates while reacting with oxygen to convert their lattices to OH^- forms. Primary minerals may be altered *in situ* to new forms or solubilized to hydroxides such as silicic acid [$Si(OH)_4$] and $Al(OH)_3$. Silicate hydrolysis is accelerated when solution pH is lowered by dissolved CO_2 (from the atmosphere and from respiration of soil organisms: $CO_2 + H_2O \leftrightarrow H^+ + HCO_3^-$) and by organic acids, also of microbial origin. Hydrolysis products, including swarms of metal ions released from lattice and ribbon structures, may migrate in soil before recrystallizing as new, **secondary minerals**. Secondary minerals are similar to primary minerals in their silica and alumina lattices and metallic inclusions.

Secondary minerals are small particles in the clay range (Table 7.1). Those less than 0.2 μm in size behave as colloidal fluid (stable suspensions) when dispersed in water. In that form, clays may be carried away in surface runoff or leached downward

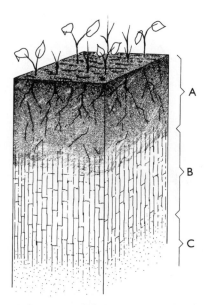

Fig. 7.2 Soil profiles showing the A layers with accumulations of organic matter; B layers with accumulations of secondary minerals; and undeveloped parent material, C.

in soil profiles. Solubilization, recrystallization, and colloid movement are continuing processes. Secondary clays form throughout profiles but clay particles and dissolved minerals are leached gradually from surface layers transforming them into distinctive **A horizons** (Fig. 7.2). A horizons are also characterized by accumulation of organic matter. Clays leached from A horizons tend to accumulate in lower **B horizons**. The B horizons are underlain by relatively undisturbed parent material (**C horizon**).

The type of secondary minerals formed in A and B horizons depends upon the primary minerals present, pH of soil solution, and amount of leaching. The **stability diagram** presented in Fig. 7.3 illustrates some clay transformations that may occur beginning with albite feldspar. Such diagrams are based on the solubility products of clays in equilibrium with the most abundant ions at various concentrations. In this case, silicic acid activity increases along the abscissa while the activity ratio of Na^+ to H^+ increases along the ordinate. The lower left of the diagram represents an acid-leached situation, in contrast to the alkaline, unleached, region in the upper right. Above a solid line, solutions are supersaturated so the indicated mineral precipitates; below the line, the solution is unsaturated so it dissolves. Beginning in the upper right: as (Na^+) and/or $[Si(OH)_4]$ declines, albite (a form of feldspar) dissolves and montmorillonite forms; montmorillonite is replaced in turn by gibbsite when silicic acid is low and by kaolinite when Na^+ is low. This stability diagram represents a very simple case with only a few species of ions. The pattern changes with temperature, and when other ions (e.g., K^+ or Ca^{2+}) or other secondary minerals are present.

Clays form amorphous, gluey gels with small amounts of water, and clayey soils are stiff and intractable to tillage when moist. Montmorillonite (in Fig. 7.3) is of special interest in this regard. It and other smectite clays shrink and swell with changes in

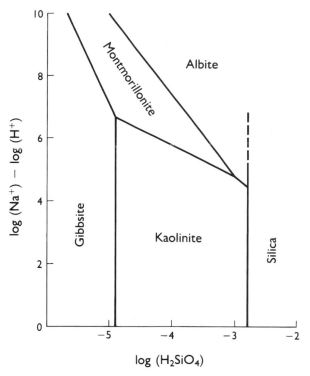

Fig. 7.3 Stability diagram for a system of clays derived from albite feldspar in equilibrium with various solutions. The diagram was constructed from solubility products and ion activities for a system, denoted by (), containing Na_2O, Al_2O_3, SiO_2, and water at 25°C. The text offers additional explanation. (Redrawn from Huang 1989.)

water content. In montmorillonite, silica plates are bound together poorly by Mg^{2+} and the clay hydrates and dehydrates strongly during wetting–drying cycles. When dried, soils containing smectite clays tend to shrink to massive structures and large cracks that appear at the surface can extend deep into the soil profile.

Characteristics of secondary clay minerals, including their very fine particle size, determine many soil properties. In addition to their role as nutrient sources, their surfaces contribute to soil structure, ion exchange, and water retention. More on each of these topics will emerge as we proceed. The properties of clay are so dominant in soil processes that soils with as little as 20 or 30% clay particles by mass are classified as clay soils.

Vegetation exerts a strong influence on clay mineralogy through recycling of elements to the soil surface and through its influence on soil water content. Grasses and deciduous trees, for example, are effective in recycling K^+, Mg^{2+}, and Ca^{2+} from deep in the profile. These ions participate in clay formation each time they leach through the profile. Residues of conifers, on the other hand, are low in those metals and serve, upon decay, as a strong source of H^+. Acid-leached soils having only small clay content are formed in that case. Na^+ is not recycled effectively by plants and it leaches from the profile over time (sea water is therefore high in Na^+). Leaching and recycling patterns result in

different secondary minerals being formed under different climate–vegetation regimes even with the same starting material. Conversely, the same class of secondary clays may form under grasslands from different starting materials. Under appropriate conditions, portions of soil profiles may develop into very rich sources of silicon, aluminum, or iron oxide minerals providing humans with abundant sources of those metals for industrial purposes.

In the humid and sub-humid tropics, warm temperatures and high rainfall accelerate weathering processes. Copious amounts of water over millions of years dissolve most of the primary and secondary minerals and leach basic cations, leaving behind **sesquioxide** minerals such as gibbsite or goethite composed of aluminum and iron hydroxides. As a result, highly weathered soils are acid and have low inherent fertility.

7.2.2   Organic matter in soils

Higher plants serve as the source of organic inputs to soil, which include litter and residues from their aerial portions, supplied to the surface, and from underground organs, mainly roots. In natural systems and managed grasslands, aboveground material decays *in situ* and only a small portion is carried into the profile by rain or insects. Roots and other underground organs are then the principal substrates for formation of soil organic matter (SOM). When crop residues are incorporated by plowing, decay takes place throughout the plow layer whereas no-till systems (Section 12.5.2), without incorporation, behave more like grassland.

Decay processes are considered in detail in Chapter 8. Fungi and bacteria do most of the work but they are accompanied and subject to grazing by a wide range of fauna including worms and protozoa. As Jenny (1980) put it: "invertebrates act as mechanical blenders [in the soil]. They break up plant material, expose organic surfaces to microbes, move fragments up and down, and function as homogenizers of soil layers." Each type of animal from earthworms to protozoa is open to predation and their residues offer new chemical combinations and new opportunities for bacterial and fungal attack. Jenny estimated (his Table 5.1) that soil might contain 2000 kg ha^{-1} of dry microbial biomass and 600 kg ha^{-1} of invertebrates.

Soil organic matter can be divided into several principal fractions: microbial biomass; plant residues; biochemical compounds (sugars, tannins, proteins, and amino acids) recently freed from living roots and decaying material; and humus. **Humus** is the final end product of decomposition. No two humus molecules are alike because they consist of heterogeneous composites of long-chain condensate polymers of aromatic phenols, aliphatic chains, sugars, organic acids, and nitrogenous compounds. These polymers have many free $-OH$ groups that are important in soil properties. These polymers are chemically stable and resistant to further attack, particularly when physically stabilized by adsorption to clays or entrapment in soil particles. Some old organic matter, perhaps the portion that is not physically stabilized, has a lifetime in soil in the order of years to decades and can be classified as "active". Another portion is much more resistant. Carbon dating reveals that the oldest fractions in many North American soils date from the last glacial cycles, 10 000 to 20 000 y ago.

The simplest method for determining SOM content is to remove undecomposed plant material from the soil by sieving and then to analyze the soil for organic C or N. Humus is the major organic component in cultivated soils compared to partially decayed material and microbial and faunal biomass. Because humus is near 6% C, C/0.6 provides a crude estimate of the organic matter content of a soil. The C/N ratio of humus in agricultural soils is near 10 to 13. If we assume that the ratio is 12, N $\times$ 20 provides an additional rough estimate of humus content. The C/N ratio of fresh residues ranges from 25 to 40 for legumes to 60 to 100 for cereal straw and maize stover. Because the C/N ratio of microbial cells is typically less than 10, microbial growth in soil is N limited as crop residues begin to decompose. As microbes oxidize reduced C to provide energy for metabolic processes, CO_2 is released and C/N ratio decreases. "Available" inorganic N as NO_3^- and NH_4^+ is *immobilized* during the initial stages of residue decomposition as microbes scavenge for the most limiting resource. When residue C/N ratio decreases to 20 to 25, net release of inorganic N occurs making it available for uptake by roots. Lignins are organic polymers in plant cell walls composed of phenolic subunits. Because phenols are recalcitrant to microbial degradation, residue decay rate is inversely related to lignin content.

Soil organic matter is usually distributed within soil profiles in about the same pattern that decay occurs, i.e., SOM content is greatest near the surface and declines more or less exponentially with depth. Some humic substances are partially soluble and other small-molecular-mass humus fractions may leach gradually within the profile as colloidal solutions. In some soils (termed spodozols), an organic layer forms at depth where humus, solubilized from the surface zone, precipitates as salts of iron and other metals.

7.2.3   Equilibrium level of soil organic matter

The amount of organic matter present in a soil reflects past balance between rates of humus formation and loss. Loss occurs because humus is oxidized slowly when exposed to air or attacked by microorganisms. Nitrogen and other elements are released in mineral forms and the process is termed **mineralization**. Mineralization is a first-order process, i.e., the rate is proportional to the amount of humus present. Because it depends on microbial activity, the rate of mineralization is near zero at 0°C and increases to a maximum at about 40°C. By contrast, rate of humus formation depends mainly on annual input of plant residues. Greatest amounts of residues are produced with a long growing season, intermediate temperatures (15–30°C), and abundant rainfall. In soils treated the same way year after year, covered with native vegetation or farmed in a certain way, rates of humus formation and loss eventually are equal and humus content comes to a steady state.

Grasslands of the North American Great Plains present an outdoor laboratory for study of temperature and rainfall influences on steady-state level of SOM. Mean annual temperature decreases from south to north while a humidity index (mean annual rainfall divided by ET_0, Section 9.2) increases from west to east. By sampling soil under native grasslands along north–south and east–west transects across the plains, Jenny (1930) identified the important generalizations illustrated in Fig. 7.4. Total soil N, which is

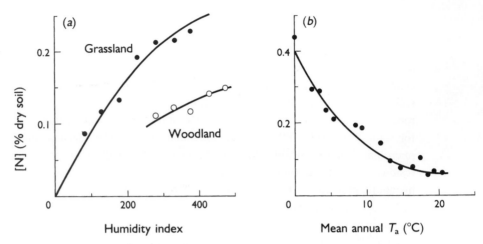

Fig. 7.4 (*a*) Average total N content in soils along a transect of increasing rainfall relative to evaporation. Data for grassland and woodland sites extending from Colorado (semi-arid) to New Jersey and with the same mean annual temperature (near 10°C) are depicted. (*b*) Average total N in grassland soils along a transect from Canada to Texas. Semi-arid sites having similar annual humidity indices are shown. (Adapted from Jenny 1930.)

a proxy for SOM content, increased in a diminishing-return relationship with rainfall and declined in a negative exponential relationship with increasing temperature. Marked differences are evident in total N of soils that develop under grasslands and deciduous woodlands. Grasses have greater production and turnover of roots and their leaf and stem residues have a greater lignin content than is found in leaves of deciduous trees. In addition, nutrient cycling is tighter with the dense fibrous root system of grasses.

Greatest amounts of soil N (and C) accumulate under grasslands in cool environments because of large plant biomass production with a high content of lignin and depressing effects of low temperatures and short season on mineralization. Quantities of C and N involved are impressive: over 30 t N and 330 t C ha^{-1} accumulate in profiles under good conditions. In contrast, humus N contents near 1 t ha^{-1} are not uncommon in desert regions where biomass production is small and high temperatures favor humus oxidation. For Earth as a whole, it is estimated that 1500 to 3000 Gt C are sequestered in soils. If C/N = 12, 125 to 250 Gt N are also held there. Sensitivity of the humus equilibrium to temperature and rainfall means that large changes can occur with a change in climate.

7.2.4 Influence of farming on soil organic matter

One consequence of tillage is accelerated breakdown of humus. That, coupled with removal of N in harvested crops, usually results in smaller steady-state levels of organic matter in cultivated soils than are found under natural vegetation. Exceptions sometimes occur where desert soils are brought under irrigation or where low fertility soils are cropped intensively with fertilizer or manure. In those situations, greater residue inputs to humus formation can offset greater rates of mineralization due to tillage and N removal. Examples of effects of agriculture on SOM levels are presented in Fig. 7.5.

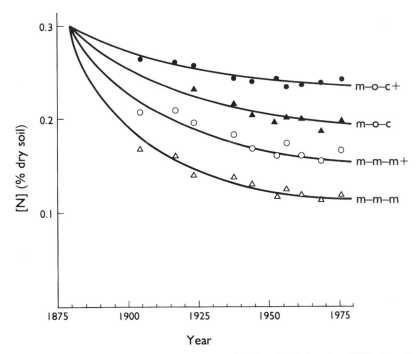

Fig. 7.5 Changes in organic N content in a Mollisol at the University of Illinois under two crop rotations (m–m–m, continuous maize; m–o–c, maize–oat–clover) with (+) addition of manure, lime, and phosphorus. Farming began in 1876. (Adapted from Stevenson 1982.)

New steady-state conditions were reached 60 to 70 y after the Illinois prairie was first put to the plow. Long-term plots in Kansas and Ohio (USA), and England have behaved in a similar way. Final equilibrium levels of SOM reflect farming practices. Lower levels result after continuous grain production than with rotations involving forage crops and manure. As has been demonstrated in long-term plots at Rothamsted (UK), steady-state levels of SOM are very low with continuous grain that does not receive fertilizer or manure inputs but can increase substantially if crops are given N amendments to increase crop productivity and return larger amounts of residues for humus formation.

Flooded soils, such as those under irrigated rice farming, tend to have the highest SOM contents among intensively cropped soils because O_2 diffusion is slow in water such that soils remain anoxic for much of the year. In the tropics and subtropics where most rice is grown, two and sometimes three rice crops can be grown each year on the same piece of land. Total annual grain yields of 10 to 15 t ha^{-1} are achieved with adequate N fertilizer, which leaves an equivalent amount of residue with high C/N ratio and lignin content. Decomposition of lignin is slow under anoxic conditions because O_2 is required by chemical reactions that open the cyclical rings of phenol subunits. As a result, humus formed in flooded soils has a strong phenolic character that reduces its subsequent rate of decomposition and leads to greater SOM content at equilibrium (Olk *et al.* 1998) (Table 17.2).

Whenever soils accumulate C, discussed further in Box 7.1, they also accumulate N and sulfur because humus molecules contain all three elements. Conversely, soil

Box 7.1 Sequestration of carbon by soils

Maintenance or increase in SOM content is required for conserving or improving soil quality. More recently, soil C sequestration has been proposed as a means to offset anthropogenic emissions of greenhouse gases (GHG). Some US farmers who practice no-till receive payments for GHG mitigation based on the assumption that no-till farming causes a net increase in SOM compared to conventional tillage with a plow or disk. Most evidence to support this supposition, however, comes from long-term field studies that neglected two key requirements: (1) measure changes in soil C in the entire root zone, not just the surface layer, and (2) sample equivalent soil mass (Baker *et al.* 2007; Vandenbygaart & Angers 2006). Most crop species have extensive root systems that explore soils to 1.5 m depth or more, as illustrated below for maize.

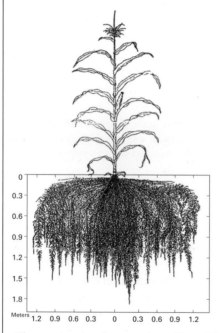

The root system of a corn plant. (From Kiesselbach 1949.)

Failure to measure changes in soil C throughout the entire rooting volume can bias comparisons of tillage treatments that influence root distribution. No-till systems leave residues as mulch on the soil surface, reducing surface evaporation, relative to conventional tillage surface. Root distribution follows these moisture patterns with relatively more roots in surface layers under no-till, and more at depth under conventional tillage. Tillage also affects soil bulk density, so comparisons of SOM must be made on an equivalent soil mass. A recent study that evaluated entire root zones of equivalent soil mass found little evidence of soil C sequestration in maize-based cropping systems of the US Corn Belt (Blanco-Canqui & Lal 2008).

management that results in loss of SOM also depletes soil stocks of N and sulfur. Increasing SOM is very costly in terms of N: every tonne of C stored as humus requires about 50 kg N as well.

7.3   Soil types and uses

Soils can be grouped with respect to their properties and suitability for specific uses. Here we deal with general principles of soil classification and consider those properties that determine suitability to support agriculture.

7.3.1   Soil classification

Systems for identifying and naming soils allow information about their characteristics and behavior to be shared over their regional distribution. Profile characteristics (chemistry and morphology) and circumstances of formation (genesis) provide the framework for classification schemes. Most classification systems assign soils into one of several great soil groups distinguished by whether formation was dominated by regional climate and vegetation leading to a **zonal** soil, or by local topography (e.g., wetlands) or parent material resulting in an **azonal** soil.

Most soils fall into one of several zonal groups. Temperate grasslands of Eurasia, N. America, and Argentina provide examples. All generate zonal **Chernozem** soils. Dokuchaev gave that name to a large group of black grassland soils characterized by a dark-colored A horizon, rich in organic matter and base metals, with accumulations of Ca^{2+} in the B horizon. The Chernozem name is still in common use around the world. Profile characteristics vary continuously within and between main groups such as Chernozems and may be further differentiated by variations in age, drainage, and parent material. As a result, a large number of specific soils are recognized within the Chernozems.

The Russian classification placed emphasis on soil genesis and the first United States Department of Agriculture (USDA) schemes followed the Russian approach. Western Europeans, on the other hand, have emphasized chemical characteristics of profiles in classification. Australian soils fit poorly into any European or American scheme (Northcote *et al.* 1975). Most Australian soils are very old **paleosoils** that are typically acid and low in nutrients. In some, shallow surface horizons reflecting genesis under recent semi-arid conditions overlie and contrast sharply in texture and structure with lower layers. Such duplex soils may have unfavorable water infiltration and holding characteristics. One theory is that the lower layers formed millions of years ago under vastly different climates.

The USDA soil taxonomy (Soil Survey Staff 1988) emphasizes profile description while retaining some attention to genesis. The system is pragmatic, focusing more on current soil properties than on formation. Soils are given a common series name as well as complex scientific name. Scientific names are based on criteria that define the soil's place among families and subgroups of 12 great **orders**. A soil of the Webster series in

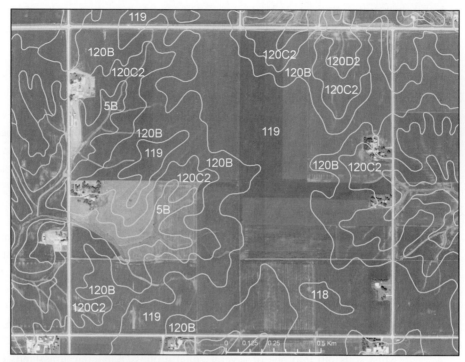

Fig. 7.6 Portion of soil map for Jasper County, Iowa, from the USDA-NRCS Soil Survey database (http://websoilsurvey.nrcs.usda.gov/). The quadrant centered in the map represents one section (i.e., one square mile, which is 640 acres, or 259 ha). Soil series mapping units are outlined and labeled with numbers so that soil data for that mapping unit can be retrieved from the database.

the US Corn Belt, for example, could be classified in the family fine-loamy, mixed, mesic soils of subgroup **Typic Haplaquolls**. Mixed indicates that no one clay family dominates the profile; mesic refers to its occurrence in a moderate temperature regime. "Typic" indicates that it is a central member of the subgroup Haplaquolls. The Haplaquolls name combines its order (**Mollisols**, ending -olls) and its great group (Hapla- for simple horizons and -aqu- for wetness). Mollisols include Chernozems and other soils with dark-colored A horizons, considerable organic matter, moderate base saturation, and good structure. They generally develop under grassland.

Detailed soil maps are now available at scales of 1:15 000 to 1:30 000 for the USA, and several European countries. Such maps reveal a high degree of spatial variability in soil distribution as shown in Fig. 7.6. To aid farmers or potential land purchasers, maps can be linked to a database that provides information about soil classification, physical and chemical properties, and land capability category (Section 12.7). Maps at coarser scales of 1:100 000 or more are available for most countries and can be used for regional and national land-use planning. A global soil map has recently been published at a scale of 1:500 000 (Batjes 2009). Direct comparisons among various systems of soil classification are difficult (see Buol *et al.* 1989; Sanchez 1976). The USDA names for soil orders are used in this book. The usefulness of USDA taxonomy has recently been improved through extensions to some African soils and through the addition of an **Andisol** order for soils that originate from volcanic ash and **Crellisol** permafrost soils in arctic regions.

7.3.2   Soils used in agriculture

Level soils with deep, well-drained profiles provide the best circumstances for agriculture. High fertility and near-neutral pH are also useful traits. Only a small portion of the world's land area satisfies these criteria; currently, only about 12% of the ice-free area is cultivated. The current utilization of land and the potential for expansion in agriculture are discussed in Box 7.2.

In retrospect, it is not surprising that many early agrarian centers developed on soils formed from young alluvium (e.g., the Nile Delta), wind-blown loess (Chinese and North American sites), volcanic ash (Indonesia, Meso-America), and in semi-arid and arid regions (the Near East), which have not been leached of their rich base-metal fertility. Many arid-zone soils are found in the order **Aridisols**. As human populations increased in Europe, agriculture spread slowly into deciduous woodlands. Two groups of soils then came into importance: **Inceptisols**, immature soils with poorly developed profiles and high in base metals; and **Alfisols**, older soils with high base saturation and clayey B horizons. Alfisols generally develop under deciduous woodland with favorable moisture.

Special knowledge or technology was needed with many soils before they could be farmed successfully. Cultivation of Mollisols formed under tall grasses is a recent phenomenon aided by the invention of steel plows. **Histosols**, highly organic peat and muck soils formed in boggy places and swamps, require drainage, while the extensive **Vertisols** of India, Australia and Texas (USA), characterized by shrinking and swelling clays, can be cultivated only over a limited range of moisture content. (The prefix *verti-* refers to churning of A and B horizons that occurs on wetting and drying of smectite clays.) Highly weathered **Ultisols** and **Oxisols** cover large areas of India, Brazil, Africa, southeast Asia, and southeastern USA. Both are common in the humid tropics and present special problems to agriculture. Although they occur in mild to warm climates with generous rainfall well suited for agriculture, these soils are very old and rich in Fe and Al oxides (hence Oxisol) that in some cases can harden irreversibly to **laterite** (brick-like structure) on drying. Low P availability is common because P reacts with aluminum and iron to form relatively insoluble aluminum and iron phosphates. Acidity and toxicities from aluminum and manganese also occur. Despite poor inherent fertility and acidity, Ultisols and Oxisols can sustain highly productive cropping systems when supplied with adequate inputs of nutrients and lime (Sanchez 1976).

7.4   Soil properties

The ability of soils to support plant growth depends upon physical, chemical, and biological properties that determine their capacity to retain and release nutrients.

Box 7.2 Land available for agriculture and implications for future food supply

Current agricultural land use is presented below. It reveals that worldwide, 38% of the global land area of 13 Gha is used for crop or pasture production but of this just 1563 Mha (12% of land area) is in cropping, including 152 Mha in permanent crops such as orchards and vineyards.

Total land area and current land use for agriculture (Mha) in relation to population

Continent	Land area	Crop land[a]	Permanent pasture	Forest area[b]	Population (10^6)	Crop land per capita (ha)
Asia	3093	573	1090	574	4167	0.14
Africa	2964	246	911	627	1033	0.24
America, Central & South	2005	162	545	908	589	0.28
Europe	2207	293	181	1003	733	0.40
America, North	1867	226	254	614	352	0.64
Oceania	849	47	393	206	36	1.31
World	13014	1554	3378	3937	6910	0.22

[a] Cropland refers to land cultivated with annual and permanent crops; double-cropped areas are counted only once. World includes Caribbean.
[b] Land under natural or planted stands of trees, whether productive or not.
Source: FAOSTAT.

While Asia has the most cropland, it also has the largest population and thus the smallest area per person. North America and Oceania have relatively large amounts of crop area per capita and are able to export large proportions of their agricultural production. Because population is expected to increase more than 2 billion by 2050, a 35% increase from current, the question arises of how much spare land is available for expansion of crop area? Some estimates that use a generous assessment of land suitability based on minimum levels of soil quality and rainfall for crop production conclude that there is more spare land than current cropland (e.g., FAO 2000). Other estimates use more rigorous requirements for soil and moisture regimes for sustainable rainfed cropping and include preservation of land for nature. Fischer and Heilig (1997), for example, estimate potential new crop land at just 300 Mha. Since 100 Mha of existing cropland will likely be lost to urban expansion and industrial development by 2030 (FAO 2002), the net amount of new land for expansion of crop and pasture area is relatively small, perhaps no more than 200 Mha. Nearly all the best farmland is currently in production, so new land suitable for cropping is of poorer quality. That, together with anticipated population growth, means that yields on all cropland must increase substantially to meet future food demand.

7.4.1   Surface area and aggregation of soil particles

Surfaces of clays and humus are very active. Because clay particles are extremely small, they possess enormous surface area per unit soil volume. One cubic centimeter of solid clay has a density near 2.65 g cm^{-3} and a surface (if cubic) of 6 cm^2. Subdivided to 1 μm cubes (coarse clay), density drops to near 1.3 g cm^{-3} because cubes do not pack tightly and about 50% of volume occurs as pore space. There are then 5×10^{11} cubelets per cm^3, each with a surface area of 6 μm^2. Total surface area per cm^3 of this "soil" is 3 m^2! The surface area of a real clay soil is less, but still an impressive 0.5 to 1 m^2 per cm^3. Part of the mass of real soils is due to larger particles (sand and silt) with much less surface area per unit weight and clay particles are usually aggregated into larger structures, termed **aggregates** or **peds**, with less surface area. Aggregated soils have a granular texture with greater porosity between peds and thus better aeration and water infiltration rates than poorly aggregated ones. In addition, they are more easily tilled.

Linkage of clay platelets and ribbons into large aggregates can occur through cation-bonding by calcium (Ca^{2+}) and "gluing" by humus. Calcium ions, with their small radius and large charge density, flocculate clays by binding negatively charged clay platelets and ribbons into larger particles.

Porosity and **aggregation** are important factors in soil structure. Structure can be evaluated through measures of soil strength (e.g., ease of tillage), air porosity, and water infiltration rate. Humus content enhances each of these aspects of structure. Farm operations affect soil structure in several ways. Traffic by animals and vehicles on wet soil, for example, compacts clays into massive structures and should be avoided except with flooded culture of rice where an impervious clay pan just below the plow layer helps prevent loss of water by deep percolation. Structure generally deteriorates with tillage because increased aeration promotes oxidation of organic matter in top soil. On the other hand, incorporation of animal manures, green manures, and crop residues through tillage helps in maintaining SOM and thus structure. **Green manures** are crops grown specifically for incorporation with soil; grasses high in lignin are most effective in maintaining organic matter. Leguminous green manures are used to add N to a soil.

Freezing and thawing, and wetting and drying, help in maintaining soil structure in most climates. In hot, dry climates, where organic inputs are small and oxidation of humus is rapid, soils are naturally low in humus, and massive structure is common. Tillage of soil when dry is one way to fracture massive structures into granules. In cases where tillage leads to compaction, ripping with chisel plows is sometimes required to shatter compacted layers. Additional energy is required for such operations, however, which reduces the energy efficiency of a cropping system (Section 15.3).

7.4.2   Ion exchange by clays and organic matter

Oxygen lattices endow clay minerals with important surface properties. Shortages of metal ions within lattices, exposure of oxygen surfaces, and conversion of some oxygen to hydroxyls ($-OH \leftrightarrow -O^- + H^+$) leave clay particles with negatively charged surfaces that are highly effective in binding cations and water molecules.

Cation bonding is ionic (e.g., Ca^{2+}: $clay^{2-}$). Various cations can be exchanged for one another through mass action. The strength of bonds depends upon the charge density of the cation involved. Protons (H^+) are very small and thus have a high charge density. Na^+, with its small charge density, is weakly held by clay and easily leached from the profile. Cations can be ranked in a **lyotropic series** according to decreasing charge density and thus decreasing strength of binding to clay: $H^+ > Ca^{2+} > Mg^{2+} > K^+ \approx NH_4^+ > Na^+$. Ion exchange and lyotropic position can be demonstrated by eluting a column of Na-saturated clay with dilute potassium solution (e.g., KCl). Sodium is displaced to the eluent solution, leaving a K-saturated clay. Because exchange depends upon mass action, an ion low in the series but present in sufficient concentration can displace a higher one.

The amount of cations that can be held by a soil defines its **cation exchange capacity** (CEC, in mol_{charge} kg^{-1} of dry soil). Soils with a small CEC can hold only small quantities of essential nutrient cations. With high proportions of H^+ or Al^{3+} on their exchange sites, soils will be both acid and infertile regardless of CEC. Young primary minerals such as montmorillonite have a relatively large CEC that does not vary with soil pH. In contrast, sesquioxide minerals in Ultisols and Oxisols have a small CEC that varies with soil pH.

Soil organic matter is even more effective in cation exchange per unit mass than are clay minerals. Metal cations exchange readily with protons of amino ($-NH_2$), carboxyl (–COOH), and hydroxyl (–OH) groups of SOM. Representative CEC values for clays range from 30 to 100 meq 100 g^{-1} whereas CEC of humus may be as great at 300 meq 100 g^{-1}. Although humus usually constitutes only 2 to 5% by mass of agricultural soils, it makes significant contributions to CEC. As with sesquioxide minerals, the CEC of SOM is pH dependent.

In arid regions, soils with alkaline pH (>7.0) prevail and often have high levels of Na^+ on the exchange complex. Na^+ hydrates to a large ion with a charge density too small to aggregate clays. That allows clays to disperse and seal soil surfaces. Sodium-saturated clays collapse into dense ("massive") structures when dry and disperse as sols when wet. Proportion of CEC occupied by Na^+ depends upon amounts of divalent cations present. An alkaline condition (pH > 8.5) develops when Na^+ is abundant and balanced by a weak anion such as HCO_3^-. Sodium status is evaluated by the **sodium adsorption ratio** (SAR) calculated from concentrations of soluble Na^+, Ca^{2+}, and Mg^{2+} in the soil solution:

$$SAR = \frac{\left[Na^+\right]}{\sqrt{\left(\left[Ca^{2+}\right] + \left[Mg^{2+}\right]\right)/2}} \tag{7.7}$$

Soil and plant problems generally begin with SAR values >10. Such sodic soils can be reclaimed by leaching with water of low sodium concentration and through additions of gypsum ($CaSO_4$). Ca^{2+} from gypsum displaces Na^+ and flocculates clays into aggregates while the acid anion helps in lowering pH. Amendment with elemental sulfur is a common treatment for alkali soils. Sulfur oxidizes to SO_4^{2-} and in association with

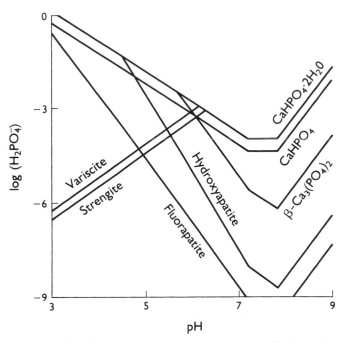

Fig. 7.7 Stability diagram for phosphate minerals in equilibrium with 10^{-3} M Ca^{2+}, Mg^{2+}, K^+, Na^+, and Cl^-. Al^{3+} and F^- are in equilibrium with gibbsite and kaolinite clays and Fe^{3+} with iron hydroxide. Free phosphate ions occur below the limiting mineral lines. See text for further explanation. (Redrawn from Lindsay *et al.* 1989.)

water provides H^+ for Na^+ displacement. Salinity and alkalinity are considered further in Chapters 13 and 14.

Both clay and organic matter can also hold and exchange anions. In humus, amino groups ($-NH_2$) that become positively charged through attraction of a proton ($-NH_3^+$) are important in anion exchange. Protonation of hydroxyl groups in sesquioxidic minerals can create substantial anion exchange capacity in Ultisols and Oxisols with acidic pH. Anion exchange capacity reduces the mobility of nitrate (NO_3^-), which in turn can reduce susceptibility to leaching.

7.4.3 Nutrient availability

Factors affecting solubility of phosphate ions (PO_4^{3-}) offer insight into the complexity of soil solutions. The ion has a dense negative charge that causes it to form sparingly soluble salts with calcium, iron, and aluminum. In addition, the oxygens of phosphate can fit tightly into clay lattices. Phosphate that is "fixed" (bound) in soils by these processes is not available to higher plants. Figure 7.7 is a stability diagram of phosphate ions in equilibrium with several minerals over a range in pH. The figure is similar to that presented in Fig. 7.3. Above a particular line, the solution is saturated for that mineral and it tends to precipitate; below its line, the mineral dissolves. At pH 4, solubilities of iron phosphate (strengite) and aluminum phosphate (variscite) restrict free phosphate ions to less than 1 μM. Formation of these minerals also contributes to iron deficiency in plants. Above pH 6, phosphate solubility is increasingly restricted by precipitation with calcium. Hydroxyapatite is the important calcium phosphate in most soils since activity of F^- and thus formation of fluorapatite is restricted by the precipitation of fluoride as CaF_2. The stability pattern is altered significantly when other ions, in particular CO_3^{2-} and SO_4^{2-}, are present.

Plant uptake also helps to restrict the amount of phosphate in solution. About 400 g P ha^{-1} are needed daily to support a crop growth rate of 200 kg ha^{-1} d^{-1} (0.002 is a conservative value for the concentration of P in dry biomass; see Table 12.1). With 2000 m^3 (2 Ml) of soil solution in the surface 0.5 m of 1 ha of drained clay soil, and phosphate at 1 μM, the solution contains only 124 g P. Crop nutrition therefore depends on continued dissolution of phosphate minerals and desorption of fixed phosphate as soil solutions are depleted of P by uptake.

Nitrogen availability is also dynamic. A typical crop demand is 4 kg ha^{-1} d^{-1} (0.02 of a daily growth rate of 200 kg ha^{-1} d^{-1}). Concentration of nitrate ions in soil solutions depends mainly on the balance between rates of mineralization of organic matter and use of nitrate and ammonium by plants and soil microorganisms (Chapter 8). Concentrations of nitrate range from 1 μM with low fertility to 1 mM in fertile soils. At 100 μM, the supply in 0.5 m of clay soil with 2000 m^3 water ha^{-1} is only 2.8 kg ha^{-1}. A substantial application of 100 kg NO_3-N with that amount of water raises the concentration temporarily to near 4 mM.

Typical concentrations for K^+ and Ca^{2+} range between 0.1 and 1 mM. They are low due to mineral solubility, strong retention by cation exchange, and aggressive uptake by living organisms. Potassium can also be preferentially fixed between clay lattice sheets of illite and vermiculite minerals. These minerals are formed through weathering and loss of K from the primary mineral biotite mica. Once soils containing these minerals become severely K depleted through exhaustive cropping without replenishment, a large portion of K from initial fertilizer applications is fixed and not available for crop uptake. With continued K application in excess of removal in harvested crop materials, a new equilibrium is established with higher [K^+] in the soil solution. Because the ammonium ion (NH_4^+) has a similar hydrated radius and charge density to K^+, it can also be fixed in soils containing these minerals.

7.4.4   Soil acidity

There are two important points to make about soil acidity and alkalinity. First, few plants grow well in soils with pH outside the range 5 to 8. Legumes are particularly sensitive to low pH. Soil pH has direct effects on plant roots, but in most cases that is overshadowed by pH-dependent toxicities and deficiencies of specific elements. In alkaline soils, high levels of sodium cause toxicities. As noted above, deficiencies of iron and phosphorus occur at both extremes of high and low pH. The main problem at pH < 5.2, however, is that aluminum minerals are hydrolyzed and soluble Al^{3+} reaches toxic levels near 10 μM. Root growth is especially sensitive to toxic [Al^{3+}] in soil solutions. Under anoxic (reduced) soil conditions, Mn^{2+} and Fe^{2+} toxicities can also occur. The second key point is that soils acidify naturally under vegetation in regions where rainfall exceeds evapotranspiration, and acidification is accelerated in pastures and fields used for production agriculture. In drier climates with less leaching, reactions involving calcium and carbonates dominate and soils tend towards neutral to alkaline pH.

Hydrogen concentration, $[H^+]$, in soil solution is controlled by complex equilibria and buffering with organic matter and clay minerals. Most soils are naturally acid due to exchangeable protons from SOM and leaching of base metals from parent materials during soil formation. Rainfall from unpolluted air contains dissolved CO_2 that buffers near pH 5.7 $\left(H_2O + CO_2 \leftrightarrow H^+ + HCO_3^-\right)$. H^+ is balanced by HCO_3^- but with heavy leaching, K^+ is lost from soil, H^+ is retained, and soil pH declines to equilibrium near pH 5.2 (Helyar & Porter 1989). Although soils may differ markedly in buffering capacity over other pH ranges, most buffer well near pH 4 to 4.5 owing to aluminum hydroxides released through the dissolution of clays.

Cycling of both C and N add or subtract H^+ to soil–plant systems through changes in base cations. The C cycle is involved with soil pH through accumulation of plant biomass, which takes up more base cations than anions from soil. Within plants, the difference is balanced electrochemically with organic anions. In terms of equivalents of inorganic cations (C_i), inorganic anions (A_i), and organic anions (A_o): $C_i - A_i - A_o \approx 0$ in biomass. The combination of inorganic cations (mainly K^+ and Ca^{2+}) with organic anions (weak acids such as malate and citrate) buffers internal pH of plants in a physiological range near pH 6.5. One consequence of the C_i–A_i imbalance in biomass, however, is a surplus of A_i relative to C_i in soil: for each organic anion created in plant metabolism, one H^+ accumulates in soil.

C_i–A_i imbalances in plant material can be determined from organic acid content or, more simply, from **ash alkalinity** (a sample is burnt to ash, dissolved in water, and titrated to neutrality with acid). When C_i–A_i imbalances are exported in harvested crops, H^+ ions remain behind in the soil. Harvest of vegetative material (hay or silage) therefore results in considerably more soil acidification than does harvest of grain for two reasons. First, vegetative tissues have much higher $[K^+]$ than grain. Thus, using ash alkalinity data from Helyar and Porter (1989), removal of 10 t legume forage ha^{-1} adds 10 kmol H^+ ha^{-1} to soils whereas production of 10 t grain ha^{-1} adds only 0.5 kmol H^+ ha^{-1}. Second, return of vegetative biomass in residue from grain crops is less extractive than complete harvest and removal of entire aboveground biomass as practiced in hay and forage crops.

Events during N cycling also contribute to soil acidification because mineralization and nitrification both release H^+ ions (see Fig. 8.2) and produce nitrate ions that can leach from the root zone. NO_3^- is a strong acid but it is accompanied mainly by K^+ and other base cations when it leaches because metal cations are far more abundant than protons at pH 4 to 7; increased $[H^+]$ in soil is balanced by the weak acid HCO_3^-. With K^+ as the accompanying ion in leachates, 1 mol H^+ accumulates in soil per mol nitrate lost. Relationships between soil acidification and nitrogen transformations are discussed further in Box 7.3.

> **Box 7.3** Summary of soil acidification related to nitrogen transformations
>
> Both natural ecosystems and agroecosystems depend on inputs of N to maintain productivity. Even with the relatively closed nutrient cycles in natural systems, a small amount of N is lost via leaching, denitrification, and ammonification. Inputs are from wet and dry deposition, biological N fixation by free-living microbes and symbionts in association with legumes as discussed in Chapter 8. The magnitude of acidification resulting from N losses depends on the form in which N originally entered the soil–plant system. In the table below, Helyar and Porter (1989) estimate that acidification is greatest with N derived from fertilizers containing ammonium (e.g., ammonium sulfate, ammonium phosphates, and urea) and from composted manure; it is smallest with N from nitrate as found in other types of fertilizer and in wet deposition.
>
> | Nitrogen enters as: | mol H^+ gained per mol N leached |
> |---|---|
> | Nitrate ion (NO_3^-) | 1 |
> | Ammonia, urea, protein (residues or manure), or by N fixation | 2 |
> | Ammonium ion (NH_4^+) (ammonium sulfate fertilizer or composted manure) | 3 |
>
> Important messages are: (a) the addition of N in any form will contribute to acidification when it leaches from the system as nitrate, and (b) the addition of more reduced forms of N cause more acidification than when the original source is nitrate. Most N inputs to agricultural systems, however, come in reduced forms, which means that soil acidification represents a substantial challenge to long-term sustainability of crop agriculture.

Cropping practices influence rates of acidification. Changes of 0.02 to 0.2 pH units y^{-1} have been observed in agricultural soils (Williams 1980; Helyar & Porter 1989). In grazed pastures, most nutrients recycle within the pasture. Milk production results in modest removals of N, Ca, and P; exports with grazing by sheep or beef cattle are less. Despite small removals, grazed lands acidify rather rapidly, owing to the large amount of forage consumed and to large amounts of N in urine and dung deposited in patches (at rates up to 700 kg N ha^{-1}) that result in localized nitrate leaching and acidification. The deposits tend to be greatest near resting and watering places.

In swidden agriculture, vegetation is burned and its N and organic acid components are combusted. The $C_i - A_i$ imbalance of the ash is returned to the soil where it has an alkalizing effect. The same phenomenon occurs when cereal stubbles are burnt. Recycling of crop residues and animal manures to the soil also has the virtue that they tend to restore $C_i - A_i$ balance but that may be offset by an acidifying effect from exchangeable H^+ of organic compounds.

Concern exists, particularly in the northern hemisphere, about the influence of *acid rain* due to contamination of air by oxides of sulfur and N. The pH of acid rain is near 4 and in extreme cases may be < 3. At pH 4, 1 kmol H^+ ha^{-1} y^{-1} is added to soil with 1 m rainfall. The effect is small relative to those of forage production but, continued for a long time, equilibrium soil pH will decline.

Correction of soil acidity is achieved through applications of lime. The Romans recognized the benefits from liming and today the practice is widespread in Europe, North and South America. The ley-farming system of southern Australia (Section 16.5) has acidified soil significantly and farmers there are now beginning to use lime. That system performs reasonably well with low pH because acid-tolerant strains of subterranean clover are used, although rhizobial activity is reduced. Without lime, however, it seems likely that clay dissolution is occurring and the acid layer of the profile may be deepening to an extent that liming will become uneconomic. In the long run, use of acid-tolerant plants instead of lime can only exacerbate such problems.

Ground calcite limestone is the most common source for pH amendment but marl and other materials high in $CaCO_3$, including waste lime from paper mills, sugarbeet factories, and other industries, are also used. Lime requirements are determined in several ways. The simplest is to slurry a soil sample in neutral salt solution and titrate it to pH 7 to 8 with $Ca(OH)_2$. A high endpoint is used because equilibration of calcium with clay minerals to a true pH endpoint may require several months. Where Al^{3+} toxicity is a problem, exchangeable cations must also be eluted from the sample and titrated so that both H^+ and Al^{3+} are considered.

Acidification processes are amplified by spatial differences in wetness, drainage, and N supply as well as by grazing intensity. As a result, buffering capacity differs spatially and liming may cause additional variation as some places are neutralized and others stay buffered at a low pH. In some cases, it pays to map the lime requirement and apply amendments according to within-field variation in soil properties and existing levels of acidity (Section 12.1.1).

7.5   Water and air components

Crop growth and survival depend upon the supply of water and oxygen to root systems in adequate amounts. Here we deal with the impacts of soil physical properties on entry, storage, and availability of water to crops and the close link that exists between soil water content and soil aeration.

7.5.1   Water in soils

The water molecule is a very strong dipole with separate, strong centers of positive and negative charges (Fig. 7.8). A proton pole of water can associate with oxygen in another water molecule or with clay and humus. Water binding by clay and humus depends on the same principles as ion exchange. Association of a water proton with the negative surface of clay leaves a negative ($-OH^-$) tail of the water molecule extending into the

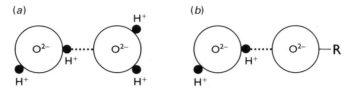

Fig. 7.8 H-bonding in water. (*a*) Oxygen atoms of two water molecules share the same proton. Such linkages extend throughout aqueous media. (*b*) One oxygen (O–R) belongs to a ligand such as humus or clay. This is the basis of matrix potential. Oxygen in R–OH, R–COOH, and in lattices of clays can enter into H-bonding.

soil solution. That tail links and thus binds with a second layer of water. Carboxyl and hydroxyl groups of humus act on water in a similar way. Such "adsorbed water" is held very tightly to the surface matrix restricting its mobility and possibility for uptake by plants. Chains of H bonding extend throughout an aqueous solution, imparting a "structure" to water that is seen in phenomena such as surface tension and viscosity.

Texture and structure determine the portions of a soil volume (V) that are pore space (V_p) and solid material (V_s). **Porosity** (V_p/V) and **packing density** (V_s/V) serve as relative measures of pore space. **Soil bulk density**, the mass of dry solids per unit volume (M_s/V; g cm^{-3}) is a more common measure of solids content. Values of bulk density range from 0.5 (volcanic ash) to 2.0 (compacted soils) g cm^{-3}; most agricultural soils have values near 1.3 g cm^{-3}. As noted earlier, the density of soil minerals is 2.65 g cm^{-3} so a bulk density near 1.3 indicates 50% porosity. That space can be variously filled with water or air and these important components of soil are reciprocally related.

Large macropores and spaces created by earthworms, past root growth, burrowing animals, and shrinking clays are major routes for root, water, and air movement. It is not uncommon, for example, to find plant roots descending through decaying residues of earlier roots. Macropores drain freely, retaining only films of water adsorbed by H-bonding. Small capillary pores, and voids within aggregated soil peds, on the other hand, are important holding spaces for water. Pores < 30 μm diameter fall in the capillary range and can hold water against the force of gravity. Most capillary water is held weakly and is easily removed by plants but the final film of adsorbed water is held very strongly. As a consequence, there are two basic ways to characterize soil water: **water content**, θ, expressed as a fraction of dry mass or volume; and soil **water potential**, Ψ_S, in pressure or energy units, which characterizes the activity of water associated with surfaces and solutes.

7.5.2   Water potential

Water potential describes the activity of water in a particular system. Individual water molecules have a velocity of movement proportional to their temperature and tend to diffuse or to escape as vapor. The collective free energy of all water molecules in a unit volume (in units of J m^{-3}) is termed water potential, Ψ. J m^{-3} is dimensionally the same as force per unit area ($1 \text{ J m}^{-3} = 1$ Newton $\text{m}^{-2} = 1$ Pascal), and Ψ is commonly expressed in either Pa or bar (1 bar $= 0.1$ MPa $= 0.987$ atm).

Water movement in soils and plants occurs along gradients of free energy, from regions where water is abundant, and thus has a high free energy per unit volume, to those where free energy of water is less. Potential of pure water is high because all molecules are free to move; it is taken as a reference state with the symbol Ψ_0. Ψ of water diluted by solutes is less than for pure water. The activity of water is further restricted by hydration reactions and by H-bonding to clays and organic matter in soils and to cell walls and proteins in plant cells. These factors have an additive effect in lowering potential of water in a soil or plant system relative to Ψ_0:

$$\Psi_{\text{system}} = \Psi_0 + \Psi_m + \Psi_\pi + \Psi_p + \Psi_g \tag{7.8}$$

By definition, Ψ_0 is set to 0 Pa and all other terms, except Ψ_p, the hydrostatic potential, are therefore negative (i.e., < 0). Ψ_g, the gravitational term, results from differences in height. It decreases 10 kPa m^{-1} height and thus is important in 80 m mountain ash and redwood trees but can be ignored in crop systems. Ψ_p, due to hydrostatic pressure, is positive in turgid plant cells but is 0 in soils. Ψ_m, the matrix potential, results from adsorption on surfaces. Ψ_π, the solute or osmotic potential, can be approximated for dilute solutions from the Van't Hoff relation derived from the gas law:

$$\Psi_p = -CRT \tag{7.9}$$

where C is the concentration of solutes (mol 1^{-1}), R is the gas constant (8.314 J mol^{-1} K^{-1}) and T is Kelvin temperature. Simplifying Eq. 7.3 by ignoring Ψ_g, Ψ_m, and Ψ_p when appropriate:

$$\text{Soil:} \quad \Psi_s = \Psi_0 + \Psi_m + \Psi_\pi \tag{7.10}$$

$$\text{Plant:} \quad \Psi_{pl} = \Psi_0 + \Psi_\pi + \Psi_p \tag{7.11}$$

Ψ_m and Ψ_π are negative and Ψ_p, although positive, is never greater than the magnitude of Ψ_π. As a result, Ψ_s and Ψ_{pl} are always less than Ψ_0, and plants can only take up water from soil when $\Psi_{pl} < \Psi_s$.

Total Ψ_s and Ψ_{pl} can be determined on samples in the laboratory with psychrometric measurements of equilibrium vapor pressure (e in kPa; Eq. 6.17). Water adsorbed to soil colloids or diluted with solutes has low equilibrium vapor pressure compared with the saturation vapor pressure of water (e^*). Water potential of soils and plant materials is related to vapor pressure as follows:

$$\Psi = [RT\ln(e/e^*)]/18 \text{ (J g}^{-1}) \tag{7.12}$$

where 18 is the relative molecular mass of water. The temperature of psychrometers must be controlled precisely so measurements are generally performed in laboratories. An alternative method used with soils involves squeezing water from a sample in a **pressure membrane** apparatus. The soil is compressed in a cylinder against a porous membrane and a graph is constructed of pressure applied (taken as Ψ_s) and water content (θ) remaining in the sample as determined gravimetrically. Examples of such

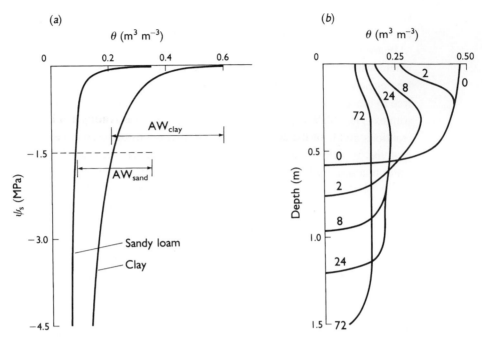

Fig. 7.9 (*a*) Moisture-release curves for sandy loam and clay soils. Initial volumetric water contents (θ), near 0.35 for the sand and 0.60 for the clay, provide a maximum estimate of field capacity. (*b*) Infiltration and redistribution of moisture with time in a homogeneous medium (slate dust). The number of hours after the beginning of the experiment is indicated for each curve. (Redrawn from Miller & Klute 1967.)

moisture-release curves are presented in Fig. 7.9*a*. In practice, standard curves are constructed for individual soils in the laboratory and used to translate field data gathered with gravimetric samples or neutron probes.

7.5.3   Water and plant growth

Several aspects of the moisture-release curves of Fig. 7.9*a* deserve emphasis. The upper limit of θ after free drainage is termed **field capacity** (FC) or just drained upper limit. Ψ_s at FC typically varies from -0.020 to -0.035 MPa depending upon several factors including soil compactness. A more consistent number, **moisture equivalent** (ME) corresponding to θ at -0.030 MPa can be obtained with a pressure-membrane or after centrifuging wet soil at $1000\times$ gravity. Near FC, loosely held capillary water can be removed from soil with only a small decline in Ψ_s. As a result, estimates of FC are never very accurate. An empirical exponential relationship, $\Psi_m = a\theta^b$, fits the curve well over most of the range of θ except near FC.

A second point from Fig. 7.9*a* is that clay, with its much greater surface area and capillary volume, holds more water than sand and its Ψ_s declines gradually as capillary water is removed. After capillary water is removed, the remainder is held very tightly in both sand and clay and Ψ_s drops dramatically with further drying due to decreasing Ψ_m.

It is obvious that plants must make dramatic adjustments in Ψ_{pl} if they are to continue to extract moisture at low values of Ψ_s. For most plants, the practical limit beyond which water uptake and growth cease is near $\Psi_s = -1.5$ MPa. Soil θ at -1.5 MPa is called the **wilting point** (WP). **Available-water capacity** (AWC) of a soil, FC − WP, defines the amount of **available water** (AW) that plants can access easily, and is approximated by the difference in θ between -0.03 and -1.5 MPa. The wet end of that range down to around -0.06 MPa, depending on crop, is referred to as **readily available water** (RAW). Given the shape of the curves in Fig. 7.9, it contains between 0.5 and 0.3 of AWC, for sandy and clayey soils, respectively. Readily available water finds practical use in the determination of irrigation frequency for high-value horticultural crops.

How plants are affected at lower Ψ_s, especially near WP, depends mainly on the relative rates of water uptake and loss. Growth and photosynthesis of plants with poor root systems and/or under high evaporative demand are restricted at Ψ_s higher than -1.5 MPa (WP). Given time for adjustment (i.e., small evaporative demand), most plants can lower Ψ_s well below -1.5 MPa. Cotton plants grown in pots of soil with limited supplies of water will wilt and their stomates close at Ψ_s near -1.5 MPa. Under field conditions, the much greater soil volume dries slowly and plants have time to adjust by increasing solute content causing a decrease in Ψ_π and thus Ψ_{pl}. Under those conditions, wilting may not occur until Ψ_s drops below -3 MPa. Extreme examples have been observed of plant adjustment to $\Psi_s = -7$ MPa. Despite these large differences in the magnitude of wilting point Ψ_s, the amount of additional water gained beyond -1.5 MPa is relatively small.

Plants extract water from soil profiles according to the distribution of their fine roots. As a result, some parts of soils may remain near FC as others approach WP. Ψ_s varies accordingly while Ψ_{pl} reflects an integration of Ψ_s over the root zone. During daytime, Ψ_{pl} varies among plant parts depending upon their position in the transpiration stream (Fig. 9.1). At night, with stomates closed, plant water content is restored and Ψ_{pl} comes to equilibrium with root zone soil. Pre-dawn values of Ψ_{pl} therefore serve as useful indicators of integrated soil water status.

7.5.4   Infiltration of water

Figure 7.9*b* illustrates the downward advance of a wetting front from a saturated zone at the surface. Wetting progresses at close to field capacity so there is generally a sharp transition in water potential between wet and dry soil. Infiltration of water influences more than just replenishment of soil moisture. Erosion (Chapter 12), leaching, and soil aeration are also affected. If the rate that water is added to the surface exceeds infiltration rate, water will pond or run off (with danger of erosion).

Pore size controls water movement in soils. In dense soils with few macropores, water flux is limited by capillary flow. The hydraulic (water) conductivity of saturated clay, with very small pore sizes, is only about 0.5 mm h^{-1} at 20°C. Temperature is specified because viscosity of water decreases (and conductivity increases) as temperature increases. Conductivity increases by approximately one order of magnitude for each step in the textural series, clay→loam→sand. The "soil" in Fig. 7.9*b* is homogeneous with depth whereas the clay content increases with depth in most soils. Lower depth therefore have smaller hydraulic conductivities than those near the surface and rapid infiltration of A horizons is followed by slower wetting of B horizons.

Wetting characteristics are affected dramatically by soil structure. Organization into structural peds leaves macrochannels that allow rapid, deep wetting followed by a slower frontal advance into peds. Marked structural organization of B horizons is rare, however, so replenishment of deep moisture is slow except with shrinking clays where macrochannels extend to depth. As these clays are wetted, however, they swell, closing channels and macropores so wetting continues through saturated flow.

Hydraulic conductivity declines when a soil dries because the remaining water is held strongly by matrix forces. The change is dramatic: conductivity declines by a factor of 10^4 between FC and WP, sharply restricting capillary flow to plant roots.

7.5.5   Soil aeration

Entry of O_2 into soils to support respiration and release of CO_2 that would otherwise increase to toxic concentrations is closely linked to soil water content. Capillary pores are open to diffusive exchange of gases only after a soil has drained following a thorough wetting event. Oxygen diffusion through macropores and channels, while more significant than in capillaries, is inadequate to support the potential biological activities of most soils. Pressure changes due to atmospheric turbulence help exchange gases between atmosphere and soil. Wetting and drying are also important. Stale air (low in O_2, high in CO_2) is displaced on wetting and fresh air is drawn into soil as macropores drain. The solubility of oxygen and its diffusion rate in water are both small causing wet soils to become anaerobic as oxygen is exhausted by respiration of roots and microorganisms. The centers of small clay aggregates, and large peds also, may remain anaerobic for extended periods after soils drain. Relative activities of aerobic and anaerobic bacteria change accordingly. Decay is slowed under anaerobic conditions and the large organic accumulations found in Histosols and lowland rice soils are one result. Rice is one of few crops that tolerate flooded soils, in part because oxygen reaches rice roots through specialized aerenchymatous tissues of stems and roots (see Section 13.1).

Respiration in soils includes the activities of animals as well as living roots and decay organisms. Trophic chains of soil organisms begin with crop residues and their respiration can be estimated from amounts of organic matter added to soils. Requirements for soil aeration are considerable. A wheat crop with 15 t ha^{-1} of aboveground production would return perhaps 9 t straw and 2 t fine roots to the soil at the end of the growing season. Assuming that residues contain 44% C, 4.84 t C ha^{-1} (11 t × 0.44) are released during complete decay. This is converted to 4.0×10^5 mol CO_2, or 1 m^3 gas m^{-2} land (Eq. 6.15). A somewhat greater amount of O_2 would be consumed. Averaged over a year, that amounts to 13.3 kg C (28 m^3 CO_2 ha^{-1} d^{-1}). The flux would occur mainly during periods with favorable temperature and moisture. Respiratory activities of roots represent an additional source of CO_2 that can amount to 10 kg C or more ha^{-1} d^{-1} (21 m^3 CO_2). As a result, soil atmospheres generally have less O_2 (18 to 20% v/v) than normal air (21%) and more CO_2 (up to 1% compared with the current 0.038%). Distinguishing what portions of total soil respiration come from plant roots, various animals, and decay organisms is impossible without isotopic labels or antibiotic treatments that suppress microbial populations.

7.6   Soil temperature relations

Soil temperature varies complexly with depth, and over daily and annual cycles, owing to gains and losses of heat energy. Soil heat flux, G, was defined in Eq. 6.12 as a term in the energy balance at the soil surface. There, contributions of heat energy come from radiative, convective, and latent heat exchanges. Those are not the only sources of heat energy, however. A small, upward flux of about 50 nW m^{-2} resulting from radioactivity and cooling of the Earth's core ensures that temperatures deep in the profile (beyond 10 m) are warm (10–15°C), thus limiting the depth of freezing during cold winters. Respiratory activities of plant roots and soil organisms add a much larger source of heat. In the example involving the decay of wheat residues, 11 t residues ha^{-1} × 17 GJ combustible heat energy t^{-1} = 187 GJ ha^{-1}. The average flux would be 0.05 MJ m^{-2} d^{-1} (0.6 W m^{-2}). Although small compared with LE and H terms in surface energy budgets, heat from decay of large amounts of organic matter will serve to warm protected beds of garden soil.

7.6.1   Soil heat flux

Surface exchanges, as modified by characteristics of the soil surface, are important components in soil heat flux. Cover provided by living plants, stubble, and residue mulches insulates soil, and the color, wetness, and texture of the soil surface affect reflectance of short-wave radiation. Dark soils absorb well; wet soils are usually more reflective than dry soils; and a granulated surface is more absorptive than a smooth one. Wetness is particularly important because +R_n then is dissipated in evaporation rather than as soil heat flux. The slope and aspect of a surface also influence its radiation balance through the cosine law (Section 6.1.2). Daily and seasonal patterns of soil temperature at various depths are influenced further by thermal conductivity and heat capacity of soil, both of which vary depending on the content of minerals, organic matter, air, and water (Table 7.2).

The magnitude of the heating pulse from R_n to bare soil varies sinusoidally over daily and annual cycles (Fig. 7.10). Peak soil temperature lags behind peak irradiance, however, because soils continue to gain heat energy from air and radiation even as the radiation load declines in the afternoon. Daily maximum soil temperature at 10 cm depth generally occurs 2 to 3 h after solar noon, while the annual maximum may be 4 to 6 weeks after the summer solstice. In temperate regions, a similar lag in soil

Table 7.2 Thermal properties of soils

Material	Water content $(m^3\ m^{-3})$	Thermal conductivity $(Wm^{-1}\ C^{-1})$	Heat capacity $(MJ\ m^{-3}\ C^{-1})$	Thermal diffusivity[a] $(10^{-6}\ m^2\ s^{-1})$
Sand	0.0	0.29	1.2	0.24
	0.2	1.76	2.1	0.84
	0.4	2.18	2.9	0.75
Clay	0.0	0.25	1.2	0.21
	0.2	1.17	2.1	0.56
	0.4	1.59	2.9	0.55
Peat	0.0	0.06	1.5	0.04
	0.4	0.29	3.1	0.09
	0.8	0.50	4.8	0.10

Note: [a] diffusivity = conductivity/volumetric heat capacity.
Source: calculated from van Wijk and de Vries 1963.

warming following the winter solstice delays planting of crops in spring. In those regions, few warm-season crops achieve full cover before peak irradiance near the summer solstice.

With full crop cover, only about 5% of R_n goes to soil heat flux (G), and soil temperatures tend to be rather stable and cool. Energy exchanges by bare soil are much larger. In both cases, the heat pulse is propagated downward with $+R_n$ and towards the surface during periods with $-R_n$. The heat pulse penetrates slowly, owing to the insulating effects of air-filled pores, and the lag time to maximum temperature increases with soil depth. When pores are filled with water, thermal conductivity increases dramatically (Table 7.2) so lags are smaller. The pulse of heat energy is depleted in proportion to the soil's heat capacity causing diurnal temperature amplitude to decrease with depth; little diurnal variation in temperature is evident below about 0.5 m. Thermal conductivity and heat capacity both increase with increasing water content (Table 7.2); the combination of these changes is seen in the parabolic change in thermal diffusivity (= thermal conductivity/volumetric heat capacity). Soil energy balance is confounded further by latent heat exchange and vapor transfer within the soil profile: water evaporates in one region and condenses in another.

7.6.2 Influence of temperature on soil and plant processes

Temperature affects physicochemical as well as biological processes in soils. The solubility of calcium phosphates (Fig. 7.7), for example, shows a marked dependence on temperature. Phosphorus deficiencies are common during cool seasons in crop grown on the Holtville soil illustrated in Fig. 7.11 but generally disappear when soils warm because the solubility of phosphate minerals increases with temperature. With the tendency towards earlier planting in the US Corn Belt to increase growing season duration (Section 17.2), small doses of "starter" P fertilizer are sometimes applied near the

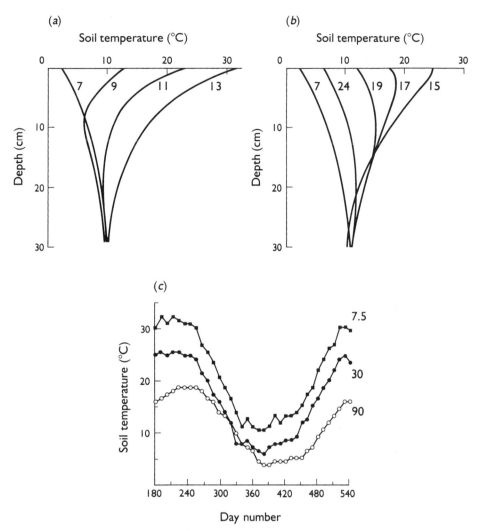

Fig. 7.10 Soil temperature patterns. (Adapted from F. A. Brooks, *c.* 1960, *An Introduction to Physical Microclimatology*, Mimeo, University of California, Davis.) (*a*) Daytime and (*b*) nighttime soil temperature patterns in bare soil at Riverside, California (34° N) for a day in late February. Numbers on the curves indicate the hour of the day. (*c*) Ten-day running means of temperatures under bare soil (Yolo fine-sandy loam) at Davis, California (38.5° N). Depths in cm are indicated. Measurements began July 1 (Day 182) and continued until June 30 of the next year.

seeds to avoid temporary P deficiency due to cool temperatures. Nutrient uptake and root growth generally follow the temperature response presented earlier in Fig. 3.12, and root permeability and thus water uptake are restricted by low temperatures. Soil temperature has a large effect on N mineralization from soil, the rate doubles for each 10°C rise in temperature, i.e., has a Q_{10} of about 2, typical of biological processes (Chapter 11).

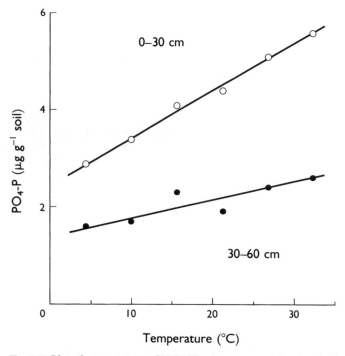

Fig. 7.11 Phosphorus content of Holtville clay loam soil that is soluble in dilute bicarbonate solution as a function of temperature. The two soil strata differ markedly in phosphorus release. (Previously unpublished data from R. S. Loomis, A. Ulrich, and G. F. Worker, Jr.)

The phase differences in times of temperature maxima and minima in air and soil present an interesting problem for plants. Photosynthate is gained as a daytime pulse following the pattern of short-wave radiation. At a particular time, user tissues, such as apical meristems of shoots and roots, may be at different positions on their temperature–response curve for growth. Some may be limited by temperature (too high or too low) while others may be in an optimum environment. As a result, phase and amplitude differences between air and soil exert control over the pattern of plant development. This topic was considered further in Chapter 5.

7.7   Review of key concepts

Soil formation and properties

- Soil formation is dominated by parent material, climate, and vegetation; similar soils are found in similar climates around the world. Over time, surface soil layers are leached, organic matter accumulates, and primary minerals weather and recrystallize as secondary minerals (clays). Minerals are only sparingly soluble. They consist of lattices and ribbons of silicon and aluminum oxides held together by metallic ions.

- Clays and organic matter exert great influence on soil properties. They provide immense areas of negatively charged surfaces that attract and hold water and ions.
- Level soils with deep well-drained profiles provide the best circumstances for agriculture. High fertility and near-neutral pH are also useful traits. This combination of traits is not common. Most favorable soils are already in use. Currently only about 1500 Mha, *c.* 12% of Earth's ice-free area of 1.3 Gha, is cultivated. A further 3400 Mha is in permanent pasture.
- Soil fertility can be improved with fertilizers and manures, salinity and acidity ameliorated by mineral amendments, and compaction alleviated by tillage. Sandy soils always have serious limitations due to low water-holding capacity.
- Soil classification systems are based on formation processes, profile characteristics, and morphology. Soils are classified in hierarchical categories (e.g., Orders, Families, and Subgroups in the USDA soil taxonomy scheme), based on soil survey data. Soil maps at various scales are available for most of the world.

Capacity to hold water and nutrients

- Organic matter is active in C and N cycling and it serves to bind soil particles into aggregate structures. Old organic matter (humus) consists of stable products of decay with C/N ratios near 12. Organic content of soils results from an equilibrium between humus formation (favored by high input of residues and thus by climates favorable for plant growth) and humus loss (favored by high temperature and moisture).
- Attraction of water to negative surfaces results in the matrix component (Ψ_m) of total soil water potential (Ψ_s). Solutes and gravity also contribute to lowering the activity of soil water. The ability of soil to hold and deliver water to plants is defined in moisture-release curves. The difference between an upper limit of water content after drainage, field capacity (FC), and a practical lower limit at -1.5 MPa (wilting point, WP), serves as a measure of plant available water capacity (AWC). At WP, water remaining in soil is held so tightly by matrix forces that little additional water can be extracted.
- During wetting–drying cycles, water and air sequentially refill soil pores. Roots and soil organisms require oxygen for respiration and so their activity strongly affects the composition of soil atmospheres. Macropores, especially, facilitate exchange of both air and water. In addition to cooling effects through evaporation at the surface, water increases the thermal conductivity and heat capacity of soil.
- Clay minerals and organic matter hold ions (plant nutrients) and exchange them with plant roots. Ability to exchange ions is measured as cation exchange capacity (CEC). Primary minerals, such as montmorillonite, have large CEC but sesquioxide minerals in Ultisols and Oxisols have small CEC that varies with pH. With a high proportion of H^+ or Al^{3+} on the exchange complex a soil will be acidic and infertile regardless of CEC.

Acidification

- Soils acidify naturally under any vegetation and more rapidly with cropping or grazing. Base cations (Ca^{2+}, K^+, Mg^{2+}) and organic acids are lost from the system in harvested material leaving H^+ in the soil. Addition of N to soil by any means, including biological fixation, also leads to acidification because NO_3^- that leaches from the profile carries base cations with it. To avoid deleterious effects of low pH, including aluminum toxicity, and to sustain productivity, pH is corrected towards neutrality with lime.

Terms to remember: acidity; aggregates (peds); ash alkalinity; bulk density; carbon sequestration; cation exchange capacity (CEC); enthalpy; entropy; equilibrium constant; field capacity (FC); green manure; humus; ion activity; lyotropic series; mineralization; minerals, primary, secondary, sesquioxides; moisture equivalent; moisture-release curve; packing density; parent material; redox reactions; sodium adsorption ratio (SAR); soil aeration, porosity and texture; soil organic matter (SOM); soil groups, Alfisol, Andisol, Aridisol, Chernozem, Histosol, Inceptisol, Mollisol, Oxisol, Ultisol, Vertisol; soil horizons A, B, and C; soil types, azonal, duplex, and zonal; soil water, available (AW), potential, readily available (RAW); soil acidity; solubility product; stability diagram; thermodynamics; weathering; wilting point (WP).

Part III

Production processes

The adequacy of agriculture as our source of food depends upon production rates in crops and pastures. Part III considers important "production processes". It begins with production and cycling of nitrogen (Chapter 8). Nitrogen has key roles in the structure of proteins and nucleic acids and thus can be considered, along with carbon, as one of the central elements of life. Nitrogen is subject to complex cycling and yet its supply is frequently limiting to performance of plant communities. Water is also commonly in scarce supply. Uptake of water from soils and its movement along the soil–plant–atmosphere continuum is the focus of Chapter 9, which gives particular attention to distinguish losses that occur through soil evaporation versus transpiration through plants. Acquisition of atmospheric carbon dioxide and its fixation and reduction (Chapter 10) are the central production processes. Photosynthesis is closely coupled with partitioning (pattern of use) of reduced carbon in respiration and growth (Chapter 11), which ultimately determines economic yield.

8 Nitrogen processes

Nitrogen is unique among the essential nutrient elements of higher plants in terms of its roles in biological systems and its complex cycling. In addition, it is the element most commonly limiting to crop production and the one most demanding of management skills. Soil organic matter (SOM) has a pivotal place in cycling N contents of crop residues and animal manures to mineral forms that are used by higher plants. In Chapter 7, we found that amount of SOM reflects the relative rates of C and N inputs to decomposition on the one hand and mineralization on the other. Nitrogen available for plant uptake comes from two sources: (1) indigenous N supply from SOM mineralization, biological N fixation, and atmospheric deposition, and (2) applied N in manure, compost, and mineral fertilizers. This chapter is concerned with several microbiological and physical processes important in N cycling and management of N in agriculture.

8.1 The nitrogen cycle

Nitrogen cycles between the atmosphere, soils, and plants through biologically mediated transformations that cover a wide range of chemical states. All N forms other than dinitrogen gas (N_2) are "reactive" in the sense they have a large impact on ecosystem structure and function. Too much reactive N causes **eutrophication** with detrimental effects on the composition and health of plant and animal communities in ecosystems. Too little N results in low net primary production and low crop yields. Understanding how N cycles through agricultural systems is important for identifying ways to improve productivity and protect environmental quality. We commence with a discussion of N oxidation–reduction states, soil N pools, and soil microbes as a basis for understanding factors governing N cycling in the biosphere.

8.1.1 Oxidation–reduction states of nitrogen

Nitrogen is present in soil–plant systems at stable oxidation states ranging from +5 (oxidized) to –3 (reduced). The arrows shown in Fig. 8.1 indicate the transformations among states that occur in N cycles. Most transformations are mediated by microbes. Nitrogen in SOM and in proteins and nucleic acids of living organisms is reduced at the –3 level. Most plants produce protein and other reduced N compounds beginning with mineral N absorbed from soil as nitrate (+5) or ammonium (–3) ions. Ammonium ions

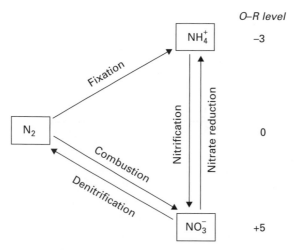

Fig. 8.1 Oxidation–reduction levels of nitrogen in nitrate (NO_3^-) and ammonium (NH_4^+) ions and dinitrogen gas (N_2) and the transformations between these important levels. Nitrite (NO_2^-) lies at +3 in most pathways to and from nitrate but has been omitted for simplicity.

enter directly into biosyntheses of amino acids and other compounds while nitrate must first be reduced to –3 (**nitrate reduction**) through the addition of eight electrons. Respiration supplies the energy and reducing power required in nitrate reduction; glucose, $C_6H_{12}O_6$, a representative substrate for respiration, yields 24 electrons as C goes from the 0 to +4 oxidation state (see Chapter 11). By contrast, humans, like most animals, do not have the ability for nitrate reduction and are dependent upon reduced N in their diets.

Soil microorganisms have evolved ways to utilize each form of N found in soils. They participate in the release of ammonium-N from SOM (mineralization) but they also compete with plants for use of N in growth. Microbial processes for oxidation of ammonium to nitrate (**nitrification**, removal of 8 e^-) and reduction of nitrate to gaseous N_2 (**denitrification**, addition of 5 e^-) are also identified in Fig. 8.1. Denitrification to N_2 (or to gases such as N_2O) is the principal way in which N returns to the atmosphere.

A number of microorganisms, including some found in symbiotic associations with higher plants, can reduce N_2 gas to the ammonium level (**nitrogen fixation**, addition of 3 e^- per N atom). Major transfers from N_2 to NH_3 also occur in the commercial production of fertilizer. The abundant atmospheric pool of N_2 gas could, theoretically, be oxidized to nitrate and enter biological soil systems in that form. The free energy change for oxidation is favorable, but the activation energy is very large. The process occurs naturally only with lightning discharges in the atmosphere and to some extent in diesel motors, together resulting in small fluxes of nitrate in rainfall ($1–5$ kg N ha^{-1} y^{-1}).

8.1.2   A generalized nitrogen cycle

The generalized N cycle presented in Fig. 8.2 includes all major pools and transfers found in cropping systems and pastures. For simplicity, nitrification is shown as a single

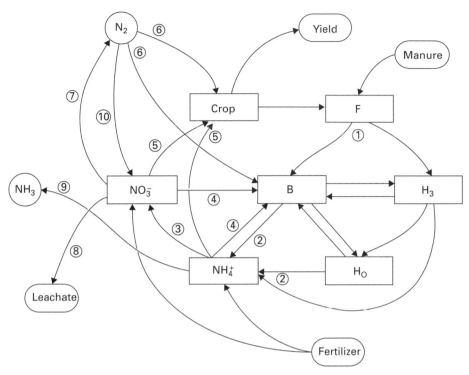

Fig. 8.2 A generalized nitrogen cycle for agricultural systems. Fresh organic matter (F) added to soil decays (1), to microbial biomass (B), active humus (H_a), and eventually to old humus (H_o). B, H_a, and H_o are subject to mineralization (2) to NH_4^+ and then by nitrification (3) to NO_3^-. The mineral forms are subject to immobilization (4) to B and then to H_a and H_o. Uptake to a crop (5), fixation of N_2 (6) to crop (symbiotic) or to B (free-living), and denitrification (7; to various N oxides as well as N_2) complete the biological transfers. Also shown: loss by leaching (8), volatilization (9) and yield; and inputs from lightning (10), manure, and fertilizer. NH_3 is also volatilized (process 9) from Crop and F (fresh residue). The major paths are emphasized by heavy lines.

process because accumulation of the intermediate nitrite form is rare. Soil organic matter is subdivided here into four pools:

F, fresh organic matter (dead and decaying material);
B, flora and fauna biomass (mainly bacteria and fungi);
H_a, young organic matter (active humus); and
H_o, old organic matter (old humus).

The distinction between the H_a and H_o pools is based on observations with C isotopes that some SOM fractions turn over more quickly than others (Paul & van Veen 1978). Active organic matter probably consists of slowly degradable intermediate products such as lignin– and phenol–protein complexes and is not distinguished from old humus in most assays. Protection of both humus pools comes from phenolic components, by adsorption to clays, or by sequestration within soil aggregates where it is inaccessible to attack by microbes. The H_a pool can be quite large (around 1 t N ha^{-1}) and its turnover is fairly rapid (10–20 y). The B and H_a pools are the principal sources of N released in

Box 8.1 Global N fertilizer use 2007–8

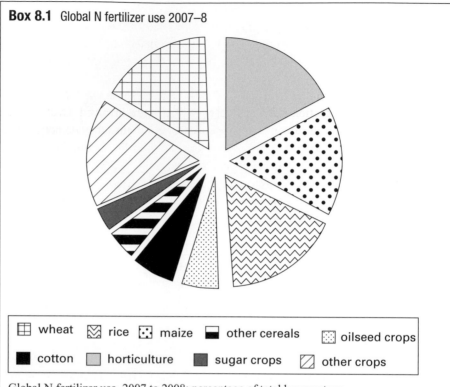

wheat   rice   maize   other cereals   oilseed crops

cotton   horticulture   sugar crops   other crops

Global N fertilizer use, 2007 to 2008; percentage of total by crop type.

The diagram above shows the proportions of 100 Mt fertilizer N applied to various crop types in 2007–8 (Heffer 2009). The dominant use is with three staple cereal crops: wheat (17.4 Mt), maize (16.9 Mt), and rice (15.7 Mt) and other cereals as a group, including sorghum, barley, oats, and millets (5.1 Mt). Together cereals accounted for 55% of total N fertilizer use. A large amount (15.7 Mt) is also used on high-value horticultural fruits and vegetables. Cotton (4 Mt) and oilseeds – including canola, oil palm, and soybean (6.3 Mt) – utilize significant amounts. A wide range of other crops and horticultural plants receive the remainder. Using data on the area of principal crops from Table 4.1, average application rates are 81, 100, and 107 kg N ha^{-1} for wheat, rice, and maize, respectively. In practice, application rates vary widely from zero to over 300 kg N ha^{-1}.

mineralization. By contrast, the old (H_o) pool is highly resistant to further decay and its half-life is typically measured in hundreds, or even thousands, of years.

Dinitrogen gas in the atmosphere is the central reservoir in exchanges of N among terrestrial and aquatic ecosystems, including oceans. Its entry to soils and crops occurs mainly through biological and industrial fixation to ammonia. Use of N fertilizer in crop production is shown in Box 8.1. The atmosphere also supplies nitrate in rainfall and ammonia that has volatilized from adjacent ecosystems. Inputs of nitrate and ammonia from the atmosphere occur by both wet and dry "deposition". Aquatic systems receive N in runoff and drainage from terrestrial systems. Sea waters are very low in N, however, and little is atomized by storms so the mineral flux from ocean to land is small; fluxes via harvest of fish by humans and birds (e.g., the guano of "Chilean nitrate") are also small. The principal avenue by which N returns to the atmosphere is through denitrification to N_2 from wet soils and aquatic systems.

Cycling of sulfur is similar to that of N in the sense that bacteria, SOM, changes in oxidation–reduction state, and gaseous phases are involved. Cycling patterns of other mineral elements are quite different, however, because they do not have gaseous forms and most have a variety of insoluble mineral forms. Potassium, for example, is subject to leaching in soils with low cation exchange capacity and to chemical and physical fixation in some clay minerals. Phosphorus cycles through organic matter and is sometimes present in organic forms such as phytic acid. In addition, phosphorus is fixed by clay and forms sparingly soluble compounds with iron, aluminum, and calcium. Oceans are the major sink for most nutrient elements and only modest amounts are recycled from there through atomization and rainfall. Without gaseous forms for recycling from oceans, agricultural systems are dependent for their main supply of most minerals on weathering of parent material and inputs of fertilizers. Principal sources for K, P, S, and Ca fertilizers, for example, are mineral deposits laid down in ancient lakes and oceans.

8.1.3   Microbial populations

Four major groups of organisms are involved in N processes: bacteria, fungi, cyanobacteria (blue–green algae), and algae. Each group is highly diverse. Cyanobacteria and algae are generally surface dwelling and autotrophic, developing their own C supplies through photosynthesis. Most bacteria and fungi are heterotrophic and obtain reduced C from organic matter. Some important exceptions are chemoautotrophic and obtain energy for C reduction by oxidizing iron, sulfur, or N.

In soils, the abundance and activity of various microorganisms change quickly as substrates and physical conditions change. Differences in species' growth rates influence patterns of succession in microbial communities cohabiting crop residues. Fast-growing microbes have reproductive rates as great as 0.5 d^{-1} while others grow very slowly. Populations are also affected by grazers such as protozoa. Paul and Clark (1989) reported average residence times for microbial biomass ranging from 0.24 y under sugarcane in Brazil to 6.8 y in a wheat–fallow rotation in Canada. Long residence times indicate that most microbes exist as resting spores or are quiescent.

Amounts of microbial biomass can be estimated by the **fumigation–incubation** method. Fresh samples of soil are fumigated with an agent such as chloroform to kill microbes and release their contents. Some organisms survive and metabolize released materials allowing CO_2 respired during a subsequent incubation period to serve as a measure of the biomass that was killed. Sizes of microbial pools found by such methods are surprisingly large: 1000 to 6000 kg dry material ha^{-1} (Jenkinson 1988). The large size indicates that supply of substrates must be a principal factor limiting microbial activity. Dry microbial biomass seems to have relatively constant proportions of C (0.45) and N (0.067) so, with a C/N ratio of microbial biomass near 6.7 (0.45/0.067), large amounts of N (67–420 kg N ha^{-1}) are held in this pool.

In addition to nutritional factors, soil microbiological processes are dependent upon temperature, water, and pH. Little activity occurs near 0°C but process rates increase with a Q_{10} around 2 as temperature increases to an optimal range near 40 to 50°C.

8.2 Decay and immobilization

Soil flora and fauna meet their energy and N requirements through decomposition of organic residues of higher plants. Because N supply plays a critical role in controlling plant growth, we must understand the processes governing N cycling. Here we examine the primary controls on decomposition, namely, decay and immobilization.

8.2.1 Decay

Carbohydrates and proteins are abundant in plant biomass and consequently in fresh organic matter also (Chapter 2). Proteins, lipids, and non-structural carbohydrates are excellent substrates for bacterial growth and they disappear quickly during decay while cellulose and hemicellulose decay more slowly. Other chemicals, including lignin, phenols, and higher-order aromatic and heterocyclic compounds, are metabolized slowly and thus increase in relative abundance as decay proceeds. Mineral elements such as K, Ca, Mg, P, and S are also released during decay. Bacteria are found throughout the succession of organisms that attack residues and they generally dominate during final stages. Fungi are important early during decomposition and with acid residues and soils; bacteria are more prominent at higher pH.

The disappearance of each organic substrate can usually be described by first-order kinetics, i.e., the rate declines in proportion to the amount of substrate remaining:

$$\mathrm{d}\chi/\mathrm{d}t = -k\chi \tag{8.1}$$

where χ is the amount of substrate and k is the rate constant for its disappearance. By integration, the amount remaining at time t can be calculated from the original supply, χ_o:

$$\chi_t = \chi_o e^{-kt} \tag{8.2}$$

Under optimal conditions in the laboratory, k is about 0.2 d^{-1} for proteins, 0.08 d^{-1} for cellulose and hemicellulose, and 0.01 d^{-1} for lignin (Paul & Clark 1989). The half-life of a material is $0.693/k$, under optimal conditions. Thus, half of the lignin in a crop residue would decay in 69 d ($0.693/0.01$ d^{-1}) and three-quarters would be gone in 138 d. Most N and a large portion of C in various substrates are transformed into microbial biomass. When substrates are exhausted, microbes sporulate or die. Materials released from dead bacteria can be used by other bacteria or higher plants whereas resistant spores may remain in soils for many years.

Under field conditions, where moisture and temperature fluctuate and generally are non-optimal, decay is slower and more variable than in the laboratory. Some of the difference in behavior is due to variation in size of residues: coarse residues decay more slowly than finely divided ones.

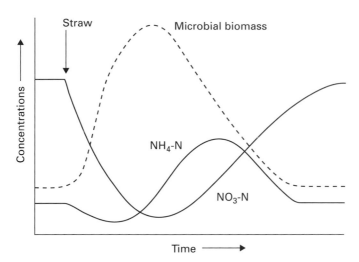

Fig. 8.3 Conceptual diagram to illustrate the time course over several weeks of immobilization of NH$_4$-N and NO$_3$-N by growth of a microbial population in response to addition of straw (an energy source of low N content) indicated by the downward arrow and then, as the straw is consumed and the microbial population decays, release of mineral NH$_4^+$ and NO$_3^-$.

8.2.2   Immobilization

Immobilization refers to the incorporation of inorganic nitrogen (NH$_4^+$, NO$_3^-$) into microbial biomass, and then more permanently into humus. Fresh residues are relatively rich in C compared with N, which gives a high **carbon-to-nitrogen ratio** (C/N = 25 to 100). Soil organisms utilize reduced C as a respiratory substrate while accumulating both C and N in proteins and the walls of their bodies. As a consequence, C/N ratios of residues decline as decay proceeds. In agricultural soils, C/N ratio of older fractions usually stabilizes between 10 and 13. In addition to N from crop residues, ammonium and nitrate ions from surrounding soil are also highly suitable substrates for microorganisms and these mineral forms are also immobilized to microbial biomass and humus.

Immobilization is particularly evident during decay of residues with low [N]. In that case, microbial growth and decay are N-limited and the concentration of free mineral N in the soil may be reduced to a very low level. Cereal straw with 0.5% N and 45% C has a C/N ratio of 90 whereas in the microbial pool, as noted above, it is near 7. Given rich sources of C, increase in microbial biomass depends on the use of mineral N from soil (including fertilizer N). Not only does this deplete the mineral pool, but continuing fluxes from mineralization are usurped also. The time-course of mineral pools during immobilization is characterized in Fig. 8.3. In aerated soils, some immobilization occurs with residues having C/N as low as 25 (i.e., 1.8% N in dry matter) but a ratio of 45 (1% N) or more is needed for a noticeable effect on supply of (available) inorganic N.

While the longer term processes of immobilization that lead to humus formation are critical to maintain SOM levels, short-term immobilization in microbial biomass can prevent adequate supply of available N for crop uptake. As much as 200 kg N ha^{-1} may become temporarily unavailable through incorporation of high-C residues. Therefore, efficient use of N inputs in crop residues and manures must anticipate the timing and amounts of mineralization and immobilization. One solution is to incorporate residues during fallow rather than just before sowing. This allows time for decomposition and has the advantage that carry-over mineral N is immobilized during decay and prevented from leaching. Another solution is to apply fertilizer N with residues, i.e., to feed decomposer populations. That capital investment in N carries the risk that some N may be lost and not recoverable from humus although the fraction of N applied and incorporated with residue that ends up in humus is relatively small. Fresh manures that have been collected on straw bedding typically cause immobilization during their initial decay in soil. That problem can be avoided by storing manure for sufficient time to allow decay to a lower C/N value before applying it to the field.

Immobilization problems are among the reasons why grass and cereal stubbles are sometimes burned or left as surface mulches (no-till systems) thereby reducing the amount of C entering soil. Burning is particularly common in Mediterranean climates where wet seasons and thus decay periods coincide with the growth of winter crops such as wheat and barley. Cycling of mineral nutrients is accelerated by burning and the amount of N lost is small (e.g., 0.005 kg N kg^{-1} straw $\times$ 3000 kg straw $=$ 15 kg N ha^{-1}) compared with savings in N fertilizer required to overcome immobilization. Other reasons for burning include disease control, especially in high-yielding cereal systems with a large amount of residue, avoidance of soil acidification, and reduction in energy and time required for tillage and seed bed preparation.

8.3   Mineralization and nitrification

Nitrogen in SOM and other organic forms is largely unavailable to higher plants. Mineral forms (NH_4^+ and NO_3^-) are the primary N source for uptake. Here we learn about the processes that transform SOM and organic N forms into mineral forms available for plant uptake.

8.3.1   Nitrogen mineralization

Release of NH_4^+ and NH_3 from organic materials is termed **ammonification**, or **nitrogen mineralization** (in contrast to release of other nutrients). Ammonification is accomplished by microbial populations during their attacks on dead bacteria, residues, and humus, and by cell-free, hydrolytic enzymes including proteinases, peptidases, and ureases. These enzymes are released from decaying plant material and bacteria and they may exist free in soil solutions or adsorbed to soil colloids.

Both extracellular enzymes and microbes are sensitive to pH, ion concentration, and temperature. In general, ammonification depends mainly on aerobic bacteria and is favored by neutral pH, moistness with good aeration, and adequate C substrate. Nitrogen fertilizer sometimes has a priming effect on ammonification through stimulation of bacterial growth. Tillage promotes aeration, disrupts soil aggregates exposing SOM previously inaccessible to microbial attack, and brings a portion of bacterial biomass to the surface where it is exposed to desiccation and other hazards. Death of those bacteria contributes to ammonification.

Rates of ammonification of humus are usually rather small in proportion to the amounts of organic matter present. In relative terms, 0.01 to 0.04% of total SOM may be mineralized annually in root zones of tilled temperate soils. Most NH_4^+ comes from residues (and organic amendments), microbial biomass, and active organic material (H_a). It must be stressed that mineralization varies not only with weather and tillage but also with amounts and types of residues and manures incorporated with soil. Another key point is that the ammonification rate is much slower than nitrification such that NO_3^- rather than NH_4^+ accumulates in soil when the supply of mineral N exceeds depletion by plant uptake or losses via other pathways.

8.3.2   Nitrification

The important process of oxidation of NH_4^+ to NO_3^-, termed **nitrification**, is mediated by chemoautotrophic bacteria of the family *Nitrobacteraceae* and by a range of heterotrophic organisms, which includes bacteria and fungi. Autotrophic nitrifiers use energy gained from oxidation of NH_4^+ (or NH_3) for fixation and reduction of bicarbonate to organic material. Nitrification takes place in two steps. First, bacteria such as *Nitrosomonas* oxidize NH_4^+ to nitrite (NO_2^-); second, NO_2^- is oxidized to NO_3^-, principally by *Nitrobacter* and *Nitrospira* (Norton 2008). Nitrous oxide (N_2O) is a minor by-product of that transformation when oxygen supply is limited and NO_2^- is used as a terminal electron acceptor. Heterotrophic nitrification, in contrast, oxidizes NH_4^+ or organic N using energy generated by the aerobic respiration of organic C sources. Nitrite is again an intermediate and many autotrophic nitrifiers also use some NO_2^- as an electron acceptor and so release N_2O (Wrage *et al.* 2001).

Oxidation of NH_4^+ to NO_2^- is usually the limiting step, and free NO_2^- seldom accumulates to high levels in soil, although exceptions occur at high pH. Nitrification is rapid in most soils used for agriculture but is strongly suppressed in many natural systems. Acid soils are a limiting factor under coniferous vegetation and in tropical Oxisols and Ultisols, but it seems that natural inhibitors of nitrification also exist. In contrast to NH_4^+, which is held by cation exchangers, both NO_2^- and NO_3^- are mobile in soil solutions. They are drawn toward plant roots by water uptake (in transpiration stream, Section 9.1) but equally can be lost by leaching as drainage water carries them below the root zone.

8.4   Loss of nitrogen

Nitrogen losses from agricultural systems are important determinants of N-use efficiency and economic benefit from N fertilizer application. They also have significant potential for detrimental impacts on the environment. Decaying organic matter, especially animal excreta, releases ammonia and other volatiles that carry unpleasant odors. Eutrophication of rivers, lakes, and estuaries occurs from NO_3^- accumulation that promotes unwanted plant growth (e.g., algal blooms) that reduces $[O_2]$ and can also be toxic to animals and humans. Finally, N_2O is a potent greenhouse gas (GHG) that has a climate warming potential 280 times that of CO_2, and also, by reaction with ozone (O_3) it reduces the ultraviolet filtering capacity of the stratosphere.

Here we deal with several processes that contribute to losses of N from agricultural systems, including denitrification, runoff, leaching, and volatilization.

8.4.1   Source materials

The popular belief that most losses of N, and pollution, occur only with fertilizer and that organic sources do not contribute to losses is not correct. Significant amounts of N are in fact lost directly and indirectly from organic materials. Once SOM and crop residues are mineralized to NH_4^+ and NO_3^-, these highly soluble mineral N molecules are subject to the same loss mechanisms as NH_4^+ and NO_3^- derived from commercial fertilizer. In cases where organic materials enter surface waters in runoff, mineralization and gasification of the N content are simply transferred to streams and lakes. As was illustrated in Fig. 8.2, the principal losses occur through leaching or denitrification from the pool of NO_3^- -N, which is supplied as much or more by mineralization of SOM as by fertilizer. Applied properly, fertilizer N enters quickly into organic forms through uptake by plants and microbial immobilization. When residues of those crops and microbial biomass cycle to older fractions of SOM, the original source of N becomes indistinguishable from other sources including legume fixation and manure. Whereas fertilizer applications can be timed and placed for efficient use by plants, mineralization of organic matter occurs whenever conditions are favorable, whether a crop is present or not. Because organic farming systems require high levels of organic inputs and high rates of mineralization to meet crop needs, they are also vulnerable to nitrate losses.

8.4.2   Denitrification

The pathway of this process, which returns NO_3^- to the atmosphere, follows the sequence $NO_3^- \rightarrow NO_2^- \rightarrow NO \rightarrow N_2O \rightarrow N_2$. **Denitrification** is carried out by various heterotrophic bacteria, including members of the genera *Pseudomonas*, *Bacillus*, *Thiobacillus*, and *Propionibacterium* (Snyder *et al.* 2009). The principal end product is N_2 but, as for nitrification also, with the possibility for leakage of N_2O. Bacteria gain energy from metabolism of C substrates using NO_3^- rather than oxygen as an electron acceptor for respiratory activity under **anaerobic** conditions. Given an abundant supply of nitrate, denitrification can remove significant quantities of N from soils that might otherwise have been used by crop plants or leached from the profile. Denitrification also increases soil pH because 1 mol H^+ is used per mol N lost.

Heavy soils with poor drainage, as well as water tables found within most soils, offer favorable anaerobic environments for denitrification. Transformations also occur in well-drained soils during brief periods of saturation and at anaerobic microsites in interior spaces of aggregates. In addition to anaerobic conditions, supplies of NO_3^- and C substrates are needed. For this reason, diurnal and seasonal patterns of dentrification are strongly episodic. Peak emissions correspond with warm temperatures, rainfall events or irrigation, and addition of fertilizer (organic or inorganic) or presence of grazing animals. Rolston *et al.* (1978) observed peak rates of loss of 70 kg N ha^{-1} d^{-1} from a wet soil heavily supplied with nitrate fertilizer (300 kg N ha^{-1}) and manure (34 t ha^{-1}). Total loss during the season was 198 kg N ha^{-1}, but this represents a case of excessive N inputs. With proper N management, NO$_3^-$ supplies are seldom large enough to support high rates of denitrification for more than a few hours and observations of seasonal rates in crops are much smaller. In fact, crops compete effectively for available NO$_3^-$ and can remove up to 10 kg ha^{-1} d^{-1}; their transpiration dries the soil and thereby improves aeration, which further reduces denitrification. Based on a meta-analysis of denitrification rates measured in a wide range of agricultural systems, IPCC-AR4 (2007) estimates average field-level denitrification at 1.0% of N inputs from both organic and inorganic sources with another 0.8% lost via "downstream" denitrification and volatilization of N in leachate and runoff. Even in high-yield cropping systems it is possible to keep denitrification losses below 1% of applied N when fertilizer is applied progressively during growth to match uptake capacity (Adviento-Borbe *et al.* 2007).

Special conditions favoring denitrification develop with flooded rice. A thin surface layer of soil is aerated by dissolved O$_2$ such that mineralization and nitrification are favored in that region. Most of the soil profile is anaerobic, however, and nitrate that leaches into the anaerobic zone is denitrified. The problem of supplying rice crops with N is resolved by placing ammonium fertilizer in the reducing zone where nitrification is inhibited, and by applying multiple small applications at key growth stages to avoid high [NH$_4^+$] in the soil–floodwater system.

8.4.3   Leaching and runoff

Ammonium ions are strongly absorbed by soil colloids with large CEC, which keeps [NH$_4^+$] relatively low in soil solution and thus reduces potential for loss by leaching. In contrast, anion-exchange capacity of soils is weak and nitrate moves rather freely with water. Downward movement of NO$_3^-$ is proportional to the downward flux of water. Conditions most conducive for NO$_3^-$ loss are thus rainy periods, permeable soils, and an abundant supply of NO$_3^-$. Loss occurs once NO$_3^-$ moves below the root zone or into an anaerobic zone and is denitrified. Nitrogen lost via leaching can become a pollutant of surface and ground waters. In addition, NO$_3^-$ leaching is important in soil acidification (Section 7.4.4). A long fallow period and tillage establish conditions favorable for mineralization but leaching is prevented if crop residues with C/N ratio >50 are incorporated by temporarily immobilizing mineral N.

Where winters are mild and rainfall adequate, **cover crops** (e.g., ryegrass) can be grown between main crops to catch the nitrification flux. That lessens the potential for NO$_3^-$ pollution but adds costs (additional tillage and seed), a new problem with residue incorporation, and uncertainty about mineral flux during the subsequent cropping period.

Less leaching occurs from natural vegetation than from pastures and cultivated fields. In pastures, dung and urine patches create locally high concentrations of N that contribute to nitrate leaching (Section 7.4.4). Steele and Vallis (1988) report examples of unfertilized ryegrass–clover pastures in New Zealand that lost over 100 kg N ha^{-1} y^{-1} through leaching from dung and urine.

It is convenient in some systems to apply **ammoniacal fertilizer** during fallows when labor is generally more available than during busy periods of land preparation and planting. Fallow applications work well in temperate climates where low temperature limits microbial activity during winter, but with mild conditions N is exposed for a long time during which nitrification, denitrification, and leaching can occur. Problems of that sort have prompted research towards controlling the rate of nitrification of ammoniacal fertilizers. Chemicals such as nitrapyrin, dicyandiamide, and dimethypirazole phospate slow ammonia oxidation and find some use in agriculture as fertilizer amendments despite their expense. Some N fertilizers (e.g., ureaformaldehyde compounds) are themselves inhibitors of nitrification and thus act as slow-release materials. Crop residues with C/N ratio >50 and manures are also slow to release N through decay processes.

Slow release is generally an undesirable trait for materials that are applied to crops during the growing season if they are expected to overcome a deficient N supply. In contrast, **controlled-release** N fertilizers use a thin porous plastic membrane around each fertilizer particle to physically restrict the diffusion of mineral N from particle to soil solution. Differences in pore size and film thickness determine the rate of release, which can be adjusted to match the duration of active crop growth and N demand. While effective in achieving greater congruence between N supply and crop demand, controlled-release fertilizers are more expensive and, to date, have only found widespread commercial use in high-value horticulture and nursery crops.

Nitrate pollution of ground and surface waters gives rise to two environmental concerns. One is that N can promote the growth of algae in surface waters. The second is that nitrite, which may be produced from nitrate by bacteria under anaerobic conditions in the digestive tracts of animals, can cause a toxic blood disorder (methemoglobinemia). In dry regions, livestock may encounter nitrite toxicity through drinking from surface waters with high concentrations of nitrate (500 ppm NO$_3$-N; 36 mM). In human populations, methemoglobinemia occurs mainly in infants and is very rare. The US Public Health Service has set a standard of 10 ppm NO$_3$-N (0.7 mM) as the maximum safe level for drinking water. The standard is conservative, being set well below the level at which health problems have sometimes been observed (Powlson *et al.* 2008).

8.4.4   Volatilization

Ammonia is a volatile gas and can be lost to the atmosphere from aqueous solutions. In water:

$$NH_3 + H^+ + OH^- \leftrightarrow NH_4^+OH^- \tag{8.3}$$

Equilibrium is strongly dependent upon the buffer pH of the soil solution. At pH 5 and below, about 0.004% of N is present as free NH$_3$ but that fraction increases approximately ten-fold with each unit increase in pH so that nearly 40% is volatile at pH 9 (Nelson 1982). As a result, gaseous losses of NH$_3$ are significant only with dry or calcareous soils and where NH_4^+ is abundant at the surface. In tilled soils, and with dispersed NH_4^+, nitrification rapidly depletes the vulnerable supply of NH_4^+ -N. Volatilization losses from ammoniacal fertilizers are restricted by incorporation with soil.

Animal manures are neutral to slightly alkaline and lose NH_3 easily through volatilization. Cattle excrete 0.1 to 0.3 kg N animal^{-1} d^{-1} of which about 50% is NH_4^+ or urea, mostly in urine. Urea is hydrolyzed readily to NH_4^+ and on drying dissipates as NH_3. Efflux of NH_3 to the atmosphere can amount to 10 kg N ha^{-1} y^{-1} from ordinary pastures and up to 45 kg N ha^{-1} y^{-1} from fertilized pastures subject to heavy grazing (Vertregt & Rutgers 1988). Animal feedlots, if kept dry, lose considerable NH_3 by volatilization; when wet, losses of N are smaller and occur principally by denitrification. Ammonia and volatile amines are also lost directly from the leaves of some plants. Ammonia losses are linked with the process of photorespiration (Section 10.1.2). The magnitude of such fluxes typically represent from 1 to 4% y^{-1} of total aboveground shoot N (Sommer *et al.* 2004).

An important property of gaseous NH_3 is the rapid absorption by any wet system including moist soils, interior surfaces of leaves, and surface waters. The negative surface charge of soils and plant materials favors adsorption. Half-life of NH_3 in the atmosphere is thus short and what is lost from one system is soon gained by another. Annual influxes of 5 to 10 kg NH_3-N ha^{-1} are not uncommon (Jenkinson 1982) and in regions with intensive dairying (e.g., the Netherlands) they can be much greater.

8.5   Assimilation of mineral nitrogen by plants

Higher plants absorb and use both NH_4^+ and NO_3^- forms of N from soils. Nitrate is the dominant form in crop nutrition because it is both more abundant (in tilled soil) and more mobile than NH_4^+. While NH_4^+ is assimilated directly into amino acids and other organic forms, NO_3^- (+5) must first be reduced to the ammonium level. That eight-electron change requires respiration of at least 0.33 mol of a glucose-level substrate mol^{-1} N (67 MJ kg^{-1} N based on 15.6 MJ kg^{-1} as ΔH_C of glucose at 20°C). Although most plant tissues are capable of nitrate reduction, it takes place mainly in the chloroplasts of leaves through direct transfer of photosynthetic reductant without intervening steps of carbohydrate synthesis and respiration. Photoreduction of nitrate does not appear to compete with CO_2 fixation (Bloom *et al.* 1989) bringing the real cost to considerably less than 67 MJ kg^{-1} N.

Despite the larger energy expenditure involved, most plants grow better with NO_3^- than with NH_4^+. Ammonium-fed plants seem to encounter difficulty with control of internal pH because H^+ remains when $NH_4^+ \rightarrow$ amino compounds (R–NH_2). Plants do not have good systems for buffering or disposing of H^+. In contrast, plants encounter surplus negative charges when nitrate is reduced ($NO_3^- \rightarrow R - NH_2 + e^-$). Those charges are easily transferred to organic acids (Fig. 7.8), which have excellent pH-buffering properties with K^+ and find service as osmotica in vacuoles.

8.6   Nitrogen fixation

After N fertilizer (100 Mt N y^{-1}), N inputs from **biological nitrogen fixation** (*c.* 50 Mt N y^{-1}) represent the greatest source of N input to agricultural systems. A wide range of bacteria (diazotrophs) is able to fix N_2. It includes **heterotrophic** fixation by bacteria that live loosely in the rhizosphere around roots and feed upon plant exudates, endophytes that live within leaves and stems of plants and fix N in an association that is clearly beneficial to both, and more complex symbiotic associations within the root nodules of legumes. First, we examine some common issues about N fixation systems.

8.6.1   Fixation systems

Reductive fixation of N_2 to the NH_3 level is the principal route by which new N enters agricultural systems. The process is accomplished biologically by bacteria and also industrially. The process is energetically expensive: it requires three electrons for reduction (in terms of glucose, 0.125 mol glucose mol^{-1} N; 25 MJ kg^{-1} N) plus significant energy for activation of the N. Symbiotic systems also encounter costs of constructing and maintaining the microorganisms. Legumes support bacteria in root nodules; theoretically, their combined cost for bacteria, nodules, and N fixation is at least 0.7 mol glucose mol^{-1} N (140 MJ kg^{-1} N). Measured costs are significantly larger, however, ranging from 1.2 to 1.3 mol glucose mol^{-1} N fixed (240–260 MJ kg^{-1} N) (Ryle *et al.* 1979; Schubert 1982). That is several times the cost of nitrate reduction by non-legumes.

A modern fertilizer plant employing the Haber–Bosch process for NH_3 production from N_2 also uses much less energy: about 45 MJ kg^{-1} N. In that process, H_2 is reacted with N_2 over a catalyst under heat and pressure to yield NH_3. Hydrogen gas is generated through the partial combustion of natural gas (or other fossil fuel) to H_2 and CO; CO is then reacted with water to yield additional H_2 and CO_2 as a by-product. Hydrogen, for the process, can also be produced by the electrolysis of water using various power sources.

Biological N-fixation systems employ the enzyme **nitrogenase**. It is composed of two Fe–S proteins, one of which carries an Mo–Fe center. In the absence of molybdenum, some microorganisms generate an alternative protein without Mo but, for practical purposes, Mo is considered an essential micronutrient for N fixation. In addition to N_2, nitrogenase can reduce H^+ to H_2 and ethyne to ethene. Wasteful H_2 evolution, with a minimum ratio of one per N_2, contributes to the high cost of fixation observed for legumes. Because ethyne is not common in soil and ethene concentration is easily measured with a gas chromatograph, ethyne reduction is used as an assay for the potential rate of nitrogenase activity. Nitrogenase is inactivated by O_2 and fixation occurs only under anaerobic conditions, or in organisms with protective systems against oxygen. Another common property of these organisms is that the ability to fix N_2 is repressed in the presence of nitrate or ammonium ions and they then live on mineral N rather than expending energy and reductant in fixation.

Table 8.1 Types and performance of N_2 fixing systems in agriculture (adapted from Unkovich *et al.* 2008)

N_2-fixing system	Range measured	Range commonly observed
	(kg N ha^{-1} y^{-1})	
Free living (soil and water)		
Heterotrophic bacteria	1–39	<5
Cyanobacteria	10–80	10–30
Associative – endophytes		
Tropical grasses	0–45	10–20
Sugarcane	0–240	5–65
Azolla – cyanobacteria	10–150	10–50
Symbiotic (rhizobia)		
Green manure legumes	5–325	50–150
Pasture & forage legumes	1–680	50–250
Crop legumes	0–450	30–150
Trees & shrubs	5–470	100–200

Three N-fixing systems that are important in agriculture are listed, together with examples of performance, in Table 8.1. The first column lists observed data, not all of which, in view of the complexity of the measurements, are authenticated. The second column presents the best current estimates of the N contribution of each N_2-fixing system to agricultural production systems.

8.6.2   Free-living organisms

Ability for N fixation is found in a range of free-living bacteria including, cyanobacteria (blue–green algae) and actinobacteria. Cyanobacteria are autotrophic for C whereas others are heterotrophic. Heterotrophic forms are able to exploit environments low in N and rich in reduced C and are common in decay processes. Free-living bacteria include *Clostridium*, *Azospirillum* and *Azotobacter*. These are mostly slow-growing and with few exceptions, the principal one being *Azotobacter*, require anaerobic or near-anaerobic conditions for fixation. Low-N residues of cereals would seem to be a good substrate for these organisms but their slow growth and aeration introduced by tillage disrupt performance.

Estimates for annual rates of fixation by free-living organisms in agricultural soils are small, in the range of 1 to 5 kg N ha^{-1} y^{-1} (Table 8.1). Some grasses develop a carbohydrate-based gelatinous sheath external to their roots and there is evidence that free-living bacteria may form loose associations within the rhizosphere of some grasses (**associative N fixation**). Cyanobacteria are found in various agricultural situations. In humid areas of Europe, for example, cyanobacteria may occur at the soil surface under cereals; rates of fixation are similar, 1 to 5 kg N ha^{-1} y^{-1}, to those with free-living bacteria. Nitrogen fixation is more prominent in irrigated lowland rice systems where cyanobacteria contribute significantly to the N economy of flooded soils. Nitrogen inputs of 30 to 50 kg N ha^{-1} from heterotrophic and associative N fixation are common in the continuous rice systems of Asia (Cassman *et al.* 1998). Indigenous N inputs of this magnitude explain why irrigated lowland rice systems have sustained relatively high grain yields for thousands of years before the advent of modern fertilizers.

8.6.3   Associative fixation

Tropical C4 grasses (see Section 10.1) are known to contain endophytic bacteria and some cultivars of sugarcane, the most important crop of the group, gain significant N-nutrition benefit from them. These are conclusions of agronomic experiments of crop N balance in Brazil where N applications to sugarcane are less than is usual elsewhere and do not account for observed uptake and extractions by crops. It is suggested that associative N fixation can account for up to 70% of crop N requirement, consistent with commonly recorded contributions of associative fixation in sugarcane of 5 to 65 kg N ha^{-1} y^{-1} (Table 8.1). The endophyte concerned has been identified as *Gluconaceto-bacter* (previously *Acetobacter*) *diazatrophicus* that enters plants through wounds, or with vesicular–arbuscular mycorrhiza, and is then transmitted widely in stem pieces used as planting material. *Gluconacetobacter* resides in stem cavities and has unusual physiological properties for a diazotroph, including tolerance to low pH, high sugar and salt concentrations, and nitrogenase that tolerates short-term exposure to ammonium (Boddey *et al.* 1991; Dong *et al.* 1994).

Cyanobacteria enter into associations with other organisms. Lichens, for example, are fungal–cyanobacterial associations. The cyanobacterium *Anabaena azollae* is important in agriculture. It occupies cavities in the water fern, *Azolla*, that is sometimes employed as an N source for rice culture in Asia, providing up to 50 kg ha^{-1} y^{-1} (Table 8.1).

8.6.4   Symbiosis with legumes

There are many natural symbiotic systems. Of these, facultative associations involving members of the family *Leguminoseae* and **rhizobia** bacteria are the most important agriculturally. For convenience, the name rhizobia is used to refer to 13 genera that have now been distinguished, mostly since the 1990s, from the original species *Rhizobium leguminosarum* identified in 1889. The most populous genera are *Azorrorhizobium* with 25 species, *Rhizobium* with 22, *Mesorhizobium* and *Enisfer* (previously *Sinorhizobium*) both with 15, and *Bradyrhizobium* with five. The genera *Rhizobium* and *Bradyrhizobium* are most prominent in agricultural studies. The associations they form with legumes are facultative since rhizobia survive in soils as free-living bacteria and legumes flourish on mineral N.

The symbiotic phase can be recognized from nodules that form on the fine roots of host plants and by the superior performance of those plants in N-poor environments. Individual species of rhizobia are distinguished by their association with specific hosts as outlined in Table 8.2. Within these groups, bacterial strains can be recognized that differ widely in their effectiveness at nodulation and/or N fixation. It is important to have

Table 8.2 Inoculation groups of legume–rhizobia symbioses

Group	Species of rhizobia	Host genera	Crops included
Alfalfa	*Rhizobium meliloti*	*Medicago, Melilotus, Trigonella*	Alfalfa, sweet clovers, fenugreek
Clover	*R. leguminosarum* (biovar. *trifolii*)	*Trifolium*	Clovers
Pea & vetch	*R. leguminosarum* (biovar. *vicia*)	*Pisum, Vicia, Lens*	Peas, vetch, fababean, lentil
Bean	*R. leguminosarum* (biovar. *phaseoli*)	*Phaseolus*	Common bean (various)
Leucaena	*Rhizobium* spp.	*Leucaena, Sesbania, Calliandra*	Leucaenas and various other tree and shrub legumes
Chickpea	*Bradyrhizobium japonicum*	*Cicer*	Chickpea
Soybean	*B. japonicum*	*Glycine*	Soybean
Cowpea	*Bradyrhizobium* spp.	*Lupinus, Arachis Vigna, Cajanus, Phaseolus, Psophocarpus*	Lupin, peanut, cowpea, pidgeon pea, lima bean, winged bean, various tropical pasture legumes

Adapted from various sources.

crop plants inoculated by effective strains of the appropriate group. That can be achieved by coating seed with inoculum produced from cultured bacteria, but the efficacy of the method is highly uncertain. Where land has not seen legumes, inoculation can be very effective. With periodic culture of legumes in rotation, however, rhizobia populations increase and strains that have lost effectiveness in fixation can be highly competitive with inoculated bacteria. No practical solution to that problem has been found; inoculation is no longer practiced in many areas of Europe and North America.

8.6.5   Factors affecting legume performance

Amounts of N fixed annually by legume crops are variable. Total N assimilation and thus fixation are proportional to biomass production by host crops and therefore influenced by any factor that causes variation in host performance. In addition to weather and mineral nutrition, rates are affected by legume species, stand density, degree of nodulation by effective strains, and supply of mineral N from soil. Biomass production and N fixation by most legumes are severely reduced by low soil pH and by low supplies of P, Ca, and K. Phosphorus fertilization is common practice for the production of legume forages and soybean (Box 17.2) on many soils and lime is necessary for acid soils. Extensive areas of Australia need both Mo and P fertilizers for legume symbiosis (as well as acid tolerant cultivars; Chapter 16). In California, P, S, and Zn deficiencies are found in the legume component of annual grasslands.

The corollary of good supplies of P, Ca, and K for legumes is that uptake and export of these nutrients is also relatively large. Examples of nutrient content of legumes and grains

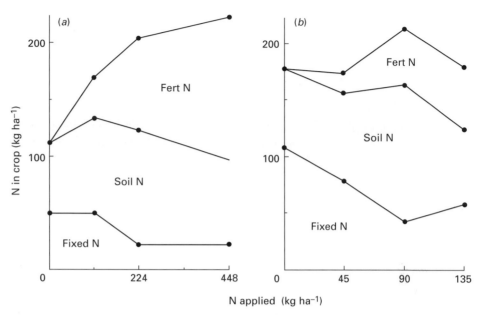

Fig. 8.4 Amounts of nitrogen in mature soybean grain acquired from fertilizer, soil, and fixation. Parts (a) and (b) are for soils of high and low N fertility, respectively. (Redrawn from Johnson *et al.* 1975; data from Diebert *et al.* 1979.)

are given in Table 12.1. An alfalfa crop yielding 20 t hay ha^{-1} each year, for example, carries with it 254 kg Ca, 308 kg K, and 56 kg P. Those removals, coupled with balancing organic anions, contribute significantly to soil acidification (Section 7.4). By contrast, good crops of maize or wheat grain remove somewhat less P per hectare (30–40 kg) and much less Ca (3–4 kg) and K (40 kg), and their production is less sensitive to pH and P supply. Likewise, grain legumes such as soybean, common bean, and cowpea have two to three times higher [P] and [K] in their seed than cereal grains. Thus removal of P and K with harvest of grain legumes is much greater than by cereals.

The facultative nature of rhizobial fixation is illustrated in Fig. 8.4 with data for soybean. Fixation is repressed in the presence of mineral N, as seen in Fig. 8a relative to Fig. 8b. On fertile soils, less than half of the N removed with legume crops may be obtained from rhizobia. Piha and Munns (1987) found that nodules were sloughed quickly from legume roots following large additions of mineral N to sand–vermiculite cultures. Under field conditions, where supplies of mineral N are not so large, nodules may simply become inactive. Several mechanisms may be involved. Carbon supplies of plants well supplied with mineral N may be monopolized in rapid growth, leaving less for nodule formation and fixation (see **functional equilibrium**, Section 11.4). While that seems to be a principal mechanism, there is also evidence that accumulation of ammonium ions and/or amino acids within plants can repress fixation. Most legumes seem to reach higher yields with mineral N than through symbiosis, owing, presumably, to smaller energetic costs of growth on nitrate. Fixation by free-living microbes is also repressed by mineral N.

Flexibility of legumes in their use of mineral and rhizobial N is viewed as a vexing problem by some but in fact it allows legumes to serve an important scavenger role in agriculture. Mineralization of SOM continues while legumes grow and, if unused, that available N might be lost through leaching or denitrification.

8.6.6   Estimates of nitrogen fixation by legumes

There is a range of methods, some ingenious but none entirely satisfactory, for determining the amounts of N fixed by legume crops. Unkovich *et al.* (2008) provide the most recent description and comparison of methods and explain their limitations. Measurements of changes in soil N content are conceptually simple but experimentally difficult because even a large addition of N is very small relative to total N content of most soils. Changes can be detected only in long-term experiments.

Nitrogen accumulation in legume biomass is a poor index of apparent fixation because the proportion acquired from soil is unknown. Legumes always acquire some mineral N and they are completely dependent on it during nodule formation on new seedlings. It is tempting to conclude that nearly all N accumulated in biomass during later years of an alfalfa sequence is fixed, but alfalfa is so deeply rooted that it can scavenge mineral N from deep in a profile, or from a water table, for many years. Comparisons of N accumulation between legume and non-legume crops and between nodulated and non-nodulated legumes also encounter problems because crops may differ in rooting habits, total N requirements, and seasonal patterns of demand.

Some methods for distinguishing the source of legume N rely on differences in ratios of natural isotopes of N in air and soil. Denitrifying organisms discriminate against ^{15}N and so soils and mineral N taken up by plants are slightly enriched in that isotope relative to ^{14}N. A problem with this method is that denitrification, and thus the isotopic ratio, varies with soil depth. Addition of **^{15}N-labeled fertilizer** (either depleted or enriched in the isotope) can be helpful in establishing what portion of legume N came from mineral sources, providing its distribution within a soil profile can be matched with the distributions of mineralization and root activity. But that is very difficult to achieve. Ethyne reduction assays are also used because they are simple and rapid. Harvested roots are placed in an ethyne atmosphere for a few minutes and while loss of ethyne says something about potential rates of fixation (independent of internal N status) it provides no quantitative estimate of actual rates. Actual rates are much smaller than ethyne values in part because fixation varies diurnally and seasonally in response to the host's status for C and N.

LaRue and Patterson (1981) and Herridge and Bergersen (1988) provide tables showing estimates of N fixation reported for a number of legume crops (as in Table 8.1). The values range from 10 to over 300 kg N ha^{-1} y^{-1}. From Herridge and Bergersen's review, it seems that 100 to 200 kg N ha^{-1} y^{-1} is a reasonable average for soybean under good conditions. Long-season forages subjected to repeated cutting tend to have the largest rates. Record annual yields of alfalfa (30 t dry matter ha^{-1} with 2.5% N) contain as much as 750 kg N ha^{-1} and it seems that fixation rates of the order of 500 kg N ha^{-1} y^{-1} are possible.

Estimates of N fixation for the world as a whole are also crude. Annual input of fixed N to agricultural lands, including cropland, pasture, and grassland savannas used for grazing, is perhaps 60 Mt, comprising about 29% of global N input of

---

Box 8.2   Global N inputs to agricultural land

Sources of N input to agricultural land include biological N fixation, fertilizer, **wet and dry deposition**, and livestock manures. Among these, annual fertilizer input of 100 Mt is most certain because it can be tracked by national production and trade statistics (Heffer 2009). Input from livestock manure is less certain but reasonable estimates can be obtained from national statistics on livestock production, manure production rates per animal, and assumptions about N losses from manure during storage. Using this approach Sheldrick *et al.* (2003) estimate N inputs from livestock manure at 34 Mt. Estimates of N fixation and deposition inputs are also uncertain. Herridge *et al.* (2008) estimate N from biological fixation at 50 to 70 Mt annually based on a global tally of all N-fixing systems in Table 8.1. About 19 Mt of N falls on agricultural land as wet or dry deposition based on estimates of global N deposition to terrestrial ecosystems of 64 Mt (Galloway *et al.* 2004). At about 30% of global total, this is greater than the 13% of total land area occupied by agriculture because most atmospheric N is NH_3, which does not travel long distances. Because farm land and livestock facilities are themselves a large source of atmospheric N, it is likely a large portion of atmospheric N from agriculture falls back on agricultural land.

The total annual N input to agricultural land is thus estimated at 213 Mt and the components are as follows:

Source	Amount (Mt N y^{-1})	Percent of total
Fertilizer	100	47
Biological N-fixation	60	28
Livestock manure	34	16
Deposition	19	9
Total	213	100

213 Mt (see Box 8.2). This corresponds to about 31 kg N capita^{-1} for the present world population compared with 2.9 kg N cap^{-1} y^{-1} as the *minimum* dictary requirement (50 g protein cap^{-1} d^{-1}). If we assume that foods actually supply 5 kg N cap^{-1} y^{-1} the apparent efficiency of agriculture is 5/31 or about 16%. Most N consumed by humans is dissipated in sewage whereas animal manures and crop residues are mostly recycled within farming systems.

8.7   Example nitrogen cycles

The magnitude of various fluxes in N cycles varies enormously among agricultural systems (Frissel 1977). To gain an appreciation of when small fluxes are important, and when they are not, and which pools are involved, we will examine agricultural systems differing in intensity of farming. An N budget for a farm in medieval Europe provides the first example and is followed by others from transition and modern systems.

8.7.1 A medieval farm

An English farm of the fourteenth century is the basis for construction of this N budget (Loomis 1978). Slicher van Bath (1963) and others provide historical data, drawn from meticulous records of abbey estates, on farming methods and yields.

This imaginary farm was operated in communal fashion by peasants under the organization of a manor. Land was abundant relative to population but farming was not easy. The farm employed an "open-field" (unfenced) system with wheat rotated in alternate years with fallow, and with pea substituted on occasion for wheat. Fallow served two basic functions: it provided weed control and it improved N supply. By cropping only in alternate years, two years of N inputs from rainfall, free-living bacteria, and leguminous weeds were accumulated for each crop of wheat. In addition, residues from previous crops were allowed time for decomposition thus avoiding complications due to immobilization. Wheat yields were low, around 1000 kg ha^{-1}. Nitrogen was probably the most limiting nutrient.

The farm also carried livestock: bullocks for tillage and haulage; other cattle for meat and cheese; and sheep for wool. Meadows and woodlands were grazed by cattle and sheep, while swine were run in the woodlands. The mild climate of England allowed some grazing through winter but the supply of hay (saved grass) was small, limiting the number of cattle and sheep that could be kept. A critical feature of this system is that animals were penned at night for protection. That resulted in accumulations of manure containing N gathered from meadows: manure that could then be spread on arable lands. With a ratio of 4 ha meadow:1 ha arable, access was gained to a significant additional supply of N. Meadows and woodlands were sustained by natural inputs of N.

Management of the farm illustrates two different applications of the non-uniformity principle for concentration of scarce soil resources. The crop–fallow sequence represents non-uniform treatment of land in time, while grazing of meadows and woodlands with manure transfer to arable land involves non-uniform use of space. Although not highly efficient in accumulation and retention of N, crop–fallow rotations still serve to supply N in some farming systems.

Construction of the **nitrogen cycle** displayed in Fig. 8.5 begins with the wheat crop. With 0.02 kg N kg^{-1} grain, a harvest index of 0.33, and 0.005 kg N kg^{-1} straw, the mature wheat crop contained 20 kg N in grain plus 10 kg N in straw for a total of 30 kg N ha^{-1}. If we assume that half of the straw was harvested for feed and bedding, then 25 kg was removed from the land and 5 kg (2.5 kg N ha^{-1} y^{-1}) was returned to soil. Other inputs to soil (free-living fixation, wet and dry deposition, and a small credit for an occasional pea crop) are estimated to total 8 kg N ha^{-1} y^{-1}. Reproductive rates for medieval farming were very small: only four times more grain was generally harvested than sown, so N input of 2.5 kg N ha^{-1} y^{-1} (5 kg N ha^{-1} crop^{-1}) from seed was significant.

To estimate the amount of N supplied from manure, we used information from Slicher van Bath (1963) that a minimum of one bullock was needed for each 4 ha of arable land,

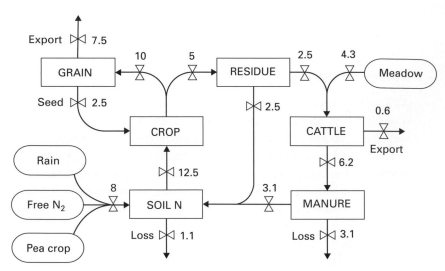

Fig. 8.5 Nitrogen transfers in a medieval farm. Major pools are identified in rectangles, transfers (solid lines with gate valves and arrows) carry numbers indicating the annual flux in kg N ha^{-1} y^{-1} (multiply by 2 to obtain the flux for the full fallow–crop cycle). (Data from Loomis 1978.)

and from Azevedo and Stout (1974) that 1000 kg dry manure with 25 kg N could be collected annually from each bullock. If manure from sheep and cows matched that of bullocks, 12.5 kg N ha^{-1} was available in manure. Only 2.5 kg of that N originated from arable land; the balance was acquired from grazed areas. If 25% of manure N was conserved, the estimated flux to arable land was 3.1 kg N ha^{-1}.

Total N input (2.5 seed + 2.5 residue + 3.1 manure + 8 fixation and rain = 16.1 kg N ha^{-1} y^{-1}) was 32.2 kg N per wheat crop compared with 30 kg N uptake by wheat. Inputs and outputs of N by dust, pollen, insects, and birds are assumed to balance, leaving 2.2 kg N to cover losses by leaching, runoff, and denitrification in the field. Leguminous weeds may have had some input; that plus the possibility of a larger supply of manure allow for possibility of larger losses.

Several basic points about farming with limited N emerge from this example. First, very small fluxes of N are significant in poverty systems. Second, yields are very low in relation to land, labor, sunshine, water, and other inputs. Third, annual fluxes of N and other nutrients from manure, rainfall, and weathering would probably have been adequate to sustain yields from this low-input system indefinitely unless soil erosion or acidification were severe. Fourth, about 20% of N input to cropland came by transfer of manure from grazing land, and the system required four acres of grazing land for every acre of crop. In contrast, a modern US maize crop produces ten times more grain yield and has a total N supply 14 times larger (Table 17.6) than wheat production in this medieval farming system.

8.7.2 Transitions

Present systems of agriculture found in most of Europe and America evolved from such medieval beginnings. Major changes were effected in the fifteenth and sixteenth centuries by Flemish and Dutch farmers. Released from feudal control, those farmers owned their own plots of land. They enclosed their fields and developed fodder systems (particularly with beet) that allowed increases in animal numbers and supplies of manure. They also rediscovered Roman practices involving legume rotations and lime. Changes in British farming that followed during 1650 to 1850 brought a revolution of "new farming" with many improved practices, new breeds of livestock, and new crop cultivars. They also brought experimental studies in plant and animal nutrition.

Further evolution between 1850 and 1940 brought mechanization, plant breeding, and gradually increasing use of fertilizers. Organic farming with proper nutrition of legumes (P and K fertilizers) and better management of manure produced much larger yields than were possible in medieval systems. By 1940, a high state of organic farming had brought average yield of wheat in Britain to near 2500 kg ha^{-1} but yields and N cycling in much of the world still approximated the medieval example. Famine was still a problem in Asia, and in the USA the dietary protein supply per capita had declined steadily as the population increased.

Much has changed since the 1940s. Nitrogen cycling still follows the same pathways and still includes legume rotations and improved methods for recycling residues and manures. Nitrogen is now also supplied in mineral fertilizers (Smil 2001) and N deficiency no longer need limit yield. The necessary external inputs of N are much greater than could be supplied by animals through manure and large amounts of N exported in harvested crops (100–200 kg N ha^{-1} are common). British wheat yields now exceed 7000 kg ha^{-1}. American maize yields, which in 1940 were static near 2000 kg ha^{-1} with organic methods, now exceed 9000 kg ha^{-1}.

Nitrogen budgets have changed significantly. Seeding rates have increased slightly but reproductive rates are much larger and contributions from seed N can generally be ignored. Free-living fixation is also a small fraction of total N supply, whereas wet and dry deposition appears to have increased significantly in some cases.

The data in Table 8.3 illustrate some of the changes in N cycling of grazed pastures and harvested forages in the Netherlands, from 1800 to 1972. One third of animals on the 1800 pasture were draft animals, whereas in 1937 and 1972 the N budgets involve only dairy cattle. Nitrogen transfers between pastures and arable lands occurred on all of these farms via harvested hay and barnyard manure. On the 1937 farm, a leguminous forage crop also contributed to N transfers. The 1972 pasture actually received more N in manure applications than was exported in animal products and hay. Many numbers are uncertain, particularly for events in the soil. The authors assumed that the large surplus shown for fertilized pasture (1972) accumulated as SOM. Small rates of mineralization and immobilization are expected in untilled pasture whereas zero values are shown in 1937 and 1972. At Rothamsted, nitrate leaching (corrected for rainfall inputs) continued at rates of 30 to 45 kg NO$_3$-N ha^{-1} y^{-1} for more than 40 y from bare, untilled soil, indicating continued mineralization of SOM (Addiscott 1988).

The 1972 farm represents an extreme example of N supply. Pasture received fertilizer and manure N considerably in excess of plant needs. Fertilizer N was applied at regular intervals during growing seasons because it promoted greater growth of grass, with higher digestibility than is obtained with manure and legumes. As a result, carrying capacity of the 1972 ryegrass pasture was 2.9 times that of the 1937 ryegrass–clover

Table 8.3 Nitrogen cycling (kg N ha^{-1} y^{-1}) in the pasture component of Dutch dairy farms 1800, 1937, and 1972

Year	1800	1937	1972
Stocking rate, animals ha^{-1}	0.5	0.9	2.5
Supplies to soil			
External supply			
Wet and dry deposition	8	14	14
Nitrogen fixation	80	120	0
Application of manure	0	0	149
N fertilizer	0	0	400
Total external supply	88	134	563
Internal supply			
Mineralization	13	0	0
Plant residues	–	26	180
Dung and urine	44	46	178
Total internal supply	57	72	358
Total supply	145	206	921
Removals from soil			
Plant uptake	71	158	450
Immobilization	13	0	–
Volatilization loss	34	7	26
Leaching loss	5	11	39
Denitrification loss	16	49	169
Total removal	139	225	684
Output			
Harvest of hay	32	50	126
Animal production	1	19	72
Losses	55	67	234
Total output	88	136	432
Input – output	0	–2	+131
Total grazing plus hay	71	111	270

Source: interpreted from J. P. N. Damen and C. H. Henkins in Frissel (1977).

pasture. Optimal rates of N application to intensively grazed ryegrass pastures in Europe range between 300 and 600 kg N ha^{-1} y^{-1}. A survey of Dutch dairy farms, cited by Vertregt and Rutgers (1988), indicates an average use of 350 kg N ha^{-1} grassland y^{-1} with a stocking rate of 3.5 cows ha^{-1}. The use of fertilizer, however, leads to problems of manure disposal because dairying is concentrated in regions with only small amounts of arable land that can utilize manure effectively.

Our analysis identifies manure as the main source of excess N on the 1972 farm. Dairy herds receive supplemental feed (saved grass, silage from arable portions of the farm, and purchased feed during winter while confined to barns). Manure in excess of what could be applied to available arable land (149 kg N ha^{-1}) was applied to pastures for disposal. That is done during winter since applications during summer cause cattle to reject forage. Summer applications also contribute to recycling of diseases and parasites. The principal merits of manure application were that large amounts of surplus organic material and N were disposed at low cost, and grass was supplied with essential mineral elements. As shown by the 1800 and 1937 farms, however, nitrate leaching occurs from concentrated dung and urine even in the absence of fertilizer or manure.

Despite excessive supply of N, the 1972 farm produced 3.8 times as much human-edible N products per hectare, and at about the same efficiency per unit N transferred, as the 1937 farm. With less manure, its N-transfer efficiency would have increased sharply. Although the 1972 pasture was a much greater polluter of ground water than the 1937 one, the ratios of leached N to product N were similar. These farms were located on wet clay soils and most losses occurred by denitrification; had they been located on light soils, nitrate pollution and acidification would have been more significant problems. Use of both manure and fertilizer allows the Netherlands to maintain high crop yields and produce much of their total food requirement from a small land area but contamination of surface and ground waters has led to intensive efforts at finding alternative uses for manure. Concerns about GHG emissions and rising prices for N fertilizer due to higher energy costs have also motivated efforts to reduce N fertilizer input to these systems.

A study in England by Wellings and Bell (1980) supports the view that manure may have been the main source of nitrate in the 1972 farm. Those workers found ten-fold greater concentrations of nitrate under a ryegrass pasture that received slurried manure than was found with an equivalent rate of fertilizer N (376 kg N ha^{-1}). Most farming systems employ much more conservative rates and methods of N supply than the Dutch, as is evident in other examples presented by Frissel (1977) and in Chapters 16 and 17. Fertilization practices will be examined in more detail in Chapter 12.

8.8 Farming with organic sources of nitrogen

Comparison of the 1937 and 1972 dairy farms illustrates the enormous advance in carrying capacity that is possible when farming systems are supplemented with N fertilizer. By 1940, farming with strictly organic sources of N had proven inadequate for meeting world food supplies. Since then, world population has tripled (2.3–6.8 billion) and most farms worldwide now meet some of their N needs with fertilizer. Systems that do not use N fertilizer have also survived, particularly in regions where pastoral dairy farms generate surplus manure, in extensive grazing enterprises in semi-arid regions, and in poor countries where farmers do not have access to other technologies. Special management issues that arise in using legumes and manures as sources of N deserve analysis and comment.

8.8.1 Legume rotations

Inclusion of legume crops in rotation with non-legume crops is a long-standing practice, involving non-uniform treatment of land in time, for supplying N to cropping systems.

In addition to satisfying much of their own need for N, legumes can also increase soil N content. A major limitation of legume rotations is that legume roots are not high in N. Carry-over N from legume crops other than pastures and green manures is generally sufficient for only moderate production by following crops unless some factor is limiting productivity, e.g., water supply or length of season. Grain legumes carry most of their N away in seed and most of that produced by forage legumes is removed in harvested hay.

When non-legumes in rotations are supplied with supplemental N from fertilizer or manure, yields increase but their residues and the soil then contain more N. That suppresses fixation by a following legume so that consumption of mineral N increases. This phenomenon and the inadequacy of legumes as sole source of N have contributed to less use of legumes in rotations as N fertilizers became available. Farmers now grow legumes more for the value of their products than for their contributions through N fixation.

Farming with legumes as sources of N is an uncertain business. Amounts of N carried over and released through decay and mineralization are variable, making quantitative predictions about supply difficult. Interesting problems relative to spatial variability also arise. Fixation is greatest in N-poor portions of fields, favoring uniformity for subsequent crops, but variability in legume stands, due for example to soil pH, drainage, or winter damage, increases spatial variability for N.

Without practical means for assessing carry-over supplies of N, experience of farmers and extension workers guides most farming practice. General experience is that legume residues have about the same value as those of other crops. Fertilizer recommendations for most states in the US Corn Belt include 40 to 50 kg N ha^{-1} residual benefit when corn follows soybean. While 60 to 100 kg N ha^{-1} can be expected as carry-over from a good four-year stand of alfalfa in California, farm advisors point to associated removals of P and K, which must eventually be replaced; as one noted, "It all went with the hay."

Major increases in soil N from legumes come only when forage is incorporated with soil (green manure) or through recycling in animal manure produced by feeding leguminous forages. Vetch, berseem clover (both cool climate), and sweet clover (warm climate) are used as leguminous green manures. Some estimates of N production by summer green manures exceed 200 kg N ha^{-1}. Leguminous green-manure residues decompose rather quickly, however, so the supply of N can exceed the needs of a following crop, and beneficial effects on soil structure are slight at best (MacRae & Mehuys 1985).

Green manures are generally uneconomic simply as a source of N. They require tillage, seed, proper soil pH, and fertility, and they consume significant amounts of water. In effect, 1.5 to 2 ha of land are cropped but only 1 ha is harvested for useful products. For the farming system as a whole, production per unit input of water, labor, and fuel is reduced by as much as half. That results in a very high cost for N that only developed-country farmers with specialty markets for organic foods can afford or resource-poor farmers in the developing world cannot avoid.

Grass–legume mixtures are used widely for pastures and forage production. In addition to supplying N, legume forages add protein content to feed. Transfers of N from legume to grass occur by several mechanisms. The principal pathway is through the dung and urine of grazing animals. Another occurs with legume residues including dead roots and nodules. Grazing or mowing interrupts C supplies needed for the maintenance of roots and nodules, resulting in nodule sloughing and root death. Nitrogen released by decay is then available to associated non-legumes. As noted in earlier discussion of Dutch pastures, however, the productivity of such polycultures may be significantly less than that of a well-fertilized grass monoculture.

8.8.2   Animal manures

Animal manures are the principal means by which nutrients are transferred among fields. Manures have two virtues in addition to their nutrient content: they benefit soil structure and return ash alkalinity (Section 7.4.4) removed in crop harvests to soil. In addition to manure, muck soil, or vegetation obtained from other fields, composted sewage sludge and various refuses also serve as organic amendments. Sewage sludge from industrial cities may be contaminated with heavy metals, however, and should not be used unless chemical analysis establishes that contaminants are below critical thresholds for concern.

In the USA, over 72% of animal manure-N comes from cattle; poultry and swine account for another 25%. In 2002, 6 Mt N was excreted but only manure from confined animals is collectible. About half the cattle manure falls on range and pasture, as does manure from sheep, goats, and horses. In contrast, poultry and hogs are produced in confined systems. Nearly all collectible manure is recycled to cropland but, because only about 50% of it finally reaches the soil due to N losses in composting and storage, manures are estimated to supply only about 2 Mt N to crop production. That number can be compared with 1.5 Mt N consumed by Americans, of which very little is recycled. Two million tonnes of N translates to about 12 kg N ha^{-1} cropland y^{-1}, demonstrating that widespread dependence on manures as nutrient sources is not possible.

Handling, storage and composition of manures Efficiency of manure recycling could be improved in most cases, but costs are high. All manures have common traits: their nutrient content is small and variable, and their behavior during storage and after application to fields is difficult to predict. Composition of representative *fresh* manures (feces and urine combined) is presented in Table 8.4. It is difficult to make any general statement about nutrient content of *collected* manure even on a single farm because decomposition and N losses during storage and handling are highly variable (Azevedo & Stout 1974). The degree of decomposition and whether bedding material (straw) is included are important factors. Caution is advised in using values given in Table 8.5 because of substantial variation. For example, N content of solid cattle manures in this table range from 2.0 to 2.7% N on a dry basis but values of 1 to 2% also are common for farm-lot manures.

About 50% of N in urine and dung of cattle is present as ammonium ions and urea, which can convert quickly to ammonia. Volatilization of ammonia is favored by warm, drying conditions and alkaline pH of voided materials; this apparently accounts for much

Table 8.4 Approximate production and nutrient contents of fresh animal manures (dung and urine combined) Numbers in parentheses are coefficients of variation.

Source	Daily manure production[a]		Nutrient content (kg t^{-1} fresh mass)			Dry matter (kg t^{-1} fresh mass)
	(kg animal^{-1})	(animals t^{-1})	N	P	K	
Beef cattle	21	47.6	5.9	1.6	3.6	146
	(0.20)		(0.21)	(0.29)	(0.29)	(0.30)
Dairy cattle	55	18.2	5.2	1.1	3.4	140
	(0.29)		(0.21)	(0.26)	(0.32)	(0.22)
Swine	5.1	196	6.2	2.1	3.4	130
	(0.28)		(0.40)	(0.56)	(0.55)	(0.57)
Broilers	0.076	13 200	12.9	3.5	3.4	250
	(0.15)		(0.26)	(0.18)	(0.16)	(0.06)

Note: [a] fresh mass production per animal and number of animals needed to produce 1 t d^{-1}.
Source: adapted from ASAE (1999).

Table 8.5 Approximate nutrient contents of aged animal manures obtained with good methods of conservation

Animal	Form	Handling	Nutrient content (kg t^{-1} fresh mass)			Dry matter	NH$_4$-N Total N
			N	P	K		
Beef	solid	air, no bedding	10	6	19	520	0.33
cattle	solid	air, bedding	11	8	22	500	0.38
	liquid	pit[a]	18	5	13	110	0.60
Dairy	solid	air, no bedding	5	2	8	180	0.44
cattle	liquid	pit[a]	11	4	11	80	0.50
Swine	liquid	pit[a]	16	5	8	180	0.72
Poultry	solid	no bedding	16	21	28	450	0.79

Note: [a] 1 m^3 of pit slurry is assumed to weigh 1 t, i.e., 1 l = 1 kg.
Source: adapted from Midwest Plan Service (1985).

variation reported in N content values. Volatilization losses are less when dung and urine are caught on straw bedding that can immobilize some of the N.

Dung decomposes rapidly during storage, bulk volume decreases, and concentration of nutrients other than N increases. Storage also helps in reducing the viability of weed seed and animal parasites passed in dung. The nature of the final product depends on conditions during storage. Under traditional dry-lot conditions, manure accumulates for 3 to 6 months before being removed and distributed to fields. Decomposition in manure heaps is an aerobic process leading to a humus-like final product that retains 50% of original N. Flies may be a problem during aerobic decay but few odors are produced. Nitrate accumulates if decomposition proceeds too far and may denitrify or leach if the manure is wetted. Potassium is also subject to leaching. Humification and N conservation are accomplished more efficiently when manure is collected frequently and allowed to age in moist conditions in well-aerated piles. High temperatures that develop within manure piles help in killing weed seed and animal parasites. Addition of crop residues and litter to manure produces compost.

When fresh manure is placed in unaerated piles or pits for storage, anaerobic conditions develop and 80% or more of the N may be conserved. Ammonium is the main form in the final product. Odors (methane, mercaptans, skatole) can be a serious problem, however. Confinement dairies and piggeries commonly handle manure as slurry. Daily removals are slurried with water and aged anaerobically in a pit or, after further dilution, in tanks. Because of dilution, N concentration after tank storage (3–5 kg N m^{-3}) is less than is shown in Table 8.5 for pit slurry. Slurries are applied to fields using tank wagons. Odor problems, even when avoided during storage, arise after surface spreading but can be greatly reduced if slurry is injected into soil.

Regardless of the methods used for handling manure, the low concentration of nutrients and the costs of facilities, labor, and fuel required in collection, storage, and subsequent application result in a high cost per unit nutrient. Slurry facilities are particularly expensive (Bouldin *et al.* 1984) and long-distance transport is generally prohibitive for any form of manure. As a result, some farmers tend to approach manure handling as a disposal problem rather than as a nutrient source. Recent increases in the cost of fertilizer are, however, helping to change this perspective.

Rates and timing of manure applications Placement and precise rates of application are not possible with solid forms of manure. Surface spreading followed quickly by incorporation with a moldboard plow seems to be best practice. Volatilization of ammonia (particularly from anaerobic manures; see the NH$_4$-N/total N ratio in Table 8.5) and runoff losses increase with the time that manure remains unincorporated. Because of limitations in labor, most farmers find timely incorporation of solid manure difficult to achieve. As was the case for the 1972 Dutch farm, manure is sometimes spread in winter; the danger then with arable land is that nutrients are lost to runoff and leaching before spring tillage begins. Slurries have an advantage in this regard because they can be injected with chisels well in advance of tillage.

The C/N ratio of decomposed manures is low enough that immobilization does not occur when they are applied to soil. The ammonium fraction of N is quickly available to plants after incorporation while the humic portion mineralizes slowly. Common values for the **decay series** for fractions of organic material in manure that is mineralized during each of the first 3 y following application are 0.3, 0.1, and 0.05.

Given the uncertain nature and behavior of manure, optimum use is difficult in a fertilizer program. There are trade-offs between the use of minimal rates with least risk of adding to pollution through leaching or runoff but with a high risk of reduced and more variable yields of fertilized crops. With manure as the sole N source, large rates are needed to ensure sufficient N mineralization per year. Using the decay series described above, manure containing 240 kg N must be applied each year to provide a mineral flux of 100 kg N y^{-1}, assuming negligible N release from residual manure after 3 y. That corresponds to 11 t ha^{-1} of beef cattle manure decomposed with bedding (Table 8.5) but supplies of P (176 kg ha^{-1} y^{-1}) and K (484 kg ha^{-1} y^{-1}) would then greatly exceed the requirements of any crop. It is immediately obvious that these amounts could not have been obtained from feed produced on a single hectare and that nutrients are being gathered from larger areas instead. Manure continues to mineralize whether a crop is present or not. Some excess nutrients would accumulate as SOM but serious losses and pollution of ground and surface waters by nitrate also occur (van der Meer *et al.* 1987).

Farmers who employ high rates of manure sometimes find it useful to grow grass during non-crop periods to capture the continuing mineral flux. In pastures, excess potassium contributes to an imbalance with magnesium, leading to metabolic disorders in grazing cows and ewes (grass tetany). In arid regions, the high salt content of cattle manures may contribute to salinity problems. The solution to these problems, and associated nutrient imbalances, reached by most farmers is to apply modest amounts of manure only to main crops. A base fertility is established for each field, which is then fine tuned by applying specific mineral fertilizers.

8.8.3   Rule-based organic farming

All farming is organic in the sense that produce is composed of organic material and most nutrients pass through a phase in SOM. Some farmers, however, practice a form of "organic" farming following special rules about source and types of external inputs, including nutrients, biocides, and seed. Their produce, under labels of "natural", "organic", or "ecological", is aimed at a market niche created by public apprehensions about biocides and other aspects of modern agriculture and food processing. The rules are generally established by associations of farmers but many governments have placed them into legal codes to support organic farmers, especially in marketing.

Methods used by these farmers can be termed **rule-based organic farming** to differentiate them from other types of farming. Use of "synthetic" materials is generally proscribed but that constraint is directed more at "modern" than at synthetic or manufactured inputs because traditional practices involving highly toxic inputs of copper salts and sulfur are permitted. Such inconsistencies are apparent in lists of permitted materials (EC 2000; IFOAM 2002). They include manufactured chelates, petroleum oil, and fermentation products such as gibberellic acid and antibiotics. Biocides are generally proscribed although Bordeaux mixture (copper sulfate and lime) for control of fungi and natural materials such as pyrethrum-, neem-, and *Bacillis thuringensis*-based (Bt) insecticides are allowed. Transgenic cultivars are banned, however, including those of cotton and maize that have an inbuilt Bt insect control mechanism.

The rationale for management of nutrient supplies in rule-based systems is neither logical nor based in scientific principles. Animal manures and leguminous green manures are favored as sources of N. Organic farms in Europe refuge N from an average 3 ha of pasture and legume forages in order to achieve acceptable yields on 1 ha of arable land (Boeringa 1980). Refuging also occurs when organic farms purchase manure composts from other farms. Animal manures purchased from integrated farms are generally accepted as "organic". Most rules permit the use of lime, rock phosphate, $NaNO_3$, and K_2SO_4 as "organic amendments" whereas KCl and beneficiated phosphate rock are "chemical" and not permitted. Such attempts to differentiate rule-based organic farms

Box 8.3 Best management practices for N fertilizer use

Optimum management will maximize economic response while minimizing losses to the environment by leaching and atmospheric emission.

General practices

- Choose the most appropriate form of N fertilizer (urea, ammonium nitrate, ammonia, etc.) for each soil–crop situation with its particular risks for loss. Special attention to wet sites and wet conditions.
- Use appropriate rates and placement to optimize crop yield and minimize residuals of NO_3-N.
- Use nutrient management plans that consider soil N supply, fixation by legumes, and nutrient content of all sources applied, especially manure.
- Avoid high pre-plant applications by applying N in doses to coincide, as practically as possible, with crop N demand.
- Use balanced fertilization to supply all other required nutrients to increase crop growth and crop N-use efficiency.
- For drill-seeded rice, establish permanent flood immediately after application of urea or NH_4-based fertilizer. Use appropriate management with any midseason N applications to enhance crop N recovery.
- On land prone to erosion, use tillage practices that reduce potential for runoff.

Equipment and application technologies

- Calibrate equipment to ensure accurate delivery of prescribed N rates and placement; subsurface placement deep enough to ensure N retention.
- Use variable rate and/or variable N source applications as technology is proven.
- Incorporate urea and urea sources by tillage, irrigation, rainfall where practical, and use controlled-release forms or urease inhibitors to minimize NH_3 loss in soils with high potential for volatilization.
- Use nitrification inhibitors with ammoniacal N sources in environments with high potential for NO_3-N leaching losses and/or N_2O emissions.

Crop management, system planning, and evaluation

- Use rotations, and specific site characteristics, to maximize N-use efficiency and minimize losses of each crop and field.
- Schedule irrigation rates to match consumptive water and nutrient demand; measure losses in tail water, use cover crops (where feasible) to retain and recover residual N.
- Use soil and plant analysis to determine status and demand of all nutrients to optimize efficiency of N use.

arbitrarily as sources of special food is generally wasteful of high-grade phosphate ores and an unnecessary expense for potassium. These irrationalities are small, however, compared with claims for alchemical and astrological bases for organic methods by some European groups (Boeringa 1980; Kirchmann & Bergstrom 2008).

In rule-based organic farming, weed control requires larger amounts of labor and fuel in cultivation than are used by most farmers. Diseases are sometimes difficult to control, especially when weather conditions are conducive to spread. Narrow reliance on organic sources of N, while seemingly wholesome, introduces several problems including large nutrient requirements of legumes and unbalanced nutrient content of animal manures. Uncertain release of mineral N from organic sources can be ameliorated to some extent by extended composting and with catch crops. It is difficult or even impossible, however, to achieve an adequate flux of mineral N from organic materials without also creating a considerable potential for leaching of nitrate and potassium. While it is true that successful organic farming requires considerable skill, the same is true of successful farming generally. Consider, for example, the considerations required in **best management practices** for use of N fertilizer (Box 8.3) that cannot be avoided if farmers are to feed the world while achieving acceptable economic return and protecting water quality.

In developed countries, rule-based organic farming fills roles in providing specialty foods for certain consumers, but costs are high and these farmers generally must depend on unique labeling of produce to obtain the higher prices they need. Greater costs are inevitable given less efficient use of land, energy, labor, nutrients, and water inherent in strict organic methods. In developing countries, where supplies of organic nutrients are limited, proponents of organic agriculture seek to attract support to their system with promise of higher prices for exported products. Economic benefits flow to a small number of producers in favored locations and environments, however, and not to the multitudes of resource-poor farmers who struggle for subsistence. For those farmers, organic agriculture offers little new. It is the system they are now forced to practice and all it offers is continuing hunger, poverty, and soil nutrient depletion.

8.9   Review of key concepts

Nitrogen in the biosphere

- The atmosphere offers an unlimited supply of dinitrogen gas (N_2) for plant and animal nutrition, but reduced N is required in living organisms. Transfer of N_2 to the biosphere depends on fixation to ammonia (NH_3) by free-living or symbiotic organisms and subsequent reduction by bacteria (*Nitrosomonas*, *Nitrobacter*) to nitrate (NO_3^-), the form used by most plants.
- Free-living microorganisms found in soils, and symbiotic bacteria, particularly rhizobia that nodulate leguminous plants, contribute most of the N that enters natural systems and about 40% of that entering agricultural systems. The balance comes with

rainfall and fertilizer. Nitrogen fertilizer is produced principally by the Haber–Bosch process employing natural gas.

- Nitrogen is the element required in largest amounts by plants, the only one not supplied by weathering of soil parent material, and that exists predominantly within the organic phase of the biosphere.

Microbial transformations

- Microbial populations of soils of many genera and diverse metabolism attack all available organic substrates for energy and N. All require N as an essential constituent of protoplasm. They survive through consumption (decay, mineralization) of plant and animal residues and SOM containing N, or by manipulating the oxidation–reduction states of mineral N. Nitrification $\left(NH_4^+ \rightarrow NO_3^-\right)$ and denitrification $\left(NO_3^- \rightarrow N_2, N_2O\right)$ are central processes in N cycles. Nitrification provides NO_3^- for higher plants; denitrification is the avenue by which N returns to the atmospheric dinitrogen pool.

- The decomposition rate of crop residues largely depends on the C/N ratio and lignin content, soil moisture and temperature. Populations of soil microorganisms rise and fall rapidly in response to the availability of substrates and upon water status. When organic matter of high C/N ratio (25–100) is available, microbial populations compete strongly with plants for soil NO_3^- and immobilize it in microbial biomass. As C/N ratio falls, populations decline and microbial N enters the mineral N pool to be available to plants. The C/N ratio of older organic fractions usually stabilizes around 10 to 13.

Management in agricultural systems

- Productivity of cropping systems is closely related to the uptake of N. The challenge for agriculture is to ensure adequate N supply in mineral forms for plants and as protein for animals in the right times, places, and quantities.

- Nitrate becomes available by mineralization of SOM or by additions of fertilizer. Nitrate exists in soil solution and is easily lost by denitrification to GHG-active nitrous oxides or by leaching with potential to pollute ground waters, streams, and estuaries.

- Humus is a resistant end-product of microbial attack that accumulates in soils conferring valuable physical and chemical properties. It is also a major reservoir of unavailable N. Management to increase humus content must attend to the substantial N content ($c.$ 5%) that it sequesters.

- Good management of N requires matching its supply to crop demand so that losses by denitrification and leaching can be minimized. This is more easily achieved with small doses of chemical fertilizers of known composition than by usually larger doses of organic fertilizers, often of unknown composition. Regardless of how N fertilizer is applied, and the care taken to avoid losses, much variability and uncertainty surround the management of N cycles in agricultural fields.

Terms to remember: ammonification; anaerobic; carbon-to-nitrogen ratio; best management practices; biological nitrogen fixation, heterotrophic, associative, symbiosis; cover crops; decay series; denitrification; deposition, wet and dry; eutrophication; fertilizer, ammoniacal, controlled-release, organic, nitrogen; legume rotations; ^{15}N-labeled; leaching; legume symbiosis; nitrification; nitrogen, cycle, immobilization, mineralization and volatilization; nitrogenase; organic farming; rhizobia.

9 Water relations

Plants grow by fixing, in photosynthesis, CO_2 that diffuses into leaves from the atmosphere through open stomatal pores in leaf surfaces. An inevitable consequence is that water vapor evaporates from wet cell walls that surround sub-stomatal cavities and diffuses through stomates to drier air outside. Water loss must be controlled or replaced if the plant is to maintain **turgor** and metabolic activity. Water is also a primary reactant in photosynthesis but the proportion of water required by plants that is chemically incorporated in their structure, or is used to maintain their water content as they grow, is very small.

Crops differ significantly in rooting habit and thus in their ability to acquire water from soil. Owing to differences in epidermal wax and in size, frequency, and behavior of stomates, they also vary in their ability to control loss of water from leaves. Control of water loss is always made at the expense of CO_2 uptake for growth. For crops, flow of water from the soil to the atmosphere through plants is accompanied by direct evaporation of water from soil, particularly when the surface is wet and unprotected by foliage. The discussion of plant and crop water relations presented in this chapter draws heavily on the information presented in Chapter 6 (Aerial environment) and is essential background to understanding the productivity and effective management of rainfed and irrigated crops presented in Chapters 13 and 14.

9.1 Flow of water through a crop

Water flows from moist soil through roots and stems down a gradient of **water potential** established by leaves exposed to air. That pathway is the **soil–plant–atmosphere continuum** and the loss of water is **transpiration (E_p)**. Plants occupy the central position in the pathway so their water status varies diurnally and seasonally depending on availability of water in soil and evaporative capacity of the atmosphere. Water also is lost to the atmosphere by direct **soil evaporation (E_s)**. If foliage cover is incomplete during a considerable proportion of their growth cycle, up to 50% of total water use may be lost by this route. The combined loss of water from soil by transpiration through the crop and direct evaporation from the soil is called **evapotranspiration (ET)**.

Stages in the soil–plant–atmosphere continuum are depicted in Fig. 9.1. Evaporation of water from wet cell walls of the leaf mesophyll lowers water content and water

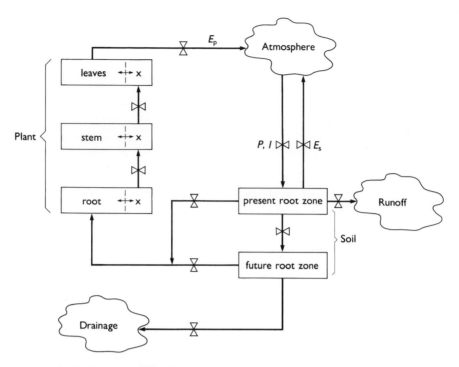

Fig. 9.1 Compartment model of water relations and water balance of a crop. The compartments define distribution (mm) of water throughout the system. In the crop, tissues of leaves, stem, and root are in hydraulic contact with the xylem (x) through which water flows from the soil to the atmosphere.

potential of leaves. In response, water moves to them passively in the liquid phase from other plant parts and ultimately from soil along gradients of water potential ($\Delta\psi$). Water moving across the root cortex through cell walls (apoplast, the non-living system) encounters a barrier at the endodermis surrounding the root's central stele. Cell walls of that tissue are rendered impermeable by waxy deposits of the Casparian strip, so passage is across membranes and through living cells (symplast).

Once across the endodermis and into the xylem of roots, there is a continuous apoplastic route through the xylem to the leaf mesophyll. Tissues of stele, stem, and leaf are in hydraulic contact with the xylem stream and exchange water with it depending upon local gradients of water potential. As a result, water status of a crop adjusts continually to changing patterns of environmental demand and water uptake.

The model presented in Fig. 9.1 considers amounts of water (mm) in each of three crop compartments (foliage, stem, and root) together with two soil layers defined as current root zone and subsoil into which roots will extend as the season progresses. The water content of crop compartments depends upon their size and on exchanges between them and the xylem pathway. Water contents of soil layers (volumetric water content $\theta \times$ depth interval Δz) are renewed intermittently from above by rainfall (P) or irrigation (I) as each is successively wetted above field capacity. E_s, runoff (RO),

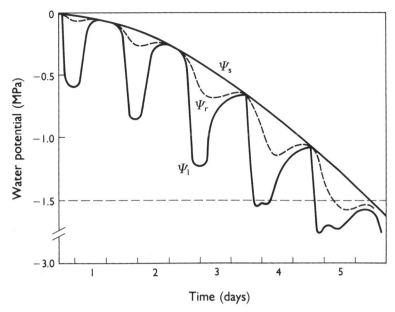

Fig. 9.2 Diagram illustrating idealized changes in water potential of leaf, root, and soil during a drying cycle of a field crop commencing with soil at field capacity. (After Cowan 1965; Slatyer 1967).

and drainage (*D*) below the root zone are included to complete a **hydrological balance**, i.e.,

$$P + I + E_p + E_s + D + RO + \Delta\theta z = 0 \qquad (9.1)$$

where $E_p + E_s = \text{ET}$.

Within the soil–plant system, flux of water (F) between contiguous compartments is proportional to differences in water potential ($\Delta\psi$) and hydraulic conductance, k. Here, resistance, which is the reciprocal of k, will be used where it leads to simpler equations for catenary flow of water in crop systems. Response of ψ in any crop compartment in Fig 9.1 to gain or loss of water depends upon cell wall elasticity, degree of osmotic adjustment, and other traits, none of which is an explicit part of this model. Capability of tissue to lose or gain water can be termed capacitance in analogy with electrical circuits. In general, crop plants have low capacitance while in trees, considerable water is exchanged between active xylem and older wood of stem and root and can exceed transpirational loss over a complete day (Running 1980).

In herbaceous crops with low capacitance, it is possible to describe water relations by ignoring internal redistribution of water. For example, a wheat crop of 9 t ha^{-1} dry mass contains around 50 t water ha^{-1} (5 mm). Under heavy evaporative demand its water content might fall to 0.9 of its water content at full turgor, i.e., its **relative water content** (RWC) would be 0.9. This represents a change equivalent to 0.5 mm or only 8% of total daily water use at $E_p = 6$ mm.

Changes in RWC lead to changes in ψ. Figure 9.2 presents an "idealized" picture of changes in root zone water potential (ψ_s) during five consecutive diurnal cycles of a crop growing in a fixed volume of initially wet soil ($\psi \sim -0.02$ MPa). Over five days, transpiration lowered ψ_s to around -1.5 MPa and the diurnal pattern of water potential in leaves (ψ_1) responded differently even though daily evaporative demand remained unchanged. During day 1, with freely available water in soil, ψ_1 fell to -0.6 MPa at midday but recovered completely, i.e., to ψ_s, by dusk. Although water uptake during daylight hours was inadequate to maintain ψ_1 close to ψ_s, the small gradient between soil and roots (ψ_s and ψ_r) (~ 0.1 MPa) reflects high conductance to flow in moist soils and accounts for rapid recovery by dusk.

As soil dried further during days 2 and 3, ψ in each part of the system fell gradually and gradients widened. Equilibration of ψ_1 and ψ_r with ψ_s was incomplete by dusk but was achieved by the subsequent dawn. On day 4, ψ_1 fell below -1.5 MPa and stomates began to close, arresting further fall in ψ_1 and the crop recovered to near ψ_s by dawn of the following day. With further drying (day 5), stomatal closure failed to maintain ψ_1 above -1.5 MPa and recovery was incomplete even by dawn of the next day. Decreasing conductance to water flow from drier soil to the root system, seen in the widening gradient of that part of the pathway, played an important role in failure of the root system to maintain water supply to the crop. Without water or additional root length, or some reduction in demand through reduced leaf area, tighter closure of stomates, or change in weather, this crop would be in danger of suffering internal desiccation and injury.

The challenge in crop ecology is to measure and interpret such responses under difficult and variable conditions of the field.

9.2   Evapotranspiration

An extensive crop of full cover, well supplied with water, exerts minimum stomatal control over transpiration and evaporates water at a rate determined by evaporation demand of the atmosphere. That rate of water use, the maximum for those environmental conditions, is termed **potential evapotranspiration** (ET*). It defines the upper bound to actual evapotranspiration (ET$_a$) that will be smaller than ET* when cover is incomplete and the soil surface is dry or when restricted availability of water causes leaf conductance to fall. ET* at full cover varies by 10 to 20% between individual crops owing to differences in color, height, aerodynamic roughness, and maximum canopy conductance (Fig. 9.3).

Short green grass, or in some places alfalfa mown to 8 to 15 cm height, is an important, special case for comparison and is included in Fig. 9.3. It is easily grown in most locations and has full vegetative cover with small aerodynamic roughness that does not change during the growing season. ET* of such vegetative cover defines a **reference evapotranspiration** (ET$_0$), a practical, widely used standard measure of environmental demand (Allen *et al.* 1998).

Penman (1948) originally defined ET* as the rate of water loss from short grass (ET$_0$) but differences in ET* between crops were not then generally recognized. Nor was it appreciated that ET* of many full-cover crops exceeds short grass (ET$_0$), especially in

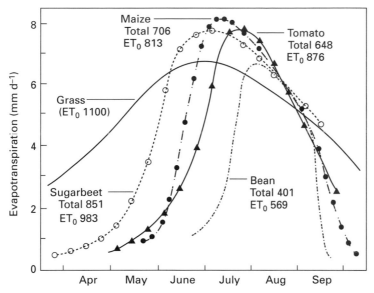

Fig. 9.3 Seasonal evapotranspiration of various well-watered crops compared with a grass reference (ET_0) under the same environmental conditions at Davis, California. (After Pruitt 1986.)

Box 9.1 Lysimeters

Lysimeters are containers of soil of depth in excess of the root zone depth of a mature crop, set within and holding part of a crop, arranged to allow measurements of gains and losses of soil water. The container, and the crop growing in it, should be large in order to minimize edge effects and be located within a representative crop stand of adequate extent (see Section 6.8 on fetch).

Accurate weighing of soil and crop in a lysimeter measures loss of water over time because gain in weight due to crop growth is relatively small compared to soil mass. Using pivot scales or electrical transducers in underground support structures, weight changes as small as 1 kg can be detected in lysimeters of 50 t total weight. Well-designed lysimeters can measure hourly ET as small as 0.03 mm. Such installations are expensive and lack mobility and are found only on research sites.

Lysimeters that are not weighed rely upon measurements of rainfall and collection of drainage water to close the water balance equation (Eq. 9.1). These installations are suited to daily or weekly estimates of ET.

---

highly evaporative conditions of arid zones. The terminology used here explains the significance of short-grass ET_0 and allows discussion of ET^* of individual crops.

Evapotranspiration can be measured aerodynamically using "eddy covariance" (Section 6.7), by partitioning energy balance using the Bowen ratio (Box 6.5), with **lysimeters** (Box 9.1), or by sequential measurements of rainfall, runoff, soil water

Table 9.1 Some equations used to predict reference evapotranspiration (ET$_0$)

Variables: r_a, aerodynamic resistance; r_c, canopy resistance; LAI, leaf area index; $f(u)$, function of windspeed; ρ_a, the density of moist air; and c_p, specific heat of air at constant pressure. See text and Section 6.5 for further explanation of terms.

Name	Equation
Penman (1948)	$ET_0 = [s(R_n + G) + \gamma\,f(u)\,(e^* - e_a)]/(s + \gamma)$
Priestley & Taylor (1972)	$ET_0 = \alpha s(R_n + G)/(s + \gamma)$
Monteith (1964)	$ET^* = [s(R_n + G) + p_a\,c_p\,(e^* - e_a)/r_a]/[s + \gamma)\,(1 + r_c/r_a)]$

content, and drainage to solve the hydrological balance (Eq. 9.1). In the latter method, changes in soil water content ($\Delta\theta_z$) are measured at intervals of at least 4 to 7 days that are needed to establish measurable change. Available methods are sampling and oven-drying soil cores, or *in situ* measurements using neutron moisture meters or time domain reflectometers.

ET$_0$ has been commonly estimated by measurements of evaporation from pans of standard shape and exposure that are a part of standard meteorological observations. Empirical pan coefficients (k_{pan}) must be established from these to estimate $ET_0 = k_{pan}E_{pan}$. Increasingly, however, ET$_0$ is calculated from a range of formulae that have, in their development, been calibrated against data collected from lysimeters.

Three equations commonly used to estimate ET$_0$ are presented in Table 9.1. Both Penman and Monteith are combination equations. Crop radiation balance (Section 6.5) provides energy for ET while the vapor pressure gradient ($e^* - e_a$) is maintained by turbulence of the atmosphere. These two weather components are evident in the equations. In contrast, Priestley–Taylor relates evaporation to radiation alone. It is derived from equilibrium evaporation (Eq. 6.18a), with parameter α (*c.* 1.0–1.3) providing correlated, additional aerodynamic contributions. Monteith includes canopy resistance (r_c) that depends upon leaf resistance (r_l) and LAI and so can describe ET* for different crops under the same environmental conditions (different r_a or minimum values of r_c). It can also describe the difference between ET* and ET$_a$ for individual crops as water supply diminishes and r_c increases. Monteith is now promoted as an international standard by the FAO, which provides methods to estimate missing energy and aerodynamic parameters (Allen *et al.* 1988).

ET$_a$ of well-watered crops with incomplete cover and intermittently dry soil surface can be estimated from ET$_0$ using **"crop coefficients"** established locally from measurements of the type presented in Fig. 9.3. Monthly coefficients suitable for irrigation scheduling of several crops at Davis, California, are presented in Table 9.2. Seasonal values of crop coefficients for many combinations of crops and locations can be found in irrigation management manuals such as Allen *et al.* (1998).

9.2.1   E$_p$ and E$_s$, the components of ET

Explanation of values of crop coefficients during development of crop cover is seen in the changing balance between E$_p$ and E$_s$. E$_p$ approaches ET$_a$ except when cover is small

Table 9.2 Monthly crop coefficients (c $=$ ET$_a$/ET$_0$) for several well-watered crops at Davis, California
Values established from measured crop water-use presented in Fig. 9.3.

Month	Crop			
	Bean	Maize	Sugarbeet	Tomato
April	–	–	0.22	–
May	–	0.17	0.41	0.12
June	0.08	0.61	1.08	0.35
July	0.45	1.22	1.15	1.15
August	1.13	1.18	1.13	1.04
September	0.40	0.79	1.06	0.79

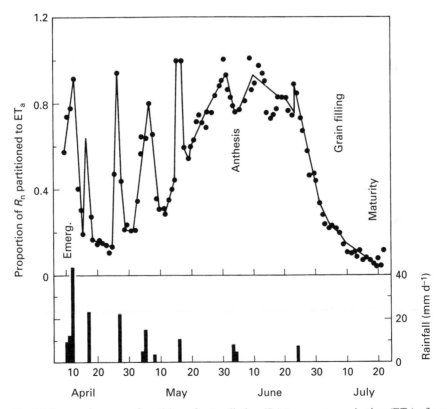

Fig. 9.4 Seasonal course of partition of net radiation (R_n) to evapotranspiration (ET$_a$) of a grain sorghum crop and rainfall at Temple, Texas. Early season peaks represent high rates of E_s following rainfall. E_p dominates late-season ET$_a$. (After Ritchie 1971.)

($< 50\%$) and soil surface is wet. This is an important point. Surfaces of exposed, wet soil dry quickly after 1 to 2 days of evaporation at the potential rate; without rewetting, ET* cannot be maintained. This is seen clearly in Fig. 9.4. As crop leaf area develops, proportion of R_n partitioned to ET$_a$ approaches unity (full partition) whenever the soil surface is wetted by rainfall, but falls back quickly to the underlying rate, that is E_p, as it dries. As full cover is achieved during May, soil is shaded and E_p builds up to equal R_n. This is a nice example of energy balance of a wet system; R_n goes predominantly to ET. During grain filling and maturation, soil surface is not rewetted and E_p declines as the root zone dries and LAI declines.

As shown in Fig. 9.3, real crops depart from ET* for a number of reasons. Incomplete cover is one. In that case, partitioning of ET between E_p and E_s depends upon exchange of net radiation by the crop (R_n) and soil surface (R_{ns}). This partitioning is a function of LAI and canopy geometry but is dominated by the same gap frequency that controls the penetration of radiation into canopies (Ritchie 1972). It can be written in the form of the Bouguer–Lambert Law (Eq. 3.4) as:

$$R_{ns} = R_n \exp(-\omega L) \tag{9.2}$$

where ω is an attenuation coefficient for net radiation and L is LAI. Separate (individual) estimates for R_n and R_{ns} allow, using one of the equations of Table 9.1, partitioning of ET* between potential soil evaporation (E_s^*):

$$E_s^* = \text{ET}^* R_{ns}/R_n \tag{9.3}$$

and its complement, potential transpiration (E_p^*):

$$E_p^* = \text{ET}^*(1 - R_{ns}/R_n) \tag{9.4}$$

9.2.2   Evaporation from a drying soil

After rainfall or irrigation, evaporation from bare soil proceeds in two successive stages (Fig. 9.5). In **stage 1**, E_s is determined by energy exchange at the soil surface and proceeds at E_s^*, of similar magnitude to ET*. In **stage 2**, E_s depends upon the supply of water and declines rapidly as surface soil dries. E_s persists until an air-dry condition, characteristic of soil type, is reached. Cumulative evaporation is small, in the range 6 to 10 mm for sandy and loamy soils respectively, so that stage 1 evaporation for exposed soils is short lived. The transition to stage 2 is abrupt and can be detected experimentally by measurement of surface soil temperature. As E_s declines, R_{ns} goes increasingly to sensible heat.

The rapid decline in E_s during stage 2 can be approximated by an inverse relationship with the square root of time (day number, t) after transition from E_s^*, i.e., $E_s = ct^{-0.5}$ (Fig. 9.5) where c is a function of soil texture with values in the range 2 to 3 mm d$^{-0.5}$. Values are established by measuring cumulative E_s during drying cycles. Rapid decline in E_s results from increasing soil depth from which water is being drawn. That distance and low hydraulic conductivity of unsaturated soil (Section 7.5.4) limit evaporation rate.

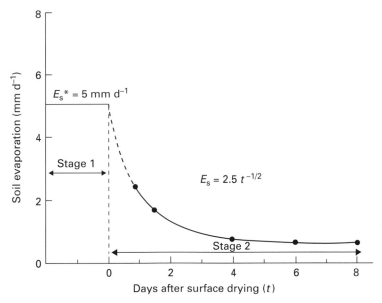

Fig. 9.5 Two stages of evaporation from a bare soil (E_s) following wetting. In this example, the energy-dependent stage 1 persists for a cumulative E_s of 10 mm (i.e., for 2 d). In the supply-dependent stage 2, E_s falls rapidly in inverse proportion to the square root of days from start of stage 2. During that time the constant of proportionality has a value of 2.5 mm $d^{-1/2}$.

9.3   Collection of water by root systems

In semi-arid regions, crops may be subjected to maximum daily ET* of 8 to 10 mm whereas rates of 5 to 6 mm are more common in humid regions with smaller (e^*-e_a). If the root system supporting a canopy is 1 m deep and has a small **root length density** (L_v, units cm cm^{-3}, commonly written cm^{-2}) of only 0.5 cm^{-2} then 5 km of root per m^2 of crop area must maintain a mean daily rate of uptake of 0.02 cm^3 cm^{-1} d^{-1} for the crop to meet ET* = 8 mm d^{-1}. However, transpiration is essentially limited to daytime, reaching a maximum rate of 1.3 mm h^{-1} near midday. Thus peak performance required of the root system is actually about three times greater, i.e., 0.06 cm^3 cm^{-1} d^{-1}.

Movement of that quantity of water from soil into roots occurs down a gradient of water potential established between root system and bulk soil midway between roots, i.e., $(\psi_s - \psi_r)$ by transpirational loss from the crop canopy. The relationship between that flow (Q, mm s^{-1}) and hydraulic resistances of two consecutive segments of the pathway in soil (R_{soil}) and across the root system (R_{root}), respectively is (Gardner 1965):

$$Q = -(\Psi_s - \Psi_r)/(R_{soil} + R_{root}) \tag{9.5}$$

Resistance of water flow from cylinders of soil surrounding roots (R_{soil}) depends upon hydraulic conductivity of soil $[k_s(\theta)]$ and distance between the roots. Therefore, it depends, within each depth interval (Δz), on the surface area of active roots that in turn is a function of root length density (L_v) and root diameter (d_r), i.e., as:

$$R_{soil} = [\ln(\pi L_v) + 2\ln(d_r)]/[4\pi L_v \Delta z k_s(\theta)] \tag{9.6}$$

R_{soil} increases markedly as soil dries from saturation to wilting point (e.g., by a factor of 10^4) because hydraulic conductivity decreases exponentially in this range. For that reason, flow of water to root systems decreases rapidly as soil dries within the available range.

The resistance of root systems (R_{root}) depends upon total root surface area and radial conductance to flow (k_r). It is usually expressed per unit root length rather than surface area, i.e.,

$$R_{root} = 1/(L_v \Delta z k_r) \tag{9.7}$$

The relative importance of R_{soil} and R_{root} (Eq. 9.5) can be estimated from measurements of E_p and ψ_r (the latter approximated from ψ_1). Such measurements show that R_{soil} becomes more important as soil dries from saturation. At high water contents, $R_{root} \approx 10^2 R_{soil}$, but as soil dries, R_{soil} increases rapidly and quickly becomes the major limitation to collection of water.

In contrast to knowledge of water conductivity of soil and complete root systems, little is known about the physical contact between roots and soil. Roots of many species secrete mucigels that assist in maintaining good contact with soil but roots have a tendency to penetrate soil through planes of weakness and to use macropores developed by previous roots. This is coupled with the fact that roots, like all plant parts, shrink diurnally and seasonally when RWC declines, which results in uncertain contact between roots and soil. Larger resistances are observed for root systems in the field than is apparent from laboratory measurements due, perhaps, to poor contact made in drying soils subject to shrinkage.

Equations 9.5 to 9.7 are the basis for analyses of performance of root systems in supplying water for transpiration as presented in Fig. 9.6 (Cowan 1965). The equations were solved repeatedly at short time intervals to account for marked diurnal patterns of E_p and strong dependence of R_{soil} on ψ_s. ψ_r was constrained to remain above a critical value of -1.5 MPa to mimic intervention of perfect stomatal control on crop water use. Calculations relate to a full-cover crop ($E_p \rightarrow ET$) with root depth of 20 cm in a medium-textured soil.

Figure 9.6a illustrates crop transpiration under three rates of ET*. At ET* = 6 mm d^{-1} the root system can supply E_p^* in wet soil ($\psi_s > -0.2$ MPa) but as ET* decreases from 4 to 2 mm d^{-1} so does the ψ_s to which E_p^* can be maintained. This latter point is explored further in Fig. 9.6b, which reveals the importance of root density under moderate ET* of 4 mm d^{-1}. As root density increases from 0.12 to 1.0 cm^{-2} and beyond, crop transpiration is maintained at a maximum rate to increasingly lower soil water potentials.

Figure 9.6c analyses sequences of daily transpiration (E_p) for crops with a range of L_v as transpiration gradually dries the profile from an originally wet condition. Since greater L_v can maintain E_p^* to lower ψ_s, E_p declines first with low L_v and progressively

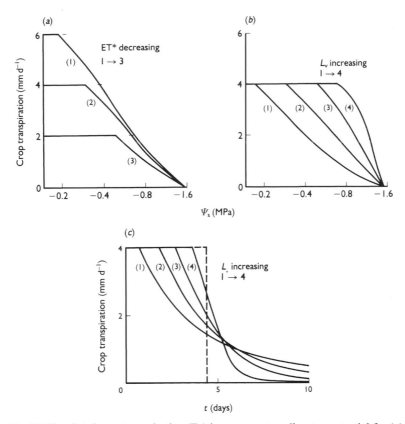

Fig. 9.6 Simulated crop transpiration (E_p) in response to soil water potential for (*a*) three levels of ET* = 6, 4, and 2 mm d^{-1}; (*b*) four levels of root-length density (L_v = 0.12, 0.25, 0.5 cm^{-2} and very dense); and (*c*) the decrease in E_p with time from an initially wet soil is shown for four levels of L_v (0.12, 0.25, 0.5 cm^{-2}, and very dense). (After Cowan 1965.)

later with greater L_v and capacity to extract water from soil. However, E_p is subsequently greater for crops of low L_v. Their more conservative use of water during early stages of drying has extended the duration of crop water use, though extraction at any Ψ_s was always slower than by crops of high L_v. This response would be amplified if stomatal control occurred at higher values of ψ_1.

Those simulations stress the dynamic nature of water uptake by crops. Availability of water depends not only upon ψ_s and how it changes with θ, but also on ET*, L_v, root depth, and patterns of ψ_1 and ψ_r. Results are consistent with observations (Fig. 9.7) on maize plants growing in pots within a field crop. Those data also show that relative transpiration (E_p/E_p^*) declined with soil water content but that form of response depended upon E_p^*. With $E_p^* = 1.4$ mm d^{-1}, E_p was maintained almost to wilting point but when E_p^* was 6.4 mm d^{-1}, a slight drying below field capacity was sufficient to lower E_p below E_p^*.

Restriction of transpiration by soil water content is less evident in the field. Thus observations of relative evapotranspiration (ET$_a$/ET*) of a number of full-cover field

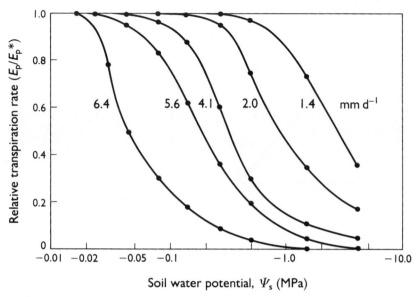

Fig. 9.7 Relative daily crop transpiration rate as a function of soil water potential for days of different E_p^* as shown on the curves. (After Denmead & Shaw 1962.)

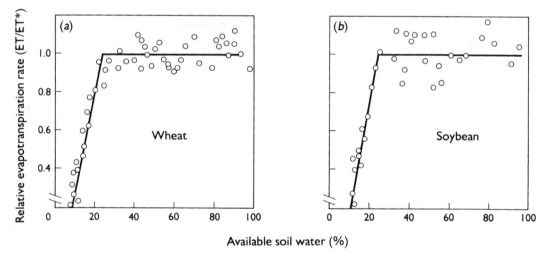

Fig. 9.8 Relative crop evapotranspiration rate (ET/ET*) as a function of available soil water content for (*a*) wheat, and (*b*) soybean. (After Meyer & Green 1981.)

crops have shown (e.g., Ritchie & Burnett 1971; Meyer & Green 1981; Shouse *et al.* 1982) that rates are often maintained until relative available soil water content falls below 0.3. This relationship (Fig. 9.8) holds across a range of soil textural classes independently of ψ_s. Below that water content, decline of ET_a/ET^* is essentially linear, an approximation widely used in models of crop water balance.

Box 9.2 A model of crop water use constructed from separate responses of transpiration and soil evaporation to atmospheric evaporative demand, crop cover, and soil water content

Step 1 Partition potential evapotranspiration (ET*) between soil and crop

ET* calculated with an equation from Table 9.1 is partitioned between potential soil evaporation (E_s^*) and potential crop transpiration (E_p^*) as in Eqs. 9.3 and 9.4 to:

$$E_s^* = \exp(-\omega L)ET^* \text{ and } E_p^* = [1 - \exp(-\omega L)]ET^* \quad (9.8)$$

where ω is an attenuation factor for net radiation (R_n) (c. 0.4) and L is crop LAI. These rates are maintained in the "energy-dependent" stage 1 (Fig. 9.5).

Step 2 Soil evaporation (E_s) and crop transpiration (E_p) at non-potential rates

E_s moves into the "supply-dependent" stage 2 (Section 9.2) falling rapidly as:

$$E_s = ct^{-0.5}\exp(-\omega L) \quad (9.9)$$

where c is a soil characteristic in range 2 to 3 mm d$^{-0.5}$ and t is day number in stage 2.

E_p^* is maintained while available water (AW) in the root zone exceeds 0.3 available water capacity (AWC) (Section 7.6), then E_p falls below E_p^* reaching $E_p = 0$ at AW = 0, i.e.,

$$E_p = E_p^* \min[1, AW/(0.3AWC)] \quad (9.10)$$

Step 3 Evapotranspiration

Under the many combinations of wet and dry surface and root zone water content, ET_a is given by:

$$ET_a = E_s + E_p \quad (9.11)$$

9.4 A model of crop water balance

Concepts introduced in this chapter concerning crop water use and its response to crop and environmental factors are drawn together in a model of crop water use in Box 9.2. It concerns a crop initially well supplied with water. ET* can be calculated using equations from Table 9.1 and be partitioned, as a function of canopy geometry, between E_s^* and E_p^* according to Eq. 9.3 and 9.4. Then as the crop dries, E_s and E_p fall below E_s^* and E_p^*, but their responses are distinct. E_s responds quickly as the surface soil surface dries, moving into a "supply-dependent" phase and decreasing rapidly below E_s^* (Fig. 9.5). In contrast, E_p^* is maintained as long as root zone water content exceeds c. 30%

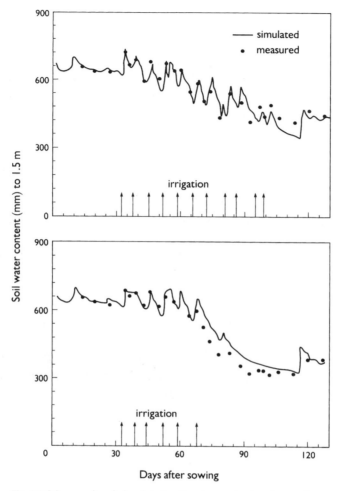

Fig. 9.9 Measured and simulated soil water content under sunflower crops with different irrigation regimes at Tatura, Australia (Connor, unpublished).

of its maximum available water capacity (AWC). Below that content, E_p falls linearly from E_p^*, reaching zero when available soil water (AW) is exhausted (Fig. 9.8). This simplification, relative to the principles of collection of water by root systems presented in Section 9.3, is made because alternatives require more detailed information on root length density and plant and soil water potentials than is usually available.

Those equations can reproduce dynamics of crop water use under intermittent rainfall and/or irrigation, variable ET*, and changing crop cover. They form the core of several published models of crop water balance. Ritchie (1972) pioneered this approach that has since been applied successfully to the analysis of many field problems. Data presented in Fig. 9.9 illustrate their use to explain observed patterns of water use by sunflower crops under two irrigation regimes at Tatura, Australia.

9.5   Responses of crops to water shortage

Drought is a meteorological term signifying a prolonged dry period; **drought resistance** is a crop's ability to withstand it. The definition of drought is purposefully broad to encompass a wide range of conditions that exist in agriculture, depending upon rainfall–evaporation balance, weather variability, soil water-holding capacity, crop type, rooting habit, and stage of development. In areas of high rainfall, a week without rain may cause significant yield loss. In semi-arid areas such "droughts" are common and cropping practices are modified to account for longer-than-usual dry spells (droughts) that can extend into months or years.

Principles controlling water flow through the soil–plant–atmosphere continuum (Section 9.1) and diurnal patterns of plant water potential presented in Fig. 9.2 emphasize that imbalance between water uptake and transpirational loss is a diurnal feature of crop water status in all environments. Difficulties arise when internal water deficit is either great or prolonged.

Three generalized patterns of water supply, depicted as relative available water contents of the root zone, emerge in relation to crop duration (Fig. 9.10). In the first, drought follows the major growth period during which water supply is adequate. Such terminal drought is characteristic of areas of mid-latitudes, i.e., Mediterranean climates, where winter and spring rainfall is followed in late spring and early summer by high evaporation and low rainfall. In the second pattern, crops suffer an initial period of drought because they are sown in anticipation of a short, reliable period of rain later in the season. This pattern is found in tropical, monsoonal areas where break of season is sharp and reliable but wet seasons are short relative to crop growth cycles. In the third case, which is characteristic of many areas (humid and semi-arid), seasonal rainfall totals may be reliable but intra-seasonal variability is large. Under those conditions, significant drought can occur at any time during growth.

Under conditions of diminishing water supply or increasing evaporative demand, maintenance of crop water status within tolerable limits depends upon crops' ability to restrict transpiration to the uptake capacity of root systems. Transpiration can be reduced either by reducing leaf area or loss per unit leaf area. Adjustments to leaf area are achieved over periods of days or weeks by less expansion and/or more senescence. In contrast, stomatal closure and leaf movement operate in minutes or hours to reduce transpiration per unit leaf area. Stomatal closure reduces leaf conductance, while leaf movements reduce radiation interception and hence temperature and saturated vapor pressure of transpiring leaves. However achieved, reduced transpiration will generally be associated with reduced growth. The common link in cases of stomatal closure and leaf movement is through reduced photosynthesis per unit leaf area. With less LAI, interception of solar radiation and photosynthesis are both reduced. This will be discussed later in Section 9.9.

The penalty of continuing water use is severe if crops fail to recover from prolonged internal deficit when water becomes again available. Where economic yield depends on critical development events such as flowering (Chapter 5), effects of transient water deficits on yield can be greatly magnified relative to effects on total growth.

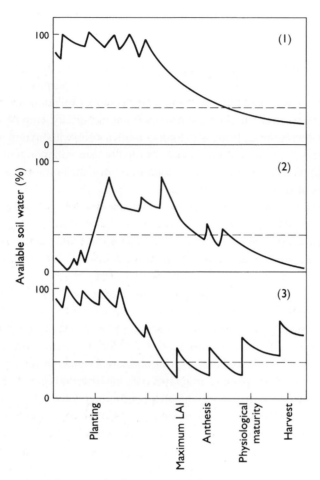

Fig. 9.10 Patterns of soil water availability identifying three types of drought: (1) terminal; (2) unreliable and terminal; and (3) unreliable. The dashed line represents the lower limit of soil water availability that will support ET*. (After Jordan & Miller 1980.)

Subsequent sections deal with the adaptations of crop plants to drought. Management of crop–water relations is discussed in Chapters 13 and 14. Variability and uncertainty of moisture supply make the design of cropping practice difficult for drought-prone regions (Chapter 13).

9.6   Adaptation to drought

Evolution has found many avenues for plant growth and survival in drought-prone areas. The challenge is to understand those responses so that crop yield can be improved by better management of existing cultivars and by purposeful breeding of better cultivars.

Table 9.3 Characteristics that enable plants to escape or resist drought

Drought escape
   Rapid phenological development
Drought resistance
(a) Drought avoidance (high plant water status)
    (i) Reduction in water loss
       high stomatal and cuticular resistance
       avoidance of radiation load
       low leaf area
    (ii) Maintenance of water uptake
       deep roots
       high root length density
       high hydraulic conductance
(b) Drought tolerance (low plant water status)
    (i) Maintenance of turgor
       osmotic adjustment
       high cell wall elasticity
       small cell size
    (ii) Tolerance of dehydration or desiccation
       protoplasmic tolerance
       cell wall properties

Plants can succeed in drought-prone regions by either escaping periods of drought or by resisting them. **Drought escape** may rely on developmental attributes that permit synchrony of life cycle to periods when there is a high probability for satisfactory growth conditions. But equally, it can also be achieved with a large supply of stored soil water per plant, as with wide-spaced plants. **Drought resistance** is achieved by various combinations of developmental, morphological, and physiological traits that allow plants either to balance transpiration and uptake and therefore **avoid** internal water deficits or to continue growth and water use and to **tolerate** them. There are many genes involved in the component traits of those adaptations, as summarized in Table 9.3.

In practice, distinctions between the classes are blurred. Drought avoiders benefit from their ability to resist drought, while for drought resistors there are many beneficial combinations of avoidance and tolerance. We need to respect evolution's well-tested solutions for survival but to realize, too, that in agriculture, production rather than survival is the major objective. Overemphasizing survival as a drought adaptation may be disadvantageous in agriculture.

9.6.1   Developmental traits

Adaptation to water-short environments requires that plants have phenological responses (Chapter 5) that provide developmental sequences that best match periods of most certain water supply. Timing and duration of phenostages are critical characteristics of adapted plants. The challenge for plant breeders is to develop cultivars that make the best use of water in dry years and yet possess flexibility for good use of wetter years.

Rate of development Rapid development can assist crops to complete their life cycle without serious water shortage. This is the behavior of many wild, desert ephemerals that also permits successful cropping in many water-short environments. Some workers have reported that water stress can hasten phenological development. Certainly, crops often develop more rapidly in dry years and this can stabilize yield. However, years that are drier than usual are often also warmer because $R_n \rightarrow H$ (see Chapter 6) so more rapid phenological development may be a thermal response rather than a direct physiological response to water stress.

Time of flowering The annual cycle of daylength (Fig. 5.2) is a reliable environmental signal and carefully tuned photoperiodic responses offer precise ways to time flower development. That is not easy, given the enormous range of sowing dates at different latitudes, however, and a common thrust in modern crop breeding has been to reduce the dependence of flowering time on daylength (Section 5.5). This widens the geographic range of individual cultivars. It may also be a valuable adaptation to variably hot and dry environments since photoperiodic requirement could restrain more rapid development in warmer conditions of drier years.

Duration of flowering In addition to timing of flower initiation, duration of flowering has important implications for seed crops in drought-prone areas. Determinate crops (Section 5.1.3), in which flowering occurs quickly and signals a rapid switch from vegetative growth to seed-filling, are vulnerable to transient stresses. A stress of even short duration at a critical time may cause mortality of a high percentage of ovules, pollen, or fertilized embryos. In such crops, there is little possibility to compensate for lowered yield potential. In some cereals such as sorghum, each tiller is determinate but tillering may extend over a relatively long period. In that case, late tillers may help compensate for losses in earlier tillers.

Crops with an indeterminate habit (e.g., soybean and field bean) maintain vegetative apices and flower from axillary buds over protracted periods. Transient stresses may cause loss of a particular age class of flowers and hence yield sites, but with return of more favorable conditions, flowering continues and yield potential can be re-established. The indeterminate habit is less efficient under favorable conditions (Chapter 5) in production of seed yield but that inefficiency is a small price to pay for the yield stability it lends in uncertain environments.

Different responses of determinate and indeterminate crops to transient stresses are summarized in Fig. 9.11. Vulnerability of the determinate crop wheat is further emphasized in Fig. 9.12, which illustrates the relative effects on final yield of stresses at various stages in development. Early stress is serious if it causes poor establishment. Reproductive development is sensitive at spikelet initiation, during the subsequent period of spike formation, and around anthesis itself.

Maize is also sensitive to water stress around flowering. In that species, water stress has been shown to delay silking, causing asynchrony between the release of pollen and the exposure of receptive silks (Hall *et al.* 1982). Under stress, this leads to reductions

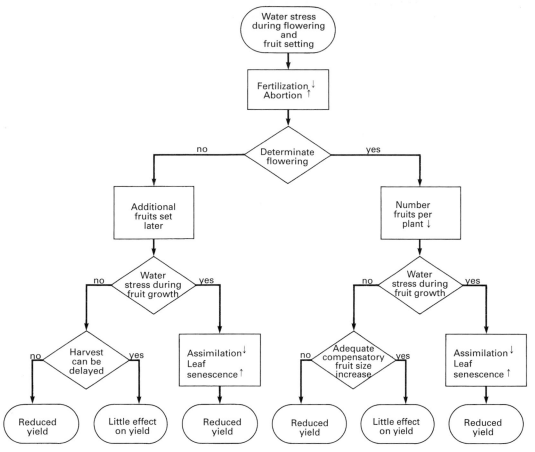

Fig. 9.11 Effects of water stress during reproductive development on grain yield of crops. The flow chart differentiates responses on grain number and grain size of determinate and indeterminate crops. (After Hsiao *et al.* 1976.)

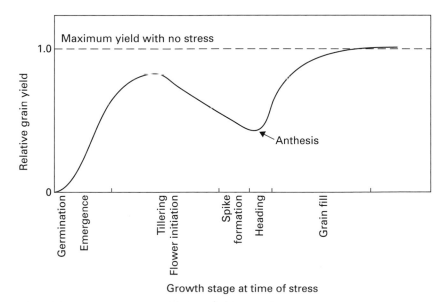

Fig. 9.12 Relative grain yield of wheat in response to water stress at various times during crop development. (After various authors.)

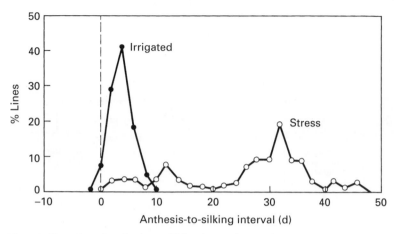

Fig. 9.13 Duration of anthesis-to-silking interval (ASI) in an elite maize breeding population subjected to water stress. Those plants that maintained a short ASI under stress also maintained yield. Yield was negligible in plants with ASI >15 d. (After Bolaños & Edmeades 1996.)

in grain set and yield. Large differences that exist among maize genotypes (Bolaños & Edmeades 1993, 1996) provide breeding material used in the development of cultivars for drought-prone areas (Banziger *et al.* 2006; Chapman & Edmeades 1999). Figure 9.13 compares the duration of the anthesis–silking interval in a number of elite maize lines grown under irrigation and under drought. Those lines that maintained a short anthesis–silking interval under stress also maintained grain set and yield. No other physiological or morphological parameter, from a wide range that was measured, was similarly highly correlated with yield under stress. Similar behavior of all lines under irrigation emphasizes that selection for this trait can only be made under conditions of water shortage. It is now known (Betrán *et al.* 2003) that the trait is largely additive and can be passed to hybrid maize from both sides of a pedigree.

9.6.2   Morphological traits

Individual traits may increase water uptake or reduce transpiration and so contribute to the control of crop water status. This section deals with the extent and behavior of exchange surfaces of root systems and canopies. The value of individual traits under water shortage depends upon total crop response, i.e., a functional balance between root system and canopy.

   Root exploration Crops can maintain high internal water status by exploring large soil volumes. Deep, dense root systems deliver water more readily and over a longer period (see Section 9.3). The value of those traits depends, of course, upon occasional replenishment of deep moisture. Although the traits are useful for productivity in some environments, their main value is for survival. The problem is that large amounts of assimilate are required for growth and maintenance of large root systems (Chapter 11).

   Grasses possess a dual root system that enables adaptation to variable conditions of water supply. Seminal roots with small L_v provide slow access to water throughout the soil profile. When the upper profile is wet, adventitious roots proliferate from nodal regions. Indeed it has been argued that observed root densities ($L_v > 10$ cm^{-2}) are excessively high and deplete soil water more quickly than is desirable for monocultures. When surface soil is dry, the few seminal roots that penetrate deeply draw more slowly upon stored water in the subsoil. A low-density root system ensures a conservative use of stored water. Perhaps high root density near the surface is in part an evolutionary adaptation to competition with unlike neighbors. It quickly scavenges surface water following rainfall and also gains access to nutrients usually concentrated there. Surface roots are an effective means to collect surface soil water that otherwise evaporates quickly to the atmosphere.

The strategy of conservative use of water has interesting ecological implications. The chaparral communities of California and Chile comprise a diversity of shrub species that show similar morphological adaptations to drought and employ a similar conservative strategy of water use (Mooney & Dunn 1970). Because no species could afford to adopt such a strategy if its competitors have equal access to soil water and continue to use it at high rates, these chaparral communities are excellent examples of co-evolution.

Different rules are possible in agricultural monocultures where water-use strategy is simplified and increasing attention is now given to it in breeding programs. Passioura (1983) outlined an interesting proposal for selection of root systems of large resistance that, under conditions of low water supply, would delay water use until the critical reproductive phase. A wheat breeding program was initiated (Richards & Passioura 1989) that reduced xylem vessel diameter of seminal roots from 65 to less than 55 μm and recorded yield increases of 8% above unselected controls in the driest environments. Consistent with theory, yield was maintained in wetter environments. The trait is now found in cultivars Kite and Cook (Richards 2006).

Attention is now turning to another conservative water use strategy based on a "maximum transpiration rate" characteristic that has been identified in a number of crop species, including sorghum (Sinclair *et al.* 2005), soybean (Fletcher *et al.* 2007), and wheat (Christopher *et al.* 2008). In these plants, leaf conductance responds to reduce, and in some cases stabilize, transpiration rate at high vapor pressure deficit. Initial analyses, confirmed by modeling (Sinclair *et al.* 2005), suggest that the trait will be most valuable for crops that rely on stored water because water saved can contribute to later use. A major problem in these approaches is that options for patterns of root growth and water use are too numerous for experimental evaluation. Modeling efforts that focus breeding efforts to the most favorable options (Chapter 13) will assist in this complex area for crop improvement.

Canopy properties Analysis of dynamics of leaf area under stress must consider component responses of **leaf initiation, leaf expansion** and **leaf senescence** if a complete picture of response is to be obtained.

Apical meristems are small and relatively isolated from water shortage with the result that leaf initiation is the last process affected by water stress. Under severe stress, leaf primordia may accumulate in meristems until relief provides conditions of assimilate supply and turgor suitable for cell division and expansion. In studies of canopy dynamics of cassava under water shortage, Connor and Cock (1981) recorded that deformed leaves

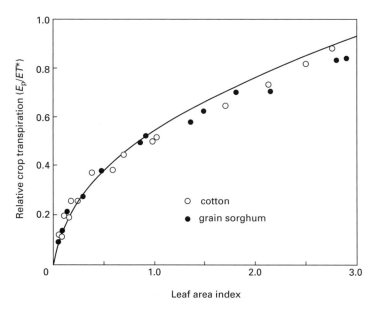

Fig. 9.14 Relative crop transpiration (E_p/E_p^*) as a function of LAI for cotton and grain sorghum crops under non-limiting conditions of soil water. (After Ritchie & Burnett 1971.)

appeared first after a 90-day period of rainfall exclusion, but were quickly followed by entire leaves.

In contrast to meristems, leaf expansion is particularly sensitive to slight internal water deficit so reduced expansion of leaves plays an important role in the adjustment of crop LAI to water shortage. As crops enter stress, transpiration and leaf expansion rates fall rapidly so measurements of leaf temperature, possible by remote infrared thermometry, and elongation provide rapid and effective means of measuring the onset and severity of water stress. Water stress reduces rate of LAI increase; this, combined with leaf senescence, may cause LAI to decrease. Both responses reduce the effect of water shortage on crop water status but at the expense of reduced LAI.

Increased rates of senescence and loss of leaves are common responses to stress. Older leaves located at the bottom of canopies generally senesce first. Ecologically, an extreme version of this tactic is the drought-deciduous behavior of many wild species. The hazard of not reducing leaf area is that complete canopies require greater amounts of substrate in maintenance respiration (Section 11.1) during a period when assimilation is severely reduced. In contrast, reduced leaf area has a lasting effect on crop water use and, with smaller respiratory load, can allow growth and survival for considerable periods. Provided crop cover is incomplete, reduction in leaf area will reduce crop transpiration although, as illustrated in Fig. 9.14 for cotton and grain sorghum, not in proportion to change in leaf area. Crops of incomplete cover experience additional radiation by reflection (shortwave) and radiation (long wave) from the soil surface particularly when it is dry. This together with direct loss of water by soil evaporation are major inefficiencies of this adaptation to water shortage. Despite this, low crop density can achieve long growth duration and is the basis for successful cropping in drought-prone areas (Chapter 13).

Interception of radiation can also be reduced by both increased leaf reflectivity and leaf movement. In some plants, water shortage increases leaf reflectivity to shortwave radiation by either increased pubescence or thickening of epidermal wax. In contrast to adjustment of LAI, leaf movement is a valuable short-term tactic from which crops can recover quickly when water again becomes available. Some species, especially legumes, have finely tuned heliotropic leaf movements that align leaves relative to the solar beam. Leaves of others, such as maize, are positioned spatially and in some cases azimuthly during development into open positions in the canopy. Leaves of some crops will only move passively, i.e., wilt or roll when they lose turgor, but these movements also contribute to reduced interception of radiation and hence to improved crop water status under water shortage.

Root–shoot balance The balance of expansive growth determines the value of morphological and physiological traits of root systems and canopies to water shortage. When water is not limiting, the abundance of active meristems in shoots can monopolize available assimilates. Root growth is then restricted owing to substrate shortage. Water stress changes activity of sources and sinks of assimilate around the plant. Stem elongation and leaf expansion are more sensitive than photosynthesis to stress. Therefore stress tends to allow accumulation of assimilate. Root systems experience higher and less variable water potential than shoots (Fig. 9.2). Given adequate substrate, root growth is generally less restricted by stress. These advantageous adjustments bring root and shoot functions into a new balance more appropriate for the current environment. Brouwer (1983) studied this process and gave it the name **functional equilibrium**. Depending upon the duration of stress there may be lasting effects following recovery.

Morphological combinations of reduced leaf area and high density of roots are important means for drought avoidance. High root/shoot ratios are a feature of many crops as well as wild species under drought. Changes in root/shoot ratios can be dramatic. Figure 9.15 illustrates profiles of L_v under weekly irrigated and rainfed sunflower. The difference is marked only in the surface layer, which was persistently dry in the rainfed treatment and generally moist under irrigation. The irrigated crop had a more extensive root system than the rainfed crop (7.8 compared with 5.2 km m^{-2} ground area). However, LAIs of the two crops were vastly different so that the rainfed crop, with 5.8 km root m^{-2} leaf area, had a greater capacity for balancing supply with demand (cf. 2.5 km m^{-2} leaf area for irrigated control).

Reproductive yield Current photosynthesis, stored assimilate, and materials mobilizable from leaves, stems, and roots serve as sources of assimilate for seed filling. Stored and current assimilates are equally available but the distinction identifies a tactic by which plants can compensate, through mobilization of stored material, for limitations to photosynthesis during stress.

The proportion of final seed yield that is attributable to preanthesis assimilate is variable. In severely stressed wheat crops it may be as great as 70% while in unstressed

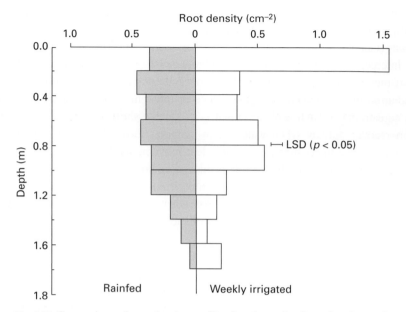

Fig. 9.15 Comparison of root density profiles five days after flowering for sunflower crops irrigated weekly or rainfed from sowing. Total root lengths were 7.8 and 5.2 km m^{-2} ground area, respectively. (After Connor & Jones 1985.)

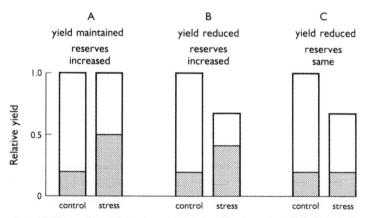

Fig. 9.16 Three alternative responses of the contribution of preanthesis assimilates to grain yield in terminally stressed cereals. In strategy A, yield is maintained under stress by increased transfer. In B, increased transfer does not maintain yield; while in C, transfer does not increase in response to stress. (After Turner & Begg 1981.)

crops it usually accounts for less than 10% of a much greater, final grain yield. Cultivars also differ in their relative contributions of preanthesis reserves to seed filling.

Three alternative responses to stored assimilate when photosynthesis is reduced by drought are illustrated in Fig. 9.16. In case A, yield is maintained by increased contribution from reserves. In case B, increased contribution is inadequate to maintain yield, while in case C there is no compensatory contribution from reserves. Even in case C, however, the proportional contribution of preanthesis assimilate to yield is increased by stress. These alternatives also raise the question of the extent and value of unused assimilate that might remain in crops at maturity. Greatest grain yields will always be achieved when reserves are completely transferred to grain. If reserves remain at maturity, then the major advantage to grain yield lies in yield stability under variable growth conditions (e.g., case A).

9.6.3   Physiological traits

Leaf resistance Rate of leaf transpiration (g H_2O m^{-2} s^{-1}) at temperature T_{leaf} ($°C$) can be expressed in the form:

$$\text{Transpiration} = [2170/(T_{leaf} + 273)](e^* - e_a)/(r_b + r_l) \tag{9.12}$$

where r_l (s m^{-1}) is leaf diffusive resistance to water vapor transfer and r_b (s m^{-1}) is the boundary layer resistance of the still layer of air associated with the leaf surface. Parameters e_a and e^* (kPa) are, respectively, vapor pressure of air and saturated vapor pressure at T_{leaf}. The ratio $[2170/(T_{leaf} + 273)]$ converts vapor pressure (kPa) to water vapor concentration (g m^{-3}) upon which diffusion depends.

Stomatal closure that increases r_l is the major mechanism by which plants can reduce transpiration per unit leaf area. This leads to higher leaf temperature and therefore closure must be nearly complete to overcome the higher vapor pressure gradient that increased temperature causes. The basic stomatal response is a feedback to low ψ_l. However, stomates of many species that maintain high water status during drought respond directly to atmospheric humidity sensed by water status of their guard cells independently of bulk ψ_l or to an unidentified hormonal signal generated in response to root water status (see Turner 1986). This gives such species a feed-forward response by which stomates close before the onset of plant water deficits. In that way they maintain high internal water status, explaining why leaf resistance in some species is independent of bulk ψ_l.

Data for well-watered and water-stressed sunflower and cassava (Fig. 9.17) demonstrate two extreme reactions of leaf diffusive resistance (r_l) to water shortage. Sunflower maintains open stomates to low ψ_l but in cassava a feed-forward response to atmospheric humidity closes the stomates at high ψ_l. Stressed crops of cassava have comparable diurnal patterns of ψ_l to well-watered crops even though their growth is severely limited by water shortage. This example demonstrates that while ψ_l does define the internal water status of a crop, it is not possible to understand the nature of response to water shortage without associated measurements of r_l.

Canopy resistance The Monteith equation (Table 9.1) can be used to assess the importance of stomatal closure to crop water use. In this case, r_c is diffusive resistance of the canopy composed of a number of leaf layers transpiring in parallel. If leaf resistance is r_l (two surfaces also in parallel) then $r_c = r_l/\text{LAI}$; r_a is the aerodynamic resistance to water vapor transport from bulk air to the effective crop surface ($= 1/K_{H_2O}$,

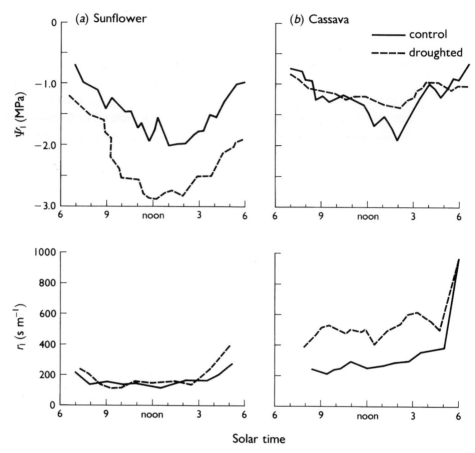

Fig. 9.17 Comparison of leaf diffusive resistance (r_l) of sunflower and cassava in response to water stress. In cassava, stomata close in response to low humidity and maintain high leaf water potential (Ψ_l) of stressed crops. In sunflower, stomata remain open and Ψ_l deviates markedly from unstressed control. In neither case is there a clear relationship between Ψ_l and r_l. (After Connor & Palta 1981 and Connor *et al.* 1985b.)

Section 6.6) that depends on crop structure and windspeed. As r_l increases due to stomatal closure, or LAI decreases, E_p will fall below E_p^* according to:

$$E_p/E_p^* = (s + \gamma)/[s + \gamma(1 + r_c/r_a)] \tag{9.13}$$

Calculations using Eq. 9.13 at $T_{air} = 20°C$ reveal that stomatal closure must be substantial to have much effect at usual aerodynamic resistances. In dense canopies, r_a is large (*c.* 700 s m^{-1}) and an increase in r_c has little effect on E_p. Such canopies are said to be "uncoupled" from the atmosphere. Stomatal closure has more effect in sparse canopies with small r_a, but in those canopies E_s may be an important part of ET and this simple equation is not suited to a complete analysis.

Stomatal closure is most effective when the temperature and hence the vapor pressure deficit are highest. Thus afternoon stomatal closure is an effective way to reduce daily crop water consumption. The extreme case of stomatal control is seen in crassulacean acid metabolism (CAM) plants, of which pineapple is the notable crop species (Ekern 1965). As that crop reaches full cover, transpiration, and hence crop water use, can be essentially suspended during daylight hours. Nocturnal uptake of CO_2 (stored temporarily in organic acids) when vapor pressure gradients and hence transpiration are least, accounts for the high water-use efficiency of this crop. The trade-off, however, is relatively low daily growth rates compared to crops with C3 or C4 photosynthesis.

Maintenance of turgor Many cellular processes respond to turgor and therefore adjustments that maintain turgor at low plant water potential are valuable attributes for avoidance of desiccation. Osmotic adjustment, increased elasticity of cell walls, and small cell size contribute to the maintenance of turgor.

Osmotic adjustment has been extensively studied in crop plants and is a common feature of those species that maintain open stomata, active metabolism, leaf and root expansion, and continued extraction of soil water at low internal water potential. Wheat is a well-studied example (Morgan 1983, 2000). Plants adjust osmotically by accumulating organic and inorganic solutes held within limiting membranes. Adjustment requires that stress develop slowly and may be reversible since recovery of some osmotic compounds does occur. In osmotic adjustment, the cost to productivity will vary depending upon how it is accomplished (Loomis 1985). Adjustment through accumulation of mineral ions (K^+, Cl^-) and organic acids has only small cost (0.5–2 mol ATP per mol osmoticum). Sugars are metabolically costly but are mobile and can be used later in growth, so actual life-cycle cost will also be low and be attributable mainly to transport. The same can be said of ordinary, highly oxidized organic acids but not of highly reduced and therefore expensive substances such as proline and betaine.

As noted earlier, the situation for natural communities and mixed agricultural stands can be very different. In both cases, if one species adjusts, others must do likewise to survive. The result is a less efficient use of the scarce resource, water. The same logic explains why plant strategies that increase a cultivar's ability to access additional water may detract from productivity of monocultures. For such communities a conservative strategy, for example closing stomata near -1.5 MPa and waiting for the next rain or irrigation, is often more appropriate (de Wit 1958).

Tolerance to desiccation is assumed to be related in some way to cellular composition, membrane stability, and enzyme activity. Examples of tolerance to low water status include lichens and vegetative parts of a group of higher plants known as "resurrection plants" (Gaff 1981). Those plants, and seeds of most plants, tolerate great desiccation. Vegetative tissues of crop plants face much less extreme water loss diurnally and during droughts, but the ability to regain function when stress is relieved is no less important.

9.6.4 Breeding for improved drought resistance

The previous discussion has emphasized the wide range of plant traits, and hence their combinations and interactions with management, that can contribute to drought resistance in crops. So far, most success has been achieved through selection and breeding

---

Box 9.3 Drought tolerant transgenic crop cultivars

Following success with insect- and herbicide-resistant cultivars, transgenic research is now turning attention to the development of drought "tolerant" cultivars. New cultivars are expected by 2015 (Marris 2008). The challenge to change morphological, physiological, and metabolic characteristics in combinations that evolution has not already found valuable over millennia seems, at present, a distant hope. This chapter has emphasized the many forms that drought can take and the many responses that can assist productivity through escape, avoidance, and tolerance. Progress so far with transgenes reveals, in contrast, hope placed in a small number of genes reported to increase photosynthesis, stabilize yield, or promote stomatal closure (Nelson *et al.* 2007; Edmeades 2008).

Two limitations to progress are evident. First, appreciation of the nature of drought and the interactions between component responses are too limited. It is a concern that the biotech literature consistently frames the search as one for drought "tolerance". This criticism is not just a matter of semantics. The new techniques are not well served by ignoring the broader background that ecology and physiology have provided on plant response to water shortage (Section 9.6). Second, while crop simulation models offer, in principle, an analytical framework to evaluate interactions of individual responses to patterns of water shortage, current models are inadequate to the task. A new generation of crop simulation models is required that deals comprehensively with physiological responses that can be linked to gene effects. These models will be intensely more physiological than those currently available and will characteristically work on short, for example hourly, time steps appropriate to the metabolic and physiological responses they portray.

Unquestionably, new drought resistant cultivars could contribute significantly to greater productivity needed to feed our expanding world population. We hope that significant gains can be achieved during the lifetime of this book but caution against excess optimism and call for greater collaboration among plant molecular geneticists, physiologists, and ecologists.

---

for phenological development suited to patterns of water availability (Fig. 9.10). Further advantages have been added by a range of, mostly individual, morphological and physiological traits that have found use in particular circumstances. Progress is seriously hampered by limited understanding of how combinations of traits might respond to management under the generally variable water supply conditions of rainfed agriculture, especially in semi-arid areas. Crop simulation models are the only known method to study such interactions. Some progress has been made but more attention is needed to the physiological content of these models if they are to contribute more. At present, the major thrust of research has moved from physiology to genetics with promises of early progress with transgenic cultivars. The hopes for this (Box 9.3) do not diminish the need to maintain a strong physiological focus because that is the level at which simulation models operate.

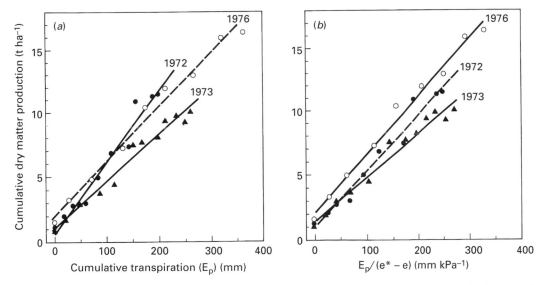

Fig. 9.18 Relation of total dry matter production in potato crops during the period of tuber growth with (*a*) cumulative transpiration (E_p), and (*b*) $E_p/(e^* - e)$. (After Tanner 1981.)

9.7   Water-use efficiency

Dry matter production per unit of water used by a crop is an important ecological and agronomic response that can be expressed with various efficiency indices. **Crop water-use efficiency** can be expressed as ratios of production of total biomass (B), yield (Y), or the glucose equivalent of those masses (G) (see Section 11.2) with evapotranspiration (ET). Those ratios are designated WUE_B, WUE_Y, or WUE_G. WUE_B and WUE_Y are most commonly used; WUE_G is of particular interest in comparisons between crops of differing chemical composition.

During the growth of an annual crop, E_s dominates ET early in the season whereas E_p is the major component once full cover is achieved. For the entire cycle, E_p accounts for 30 to 85% ET. These considerations are important because the best chance for a causal relationship exists between growth and transpiration (E_p), i.e., within corresponding **crop transpiration efficiencies** (TE_B, TE_Y, or TE_G). This derives from the functional relationship between photosynthesis and transpiration of individual leaves (Eq. 10.5) that extends collectively to entire crops. Strong correlations between crop growth and ET only develop when the surface soil is dry and E_s is small, or when $E_p \rightarrow$ ET, as occurs in full-cover crops.

9.7.1   Relationships between crop growth and water use

There are many data to demonstrate linear relationships between growth and transpiration and they have been variously summarized (e.g., de Wit 1958; Arkley 1963; Hanks 1983). An example of a relationship between cumulative biomass (*B*) and cumulative transpiration (E_p) of individual potato crops is presented in Fig. 9.18. In those data,

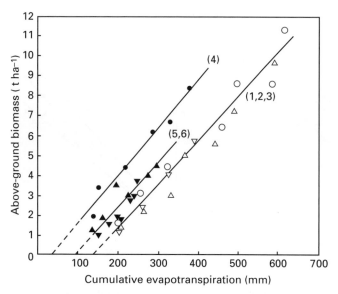

Fig. 9.19 Growth and progressive evapotranspiration (ET) of sunflower crops subjected to different irrigation sequences. Three linear relationships describe the responses. Transpiration efficiencies (TE$_B$, gradients of the lines) were comparable for all crops but soil evaporation (E$_s$) (intercepts of lines with the abscissa) were different because of differences in wetting patterns and exposure of the soil surface. Differences in evapotranspiration efficiency (ΔB/ΔET) were due to differences in E$_s$. (After Connor *et al.* 1985a.)

B of three crops is plotted against E$_p$ for individual periods of 3 to 4 months. B is plotted directly against E$_p$ in Fig. 9.18a and against E$_p$/($e^* - e_a$) in Fig. 9.18b. Both graphs display strong linear relationships but, for the latter, there is more consistency between years. That arises because inclusion of $e^* - e_a$ accounts for one important source of year-to-year variation in TE, i.e., variation in evaporative demand.

For individual crops, larger differences can be expected between geographical locations. Gradients of the relationships in Fig. 9.18b are the most generally applicable estimates of TE$_B$ from these experiments. They can be applied to other potato crops at the same or comparable locations for which ($e^* - e_a$) is available.

The functional link between B and E$_p$ also dominates relationships between B and ET. Figure 9.19 compares progressive B and ET for six sunflower crops grown simultaneously at a single site but under different irrigation regimes. Here, gradients of the lines approximate TE$_B$ so that the intercept estimates E$_s$ for the growing season. They reveal that TE$_B$ did not vary between irrigation regimes and that differences in WUE$_B$ resulted from differences in E$_s$/ET. E$_s$ was greatest in crops that were irrigated frequently (treatments 1, 2, and 3) because in those crops surface soil remained moist for a great part of the growing season. E$_s$ was smallest in crops that received no irrigation (treatment 4) because the surface soil was mostly dry. It was intermediate in other treatments (5 and 6) that received various patterns of infrequent irrigation.

Table 9.4 presents field measurements of TE$_B$ and TE$_Y$, and the same data adjusted to vpd ($\Delta e = e^* - e$) as TE$^{adj} = B/(E_p/\Delta e)$, for a number of crops for growth periods

Table 9.4 Median (and range) of field-measured TE_B (g aboveground dry matter kg^{-1} water) and TE_Y (g marketable dry matter kg^{-1} water) for various crop species, and the values adjusted (TE^{adj}) to mean daytime vapor pressure deficit (Δe) = 1 kPa

Data compiled by Patricio Grassini.

Species	n_B	TE_B	TE_B^{adj}	n_y	TE_Y	TE_Y^{adj}
C$_4$ grasses						
Grain sorghum[a, b, c]	11	4.2 (2.9–5.7)	10.0 (7.1–11.8)	10	1.9 (1.2–2.2)	4.6 (3.0–5.4)
Sugarcane[d]	1	7.0	8.7	0	–	–
Maize[a, e, f, g]	35	4.8 (1.7–8.3)	8.5 (5.5–13.8)	23	2.3 (1.3–3.6)	4.4 (2.6–7.4)
Pearl millet[h]	3	3.9 (2.1–4.6)	7.0 (6.3–7.9)	0	–	–
C$_3$ grasses						
Barley[i, j]	9	4.7 (3.0–5.3)	4.8 (3.2–6.5)	5	2.1 (1.7–2.5)	1.7 (1.2–3.4)
Wheat[e, i, j, k]	50	4.5 (2.9–8.6)	4.8 (2.7–6.0)	16	2.2 (1.0–3.0)	1.7 (0.9–2.1)
Oat[j]	6	4.1 (3.4–5.4)	3.6 (2.9–4.2)	6	1.5 (2.5–3.5)	1.2 (1.1–1.6)
Rice[l]	8	3.0 (2.2–4.0)¶	3.3 (1.3–5.0)¶	0	–	–
C$_3$ crops						
Potato[a, j]	4	4.9 (3.6–7.8)‡	6.4 (5.4–7.1)‡	4	4.1 (3.3–6.2)	5.4 (4.7–6.0)
Sugarbeet[m]	5	7.2 (5.4–9.7)‡	4.7 (4.0–5.2)‡	0	–	–
Sunflower[c, p, o]	17	2.2 (1.9–3.5)†	4.5 (3.5–5.8)†	14	0.8 (0.4–1.0)	1.6 (0.8–2.2)
Cotton[r]	1	1.9*	4.1*	1	0.8	1.8
C$_3$ grain legumes						
Soybean[a, f]	5	2.6 (1.4–2.8)*	4.4 (3.4–5.3)*	5	1.1 (0.7–1.3)	1.9 (1.6–2.2)
Peanut[h, s]	4	2.2 (1.9–4.1)*	4.1 (2.3–4.5)*	0	–	–
Field peas[j]	1	6.0*	3.8*	1	3.0	1.9
Chickpea[c]	1	3.0*	3.8*	0	–	–
Faba bean[j]	1	3.9*	3.1*	0	–	–

Notes: ‡ plants grown in containers; TE_B estimates include belowground organs; † based only on pre-anthesis growth; * estimates may be biased due to protein and oil synthesis costs during reproductive stages.
Sources: [a] Tanner and Sinclair (1983) and references cited therein; [b] Garrity *et al.* (1982); [c] Steduto and Albrizio (2005); [d] Inman-Bamber and McGlinchey (2003); [e] Kremer *et al.* (2008) and references cited therein; [f] Suyker and Verma (2009); [g] Otegui *et al.* (1995), unpublished results; [h] Squire (1990) and references cited therein; [i] Kemanian *et al.* (2005) and references cited therein; [j] Ehlers and Goss (2003) and references cited therein; [k] Abbate *et al.* (2004); [l] Haefele *et al.* (2009); [m] Clover *et al.* (2001); [p] Connor *et al.* (1985a); [o] Soriano *et al.* (2004); [r] Orgaz *et al.* (1992); [s] Collino *et al.* (2000).

of one to several months. The data illustrate two important points. First, values of TE^{adj} relative to TE reveal that measurements were taken over a wide range of $1 > \Delta e > 2$ kPa and that warm season crops generally experienced greater Δe than cool season crops (cf. C4 grasses vs. C3 grasses; and peanut, potato, and sunflower vs. sugarbeet and field pea). Second, when adjusted to comparable evaporative environments, here to 1 kPa, C4 grasses have greater TE^{adj} than C3 grasses. The larger TE^{adj} of C4 species arises from greater photosynthesis rates in the warmer environments to which they are adapted (Section 10.2.3). Third, comparison between grain legumes and other C3 crops suggests a smaller TE^{adj}. One reason for low TE^{adj} of legumes is that symbiotic N fixation has a metabolic cost that reduces the growth of host plants (Chapter 8).

The value of these numbers is that they define the terms of trade under which terrestrial plants exchange water with the atmosphere for CO_2 for biomass production and for yield. The low rate of exchange is the inescapable result of the anatomical–physiological solution found by evolution to conquer the terrestrial environment. A more effective solution would be a membrane differentially permeable to water and CO_2, but that has never eventuated so all plants rely on stomates for control of excessive water loss on stomates.

9.8   Review of key concepts

Soil–plant–atmosphere continuum

- Crops inevitably lose water by transpiration when leaves absorb CO_2 through open stomates in photosynthesis. Leaf water potential (ψ_l) falls and establishes a flow of water from soil when soil water potential (ψ_s) is greater than that in roots. Continuing uptake of water is essential to maintain internal conditions suitable for growth. In this sense, plants grow by exchanging H_2O for CO_2 with the atmosphere.
- Amount of transpiration depends upon evaporative demand of the atmosphere. The result is a wide range of evolved mechanisms, mostly associated with leaf area, stomatal density and distribution, and response to internal and external factors that seek to maintain internal water status within tolerable limits.

Crop water use (evapotranspiration)

- Crops also lose water by direct evaporation from soil surfaces and by drainage below the root zone. Separate identification of transpiration and evaporation components is necessary to explain water use of crops of incomplete cover, to determine relationships between crop evapotranspiration (ET_a) and growth, and to understand the response of crops to water shortage.
- When water is freely available and crops have large LAI, most water use is by transpiration. Under those conditions, evapotranspiration is determined by R_n and evaporative demand of the atmosphere. That potential rate (ET^*) sets the upper boundary to water use at which it presents no restriction to crop growth.
- There are small but significant differences in ET^* between various crops but the major determinants of ET^* under the same environmental conditions are duration of growth cycle and rapidity in achieving complete foliage cover.

Effect of water shortage on productivity

- Water shortage will always reduce growth of crops not restricted by other environmental conditions. The impact will be seen in smaller and fewer vegetative and reproductive organs.
- Effect of water shortage on reproductive yield depends upon timing of growth restrictions relative to the phenological cycle of flower initiation, flowering, and growth of

reproductive organs. Prolonged flowering (indeterminate habit) allows crops to compensate for loss of flowers during a short, stressful period. If stresses can be avoided during flowering, however, that strategy is inferior to determinate flowering in which an optimum balance can be achieved between vegetative and reproductive growth for a given availability of water and crop cycle length.

Water-use efficiency

- The close link between transpiration and growth can also extend to crop water-use efficiency, i.e., growth relative to ET, and even to yield, especially for crops with a relatively small component of soil evaporation.
- Management to improve crop water-use efficiency relies on reducing losses to soil evaporation, maximizing transpiration, and improving transpiration efficiency by growing crops under conditions of low evaporative demand. Mostly, however, maximum water-use efficiency is not an adequate goal. Rather it is necessary to combine that with the use of all available water to maximize production.

Drought resistance

- In crop response to drought, it is possible to distinguish escape mechanisms mediated mostly by phenological response from resistance mechanisms dependent upon morphological and physiological characteristics. A major distinction within resistance is made between those mechanisms that restrict water loss relative to water uptake and so avoid internal water deficit, and those that confer ability to tolerate low water content and therefore to continue activity. The value of these strategies in the field depends upon expectancy of rainfall and behavior of competing species.
- Since improvement of performance under water shortage can be achieved in many ways, progress in genetic improvement for drought resistance requires careful definition of phenological and physiological mechanisms and target environments for which alternative strategies are required.
- Most success in developing drought-resistant crop cultivars has been achieved by adjusting phenological responses, i.e., by selecting the growing season to suit the period and amount of water that is available through rainfall or can be made available by management.

Terms to remember: crop, coefficient, transpiration efficiency, and water-use efficiency; drought, avoidance, escape, resistance, and tolerance; evaporation and transpiration; evapotranspiration, actual, potential, and reference crop (ET_0); functional equilibrium; hydrological balance; lysimeter; plant or soil water, content, and potential; root, exploration, length density; soil–plant–atmosphere continuum; soil evaporation, stage 1 and stage 2; turgor.

10 Photosynthesis

Photosynthesis is the primary process in crop production. It supplies reduced carbon for the construction of biomass and as the source of energy in metabolism. Leaves are the functional units of crop photosynthesis; their efficiency in capture and utilization of solar energy determines productivity. The transport of CO_2 from the atmosphere to sites of fixation in leaves is limited by slower moving air within canopies, boundary layers of still air surrounding leaves, stomatal pores in leaf epidermis, and by the interior structure of leaves.

The area (LAI) and arrangement of foliage, i.e., **canopy architecture**, determine the interception of solar radiation by individual leaves of a crop. Leaf area and arrangement change during crop growth and, by leaf movement, during each day. Maximum crop production requires complete capture of solar radiation and supporting levels of water and nutrients. When water or nutrients are in short supply, productivity is reduced by incomplete capture of radiation and/or less efficient utilization of it.

This chapter begins with a discussion of photosynthesis and photosynthetic responses of leaves progressing to analyses and explanations of spatial and temporal variation of photosynthesis of crop canopies.

10.1   Photosynthetic systems

The central processes of photosynthesis are common to all plants but variants have evolved ancillary chemical, morphological, and physiological mechanisms that result in three photosynthetic systems with important ecological adaptations.

10.1.1   The central processes

The fundamental feature of photosynthesis is incorporation of inorganic substrates into organic products. The central reaction, reduction of CO_2 to carbohydrate (CH_2O), the principal first product, can be summarized as:

$$CO_2 + 4e^- + 4H^+ \rightarrow (CH_2O) + H_2O \qquad (10.1)$$

Electrons and protons are liberated by solar energy in the photolysis of water:

$$2H_2O \rightarrow 4H^+ + 4e^- + O_2 \qquad (10.2)$$

These two processes, respectively the **dark** and **light** reactions of photosynthesis, occur in chloroplasts of green leaves. NO_3^- and SO_4^{2-} are also reduced in chloroplasts but they are minor substrates for the dark reactions of photosynthesis that are principally concerned with reduction of CO_2. Plant biomass consists of around 45% C but only 2% N and 0.2% S, and there are other sites within plants where NO_3^- and SO_4^{2-} are reduced.

Equations 10.1 and 10.2 explain that photosynthesis can be measured by the uptake of CO_2, evolution of O_2, and by the production of carbohydrate. The latter can be expressed as mass or as its heat of combustion. Carbon dioxide exchange offers a practical way of measuring photosynthesis of crops with infrared gas analyzers (IRGA) that allow easy and accurate measurement of $[CO_2]$ in air streams. As their name suggests, they measure $[CO_2]$ by its absorption of infrared radiation (Fig. 6.5). There is no equally convenient method to measure photosynthesis by O_2 exchange.

These equations also establish the maximum energetic efficiency of photosynthesis because the minimum number of quanta in the photosynthetically active waveband (also known as photons) required to generate sufficient reductant for a molecule of CO_2, the **quantum requirement,** is around 12 to 16. Its reciprocal, termed **quantum efficiency**, has a maximum value in the range of 0.06 to 0.08 CO_2 per quantum.

10.1.2   Photosynthetic groups

Plants have evolved three different chemo-anatomical systems (**termed C3**, **C4**, and **CAM**) that provide suitable internal environments for the light and dark reactions. Terrestrial plants evolved from algae with C3 photosynthesis and retain it. Some have since evolved to either C4 or CAM variants. Floristic distribution of C4 types (Evans 1971) suggests that they have evolved separately on more than one occasion. Crop plants, like plants generally, are dominated by C3 species. Table 10.1 records the distribution of some important crop plants according to their photosynthetic system.

C3 photosynthesis In the presence of the universal photosynthetic enzyme rubisco, a five-carbon sugar, ribulose bisphosphate (RuBP), accepts CO_2 and the unstable product splits to form three-carbon phosphoglyceric acid, hence the name C3. While CO_2 fixation (carboxylation) is its major activity, rubisco also behaves as an oxygenase. In light, O_2 competes with CO_2 for active sites on the enzyme and catalyzes oxidation of RuBP. Subsequent metabolism, in a process called **photorespiration**, recycles the oxidation products and releases CO_2. Internal leaf tissues have high $[O_2]$ because atmospheric concentration is high (21 % O_2 v/v) and also because it is produced by photosynthesis. In consequence, photorespiration is an inescapable process that reduces net CO_2 gain in C3 plants.

In C3 plants, CO_2 from interior leaf spaces is adsorbed into cell walls of the mesophyll as dissolved CO_2 and as HCO_3^-. The enzyme carbonic anhydrase hydrates dissolved CO_2 to form HCO_3^-, thus increasing the effective $[CO_2]$ at chloroplasts. However, low conductance of mesophyll cells to CO_2 strongly restricts the flux of CO_2 from the atmosphere to chloroplasts such that $[CO_2]_i$ is high in substomatal spaces, ranging from 200 to 260 ppm ($\mu L\ L^{-1}$).

Table 10.1 Some important crops classified according to product type and photosynthetic system

Product	Photosynthetic system		
	C3	C4	CAM
Beverages and drugs	Tea Coffee	–	Agave
Root crops	Potato Cassava	–	–
Grains	Wheat Barley Rice	Maize Millet Sorghum Amaranth	–
Sugars	Sugarbeet	Sugarcane	–
Fibers	Flax Cotton	–	Sisal
Oils	Soybean Sunflower Rapeseed	–	–
Fruits	Apples Banana	–	Pineapple *Opuntia* spp.
Forages	Ryegrass Alfalfa Clovers	Panic grass Rhodes grass (no legumes)	–

C4 photosynthesis This system occurs mainly in tropical members of the family Gramineae but is also found in a few floristically diverse dicots. It is an adaptation that enables a higher photosynthetic rate than does C3 under high atmospheric $[O_2]$. In C4 plants, CO_2 is fixed in mesophyll cells by the enzyme phosphoenolpyruvate (PEP) carboxylase to form C4 acids after which the system is named. The activity of this enzyme is not affected by O_2. The C4 acids are transported to specialized chloroplasts of bundle sheath cells that surround vascular traces. There, CO_2 is released from C4 acids and enters C3 photosynthesis through rubisco. Phosphoenolpyruvate carboxylase has a high affinity for CO_2 so this combination of chemistry and anatomy concentrates CO_2 in bundle-sheath chloroplasts, thereby enhancing the performance of rubisco. C4 leaves reveal no photorespiration because rubisco is isolated in interior cells and CO_2 released internally by photorespiration is readily recycled through PEP carboxylase in mesophyll cells. As a result, they have higher rates of photosynthesis than C3 leaves, especially at high irradiance and high temperature.

CAM photosynthesis CAM, or crassulacean acid metabolism, is named after the family Crassulaceae where it prevails. It occurs only in succulent plants of which few are grown as crops (Table 10.1). Crassulacean acid metabolism photosynthesis is also of recent evolutionary origin and like C4 is an adaptation of C3. The CAM system has the same additional four-carbon sequence of C4 but not its spatial compartmentation within leaves. In CAM, separation of the two systems is largely temporal. Under drought conditions, CAM plants maintain open stomates only during nights and CO_2 accumulates in C4 organic acids. These are stored in large aqueous volumes of their succulent tissues. During daytime with stomates closed, stored CO_2 is released into C3 photosynthesis. Daytime stomatal closure greatly reduces transpiration so that the ratio of CO_2 gain to H_2O loss by transpiration over a daily period is much greater than in C4 and C3 systems. Photosynthetic capacity is limited, however, by capacity for acid storage.

When water is freely available, many CAM plants, such as pineapple, open their stomates during the day and behave chemically as C3 plants. In this way some CAM plants are capable of very high production. For pineapple, this behavior is best expressed in regions with warm days and cool nights where nocturnal accumulation of CO_2 and daytime C3 system are both active. Under appropriate conditions, commercial crops of pineapple approach the carbohydrate productivity of C4 sugarcane (Bartholomew 1982).

Intermediate photosynthetic types Recent surveys have identified plants with structure, enzyme distribution, and photosynthetic behavior intermediate between C3 and C4. The agriculturally important grass genus *Panicum* contains C3 and C4 as well as intermediate types (Holaday & Black 1981). Cassava is now a well-studied example. It possesses substantial quantities of enzymes of the C4 pathway but without the typical leaf anatomy. Many of its gas exchange characteristics are intermediate between C3 and C4 species (Cock *et al.* 1987) and leaf photosynthesis rates compare favorably with C4 species when grown in warm, humid conditions (El-Sharkawy 2007; El-Sharkawy *et al.* 2008).

10.2 Leaf photosynthesis

Leaf photosynthesis (P_n) can be measured in laboratory and field by CO_2 uptake in transparent chambers. The technique requires the control of temperature and humidity. In addition, some leaf chambers are fitted with light sources that enable *in situ* construction of P_n vs. PAR responses. The main environmental and physiological influences on P_n are incident PAR, atmospheric [CO_2], photosynthesis type, and leaf diffusive resistance to CO_2 transfer. Water stress operates initially by increasing leaf diffusive resistance through stomatal closure but, if severe, it also affects carboxylation. Temperature affects internal [CO_2] through respiration within the normal range and reduces carboxylation at extremes. Nitrogen deficiency reduces RuBP and PEP concentration, which reduces P_n.

10.2.1 Light

Net photosynthesis responds to increasing irradiance (I) according to:

$$P_n = [P_n^*(I - I_c)\alpha]/[P_n^* + (I - I_c)\alpha] \tag{10.3}$$

as represented in Fig. 10.1. At $I = 0$, the leaf evolves CO_2 by **dark respiration** at the rate R_d that can be estimated as $-\alpha I_c$. At $I = I_c$, the **light compensation point**, $P_n = 0$ because photosynthesis exactly balances loss by R_d. As I increases above I_c, P_n increases rapidly to become **light saturated** at a maximum rate of P_n^*. In this example (Fig. 10.1),

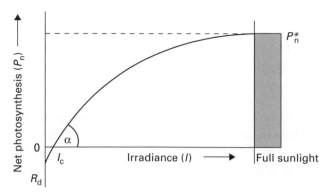

Fig. 10.1 A typical response curve of leaf net photosynthesis (P_n) to irradiance (I) at ambient CO_2 concentration. R_d is dark respiration, I_c light compensation point, α radiation-use efficiency at low light, and P_n^* is photosynthesis rate at light saturation (see Eq. 10.3). The shaded section emphasizes that "full sunlight" is a range of irradiance not a single value.

P_n becomes light saturated well below full sunlight, a characteristic of most C3 species. P_n^*, under optimal environmental conditions, varies greatly between and within C3 and C4 groups whereas the initial slope (α) of the response curve is relatively constant for all plants. The term **gross photosynthesis** (P_g) is used to describe the sum $P_n + R_d$.

Radiation-use and quantum efficiency Radiation-use efficiency of leaf photosynthesis (RUE $= dP_n/dI$ of the response in Fig. 10.1) decreases from its maximum value α with increasing irradiance. An important consequence is that greatest efficiency for foliage canopies occurs when most leaves receive low irradiance and operate near maximum radiation-use efficiency. Canopy architecture (Section 10.5) determines the distribution of irradiance over photosynthetic surfaces and hence, relative to leaf photosynthetic response, the possibility for high canopy radiation-use efficiency.

When I is given in PAR units (μmol m^{-2} s^{-1}) and corrected for losses due to reflection and transmission, the initial slope is the quantum efficiency of photosynthesis. It provides a basis for comparing intrinsic efficiency of photosynthetic systems.

In theory, C4 plants have smaller quantum efficiency than C3 plants due to additional carboxylation and membrane transfers of intermediate compounds (Osmond *et al.* 1982; Amthor 2007). However, photorespiration reduces quantum yield of C3, depending upon temperature and [O_2], so that in the range 20 to 25°C and 21% O_2, both C3 and C4 have **quantum efficiencies** near 0.06 CO_2 per quantum in low light. As a result, productivity of C3 and C4 species is generally comparable in low-radiation environments. C4s are markedly more productive at high irradiance, however, because they are less light-saturated. At lower temperatures, C3 performs best, and at higher temperatures, C4 are generally better. Experimental observations of quantum efficiency of a range of species (Fig. 10.2) are consistent with these predictions.

Maximum rates Maximum rates of leaf photosynthesis (P_n^*) are achieved at high irradiance under optimum conditions of nutrition, water supply, and temperature. Photosynthesis is then limited at ambient concentrations by CO_2 transfer through stomates in all species and by O_2 diffusion in C3. Maximum rates of C4 plants are generally

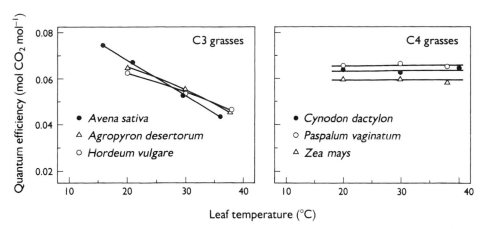

Fig. 10.2 The relationship of quantum efficiency of leaf photosynthesis to temperature for C3 and C4 grasses. Measurements made under normal atmospheric conditions. (After Ehleringer & Pearcy 1983.)

greater than for C3 plants (Fig. 10.3a). P_n^* values vary widely among species within both groups, providing hope for a route to greater photosynthetic productivity (Box 10.1).

The search for higher rates large enough to impact on field performance has been difficult. Extensive searches for germplasm with high P_n^* that might contribute to greater yield were unsuccessful for rice (Evans *et al.* 1984). However, more recent analyses of long-term progress of breeding programs of soybean (Morrison *et al.* 1999) and wheat (Fischer *et al.* 1998) have identified increase in P_n^* in sequences of cultivars released over periods of 58 and 26 years, respectively. In both cases, greater P_n^* was associated with greater leaf conductance rather than changed metabolism but there is no evidence of greater biomass production. Interpretation of present day comparisons of historical cultivar sequences is complicated. As was shown with rice (Peng *et al.* 1999), earlier cultivars may not demonstrate their past performance in present day comparisons due to changes in environmental conditions. The basic components of chloroplast systems appear to be highly conservative in a genetic sense (Richards 2000) leaving possibility for improvement in P_n^* in physical aspects of stomatal frequency and the greater sink strength of high-yielding cultivars that reduces the possibility of feedback responses on P_n (see Section 10.2.7).

Photoinhibition Low temperature, salinity, water stress, and acclimation to shade can predispose primary photosynthetic sites and various "protection" mechanisms (Sage & Reid 1994) to damage by excess light. This damage is known as photoinhibition and is best known from studies of reactions of shade plants to high irradiance and from plants exposed experimentally to low [CO_2] at high light (see Evans *et al.* 1988). Damage is associated with the dissipation of excitation energy not consumed in carbon reduction and may show its effect as a decline in either quantum efficiency or P_n^*.

The importance of photoinhibition and of protection mechanisms to productivity of crop plants is largely unknown, although it does seem likely that they have evolved to

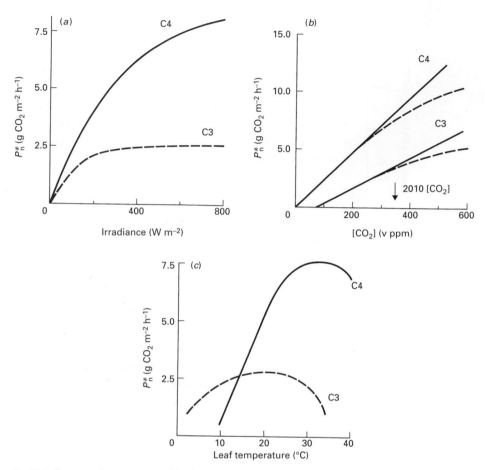

Fig. 10.3 Comparative response of leaf photosynthesis of C3 and C4 species to (a) irradiance at ambient [CO_2] and optimum temperature, (b) [CO_2] at high irradiance, and (c) temperature at high irradiance and ambient [CO_2].

resist it. Stresses that cause photoinhibition, for example low temperatures and severe water deficit, also affect photosynthesis directly. Ludlow and Powles (1988) recorded photoinhibition induced by water stress in grain sorghum. They concluded, however, that its additional effect to the severe water stress that induced it is unlikely to have agronomic significance.

10.2.2　Carbon dioxide supply

P_n responds to [CO_2] over a wide range that covers present and expected future changes in atmospheric composition. Figure 10.3b compares P_n^* from 0 to 600 ppm [CO_2] with saturating light. For C3 leaves, P_n^* responds linearly to [CO_2] but C4 saturates at [CO_2] not much above present levels. P_n^* of C4 leaves is positive down to around 5 ppm CO_2 but C3 leaves leak CO_2 (P_n is negative) in air with less than about 50 to 100 ppm. The [CO_2] at which $P_n = 0$ is called the **CO_2 compensation point**.

Box 10.1 Transferring C4 photosynthesis to C3 rice

Attention has turned once again to the challenge of transferring C4 photosynthesis to C3 crops to increase P_n, especially at high temperatures and with a smaller N requirement. Rice is in center stage because of its importance as the premier world food crop and the depressing effect that global warming would have on productivity (Section 11.3). Previous attempts at cross-breeding C3 and C4 species have only served to demonstrate the many features needed to convert C3 plants to C4. Now, however, new biotechnological tools of gene identification and transfer are available and also large sums of money to attempt the incorporation of these additional C4 components in rice. There is a current argument that because C4 photosynthesis has evolved on more than one occasion in various plant families, and that intermediate types can be found in their lineages, it should now be possible to identify the gene combinations required to move C4 traits to C3 plants. A counter argument is if C3 to C4 conversion occurs so readily via natural selection due to advantages in productivity and fitness, why has C4 not developed in one or more of the many wild rice species adapted to warm climates?

The new tools do not, however, reduce the complexity of the task, even when simplified to a four-step challenge. First, identify the genes that control the anatomical, physiological, and metabolic characteristics of C4 photosynthesis; second, transfer those genes to C3 rice to produce C3–C4 intermediates; third, manipulate expression of those genes to optimize their performance towards full C4 behavior; and fourth, intercross the C4 characteristics so obtained into locally adapted cultivars. Quite apart from the complex biotechnological issues in steps 1 to 3, step 4 takes the process back to the beginning and the lack of success in early attempts to breed C4 into C3 backgrounds.

What a pity that cassava does not share the world-food limelight with rice. This species has the most, and best studied, intermediate photosynthetic types, and beneficial growth and yield responses have been demonstrated in them (Section 10.1.2). The pathway to success ought to be shorter for this crop. It would be exciting to see progress in the search for this current "holy grail" of biotechnology during the lifetime of this book.

The linear relationship with $[CO_2]$ holds because P_n^* is strongly related to the gradient of $[CO_2]$ from bulk air to sites of fixation in leaves as expressed by the diffusion equation for photosynthesis:

$$P_n^* = (C_a - C_i)/[1.6(r_b + r_l)] = (C_a - C_c)/[1.6(r_b + r_l) + r_i'] \qquad (10.4)$$

where C_a, C_i, and C_c are $[CO_2]$ in air, internal leaf spaces, and chloroplasts, respectively. The factor 1.6 converts diffusive resistances to water vapor, r_b in boundary layer and r_l

through leaf (Eq. 9.12) to CO_2 according to relative molecular mass. r'_i is mesophyll resistance associated with transfer of CO_2 from interior leaf spaces to sites of fixation in chloroplasts. The r'_i pathway is complicated because CO_2 does not move along it by gaseous diffusion and respiration releases CO_2 within leaf tissues.

This form of relationship will hold provided there are no other responses to $[CO_2]$. It is known, however, that stomates respond to $[CO_2]$ with the effect of maintaining constant C_i in some species. That was observed with maize whereas in sunflower, r_l remained constant and C_i varied proportionately with C_a (de Wit *et al.* 1978).

10.2.3   Temperature

Leaves vary widely in their photosynthetic response to temperature. In general, C4 species perform well in warm climates and are tolerant of high temperature (Fig. 10.3*c*). Few C4 species perform well at low temperature and most suffer irreversible damage to membranes (**chilling injury**) at temperature around 10 to 12°C. Many C3 plants such as cotton and sunflower perform well at high temperature (30–40°C) and some warm-climate C3s (e.g., banana) are sensitive to chilling. Most C3 leaves, however, can withstand temperatures down to 0°C, and many to below it.

The generally broad optimum of photosynthetic response to temperature contrasts with the narrower response of growth found with most species. This emphasizes the importance of other physiological and environmental factors that control photosynthetic gain and utilization of assimilate. The relationship between photosynthesis and temperature gives C4 productive advantage at temperatures above about 25°C and is consistent with observed relationships between quantum efficiency and temperature (Fig. 10.2) and maximum production rates of C3 and C4 crops (Table 3.1).

10.2.4   Nitrogen

C4 leaves generally have a lower N content than C3 leaves because PEP carboxylase is a smaller molecule than rubisco and contains less N. In C3s, rubisco accounts for around 25% of leaf N. As a result, C4 plants have greater photosynthesis per unit leaf N and also develop more operational leaf area per unit N supply than C3 plants (Fig. 10.4).

Legumes characteristically have large N concentrations in reproductive and vegetative parts. For this reason their seeds are valuable food for humans, poultry, and swine, and their biomass is valuable fodder for grazing animals. In contrast, the low N content of C4 grasses is an important reason why they are lower quality fodder for grazing animals than are C3 grasses. C4 grasses also have higher contents of indigestible lignin because a greater proportion of their leaf anatomy is given to vascular traces.

10.2.5   Stomates and transpiration efficiency

Stomates give plants the ability to control transpiration but also the opportunity to control **transpiration efficiency** (*TE*, g CO_2 g^{-1} H_2O), the ratio of carbon gain by photosynthesis to water loss by transpiration (Section 9.6.3). Combining equations for

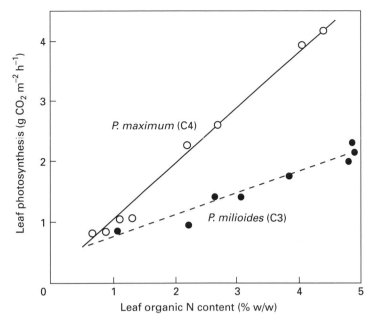

Fig. 10.4 Relationship between leaf photosynthesis rate and organic nitrogen content of the pasture grasses *Panicum maximum* (C4) and *Panicum milioides* (C3). (After Bolton & Brown 1980.)

leaf transpiration (Eq. 9.12) and the complete pathway of photosynthesis (Eq. 10.4) gives:

$$TE = (C_a - C_c)(r_b + r_l)/[(e^* - e_a)(1.6(r_b + r_l) + r_i')]$$ (10.5)

The equation shows that the effect of stomatal closure (increasing r_l) depends upon magnitudes of both r_b and r_i'. Given that r_b is very small except in still air and in C3 species, r_l' is a significant proportion of total resistance for CO_2. Therefore an increase in r_l has a larger effect on the numerator (resistance to water loss) than on the denominator (resistance to CO_2 uptake). As a consequence, TE increases when rates of photosynthesis and transpiration are restricted by stomatal closure.

Evaporation demand and photosynthetic potential vary diurnally and from day to day so if a leaf is to maximize TE, then optimization theory requires that leaf resistance vary with environmental conditions such that marginal gain of CO_2 for each unit of water lost maintains a constant maximal value (Cowan 1982). Species that have evolved strong feed-forward stomatal responses to humidity (e.g., cassava) may have this possibility.

The same argument can be extended from individual leaves to crop canopies and include other factors that influence photosynthesis and/or transpiration such as leaf movement, leaf expansion, leaf loss, osmotic adjustment, and root extension (Cowan 1986). In the field, however, water is not the only, and not necessarily the most important, resource to be used optimally. Furthermore, criteria of optimal water use are not restricted to just achieving maximum photosynthesis or growth but also to success in completing the life cycle successfully and in competitive relationships between neighbors of the same or different species. We will return to principles of optimal strategy of crop response in agricultural systems in Chapters 13 and 16.

Carbon isotope discrimination Atmospheric CO_2 contains a mixture (99:1) of two stable isotopes, ^{12}C and ^{13}C. Photosynthesis discriminates against the heavier isotope, ^{13}C, both in diffusion through stomates and in reaction with rubisco. Consequently, all plant biomass, all animals, all soil organic matter, and all fossil fuels are uniquely labeled isotopically with ratios of $^{13}C/^{12}C$ different from the atmosphere and from carbonates in oceans. Soil carbonates, being mainly derived from respiratory CO_2, have ratios more like the plants that are grown in them than carbonates formed in the oceans.

Discrimination is less in C4 plants than in C3s so measurement of $^{13}C/^{12}C$ ratios of plant material is a useful way to identify the photosynthesis pathway (see Farquhar *et al.* 1982). The ratio can also be used to measure transpiration efficiency of C3 leaves under conditions of limiting water supply that cause stomatal closure (Farquhar & Richards 1984). That possibility arises because stomatal closure causes C_i to fall so that discrimination against ^{13}C is less. Carbon isotope analysis provides a screening technique for plant breeders in search of genotypes with intrinsic high transpiration efficiency. For example, the low $\Delta^{13}C$ trait was identified in wheat cultivar Quarrion (Rebetzke *et al.* 2002) and its inheritance evaluated in crosses with cultivar Hartog. Two new cultivars "Drysdale" and "Rees", released for Australian growers, have demonstrated significant yield advantages over previously recommended cultivars, with the advantage increasing at low-rainfall, low-yield sites (Richards 2006).

10.2.6   Leaf structure

Leaves presumably have evolved as planar organs because this maximizes light interception and photosynthetic return per unit material invested in leaf structure. Their large area/mass ratio ensures efficient illumination of chloroplasts as well as a short pathway for diffusion of CO_2 to sites of fixation.

Striking differences are found between anatomy and photosynthetic response of leaves of species adapted to shade and sun. Leaves of shade plants are generally thinner with a lower chlorophyll content than those of sun plants. Shade plants are incapable of the high photosynthetic rates of sun plants, but they perform efficiently at low irradiance.

Specific leaf mass (SLM) is a useful index of leaf structure related to thickness and density, and hence to the balance between radiation capture and provision for diffusion through air spaces within leaves. Leaves of herbaceous plants are commonly around 0.2 mm thick with an SLM in the range 30 to 70 g m^{-2} (specific leaf area, SLA = 1/SLM, 330–140 cm^2 g^{-1}) and P_n^* is often strongly correlated with SLM (Fig. 10.5).

Most species are able to adjust the structure of their photosynthetic systems to a range of ambient conditions because leaf growth and development are plastic in response to the environment. A typical response is to produce thinner leaves with fewer mesophyll cells and less chlorophyll in shade than at higher irradiance. This **acclimation** is significant to both current and subsequent photosynthetic responses. Leaves formed in shade at low canopy positions have low maximum photosynthetic rates if exposed to high irradiance by harvesting or grazing of the upper canopy. Leaves formed under high irradiance at the

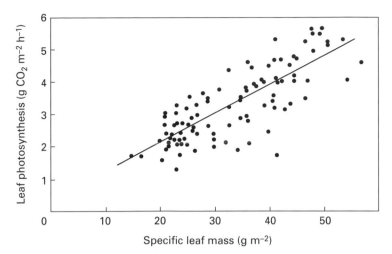

Fig. 10.5 A relationship between leaf net photosynthesis of alfalfa and specific leaf mass (SLM). (After Pearce *et al.* 1969.)

tops of canopies may be poorly adapted to low light levels below as they are gradually submerged there by the growth of new leaves.

Expanded leaves are also able to adjust to changes in radiation environment. Osmond *et al.* (1988) studied the physiological and morphological aspects of acclimation of sunflower photosynthesis to irradiance. Plants were grown in sun and shade and then transferred to the alternative condition. Quantum efficiency (dP_n/dI at low light) of both groups was similar (0.072 mol CO_2 mol^{-1}) at the start of treatments and did not change during acclimation. There were, however, large responses in P_n^*, which, as reported in previous studies (see Björkman *et al.* 1981), are related to changes in enzyme activity, frequently measured as leaf N content, but here recorded as SLM (cf. Fig. 10.5). In this case, SLM of leaves grown in sunlight decreased over 8 d in shade from 54 to 22 g m^{-2} while P_n^* decreased by 50%. In contrast, shade-grown plants commenced acclimation to sunlight with SLM = 22 g m^{-2}, increasing to 68 g m^{-2} over 10 d while P_n^* increased three-fold.

10.2.7   End-product inhibition of photosynthesis

Although leaves are well designed to capture radiation and CO_2, they have little volume to store the products of photosynthesis. Leaves of sunflower or maize, for example, have the capacity to assimilate 30 to 40 g DM m^{-2} d^{-1}, which is close to their SLM of 50 to 60 g m^{-2}. It is a general phenomenon that non-structural carbohydrates may accumulate in leaves and other parts when growth is restricted by removal of sinks or by environmental stress such as low temperature or shortage of N and other nutrients. The simple explanation is that growth is more sensitive to those stresses than is photosynthesis. Eventually, as storage sites in leaves are filled, photosynthesis declines, and this has been interpreted as end-product (feedback) inhibition (Neales & Incoll 1968).

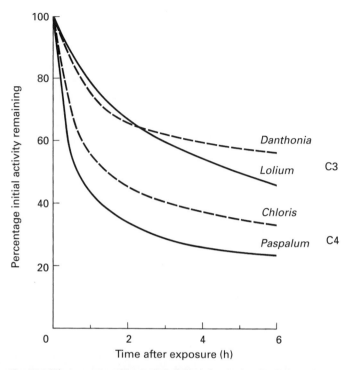

Fig. 10.6 Time course of loss of radioactivity during the light at 25°C from leaves of C3 and C4 grasses previously exposed to $^{14}CO_2$. (After Lush & Evans 1974.)

Feedback control could occur in several ways. Though chemically isolated, chloroplast starch can slow photosynthesis by interfering with penetration of light and CO_2 to active sites and, when in large amounts, by physically distorting chloroplast membranes. Studies suggest that high levels of sugars in leaves sequester a large part of inorganic phosphorus in hexose phosphates and that unavailability of phosphorus also slows photosynthesis (Foyer 1988).

Translocation of assimilates from leaves is mainly as sucrose and depends on the activity of bundle sheath cells surrounding vascular traces, which serve in loading assimilates into phloem cells for transport. Concentration of chloroplasts in bundle sheaths of C4 species in close proximity to phloem may confer an advantage by reducing the possibility of end-product inhibition. Experiments with a range of pasture species have shown that translocation rate of assimilate may be greater from C4 leaves than from C3. This is shown in Fig. 10.6, which illustrates the persistence of radioactive ^{14}C introduced into leaf photosynthesis early in the photoperiod. This label is lost more rapidly from C4 leaves. Greater translocation is one explanation, but there are complicating factors. C3s have greater respiration than C4s in light (photorespiration) and this could account for some difference. Equally it could reflect different patterns of storage and export of assimilate. To be an effective marker, a label must mix freely in the assimilate pool so that its dilution defines input–output balance. If, for example, it became embedded in inner layers of starch granules it could no longer serve this

Table 10.2 Comparison between C3 and C4 photosynthesis

	C3	C4
Leaf photosynthesis maximum rate ($g\ CO_2\ m^{-2}\ h^{-1}$)	1.5–4.5	3.5
PAR for saturation	low to moderate	very high
Temperature optimum	low to high	high
Maximum leaf conductance	similar	similar
Mesophyll conductance	low	high
CO_2 compensation point (vppm)	50–100	<5
CO_2 evolution in light	high	none
Transpiration efficiency	low	high
Nitrogen-use efficiency	low	high
Canopy photosynthesis ($g\ CO_2\ m^{-1}\ h^{-1}$)	3.5–9	5–10

purpose. Diurnally, the residence time of assimilate in leaves does display such "first in, last out" dynamics.

10.2.8 Comparison of C3 and C4 photosynthesis

Various attributes of C3 and C4 photosynthesis are summarized in Table 10.2. Their comparison raises a number of important evolutionary and ecological questions.

One important ecological difference between the two groups is that enzyme systems of C4s are poorly adapted to temperatures below 10°C whereas C3s maintain function to near 0°C. As a result, C3 species are widely distributed while C4s are in general restricted to warm climates. Further, because the dominant group of C4s is grasses, the major distribution of C4 is in tropical and semitropical grasslands. However, many C3s can also be productive at high temperatures, a fact attested to by the productivity of C3 crops such as rice and cotton in hot climates. That arises because adaptation to temperature among C3s is wide. It is easy to find published examples of distinct temperature optima between C3 and C4 species but proper comparisons of effect of temperature on C3 versus C4 photosynthesis will compare plants of similar temperature adaptation.

The second major difference between the two groups occurs in their response to O_2. In the absence of O_2, they perform similarly, but with ambient (21%) O_2, P_n of C3 plants is depressed by photorespiration. It may be that C3 photosynthesis represents an adaptation to the Carboniferous Period when terrestrial plants evolved and $[CO_2]$ was high and $[O_2]$ was low. Unless photorespiration is found to have an important beneficial function, the ecological conclusion is that high $[O_2]$ of Earth's atmosphere is detrimental to C3 species. C4s are clearly better suited to our present atmosphere that is low in CO_2, whereas C3s will benefit more from currently increasing $[CO_2]$.

C4 photosynthesis may also be an adaptation to low N availability (Brown 1978). The difference in leaf N content between C3s and C4s is striking as also is the absence of C4 members in the large and varied plant family Leguminosae, which fixes N symbiotically.

Superior performance of C4 raises an important possibility that the trait could be found or introduced into C3 plants. Neither screening of C3 crop plants (Menz *et al.* 1969) nor hybridization of C3 and C4 from the same genera (e.g., *Atriplex*, Björkman *et al.* 1969) was successful. Attention has now turned to a process-by-process approach and theoretical possibilities for modification. Amthor (2007) recently reviewed the links between photosynthesis types and potential crop productivity, and described possible genetic approaches to introduce C4 photosynthesis into C3 plants. The International Rice Research Institute (IRRI), along with international collaborators, have embarked on the quest to bio-engineer a C4 rice (Anon. 2009). The challenge is enormous because it involves a number of genes and morphological structures, which means progress will be inevitably slow (Box 10.1).

10.3   Canopy photosynthesis

Crop canopies have more complex photosynthetic responses than their component leaves but a useful parallel can be drawn between photosynthesis of crop surfaces and of single leaves. By analogy to Eq. 10.4:

$$P_{n(crop)} = (C_a - C_l)/(1.6r_a) = (C_l - C_i)/(1.6r_c) \tag{10.6}$$

where $P_{n(crop)}$ is canopy net photosynthesis, C_a, C_l, and C_i are [CO_2] of bulk air, at the effective crop surface, and within the crop, respectively. The terms r_a and r_c are the aerodynamic and canopy (all leaves acting in parallel) resistances to water vapor transfer from the bulk air to leaf surface and from leaf surface to substomatal cavities, respectively. As previously, the factor 1.6 converts these resistances to equivalent resistances for CO_2 transport.

In dense crops of high photosynthetic rate, C_l can fall to 250 or nearly 100 ppm below C_a on windless days with little atmospheric mixing (Fig. 6.12). Under those conditions, [CO_2] strongly restricts leaf photosynthesis within canopies. In contrast, for sparse canopies and windy conditions, depletion of CO_2 is much less.

Techniques are available to study the role of canopy architecture in crop photosynthetic responses. These are discussed in Section 10.5 but first information and discussion of some established responses of field crops.

Seasonal photosynthesis of wheat crops   Data in Fig. 10.7 record daily (24 h) net CO_2 gain of wheat crops made with field chambers (Box 10.2) on many individual days beginning 110 days after sowing (DAS) when crops were tillering. Three crops had different patterns of water supply; R was rainfed, Wl was irrigated weekly, and W2, biweekly. The upper curve defines seasonal assimilation of Wl. In this crop, daily assimilation rose to around 75 g CO_2 m^{-2} d^{-1} at flowering and then declined to zero at maturity (185 DAS) as crop leaf area senesced. It showed no evidence of transient water shortage between irrigations. Measurements for W2 define effects of drying cycles between irrigations. From 140 DAS, photosynthesis of W2 decreased with water shortage during the second week following each irrigation. Comparison of those crops with Wl, however, reveals full recovery after irrigation. The rainfed crop had significantly depleted

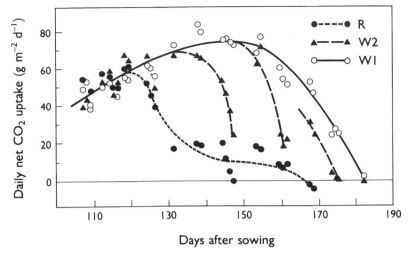

Fig. 10.7 Daily (24 h) net CO_2 uptake of wheat crops in response to three conditions of water supply. R rainfed, W2 irrigated at two-weekly intervals, and W1 irrigated weekly. (From Whitfield 1990.)

Box 10.2 Field chambers for measurement of crop photosynthesis

Transparent field chambers are commonly used to measure crop photosynthesis. Short-term measurements can be used in analyses of community architecture and in studying the effects of changes in leaf photosynthetic characteristics and environmental and management factors on canopy performance. Longer-term measurements maintained for hours or days assist in the interpretation of crop growth by estimating total assimilation from which new biomass is constructed.

Three types of transportable assimilation chambers that enclose segments of a crop are used in measurements of crop photosynthesis. In **closed systems**, photosynthesis is determined by the rate of depletion of CO_2 in a known chamber volume. To provide continuous measurements, such chambers are sequentially opened and closed. In **open systems**, air flows continuously through the chamber fast enough to maintain $[CO_2]$ close to ambient but slow enough to provide a measurable $[CO_2]$ depletion between inlet and outlet air. Rate of CO_2 exchange is the product of airflow through the chamber and depletion in $[CO_2]$ across it. In **closed-compensating systems**, chamber $[CO_2]$ is maintained at ambient level by injecting CO_2, or CO_2-enriched air, to balance the uptake by photosynthesis. Rate of injection measures crop photosynthesis rate.

It is possible to offset inevitable heating in open and closed-compensating systems and to operate all chambers with $[CO_2]$ and humidity close to ambient levels. Each chamber type requires forced mixing of air inside the chamber, however, so that CO_2 uptake can be measured accurately. Mixing increases transfer rate above that determined in the field by eddy diffusivity (Eq. 6.22) and so may increase canopy photosynthesis above actual rates in adjacent open fields.

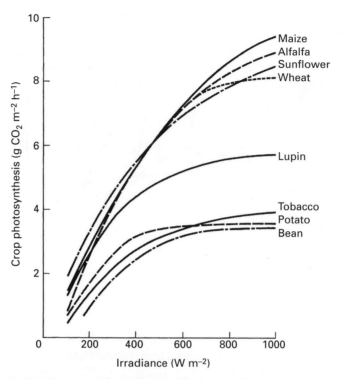

Fig. 10.8 Response of crop photosynthesis to irradiance in full-cover C3 and C4 crops well supplied with water and nutrients. All measurements made with field assimilation chambers. (Data from Connor *et al.* 1985b; Sale 1974, 1975, 1977; Whitfield *et al.* 1980, 1986).

soil moisture by 120 DAS and thereafter assimilation fell rapidly until photosynthesis ceased at 170 DAS.

10.3.1   Comparative photosynthesis of C3 and C4 crops

Photosynthetic responses to irradiance for a range of full-cover crops are presented in Fig. 10.8. These data were assembled from field-chamber measurements. Each is a diurnal response to ambient irradiance under optimum conditions of water and nutrients. Temperature regimes varied from crop to crop but were within optimal ranges for individual species. Comparison of these data with many other responses available in the literature indicates that maximum rates are in the medium to high range for each crop considered.

   The responses approach a saturation rate asymptotically. In the case of crops, as opposed to individual leaves, linear responses of photosynthesis to full sunlight are not restricted to C4 species. In these data, wheat, sunflower, and alfalfa crops also display strongly linear responses up to 600 W m^{-2} shortwave irradiance. Such responses are achieved when there is little light saturation of individual leaves in canopies. Maize, and to a lesser extent sunflower, can achieve this independently of canopy structure by virtue of leaf responses that do not saturate to full sunlight (cf. Fig. 10.3 for C3 versus C4 comparison). In contrast, wheat and alfalfa rely upon high leaf angles, and in the case of alfalfa, small leaves also, to distribute light evenly at low irradiance over leaf surfaces. Strongly asymptotic responses are typical of C3 species with leaves that saturate at low irradiance and are displayed in canopies of low leaf angle (here tobacco, bean, and potato). In those cases, a significant proportion of upper canopy photosynthesis is light saturated.

10.3.2   Radiation-use efficiency

Photosynthetic rates of crops depend on the quantity of radiation intercepted and the efficiency achieved in utilization. Thus radiation-use efficiency (RUE) can be defined relative to incident or intercepted radiation, two values that converge as crops achieve full cover and the fraction of intercepted radiation (R_i) approaches unity. To provide a general analysis, R_i can be approximated by crop cover across all LAI (L) by the expression:

$$\text{Cover} = 1 - \text{gap} = 1 - e^{-kL} \tag{10.7}$$

derived from the Bouguer–Lambert analysis (Eq. 3.4) of radiation interception by canopies with randomly oriented and randomly dispersed foliage. Variable k is an attenuation coefficient relating leaf angle to interception ability of unit leaf area.

The equation is readily extended to describe canopy gross photosynthesis in terms of RUE (ε) of intercepted radiation as:

$$P_g = \varepsilon I_0 (1 - e^{-kL}) \tag{10.8}$$

in which I_0 is incident irradiance, e^{-kL} is gap, so that $I_0(1 - e^{-kL})$ is interception. If radiation were distributed over canopy surfaces at low intensity, ε would approximate maximum RUE established from leaf photosynthesis–irradiance response as presented in Fig. 10.1 and Eq. 10.3.

Observation has shown that $P_{n(crop)}$ is often closely related to intercepted radiation in this way. Figure 10.9 illustrates this for cotton and sunflower crops of a range of canopy cover. In these data, RUE from instantaneous photosynthesis measurements on cotton is 2.3 g CO_2 MJ^{-1}. For sunflower, it is 5.3 g CO_2 MJ^{-1} for daily totals. The two values are not strictly comparable because they were made by different techniques over different timescales. However, the larger value for sunflower does reflect the unusually high leaf photosynthetic rate of that C3 species.

Comparisons of RUE for eight crop species are presented in Table 10.3. They were derived from crop photosynthetic responses of full cover crops presented in Fig. 10.8. Values range from 1.2 to 4.2 g CO_2 MJ^{-1} with generally higher values at low (330 W m^{-2}) rather than at high (800 W m^{-2}) irradiance, reflecting progressive light saturation of component leaves as irradiance increases. This effect is greatest in crops that have horizontal leaves and leaf photosynthetic responses that saturate at low irradiance, e.g., tobacco and potato.

Table 10.3 Radiation-use efficiencies of canopy photosynthesis (g CO_2 MJ^{-1} intercepted shortwave radiation) for crop responses presented in Fig. 10.8

Crop	Efficiency at	
	300 W m^{-2}	800 W m^{-2}
Maize	3.2	3.1
Alfalfa	3.3	2.8
Sunflower	3.3	2.7
Wheat	4.2	2.8
Lupin	3.2	1.9
Tobacco	1.9	1.3
Potato	2.7	1.2
Bean	1.9	1.2

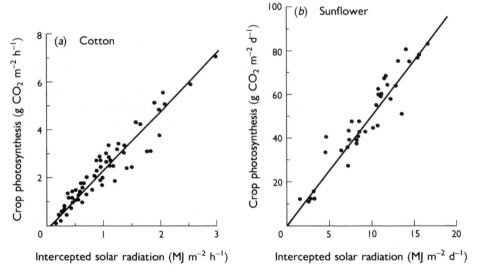

Fig. 10.9 Relationships between crop net photosynthesis and canopy interception in (a) cotton (after Baker & Meyer 1966), and (b) sunflower crops (after Connor *et al.* 1985b).

10.4　Modeling canopy photosynthesis

The exponential profile (Bouger–Lambert law) of average irradiance within plant communities (Eq. 3.4) retains utility in analyses of crop photosynthesis when combined, layer by layer, with leaf photosynthesis response to irradiance (Eq. 10.3) as described in Box 10.3 and used in later examples (Section 10.5). At the same time, much variation in RUE measured for entire crops can be laid to complex canopy geometry. At first inspection, canopies of leaves may seem hopelessly complex, but their geometry can be described in relatively simple ways and consequent photosynthetic behavior can be explained through mathematical analyses.

Box 10.3 A model of crop photosynthesis constructed by combining the exponential profile of light penetration into a canopy with a leaf photosynthesis–light response function

Photosynthesis of individual layers

At depth L in a canopy (LAI units from the top), the mean irradiance of the downward flux on a horizontal sensor is $I_0 e^{-kL}$ falling at depth $(L + \Delta L)$ to $I_0 e^{-k(L+\Delta L)}$. The difference is the mean irradiance absorbed by the layer ΔL and is $k I_0 e^{-kL}$, so irradiance incident on the leaves is $k I_0 e^{-kL}/(1 - m)$, where m is the leaf transmission coefficient (Saeki 1963). There is also an upward flux, which can be reasonably related to downward flux by the leaf reflection coefficient (φ) (Thornley 1976) to estimate the mean irradiance additively incident on both surfaces of leaves in the layer as $(1 + \varphi)\, k I_0 e^{-kL}/(1 - m)$.

Mean net photosynthesis per unit leaf area ($P_n(1)$) of this increment of leaf area at depth L can be obtained by substitution in Eq. 10.3 to give:

$$P_n(I) = [P_n^*(\beta k I_0 e^{-kL} - I_c)\alpha]/[P_n^* + (\beta k I_0 e^{-kL} - I_c)\alpha] \tag{10.9}$$

where $\beta = (1 + \phi)/(1 - m)$ will typically have a value of 1.2 since $m = \phi \approx 0.1$ for radiation incident on most leaves. Omission of this term will not have a serious effect on performance of the model since its effect can be taken up in k.

Photosynthesis of a canopy

Photosynthesis of the entire canopy per unit ground area ($P_{n(crop)}$) is the sum of component layers best made by numerical integration to allow for changing parameter values for individual layers.

$$P_{n(crop)} = \Sigma P_n(l) \tag{10.10}$$

This equation describes crop photosynthesis in **response** to environment (I_0) canopy structure (LAI, k, m, ϕ), and leaf CO_2 exchange (P_n^*, α, I_c) as in Eq. 10.9 above. Even simple models such as this that combine few essential non-linear functions can describe major responses of complex systems.

Characteristics of more complex geometrical models are depicted in Fig. 10.10. This illustrates two aspects of interaction between leaf angle and solar altitude on penetration and irradiation of foliage by direct solar beams. Figure 10.10a shows the relationship between k and solar altitude for horizontally continuous canopies in which leaves are inclined at a range of angles to the horizontal and are randomly dispersed and oriented. Figure 10.10b shows how incident radiation can be spread over a larger leaf area, to a maximum sunlit LAI of near four, as leaf angle and solar altitude increase. With leaves arranged obliquely to the Sun's rays, sunlit leaf area index increases and their mean irradiance declines.

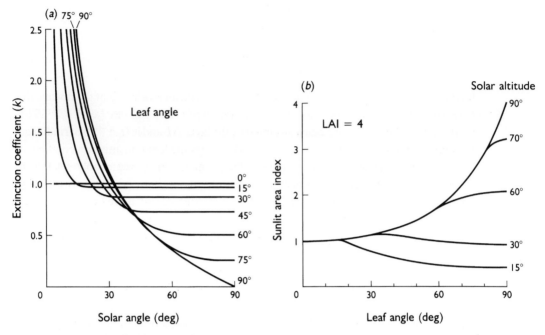

Fig. 10.10 Components of geometrical models of canopy photosynthesis. (*a*) The relation of extinction coefficient (*k*) to solar altitude and leaf angle; and (*b*) sunlit area index of a canopy of LAI = 4 in relation to leaf angle and solar altitude. (After Warren Wilson 1967.)

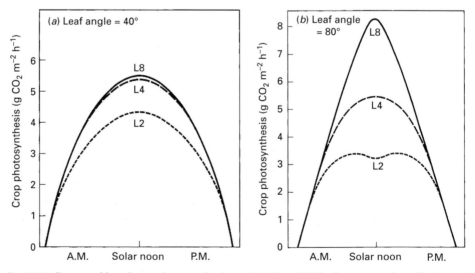

Fig. 10.11 Computed hourly net photosynthesis on July 30 at 38° N of crop canopies of LAI equal to 2, 4, or 8 in response to leaf angle: (*a*) all leaves at 40°; and (*b*) all leaves at 80° (near vertical). (After Duncan *et al.* 1967.)

Relationships in Fig. 10.10*a* deal with direct solar radiation but can be extended to diffuse radiation, originating from many point sources over the sky. Using the same underlying equations (Hanau in Duncan *et al.* 1967), photosynthesis of crops irradiated in complex patterns of light and shade by direct and diffuse radiation can be calculated as the sun tracks across the sky (Box 6.1). Calculations can be made for classes of foliage, profile layers, or for entire canopies. The multitude of calculations required is readily made by computers, but it is also possible to calculate sets of general solutions from which behavior of particular structure–irradiance–leaf response combinations can be estimated by interpolation.

Figure 10.11 presents examples of simulations of canopy photosynthesis of maize. As leaf angle increases, more foliage can be irradiated and lower irradiance over foliage leads to higher radiation-use efficiency. At low leaf angle, increases in LAI do not lead to more photosynthesis because most radiation is intercepted by upper leaves, which are generally light saturated.

10.5   Canopy structure for productivity and competitiveness

The display of leaves in crop canopies is determined by plant density, planting pattern, and morphological characteristics of component species. Two aspects of canopy display dominate productivity and persistence of individual species. First is rapid attainment of optimum cover for the environment. Second is the advantage that height confers on competitive ability in mixed communities.

10.5.1   Canopy dynamics

Development of foliage canopies involves a number of overlapping stages related to crop phenology (Chapter 5). First is initiation of new leaf primordia at vegetative apices. Second is expansion of leaves to take their place as the youngest, generally best located, and most productive in the canopy. The third, and last stage, is senescence, characterized by a gradual loss in assimilatory activity and mobilization of cell contents to other sites. That process sometimes ends with leaf abcission.

Full cover is achieved most rapidly when population density is high, when seedlings partition a significant proportion of new assimilate to leaf development, specific leaf mass is small, and leaf angle is low. Interception of radiation and early productivity are then maximum and crops are also able to shade out shorter weed competitors.

Highly productive canopies have characteristics of full cover and erect leaves, at least in the upper canopy, e.g., sugarbeet. Best designs, however, depend upon radiation environment and aspects of photosynthesis and respiratory responses of component leaves. In a simple case, all leaves with the same photosynthesis- and respiration-response functions, maximum crop photosynthesis will be achieved when all leaves receive the same low light flux at a level where radiation-use efficiency is greatest. This would require an attenuation coefficient (k) that changes continuously with canopy depth. If response functions vary with depth, or if plant morphology restricts leaf display, then

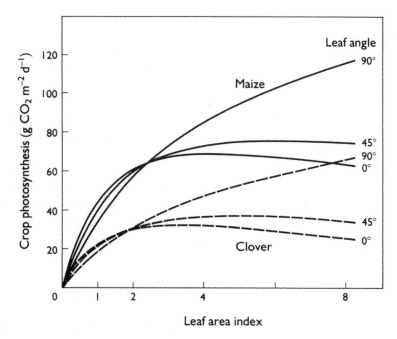

Fig. 10.12 Simulations of daily crop photosynthesis for maize and clover communities of various LAI and leaf angle combinations. Solar and skylight data are for latitude 38° N on July 1. (After Loomis & Williams 1969.)

the best achievable design will be different. Simulations of crop photosynthesis have shown themselves to be an effective way to investigate such questions of canopy design.

Examples presented in Fig. 10.12 summarize several important points about canopy architecture by comparing photosynthesis of maize (C4) and clover (C3) canopies over a range of LAI and leaf angle. The responses were generated by the simulation model of Duncan *et al.* (1967). With LAI < 2, canopies of horizontal leaves are most productive. With intermediate LAI of 2 to 4, leaf angle has little influence on productivity, but increases in LAI beyond 4, given erect leaves so that available radiation is spread over more leaf area, lead to progressively greater assimilation. The high productivity of maize arises from these C4 traits: high unit leaf rate and limited saturation at high irradiance (Fig. 10.2). During recent decades of plant breeding their advantage has been increased by the development of more vertical leaves high in the canopy and the location of cobs low in the canopy where they interfere less with light interception and photosynthesis by the canopy.

Leaves that have become acclimated to low irradiance, but have not begun to senesce, may be able to respond to management that places them once again in a high-irradiance environment. Thus productivity of a canopy following removal of some leaf area (pruning, harvesting, grazing, or insect or disease attack) depends upon which leaf area is removed because that determines what productive structure remains. Design of improved management strategies for canopies requires an understanding of the dynamics of leaf formation, expansion, longevity, and in the case of susceptible crops such as cotton, the impact of insect depredation also (e.g., Reddall *et al.* 2007; Sadras 2002).

10.5.2   Nitrogen distribution for maximum canopy photosynthesis

Large amounts of N are required in the construction of dense canopies, especially of C3 species. Canopies receive diminishing photosynthetic gains for each increment of leaf area above full cover and so, without internal redistribution, considerable additional N is required. C3s such as sugarbeet and wheat contain about 12 kg N ha^{-1} per unit LAI (SLM $= 40$ g m^{-2} with 30 g N kg^{-1}). Legume leaves generally have greater [N], ranging from 30 to 50 g kg^{-1} DM. Even C4 maize, which needs much less N for maximum photosynthesis rate, has around 10 kg N ha^{-1} per unit LAI (SLM $= 50$ g m^{-2} with 20 g N kg^{-1}).

When N is in short supply, crop photosynthesis may be limited by both leaf area and leaf capacity for photosynthesis. Given leaf photosynthetic response functions to irradiance and [N] in leaf tissue of the form presented in Figs. 10.2 and 10.4, the question arises how, given the same amount of leaf area and N, an additional unit of N should be distributed between growing new leaves and increasing N content of existing leaves? The optimum solution, that will maximize photosynthesis per unit N in the total canopy (one definition of **nitrogen-use efficiency, NUE**), depends upon N supply, canopy architecture, solar track, leaf longevity, and other factors.

Hirose and Werger (1987) compared distribution of leaf [N] through a canopy of *Solidago* sp. with optimum distribution simulated with a canopy photosynthesis model comparable to that presented in Box 10.3. They concluded that observed distribution of leaf [N] provided 20% greater canopy photosynthesis than would occur with a uniform distribution, but that it was 5% less efficient than the optimal distribution in which leaf [N] decreased exponentially with canopy depth. The same approach was used to evaluate efficiency of N-withdrawal patterns from sunflower canopies during grain filling of crops grown at different levels of N supply (Connor *et al.* 1995). Analysis compared simulated crop photosynthesis under observed N distribution to that with the same total N content distributed so as to maximize crop photosynthesis. Nitrogen was withdrawn simultaneously from leaves in all canopy positions. Observed profiles were suboptimal except at mid flowering in a low-density, low-N treatment. Differences were most marked in high-N treatments that retained "excessive" N in lower leaves, accounting for a 4% loss in $P_{n(crop)}$. Analysis also revealed that, as canopies matured, N was mobilized to grain, and leaf area index decreased, the optimal N profile changed shape from exponential to linear.

10.5.3   Maximum yield at less than maximum assimilation

In many crops, maximum yield and/or quality is obtained with less than maximum canopy photosynthesis. For total biomass, this occurs because the last unit of photosynthetic gain does not return the investment in leaf area and associated stem. There are, however, issues in crop production other than maximum biomass. In field crops, for example, light is required low in canopies to assure leaf quality in Virginia tobacco and boll retention in cotton. In horticultural crops, stone and pome fruits often require high irradiance to develop the color and size that determines quality. In addition, next year's reproductive capacity of many perennial fruit trees depends upon current assimilate supply to meristems distributed throughout the canopy. Canopies that allow these responses usually have incomplete capture of radiation and assimilation is less than maximum.

Control of canopy shape to improve yield and quality is well known in the trellis- and hedgerow-production systems common in horticulture. Interest has grown recently in hedgerow fruit production systems (e.g., Annandale *et al.* 2004; Olesen *et al.* 2007; Oyarzun *et al.* 2007) because of advantages to management (irrigation, fertilization, and mechanization) as well as control of illumination to improve production. Some crops, such as pome fruits and vines for wine, have been grown and managed in this way for centuries but the methods are now being applied to other crops. Current intensification of olive production in traditional areas of the Mediterranean and in the New World (Argentina, Australia, Chile, and South Africa) is made competitive by the high productivity and low management cost of hedgerow systems. The question of optimal design, including issues of productivity, quality, and manageability, is largely unknown.

A starting hypothesis is that productivity and quality will be maximized when the lowest parts of canopy walls receive sufficient irradiance for the limiting process in fruit production, be it flowering, fruit set, or fruit filling. A theoretical analysis of north–south (NS) olive hedgerows of various heights, widths, and wall angles (Fig. 10.13a) quantifies key parameters of optimal structure as the ratio between hedgerow height and free alley width. Hedgerow width contributes separately to productivity because it determines total length of hedgerow that can be established per unit area. That ratios, and not absolute sizes, are important is consistent with current developments in the industry. Hedgerow heights vary widely from 2 to 6 m and row (not alley) spacing from 3 to 8 m. These differences can be accommodated by machinery of appropriate size although advantages apply to short hedgerows that can be managed more easily and more cheaply with equipment readily adapted from vineyard use.

The challenge is to optimize the structure of new olive orchards, modify existing suboptimal orchards, and develop canopy management systems to maintain optimal structure once defined. A combination of observation and modeling appears to offer most rapid progress. Observations on profiles of stem growth, fruit set, fruit size, fruit oil content, and oil quality are needed to establish the response functions required by simulation models for complete analysis of productivity and quality. Available data suggest that profiles of fruit size and oil content (Fig 10.13b) can be related to radiation interception patterns on canopy walls but that prediction of fruit set is more difficult because it responds more strongly to factors other than irradiance (extreme temperatures, wind, hail, etc.; Connor *et al.* 2009). The question of hedgerow orientation is also important because of distinct patterns of seasonal irradiance. Gómez-del-Campo *et al.* (2009) have presented some preliminary data on productivity of east–west (EW) hedgerows from which it is clear that hedgerow porosity is an important determinant of productivity of the shaded sides of these hedgerows. More complex models are needed for asymmetrically illuminated EW hedgerows than their symmetrically illuminated north–south (NS) counterparts.

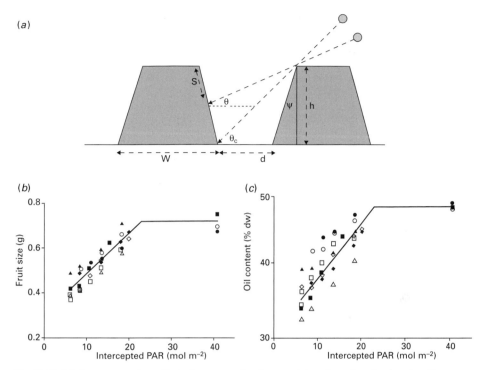

Fig. 10.13 (*a*) A cross-section of a hedgerow orchard normal to solar azimuth, showing the parameters row height (h), row width (w), and canopy slope angle (ψ) that can define a range of rectangular, triangular, and truncated-triangular shapes. Alley width is d. The diagram also shows the sunlit length (s) on the illuminated canopy wall at solar elevation (θ) together with the minimum (critical) angle of solar elevation (θ_c) that just illuminates to the canopy base (from Connor 2006). Relationships between profiles of (b) fruit size and (c) oil content with daily intercepted PAR in October by various NS hedgerows for eight cv. Arbequina orchards in Spain (from Connor *et al.* 2009).

Trellises are rarely practical for field crops and optimum canopy design must be achieved by genetic means and/or by spacing. There are, however, examples to which principles of trellising apply. Climbing beans are such a case. Their small investment in stem means that they cannot display their canopy without mechanical support. Given this by canes, strings, or wires, yield is advantaged by the small diversion of assimilate to stem. Climbing beans are commonly grown as an overlap crop in the tropics supported by stalks of maturing maize and cassava. In those polycultures, issues of canopy display and productivity are further complicated by the presence of additional crop species.

10.5.4   Polycultures

Principles of competition for light and the role of canopy structure in polyculture were developed from studies of light relations and productivity in pastures, particularly by Donald and coworkers at Adelaide, South Australia, who concentrated on simple mixtures of annual ryegrass and subterranean clover. This combination of tall grass and short legume exemplifies the structure of most grass–clover mixtures. Important exceptions occur, however, in improved tropical pastures, which often combine climbing legumes with grasses.

The stability of ryegrass–subterranean clover mixtures depends upon mutual shading by the two species. Under low N supply or heavy grazing, there is little shading of clover by grass, but as fertility increases over the life of the pasture, or if pastures are fertilized with N or excluded from grazing in preparation for harvest, clover growth suffers while grass flourishes. Suppression and gradual loss of clover from pastures with high N supply is caused by shading from taller grass. This was shown in studies of Stern and Donald (1962) who investigated above- and below-ground components of competition between annual ryegrass and subterranean clover. Clover production fell sharply as N supply increased and the effect could be explained entirely by light relationships in the canopy (Fig. 10.14). A close correlation was established between growth of clover and radiation that penetrated the upper grass canopy to the clover. The studies also demonstrated an important feature of interspecific competition that was introduced in Chapter 3. At high N, growth of clover is suppressed sufficiently to lead to mortality. Over time, low-growing species may be excluded from mixtures. On farms, grass (and pasture) productivity subsequently declines as N is depleted by grazing and reseeding may be necessary to re-establish clover in the pasture.

Arable crops may also be grown in combination on the same land. Such polycultures aim to achieve LER >1 (Section 3.4.3) and are usually successful only where niche differentiation occurs for limiting soil resources or because duration of green cover is extended, leading to better use of the growing season. Careful evaluation of intercrops is essential. To establish the real LER, polycultures must be compared with pure crop stands that are also optimally managed with regard to density, planting pattern, and duration.

The classical mixed crops of the tropics are legume–non-legume combinations of annual and perennial crops grown with limiting N. Maize and beans as annuals and coffee or cocoa under leguminous trees as perennials are the main examples. The range of possibilities in mixed cropping, herb–herb, shrub–herb, tree–herb, legume–non-legume, deciduous–non-deciduous, is enormous. Even where water and/or nutrients are limiting, attention is needed to the light relationships and impact on productivity in these mixtures. The planting pattern is one part of such management but manipulating canopy shape and extent is another. A feature of species mixtures, especially mixtures of trees and herbaceous plants (agroforestry systems), is the great morphological and physiological differences that exist between components. Difference in stature plays a dominant role in the competition for light between component species (Connor 1983).

In productive agroforestry systems there will always be competition for light. Models of canopy photosynthesis can play a major role in making preliminary assessments of likely outcome of species combinations, planting patterns, and management practices (e.g., Van Noordwijk & Lusiana 1999). Even given this start, much experimental work is then required for the development and management of these complex systems.

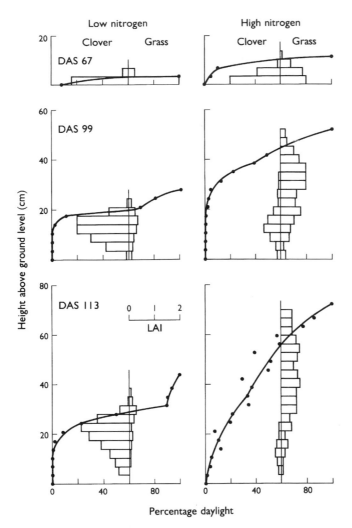

Fig. 10.14 Vertical distribution of light flux (lines) and leaf area (bars) within grass–clover mixtures grown at two levels of available nitrogen on three occasions (days after sowing, DAS) during the season. (After Stern & Donald 1962.)

10.6   Review of key concepts

Photosynthetic types

- Primary reactions of photosynthesis form the C3 compound phosphoglyceric acid (PGA), the substrate for synthesis of C6 sugars. In most plants, C3 types, CO_2 diffuses directly to chloroplasts where these reactions occur.
- Evolution has found two mechanisms to improve photosynthetic performance, both involving C4 acids that concentrate CO_2 at sites of fixation. In C4 plants, these acids are formed in mesophyll tissue and transported to chloroplasts within bundle-sheath cells where CO_2 is released. In succulent (CAM) plants, the accumulation of C4 acids

during nighttime can supplement or replace CO_2 uptake during daytime (when the stomates remain closed). In both cases, photosynthesis is favored and transpiration efficiency is increased.

- Tropical grasses are commonly C4 species but many important tropical plants, such as rice and banana, and all tropical legumes, are C3. The most important crop with CAM photosynthesis is the pineapple. Some crops, including cassava, have intermediate C3 to C4 characteristics and photosynthetic performance.

Leaf photosynthesis

- Leaf photosynthesis depends upon photosynthesis type, irradiance, N content, and temperature. C4s saturate at higher irradiances with higher rates than C3s. C4s perform better at high temperatures and lower leaf [N] than C3s. All perform similarly at low irradiance because then the basic C3 photosynthesis process is not assisted by chemical and morphological advantages of C4 or CAM.
- The advantage of high photosynthesis rates in C4, and often plant survival, is restricted to warm environments. C3s are more productive in many temperate regions.

Canopy photosynthesis

- Photosynthesis rates of canopies are determined by canopy structure as well as environment and photosynthetic characteristics of individual leaves. Structure includes size, angles, and vertical distribution of leaf area within the canopy. Because leaves are generally short-lived relative to lifespan of crops, the dynamics of initiation, expansion, and senescence determine the age structure of canopies, a major determinant of physiological status.
- Under conditions of adequate water and nutrients, light capture and utilization are ultimate limiting factors for survival and productivity. When water or nutrients limit growth, optimum canopy design must balance canopy productivity with water and nutrient supply required to produce a canopy and maintain its physiological activity.

Photosynthesis and productivity

- Linkage of photosynthesis to productivity is strong as revealed by simple quantitative analyses that relate crop growth to intercepted radiation and its efficiency of utilization (RUE). At the same time, more complex (computer) models can extend analyses to interactions between leaf display, leaf photosynthesis response, and canopy performance.
- Models are available to study interactions in continuous and discontinuous canopies and in monocultures and polycultures. They have already had an impact on the design

and management of crop canopies and offer an efficient way to screen many future possibilities, particularly those complex arrangements involved in polycultures.

Terms to remember: acclimation; canopy architecture; chilling injury; carbon isotope discrimination; light and dark respiration; compensation point, CO_2, light; nitrogen-use efficiency in photosynthesis; photoinhibition; photorespiration; photosynthesis reactions, dark and light; photosynthesis types, C3, C4, and CAM; quantum requirement, efficiency of photosynthesis, and radiation-use efficiency; specific leaf area and mass.

11   Respiration and partitioning

Assimilates from photosynthesis serve as substrates for respiration and growth. Sucrose is the principal transport form in crop plants and sucrose and starch are the main storage forms. Assimilates are consumed in respiration providing energy to maintain cellular processes and also for biosynthesis of new materials. It is the partition of those materials, also originating from the same pool of assimilates, that changes the size and morphology of plants during a growing season. This is evident in the changing numbers of stems, leaves, and reproductive structures and its control is an objective of crop production. Here we begin the discussion of partitioning with assimilate use in respiration and biosynthesis.

11.1   Carbon use in respiration and synthesis

Respiration furnishes energy for new construction and for maintenance of existing structures. The portion linked with growth is termed **growth respiration, R_g**. The magnitude of R_g varies with the chemical nature of newly constructed biomass. **Maintenance respiration, R_m**, also depends on tissue composition and has precedence over growth for assimilate; together R_m and R_g ordinarily consume 30 to 50% of gross photosynthesis. Respiration and chemical composition of new biomass, then, are important aspects of carbon partitioning.

11.1.1   The respiratory process

Respiration occurs in the mitochondria of all living cells and is termed "mitochondrial" or "dark" respiration, thus avoiding confusion with photorespiration (Section 10.1.2). Carbon enters mitochondria as organic acids derived in the cytosol from protein, carbohydrate, or lipid. Within mitochrondria, the tricarboxylic acid (TCA) cycle and electron transport chain accomplish the chemical transformations. The carbon substrates are oxidized through removal of H^+ and e^- and release of CO_2. The electrons are transferred to O_2, forming water, while the released chemical bond energy is retained in reducing agents (nucleotides such as NADH and NADPH) and in energy carriers such as ATP (adenosine triphosphate). Box 11.1 provides a summary of respiratory processes.

Respiration normally is closely coupled with the availability of ADP and oxidized nucleotide (NAD^+). If those carriers are scarce, i.e., if ATP and reduced nucleotides

---

Box 11.1 Energy and CO_2 release in respiration

Reduced nucleotides and ATP act as energy carriers within cells, supporting biosynthesis and maintenance processes. Reduced nucleotides can serve also as intermediates in ATP production. Carbon substrates are transported between cells but energy carriers are not. A total of 24 e^- is released from complete oxidation of a glucose molecule (taken here as the standard substrate). That is sufficient for the production of 12 reduced nucleotides or 36 molecules of ATP. Respiration of glucose can be summarized as follows:

$$C_6H_{12}O_6 + 6CO_2 \rightarrow 6CO_2 + 6H_2O, \text{ and}$$

$$36(ADP + P_i) \rightarrow 36ATP, \text{ or} \qquad (11.1)$$

$$12(NAD^+ + H \text{ sources}) \rightarrow 12(NADH + H^+)$$

where ADP is adenosine diphosphate, P_i is inorganic phosphate, and NAD^+ and $NADH + H^+$ (or $NADH_2$) are the oxidized and reduced forms of nicotinamide adenine dinucleotide, respectively. Organic acids, which are intermediates in the flow of C, serve as H sources. ATP and $NADH_2$ are the main energy carriers in plants. In this equation with glucose, the **respiratory quotient** (RQ; CO_2 released/O_2 consumed) is 1.0. When a more reduced material such as lipid serves as the substrate, as in the germination of oilseeds, RQ is only about 0.7.

accumulate, respiration ceases. Energy released by complete oxidation of glucose is near 2.80 MJ mol^{-1} (15.6 kJ g^{-1} at 20°C). Bond energy associated with reduction of nucleotides and formation of ATP varies with the concentrations of their reactants and products in the cells. Approximate values are 0.22 MJ mol^{-1} nucleotide and 0.053 MJ mol^{-1} ATP; approximate efficiency for formation of reducing power [(12 NADH × 0.22)/2.80 = 0.94] is greater than for ATP formation [(36 ATP × 0.053)/2.80 = 0.68]. The remaining energy is lost as heat.

11.1.2   Respiration related to growth and maintenance distinguished

The close coupling of respiration allows us to proceed directly to questions about the use of reductants and ATP. Bacteriologists and animal scientists routinely distinguish between biosynthesis and maintenance activities as causes of total respiration (R) but plant scientists came to a similar view only after pioneering work by McCree (1970). The partition can be written:

$$R = R_g + R_m \qquad (11.2)$$

The terms can be expressed as mass of C, CO_2, O_2, or glucose, per unit time. As with photosynthesis, R is most easily measured as a CO_2 flux in closed or open systems (Box 10.2) because it is more difficult to measure disappearance of substrate and release of O_2.

Respiration by mature seed is presumably all R_m, because they are not engaged in biosyntheses. Growth and maintenance activities occur simultaneously, however, in most vegetative tissues. Mature leaves, for example, import nitrate ions and synthesize amino acids that are then exported and used elsewhere in growth. Therefore, R_g related to new construction actually occurs throughout the plant. Special accounting is required for such mobilized materials. The first synthesis of a protein, for example, embodies the large cost of N assimilation from nitrate or N_2 gas (Section 8.6.1). When protein is mobilized, costs of hydrolysis, transport as amino compounds, and synthesis into a new protein at another site are also properly assigned to R_g. Those costs are less, however, than are needed to acquire additional N and synthesize a new amino acid, and the plant has conserved both carbon and N resources through recycling. Protein content of the plant remains unchanged while R_g has increased.

Two principal methods, neither entirely satisfactory, are used to distinguish R_g and R_m in growing plants. In one, plants are held in a controlled environment at their light compensation point where gross photosynthesis equals dark respiration (i.e., $P_n = 0$; Fig. 10.1). Total R is measured during brief periods of darkness. With photosynthesis balanced exactly by respiration, we can assume that no assimilate is being used in forming new biomass, that R_g is therefore zero, and $R = R_m$. In the second, "starvation", method, plants are placed in the dark so that free assimilates are consumed and growth ceases. Total R declines to a minimum value (after about 48 h in some experiments) that serves as an estimate of R_m. In both methods, the difference between normal respiration and R_m provides an estimate of R_g. The problems with these methods are that maintenance activity may be greater in rapidly growing plants than in non-growing ones, and that R_m may be substrate dependent, proceeding faster in unstarved plants.

11.1.3   Maintenance activities

Maintenance activities in higher plants include turnover of proteins and lipids and maintenance of electrochemical gradients across membranes (Penning de Vries 1975). Proteins and lipids are subject to slow rates of breakdown and ATP is then consumed in resynthesis at the same site. Because net increases in protein or lipid content do not occur and, unlike the senescence–mobilization phenomena, growth does not take place, turnover is classified as maintenance. Some enzymes turn over rapidly but rates for the larger pool of structural proteins and lipids are small. Hence, an equal or larger portion of R_m is likely due to maintenance of electrochemical gradients. Cells concentrate solutes such as K^+ and sugar within the plasmalemma and tonoplast compartments, but these materials may leak to the apoplast or to soil solution. ATP is then expended to transport solutes back across the membranes.

Specific R_m (R_m per g dry mass, sometimes termed the maintenance coefficient), increases with temperature and declines with plant age as illustrated in Fig. 11.1. These effects are important ecologically. The exponential increase in R_m with temperature up to an injury point at 40 to 50°C (depending on species) has a Q_{10} near 2 (rates double with each 10°C increase) as is characteristic of chemical reactions. Leakage and turnover presumably increase as molecular activity increases with temperature. The decline of R_m

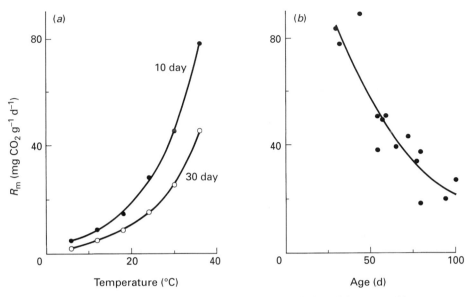

Fig. 11.1 (*a*) Variation in specific maintenance respiration of whole radish plants with temperature. The 10-day-old plants had a significantly greater specific maintenance expenditure than 30-day-old plants. (Redrawn from Lopes 1979.) (*b*) Variation of specific maintenance of whole sorghum plants with age at 30°C. (Redrawn from Stahl & McCree 1988.)

with increasing age correlates in most species with an increase in proportion of mature tissues. Mature tissues have large proportions of materials such as cellulose, lignin, and starch that do not require maintenance. In contrast, R_m rates increase linearly with N content. Nitrogen content reflects not only the amount of protein subject to turnover but also the total amount of protoplasm, and thus surface area of membranes and solute content, as opposed to wall material and starch.

In general, larger portions of new assimilates are required for maintenance in warm climates, or with N-rich plants, than in small plants, cool environments, or with N-poor plants. As plants increase in size, specific R_m declines (Fig. 11.1*b*) but that is offset by an increase in the proportion of non-photosynthetic tissues requiring maintenance. As a result, a greater portion of new assimilates goes to maintenance. This is particularly evident in crops after they achieve full cover: photosynthesis continues at a constant rate while biomass and R_m continue to increase.

Specific maintenance rates observed with different crop plants vary considerably depending on the way the plants are grown and how R_m is measured (Amthor 1989). Values obtained by the starvation method seem least variable. Typical rates at 20 to 25 °C for vegetative tissues measured by that method range between 15 and 50 mg CO_2 g^{-1} DM d^{-1}. Expressed as a ratio of the masses of carbohydrate and dry matter, specific rates vary between 0.01 and 0.035 d^{-1}. Values reported for root systems (up to 0.15 d^{-1}) are much larger than expected from their N content. Those observations are mainly from culture in nutrient solutions, however, where ion leakage may be greater than occurs in soil. Extended to field conditions, they would cause much greater CO_2 fluxes from soil than are observed.

Reduction of maintenance costs remains one of the least explored possibilities for improvements in biological efficiency. In one example, ryegrass selections having small R_m rates for mature leaf tissue produced greater forage yields under field conditions than their parent population (Wilson & Jones 1982). The lines had similar N content but whether they differed in rates of turnover and gradient maintenance was not determined. Related work by Robson (1982) indicated that yield advantage was due in part to greater leaf longevity and thus greater leaf area in the selections than in the parent population.

11.2   Growth respiration and growth yield

Growth requires substrate carbon both to form new chemical structures and as a source of reducing power and energy for synthesis. Denoting substrate that enters into the structure of new product by S, the mass ratio of product formed (B) to total substrate used ($S + R_g + R_m$) during the same time interval is termed the **apparent growth yield** (Y):

$$Y = B/(S + R_g + R_m) \qquad (11.3)$$

S, R_m, and R_g are amounts of glucose substrate and the usual units for Y are g product g^{-1} glucose. Consideration of just the substrate used in growth ($S + R_g$) allows calculation of the **true growth yield** (Y_g):

$$Y_g = B/(S + R_g) \qquad (11.4)$$

Bacteriologists handle the problem of distinguishing between R_g and R_m neatly in chemostats (Pirt 1965), and animal scientists have found ways to calculate net energy contributions of feeds for gain or maintenance (Section 2.3.2). An independent means for measuring or calculating Y_g, S, and R_g was needed for plants, and two approaches emerged. One approach depends upon analysis of proximate biochemical composition and calculation of S and R_g from biochemical pathways. (Proximate analysis was introduced in Chapter 2.) The other method begins with elemental analysis of biomass.

11.2.1 Y_g and R_g from proximate analyses

Starting from basic materials such as glucose and nitrate, it is possible to trace the metabolic pathways by which plants synthesize various organic materials and thus calculate Y_g and R_g. An overall reaction for synthesis is constructed first. Inputs are then expressed in terms of glucose for calculation of Y_g. In this method, estimate of true growth yield is termed **production value** (PV) and its inverse, the amount of glucose consumed to form the product, is the **glucose requirement** (GR).

Penning de Vries *et al.* (1974) made calculations by this method for a large number of organic materials including organic acids, lignin, lipids, proteins, and carbohydrates. Their key assumption was that plants use the least costly of any alternative pathways for biosynthesis. They sometimes included additional respiratory costs covering turnover of

Table 11.1 Glucose use in biosynthesis of 1 g of product of major classes of organic compounds obtained by calculations for type compounds in each class

The units are: production value (PV), g product g^{-1} glucose; glucose requirement (GR, the inverse of PV), g glucose g^{-1} product; and R_g, g CO_2 g^{-1} product.

Class of compound	PV	GR	R_g
Carbohydrate	0.83	1.21	0.12
Protein (with reduced N)	0.62	1.62	1.67
Protein (with nitrate N)	0.40	2.48	0.42
Lipid	0.37	2.71	1.61
Lignin	0.52	1.92	0.58
Organic acids	1.10	0.91	-0.04^a

Note: [a] a net uptake of CO_2 occurs.
Sources: adapted from Penning de Vries *et al.* (1983); Vertregt & Penning de Vries (1987).

enzymes ("tool maintenance"), ion uptake, and membrane transport between cells as a part of the glucose requirement for growth.

Estimates of PV obtained for various compounds within a given biochemical class, such as lipid, are generally very similar to each other. Therefore, each class of compounds can be represented by a single GR value (Table 11.1) that can be used with proximate analyses to estimate GR and R_g for complex biomasses.

11.2.2 Y_g and R_g calculated from elemental analyses

The cost of synthesizing organic compounds can also be estimated more directly from elemental analyses, without knowledge of biochemical composition or pathways (McDermitt & Loomis 1981). Calculations depend on the conservation of mass and electric charge and are demonstrated in Box 11.2 with the synthesis of glutamic acid using glucose and nitrate as substrates.

Formation of some compounds involves additional production or consumption of ATP, however, and GV does not always equal PV. For complex biomass with an unknown biochemical composition, an empirical relationship exists:

$$PV_{biomass} = E_g GV \tag{11.5}$$

where E_g is the fraction of total substrate e^- retained in the product. E_g for biomasses appears to be reasonably constant in the range 0.84 to 0.89, indicating that nearly 90% of free energy of glucose is retained in products of biosyntheses (Lafitte & Loomis 1988a). (Biochemistry is reduced to a constant!) This leads to an important conclusion that there is little promise in genetic selection aimed at improving efficiency of biosynthesis in crop plants. Respiration associated with growth and heat of combustion of the product biomass can also be predicted from glucose equivalent (GE) (Lafitte & Loomis 1988a).

Box 11.2 Glutamic acid synthesis as example of elemental analysis

The molecular formulae and relative molecular masses (M_r) of glucose and glutamic acid are:

Glucose	$C_6H_{12}O_6$	$M_r = 180$ g mol^{-1}
Glutamic acid	$C_5H_9O_4N$	$M_r = 147$ g mol^{-1}

Evidently, 5/6 moles of glucose are needed to provide the C skeleton of glutamic acid. In addition, C in glutamic acid may be more oxidized or reduced than in glucose. The oxidation numbers (valences) of H and O are fixed at +1 and –2, respectively. Organic N is at the –3 level while nitrate N is +5. By contrast, C may vary between –4 and +4. The oxidation–reduction state of a compound is assumed to balance to 0, allowing the reduction level (r) of C in the compound to be predicted from:

$$r + h - 2x + kn + ms = 0 \qquad (11.6)$$

where h, x, n, and s are the moles H, O, N, and S in the compound ($C_cH_hO_xN_nS_s$); k is the oxidation number for N taken to standard state ($k = 5$ for nitrate); m, the oxidation number for S, is normally $+ 6$; r is 0 for glucose. For formation of glutamic acid from nitrate:

$$r_{glutamic} = -9 + (2 \times 4) - (5 \times 1) - 0 = -6 \text{ mol e}^- \text{ mol}^{-1} \text{ glutamic acid}$$

Thus, C in 1 mol of glutamic acid has 6 mol electrons more than C in glucose, i.e., it is more reduced. Oxidation of a mole of glucose yields 24 mol e$^-$, therefore an additional 6/24 mol glucose must be respired to supply reducing power for the synthesis. A summary of these steps defines the amount of glucose required (**glucose equivalent**, GE):

$$GE = c/6 - r/24 \text{ (mol glucose mol}^{-1} \text{ product)} \qquad (11.7)$$

amounting to 1.083 mol glucose mol^{-1} glutamic acid ($c = 5$ and $r = -6$).

Estimate of true growth yield obtained by this method is termed **glucose value** (GV).

GV (equivalent to PV) = 147 g product/(1.083 mol $\times$ 180 g mol^{-1} glucose)
= 0.75 g product g^{-1} glucose.

Calculations usually proceed from glucose and nitrate but other substrates such as amides may be employed. Analyses for C, H, N, and S are easily and accurately obtained by the Dumas pyrolysis method. That provides direct measures of net photosynthesis (P_n) in carbon terms as well as of amounts of N in biomass. Unfortunately, a satisfactory direct method for measuring organic O does not exist. Oxygen can be obtained by difference after measuring C, H, N, S, and all mineral elements (a laborious procedure). That method was used to obtain the empirical formulae for grain sorghum biomass presented in Table 11.2.

Table 11.2 Growth yields and R_g values derived from empirical formulae for biomass of young grain sorghum plants grown with an adequate supply of N[a]

Elemental formula for the organic materials in 100 g biomass[b]

$C_{3.59}H_{5.74}O_{2.68}H_{0.126}S_{0.002}$

r (mol e$^-$ per 100 g)	GE (mol glucose per 100 g)
$2(2.68) - 5.74 - 5(0.126) - 6(0.002) = -1.02$	$3.59/6 - (-1.022/24) = 0.64$

GV (g g^{-1} glucose)	PV (g g^{-1} glucose)	GR (g glucose g$^{-1)}$	R_g (g CO_2 g^{-1})
0.87	0.73	1.37	0.43

Notes: [a] data are from Lafitte and Loomis (1988a) for samples collected 61 d after planting. Here, fewer significant figures were carried in the calculations causing slight differences in results.
[b] 100 g is the conventional "molecular mass" for unknown materials. Samples contained 6.45 g minerals including nitrate, sulfate, and silica.

Table 11.3 Construction costs of maize and soybean grains compared using glucose requirements from Table 11.1

The calculation is done as the sum of products of composition × the individual GR values.

	Composition in g per 100 g grain × (GR)					
Source	Crude fiber[a]	Carbohydrate	Lipid	Protein	GR per 100 g	PV
Maize	2.2(1.32)	+ 82.2(1.21)	+ 4.4(2.71)	+ 10(2.48)	= 139.1	0.72
Soybean	5.6(1.32)	+ 27.2(1.21)	+ 20(2.71)	+ 42.1(2.48)	= 198.9	0.50

Note: [a] crude fiber is assumed to be 85% cellulosic material and 15% lignin; 1.32 is the weighted average of their individual GR values.
Source: proximate composition was obtained from the National Research Council (1982).

11.2.3 Variations in R_g

In contrast to R_m, specific R_g depends entirely on biochemical composition of new biomass, and substrates employed, thus is independent of any direct effect of temperature. Actual rates of R_g vary indirectly with temperature, however, through effects on growth rate and the balance of photosynthesis and R_m.

Effects of biochemical composition on R_g are dramatic. Of vegetative structures, leaves have the largest contents of both protein and lipid, which makes their construction costs much larger than for stems and roots. Differences in composition have large effects as illustrated in Table 11.3 with calculations of growth yields for maize and soybean grains. It is obvious, given equal quantities of photosynthate for grain production, that the mass yield of soybean with its high contents of lipid and protein can never be as great as for maize. Because the two crops grow in the same climate zone and have similar growth duration, the price received by farmers for soybean grain must be more than twice that of maize if they are to choose it as a crop in place of maize. In practice, the yield difference is widened because C4 maize is more productive of photosynthate than C3 soybean. In the US maize–soybean region, the usual ratio of grain yields (soybean/maize) is near 0.33. Maize production is somewhat more costly due to N fertilizer, however, and the price ratio is usually near 5:2.

Table 11.4 Glucose requirements (GR) calculated for several seeds and grains from their principal constituents beginning with ammonium or nitrate ions or atmospheric N_2

A value of 3.43 was used for GR for protein by N fixation based on the minimum theoretical costs of nitrate reduction and N fixation (67 and 200 MJ kg^{-1} N, respectively).

Species	Composition (% dry matter)			GR (g glucose g^{-1} biomass) with N from		
	(CH_2O)	Protein	Lipid	NH_4^+	NO_3^-	N_2
Rice	88	8	2	1.26	1.32	1.46
Corn	84	10	5	1.33	1.42	1.58
Wheat	82	14	2	1.28	1.40	1.63
Chickpea	68	23	5	1.35	1.54	1.92
Safflower	50	14	33	1.83	1.95	2.18
Soybean	38	38	20	1.68	2.01	2.64
Rape	25	23	48	2.13	2.33	2.71
Peanut	25	27	45	2.10	2.33	2.78
Sesame	19	20	54	2.19	2.36	2.69

Table 11.4 illustrates the effects of composition and N source on growth yields and glucose requirements for a range of crop seed and fruits. Costs were calculated beginning with nitrate, ammonium, or N_2 gas. A conservative value of cost of N fixation was used in these calculations (see Section 8.6). The ammonium basis excludes costs of nitrate reduction (or N fixation) from other aspects of fruit growth and assigns them to vegetative tissues where they generally take place. In this table, average glucose requirements were 11% more with nitrate and 32% more with N fixation than with ammonium. The implications are seen clearly with soybean that has potential for producing 1.3 times as much yield without N fixation. The greater yield of legumes with mineral N has been demonstrated in experiments (Silsbury 1977; Piha & Munns 1987). The small growth yields of expensive compounds explain observations by plant breeders of inverse relationships between yield and protein or lipid content of cereals and grain legumes.

11.3   Seasonal patterns of crop respiration

Few measurements exist of seasonal courses of respiration in crops because such experiments are difficult and costly. Among the problems: daytime respiration cannot be estimated accurately and R_g and R_m are not distinguished easily. Work by Thomas and Hill (1949) with alfalfa, Biscoe *et al.* (1975) with barley, and Katsura *et al.* (2009) with rice define the general pattern of total respiration, R and its components R_m and R_g. Thomas and Hill (1949) and Katsura *et al.* (2009) enclosed parts of their crops in assimilation chambers (Box 10.2) whereas Biscoe *et al.* (1975) left their crop open to the air and estimated CO_2 fluxes by aerodynamic methods Section 6.7. Soil CO_2 flux (sum of root and microbial activity) was measured in small enclosures while daytime respiration was calculated, with adjustments for temperature, from nighttime values. Thomas and Hill (1949) found that about 40% of gross photosynthesis (P_g) of vigorous alfalfa crops

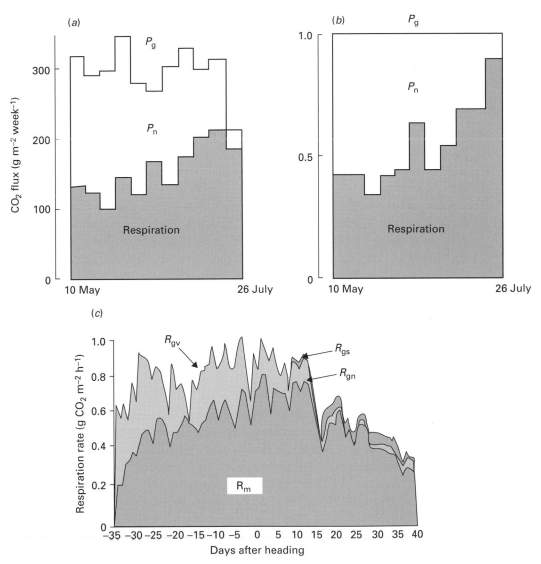

Fig. 11.2 (*a*) Weekly CO_2 fluxes relating to gross photosynthesis (P_g) and net production (P_n), and respiration for a barley crop grown in the field at Nottingham, England. The crop was sown March 18 and harvested August 21. (b) Data from (a) presented on a relative basis. (After Biscoe *et al.* 1975.) (c) Seasonal changes in rates of maintenance respiration (R_m), constructional respiration related to synthesis of vegetative organs (R_{gv}), synthesis of grain from newly assimilated photosynthate (R_{gn}), and synthesis of grain from carbohydrates stored in vegetative tissue (R_{gs}). (Adapted from Katsura *et al.* 2009.)

was expended in respiration during the season. Respiration was about equally distributed between roots and shoots. Root respiration dropped dramatically when shoots were harvested and then increased during regrowth, reflecting the dependence of growth and N fixation on carbohydrate supply. In the barley study (Biscoe *et al.* 1975), crop respiration ranged from 33 to 87% of P_g, increasing during the season as biomass accumulated. Those results are illustrated in Fig. 11.2*a, b*.

In a study with rice (Katsura *et al.* 2009), measurements of CO_2 exchange were made every 30 min from 35 days before, to 40 days after, panicle emergence. Mean values of daytime net photosynthesis (P_n) and nighttime respiration (R_{night}) were 25 and 6 g CO_2 m^{-2} h^{-1}. If respiration proceeded at the same rate during the day as at night, then daily respiration would be 40% of P_g. The authors combined gas-exchange data and harvest data to distinguish R_m and R_g with a respiration model. Results for cv. Liangyoupeijiu, presented in Fig. 11.2c, reveal distinct seasonal patterns related to maintenance and growth of vegetative material and grain from both newly assimilated photosynthate and from carbohydrates stored in vegetative tissue. Seasonal trends of respiration differed between cultivars but could be explained by differences in biomass production and N content.

These results also contribute to understanding of impact on yield of increasing night temperatures that have been observed in rice-growing regions. Earlier studies identified yield reduction of 6 to 10% $°C^{-1}$ (Peng *et al.* 2004; Sheehy *et al.* 2006). By extending their model analysis, Katsura *et al.* (2009) were able to estimate that greater R_m might account for *c.* 3% grain loss $°C^{-1}$, suggesting importance of factors in addition to R_m in measured yield losses.

Loomis and Lafitte (1987) used elemental analysis to calculate R_g and P_n of maize crops during flowering and grain-filling. R_m was estimated using a published maintenance coefficient with corrections for diurnal temperature and N content. P_g was found as sum of P_n and calculated values of R_g and R_m. In that study, 14% of P_g went to R_g and 34% to R_m. Those proportions were unaffected by moisture stress or by a doubling of the ambient CO_2 level.

Another approach for exploring crop respiration is to incorporate knowledge of maintenance and growth respiration coefficients and their dependence on tissue composition and temperature into crop simulation models where photosynthesis and respiration can be simulated independently. Ng and Loomis (1984) did that for potato (Fig. 11.3). Simulated dry matter production by this crop agreed closely with field observations. Simulated R_g exceeded R_m for small plants early in the season while the reverse was true later. Daily R fluctuated widely depending upon current weather. For the season as a whole, close to 40% of simulated gross photosynthesis was expended in respiration. That was about equally distributed between maintenance and growth.

It appears from sketchy available evidence, then, that respiration may account for 40 to 50% of the carbon assimilated by crops under good conditions. With high temperature that fraction may be much greater, which implies reduced crop yield potential in current agricultural areas under climate change scenarios with a large increase in temperature. Respiration clearly represents a major drain from carbon supply that can be partitioned to plant growth.

11.4   Morphological aspects of partitioning

Multicellular organisms face special problems in the coordination of growth and development of their various parts. Development in mammals, for example, is highly

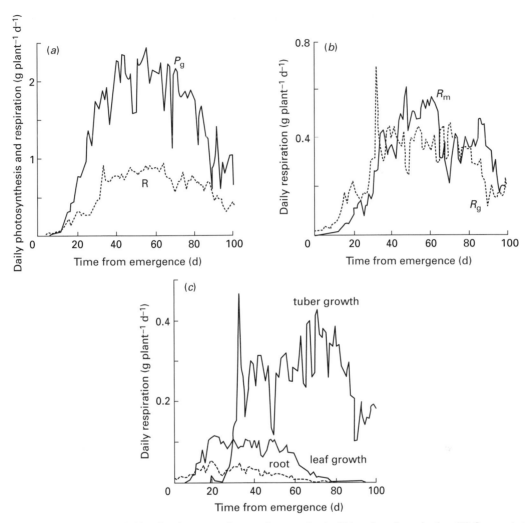

Fig. 11.3 (*a*) Simulated course of gross photosynthesis (P_g) and total respiration (R) for a potato crop at Aberdeen, Idaho, beginning with emergence on June 9. (*b*) Total respiration partitioned between R_m and R_g. (*c*) R_g partitioned among the costs of growing fine roots, leaves, and tubers. (After Ng & Loomis 1984.)

constrained to a usual form and size, in part because embryonic development continues for an extended period under controlled conditions. Throughout development, powerful control also is exercised by hormones. Embryo development in higher plants, by contrast, is minimal; a central hormone center for coordination does not exist, and subsequent growth is highly plastic. Plasticity is possible because plants have numerous meristematic centers (apical, axillary, intercalary, etc.) that continue activity throughout most of the growth cycle. Each is capable of expanding some dimension of the organism. The collective potential use of substrates by competing meristems can easily exceed supply and it is important how resources are partitioned among them. Without coordination of what grows when and how rapidly, growth could be chaotic and inefficient, if not fatal. Such coordination does occur, however, and usually leads to a plant form well suited to its environment (Trewavas 1986). Understanding qualitatively and quantitatively how such activity is coordinated remains as one of the most significant problems facing plant scientists and we now explore the various ways we can begin to do this.

11.4.1   Controls of partitioning

The principal control over partitioning rests in the capacity of meristems to grow through cell division and enlargement. Intrinsic, maximum rates of division and maximum extent of enlargement determine whether meristems can play vigorous roles in partitioning. When it grows, and to what extent, depends on internal and external milieu. Internally, growth substances serve as on–off switches of activity while substrate supply controls the rate and extent of growth. (We avoid the term hormone, which implies long-distance transport; growth substances of plants, in contrast, may act at a distance or where they are produced.) The major environmental factors that affect growth are temperature, radiation, and the supply of nutrients and water.

There are two bodies of knowledge about controls of partitioning: one focused in a reductionist way on growth substances and nuclear control, and the other in an integrative way on control by substrate and environment. Unfortunately, there are few linkages between these approaches. Some single-gene controls of partitioning are known but most aspects appear to behave as quantitative traits under multigenic control.

Control by growth substances A general theory of coordination based on growth substances alone has yet to be constructed. One problem is that the principal growth substances, auxins, cytokinins, gibberellins, and abscisic acid, are produced by most tissues (no central point) and their effects in controlling growth and development differ among tissues in confusing ways.

Growth substances are involved in a wide array of phenomena ranging from apical dominance, and control of cell differentiation, to control of flowering by photoperiod and low temperature. Evidence for hormonal involvement in flowering comes from transmission of induced states through grafts. Control by growth substances is seen most clearly in **apical dominance**. In that, development of axillary meristems is suppressed by a flux of auxin from the apical meristem. Removal of the apex removes that control and axillary meristems can then develop, establishing a different pattern of assimilate utilization. Apical dominance is very strong in plants poorly supplied with carbon substrates, because of crowding or other factors, but it has much less influence when substrates are adequate.

Nutritional control This theory of control is based on the fact that finite amounts of substrate are required for cell division and enlargement and other processes involving chemical synthesis. Amounts used are predicted by growth–yield calculations. The conservation law imposes absolute limits on partitioning because substrate used in respiration or embodied into new construction cannot be used elsewhere. In addition, rates of synthesis and growth vary with substrate level. These principles couple operationally

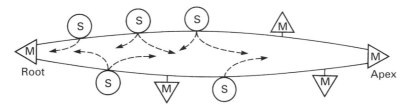

Fig. 11.4 Conceptual diagram of source–sink relations in plants with source leaves (S) and meristematic sinks (M) arranged along a common channel. Unloading by sinks is a driving force in transport and a major factor in competition among sinks for transported substrates. The channel offers some resistance to transport with the result that remote sinks can be disadvantaged in the competition.

to carbon balances from photosynthesis and respiration and provide a powerful basis for interpreting ecological phenomena. Most crop simulation models are based on the nutritional control of partitioning.

Organs capable of meristematic activity or storage serve as importers (**sinks**) for substrates transported from other organs (**sources**). Both assimilate sources and sinks can be identified by supplying individual photosynthesizing leaves with $^{14}CO_2$ and tracing the movement of labeled sugars. Such experiments reveal the existence of "normal" patterns of transport (this leaf to that sink), but they also show, after excision of various sources and sinks, that patterns are very flexible. In the absence of other sources, any leaf can supply labeled assimilate to any sink. Young growing leaves are generally sinks for carbohydrate and older leaves are sources, while stem tissues may be either sinks in which growth occurs and starch accumulates, or sources of mobilized starch. An organ that is currently a source of one substance can at the same time serve as a sink for another. Leaves, for example, are sinks for nitrate absorbed from soil and sources of newly formed amino acids.

A "channel" analogy (Fig. 11.4) accounts, operationally, for "distance effects" in assimilate use seen among competing sinks. In this, a number of growing organs are distributed along a common channel (phloem tissue) fed by source leaves. If flow of substrate along the channel is restricted by phloem conductance or depleted by unloading to intervening sinks, supplies to remote sinks will be less than for nearby ones. Such effects are seen in experiments where roots cease growing when leaves are shaded or removed because source activity is less than the capacity of intervening sink tissues to use assimilates.

In crop simulation models, distance effects can be modeled crudely by introducing transport resistances between assimilate sources and sinks. Alternatively, competing sinks can be assigned different "priorities", e.g., leaves > stems > fine roots, over a common assimilate pool with a diminishing return to substrate supply (Fig. 11.5a). Both approaches recreate observed patterns of partitioning. Some priority rules such as those related to flowering are genotype specific, while others can be stated as general rules for all crops. Maintenance respiration, for example, appears always to have priority over growth and storage.

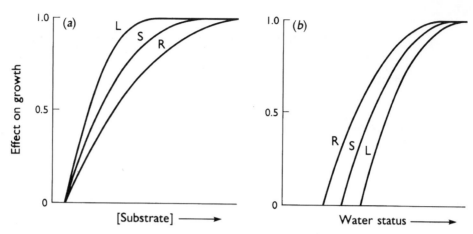

Fig. 11.5 Priority schemes useful for understanding partitioning of a common pool of substrate among several sink tissues. The schemes are based on patterns observed with varying supplies of substrate and water. The order of the lines accounts, in most cases, for nearness to source and other mechanisms that influence partitioning. (*a*) Dependence growth of leaves (L), stems (S), and roots (R) on carbon substrate level. (*b*) Dependence on plant water status.

11.4.2   Sink capacity and activity

Sink capacity of an organ is defined by the maximum rate at which it can use substrates. That depends upon its size (B), the fraction capable of growing (F_c), and the maximum specific growth rate of that fraction (μ):

$$\text{Capacity} = \mu F_c B (\text{g organ}^{-1} \text{ day}^{-1}) \tag{11.8}$$

Capacity of a single sink organ can be expressed in glucose equivalents by replacing μ with μ/Y_g, where Y_g is true growth yield (or an estimate of it such as PV or GV). Actual growth rates approach capacity when temperature is optimal and supplies of substrates, nutrients, and water are not limiting. Those conditions sometimes can be achieved experimentally by pruning competing sinks from the plant. Sink activity refers to growth or storage rates actually achieved. Operationally:

$$\text{Activity} = \text{Capacity} \times \text{Limit} \tag{11.9}$$

where Limit is a function that describes effects of limiting factors such as unfavorable temperature or substrate supply. Examples of Limit functions for effects of water stress on leaves, stems, and roots are illustrated in Fig. 11.5*b*.

The diurnal course of restrictions on sink capacity illustrated in Fig. 11.6 were produced by a substrate-dependent model of the potato crop constructed with Eq. 11.9, using priority rules of the type illustrated in Fig. 11.5. Young potato plants began with an abundant substrate supply from seed pieces, and leaf and root growth was limited only by temperature. By day 45 (Fig. 11.6*a*), plants were dependent on their own photosynthesis but, as usually occurs in a stand with competition from neighbors, substrate supply was small and growth of newly initiated tubers was limited more at night by substrate than

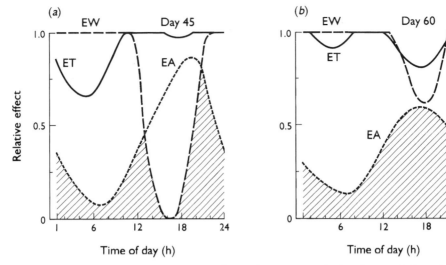

Fig. 11.6 Simulated diurnal patterns of the effects of temperature (ET), substrate (EA), and water status (EW) limitations on tuber growth of potato. The shaded area represents the net effect (Liebig's law) on tuber growth. (From Ng & Loomis 1984.) (*a*) Day 45 after emergence. Young tubers are limited by carbon substrate during the dark period and by water status during the day. The tubers are insulated by the soil and escape the effects of low night temperatures. (*b*) Day 60. Fine roots have increased and are now better able to meet the evaporative demand. Photosynthate production has also increased but intense competition among several active tubers results in a limitation by carbon substrate at all hours of the day.

by temperature. As is the case for most competitively grown crops, these plants became "sink dominated" and "source limited". In addition, transpiration exceeded capacity of fine roots for water uptake on this day and tuber growth was strongly limited at midday by low plant water status. By day 60 (Fig. 11.6*b*), leaf area and photosynthesis had increased further but with a larger root system water stress was much less. Tuber growth then was limited throughout the daytime by substrate level. Hourly integration with this model provides a realistic diurnal simulation of the interactive effects of temperature, water, and substrate supply on partitioning among leaves, roots, and tubers.

11.4.3   Growth correlations

Growth–differentiation balance When the growth of vegetative organs and fruiting bodies is less than the current rate of assimilate production due to stress, carbohydrates accumulate in temporary pools in leaves and stems, or find use in "differentiation". Differentiation, including increases in secondary products (oils, alkaloids, etc.) and wall materials (lignin, hemicellulose), is favored by conditions such as N deficiency and moisture stress that restrict growth. Trade-offs between growth and differentiation are an important feature of partitioning (Loomis 1932). Lorio (1986), for example, employed growth–differentiation balances to explain interannual variations in beetle attacks on southern pine.

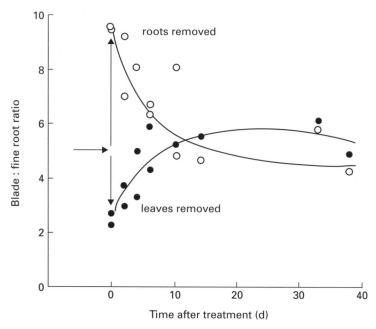

Fig. 11.7 Time course of a functional equilibrium between leaf and fine root growth in sugarbeet plants. When the plants were 30 d old, 0.5 of the leaf blades were removed from one group of plants and 0.5 of the fine roots were removed from another. Control plants (omitted for clarity) maintained a ratio of leaf blade mass to root mass near 5. Blade/root ratio of treated plants returned over time to that of untreated controls. (Redrawn from Fick *et al.* 1971.)

Functional equilibrium between roots and shoots Important examples of control of partitioning are found in the interactions between roots and shoots. Leaves are dependent upon roots for water and nutrients and roots in turn require photosynthate from leaves. When plants well supplied with nutrients and water are stressed for carbohydrates, by shading for example, carbon partitioning follows the priorities in Fig. 11.5*a* (leaves > fine roots). Given adequate carbohydrate but only limited water or nutrient, priority in carbon use is reversed to favor root growth. One reason is that root tips extending into moist soil do not experience equally severe stress to shoots. Xylem differentiation begins some distance behind the root apex and the apex is isolated somewhat from water deficits generated by transpiration. Brouwer (1983) referred to such partitioning patterns as examples of "functional equilibria" between plant organs in response to a limiting resource.

Results from an experiment with root and leaf pruning of sugarbeet plants grown in a constant environment are illustrated in Fig. 11.7. Removal of fine roots caused shoot growth to slow relative to root growth as proportions of root and shoot returned towards the starting value near 5. Similarly, leaf removal resulted in a temporary reduction in root growth. Nutritional theory adequately explains this example of internal homeostasis. The plants were grown in nutrient culture and uptake of water and/or nutrients became limiting to leaf growth after roots were removed. Note that while mass of organs serves

Table 11.5 Variation in yield components of "Insignia" wheat with planting density at Glen Osmond, South Australia

	Planted density (plants m^{-2})				
	1.4	7	35	184	1078
Density at maturity	1.4	7	35	154	447
Tillers per plant	41	30	14	7	3
Spikes per tiller	0.71	0.63	0.50	0.28	0.23
Spikes per m^2	1	132	245	302	308
Grains per spike	33	38	30	22	19
Mass per grain (mg)	34	35	33	33	33
Grain per plant (g)	33	25	7	1.5	0.4
Grain yield (kg ha^{-1})	460	1730	2470	2340	1850

Source: adapted from Puckridge and Donald (1967).

as the morphological basis here, leaf area × evaporative demand and root surface × uptake capacity are the mechanistic bases.

A similar explanation applies to partitioning between roots and shoots with limiting supplies of N. Roots acquire new supplies as nitrate but leaves are the principal sites of nitrate reduction. As a result, shoots generally enjoy a priority over roots in use of N. That priority is reversed in N-deficient plants, however, and the ratio leaves/roots is smaller for such plants. Roots of N-deficient plants are well supplied with carbohydrate and Radin (1977, 1983) proposed that the small capacity of roots for nitrate reduction serves to ensure that root apices capture adequate N despite a limited supply. Root/shoot ratios are also sensitive to plant phosphorus supply, but not to potassium (Brouder & Cassman 1994).

11.4.4   Environment-dependent variations in yield components

Morphological plasticity also is seen from the examination of variations in yield components with density. We considered changes in numbers of branches or tillers as a response to changes in density in Chapter 3 (Fig. 3.7). Numbers of inflorescences per tiller (0–1), grains per inflorescence, and grain mass also vary in turn, as is illustrated for wheat in Table 11.5. In that experiment, grain mass varied over only a small range. Examples of responses of determinate and indeterminate field bean to variations in density are presented in Table 11.6. Of yield components, numbers of branches and pods plant^{-1} were strongly affected by density whereas seeds pod^{-1} were affected less and mass per seed was unchanged.

Sequential responses to the environment are the basis of partitioning patterns in both beans and wheat. Branching (and tillering) occur early during vegetative growth and take place only if nutrients and substrates are adequate and apical dominance is overcome. As growth continues in bean plants, the numbers of axillary flower sites, flowers, seed per pod, and finally mass per seed are determined in sequence. Considerable plasticity exists

Table 11.6 Variations in yield components of field bean with variations in spacing

Determinate (D) "Taishō-kintoki" and (S) semideterminate "Gin-tebō" are commercial cultivars in Japan.

Attribute	Cultivar	Density (plants m^{-2})		
		64	16	4
Nodes per main stem	D	5.4	5.8	5.8
	S	14.3	19.9	24.6
Branches per plant	D	1.6	3.9	4.1
	S	0.2	1.8	9.0
Pods per plant	D	4.4	10.1	17.2
	S	8.0	28.1	70.4
Seeds per pod	D	3.0	3.3	3.6
	S	3.7	4.3	5.4
Mass per seed (mg)	D	376	345	334
	S	364	336	348
Seed yield (g plant^{-1})	D	4.8	11.5	20.7
	S	10.8	40.6	132.3
Crop yield (g m^{-2})	D	307	184	83
	S	691	650	529

Source: data from Tanaka and Fujita (1979).

at each stage: an unfavorable stress during the branching phase of beans, for example, usually can be compensated later in number of pods or seeds per pod if conditions improve. Determination of inflorescence size prior to anthesis is the critical step in small grains. Thereafter, plasticity is limited mainly to reductions of grain number. In most plants, seeds and grains are subject to relatively small variations in mass.

Some species, including bean and rice, are notorious for the numbers of flowers they abort, even under good conditions, as reproductive effort is balanced to resources. We are tempted to wonder what yields might be if flowers were retained and more seeds were produced. The answer is that seed number generally balances rather well with photosynthate supply. Abortion serves to bring retained flowers into balance with photosynthate supply. Plants lose little through abortion and pollination failures because flowers generally represent small costs in C and N resources. In fact, hybrid rice lines that give a 10% yield boost compared to the best inbred cultivars also have much larger numbers of unfilled spikelets, which attests to the small cost of ensuring adequate potential sink size.

Other species adjust numbers of flowers and seeds in less obvious ways. Distributive timing of anthesis in wheat and rice (from middle towards basal and acropetal regions of the ear) and sunflower (centripetally from the outer flowers towards the center of an indeterminate head) offer opportunities for adjusting numbers of grains within inflorescences. A series of "decision" stages is evident in maize. The results depicted in Fig. 11.8

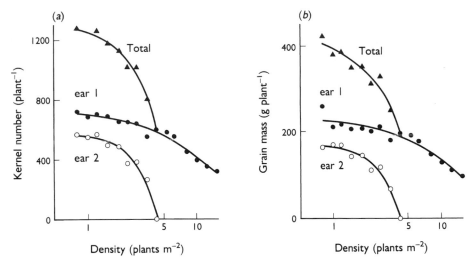

Fig. 11.8 Reproductive plasticity in maize in response to crowding for space. (Data from Tetio-Kagho & Gardner 1988.) (*a*) Number of grains per ear for the first and second ears produced on each plant. (*b*) Mass of grain for the two ears.

were obtained by varying plant density to cause variations in substrate supply per plant. Ears (female inflorescences) are indeterminate with progressive apical development of additional female spikelets. Silks (stigma) of distal spikelets emerge later than those of basal ones. When substrates are limiting, silks grow slowly, many distal ones fail to emerge and be pollinated, and grain number is lower. When plants are well endowed with substrates, those silks emerge and are pollinated; in addition, ears then develop with more columns of spikelets than occurs with limited substrate. As a result, more kernels develop per ear on widely spaced plants than on densely spaced ones. A hierarchy is also evident among successive ears. The upper ear is strongly dominant and development of axillary buds at lower positions into additional ears is very dependent upon resources.

Conservatism in seed size is not a matter that seed cannot be small (some orchid seeds are as small as 2 µg) or large (a coconut may weigh >1 kg). Differences in timing of pollination effect small variations in seed size. First pollinated spikelets on the ears of wheat and corn, and in rice panicles, are the largest seed while the last pollinated spikelets are smallest. In alfalfa, as may be the case for most legumes, ovaries generate as many as 20 ovules but only a few develop as seed. Those few are usually near the stigma end of the ovary and presumably were reached first by pollen tubes.

11.4.5 Phenotypic variations in partitioning

Differences in patterns of partitioning determine a plant's visible characteristics, i.e., its phenotype. Whether the phenotype of a bean cultivar has prostrate or erect growth, whether a maize line is "prolific" (more than one ear) or not, and whether a wheat is tall or

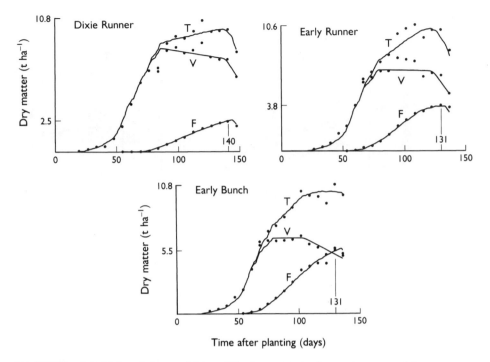

Fig. 11.9 Simulated (−) and observed (•) partitioning patterns of several peanut cultivars grown at Gainesville, Florida. Cumulative masses of leaves and stems (V), fruit (F), and total biomass (T = V + F) are shown. (Redrawn from Duncan *et al.* 1978.)

short in stature are important attributes. Alteration of patterns of dry matter partitioning towards a larger harvest index has been an underlying principle during domestication and continuing improvement of crops (Donald & Hamblin 1976, 1983; Peng *et al.* 2008, Chapter 4).

Such phenotypic differences are illustrated in Fig. 11.9 as a time-course of dry matter accumulation by several peanut cultivars. The cultivars represent a genetic advance in which seasonal dry matter production changed little while the proportion that accumulated as economic yield increased. That increase apparently resulted from restrictions on stem growth coupled with advances in fruit size and fruit growth rate as similar changes in the simulation model of Duncan *et al.* (1978) allowed it to match the patterns of real plants. We can hypothesize that competitiveness of fruit increased compared to that of stem apices. In other words, nutritional limitations led to a semi-determinate habit of stem growth. Less stem mass also means less R_m, perhaps compensating for smaller growth yield (Y_g) of fruit relative to stems.

White (1981) examined partitioning in a wide range of field bean cultivars. Dramatic differences were evident between "Type I" and "Type II" cultivars. In Type I (Fig. 11.10*a*), the main stem was determinate and the bulk of yield occurred on branches. In Type II (Fig. 11.10*b*), flowers occurred mainly on an indeterminate main stem. These bean cultivars also demonstrate strong compensatory relations among yield components. A negative association between seed number and mass is apparent in Fig. 11.11. Seed

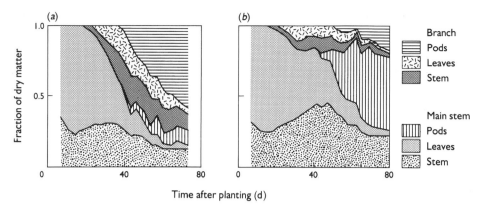

Fig. 11.10 Relative dry matter distribution during the season for two types of dry beans grown at Cali, Colombia. (Redrawn from White 1981.) (*a*) "Type I" plants have a determinate pattern of mainstem growth. Most of the pods are produced on branch stems. (*b*) In "Type II" plants, with an indeterminate main stem, only a few pods appear on branch stems.

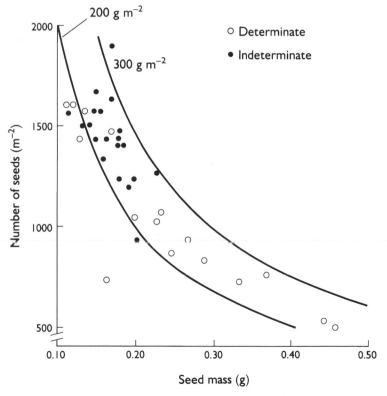

Fig. 11.11 Relation between number of seeds produced m^{-2} land and individual seed mass for a large number of determinate and indeterminate dry-bean cultivars grown at Cali, Colombia. The yield isoquants reveal that similar yields were achieved over wide ranges in seed number and size. (Redrawn from White 1981.)

size was relatively constant within a cultivar but differed widely among cultivars. Given the same seasonal supply of photosynthate, the conservation law makes such negative correlations inevitable. As a result, improvements in yield are seldom achieved through selection for an increase in only one yield component. Instead, the yield isoquants in Fig. 11.11 demonstrate a wide range in yields among cultivars, independent of seed size.

Some interesting examples of phenotypic plasticity are found with root crops. The species *Beta vulgaris*, for example, includes chard (silver beet) grown for its large edible leaves. Chard has a small storage axis in contrast to root-crop forms (sugarbeet and fodder beet) that have much larger roots and smaller leaves. Those differences result from chard having a much smaller cell size (but not fewer cells) in storage roots, and a smaller leaf initiation rate (Rapoport & Loomis 1986). That combination of a small root-sink capacity coupled with fewer leaves results in chard having a much greater substrate supply per growing leaf and ultimately larger leaves than sugarbeet.

11.5   Ideotype concepts

Plant breeders must choose their breeding objectives to address a wide array of alternative opportunities and constraints. Unless objectives are chosen carefully, the specialized, time-consuming efforts that embody the breeding process may go for naught. Traits such as resistance to insects, disease, or lodging are usually self-evident and may demand priority. They are also easier to approach than more complex traits such as yield and efficiencies in use of nutrients, water, and light. As adequate crop nutrition became a reality during the past 50 years because of availability of commercial fertilizers, plant breeders have given more attention to the biological efficiency of crops, crop photosynthesis, and optimum partitioning patterns.

Colin Donald (1968) provided an approach to the problem of optimum partitioning with his concept of an **ideotype**, a model of an ideal phenotype. "Ideal" embraces both morphological and physiological features that would suit a phenotype to a particular cropping system. Ideal ideotypes partition a minimum proportion of growth to "resource-foraging" structures, e.g., stem height to capture light or root density and depth to scavenge for water and nutrients. Partition of growth to these structures in greater proportion than is needed to maximize yield in crops composed of these ideotypes is "redundant" (Zhang *et al.* 1999). Most plants have evolved with intense interspecific competition. There, evolution favored high intrinsic productivity, and competitive ability. This explains why breeding has achieved more by changing partitioning than by increasing photosynthesis, which has long been under intensive natural selection (Denison *et al.* 2003; Box 4.3).

Donald's ideotype ideas came at a time when the international centers, CIMMYT (International Maize and Wheat Improvement Center) and IRRI, were releasing the first green revolution cultivars of wheat and rice that would subsequently provide large increases in grain yield. Most of Asia had just emerged from a severe famine in which millions had died of starvation. Donald's wheat (Fig. 11.12) was designed for an intensive

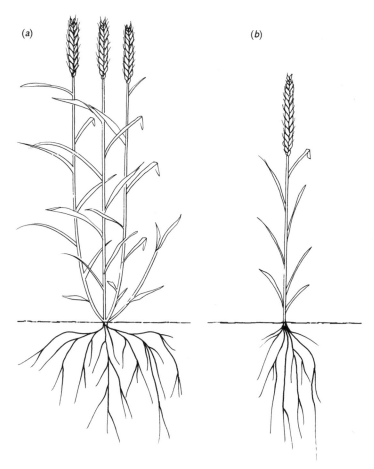

Fig. 11.12 Comparison between standard wheat (*a*) and C. M. Donald's (1968) ideotype of wheat (*b*) for high-density monocultures with soil resources not limiting. The ideotype is characterized by short stiff straw, minimum number of erect leaves, and large spike. These traits aim for non-competitive behavior, large harvest index, and maximum community performance.

monoculture system with nutrients and water supplied in adequate amounts and weeds controlled. His ideotype would achieve a large harvest index through strong restrictions on plasticity and competitive ability. Table 11.5 reveals that many tillers of traditional wheat landraces may fail to produce grain. While reduction of tillering, for example, would require greater precision in planting to assure adequate plant density, it would avoid waste of carbon and associated N required for excess tillers and leaf area. Optimization of stem growth and early flowering followed by a long grain-filling period also would help to maximize harvest index. Donald and Hamblin (1983) elaborated further on this theme, arguing that all annual plants will have similar ideotype rules. A subsequent analysis, which used gaming theory to explore ideotype design in water-limited conditions, supports this contention (Zhang *et al.* 1999).

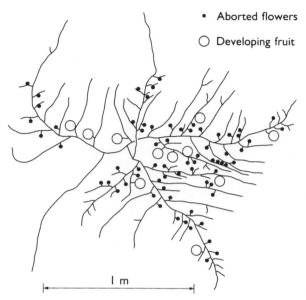

Fig. 11.13 Stem growth and fruiting patterns of an indeterminate netted melon (muskmelon) plant grown with wide spacing at Davis, California. (Redrawn from McGlasson & Pratt 1963.)

Donald's wheat fits with intensive production schemes in Europe and Asia but there are still areas in the Americas, Australia, and elsewhere geared to extensive cultivation by lower cost methods. Precision drills are now used in those areas, but with spacings that depend on good tillering for optimal performance. Ideotype concepts for wheat and other crops continue to be advanced (Bingham 1972; Loomis 1979; Smith & Banta 1983; Rasmusson 1987; Peng *et al.* 2008).

A map of netted melon, grown without competition, serves to illustrate ideotype design (Fig 11.13). The indeterminate habit is useful in gardens because it extends fruit production over a long season. But need for several hand-harvests creates a problem in commercial production. Even with repeated harvests, only a few ripe fruit are obtained per plant. As was the case for peanut, a morphology with a greatly restricted potential for stem growth and flowering sites would be advantageous in a melon ideotype for commercial production. Commercial breeding goals are thus defined as selection for a determinate plant type while maintaining or improving easily measured integrative traits such as yield, quality, and harvest index.

Ideotype questions are usually more complex than the melon example, and considerable empirical testing may be required to match each new phenotype to its optimum environment before legitimate yield comparisons can be made. Optimum space relations for bush- and vine-type plants, for example, may be quite different. As a result, most breeders seek ideotypes for existing systems (i.e., standard spacings) rather than seeking new designs for alternative systems. The point made here is that optimum designs can be defined for each system.

Ideotype questions extend beyond morphology to embrace optimal physiological traits (e.g., capabilities of roots for nutrient uptake) and optimal use of scarce resources. As an example, leaves are expensive to construct because of their high contents of protein and lipid. An optimal canopy of leaves, then, should have no more photosynthetic apparatus in each leaf than is needed in its light environment (Section 10.5.2), nitrogen content of leaves should decrease with increased shading because leaf area, rather than leaf N, becomes the limiting factor.

Ideotypes are useful in focusing attention on critical traits, integrative approaches, and optimization. Progress in designing and testing ideotypes remains slow, however, because it requires integration of many traits. Single genes may control the main features of development, but alteration of partitioning, including relations between growth and differentiation, results in changes in the carbon balance with immediate, quantitative feedback to many phenotypic traits. The sugarbeet–chard example cited above, where the "gene(s)" for large leaves is effective through changes in root morphology, is an example. The fundamental issue is that conservation of mass in partitioning confounds genetic expressions. Plant breeders continue to receive too little training in physiology, morphology, and crop ecology to approach such problems effectively. Unfortunately, the interest, skill, and funding of physiology to better understand this dynamic integration are still poor (Struik *et al.* 2007).

Integrative simulation models offer a means for dealing with such complexity. Through repeated simulations under various climates, promising ideotype concepts can be compared in a quantitative way before commitments are made to large breeding programs. The peanut example presented in Fig. 11.9 illustrates the approach; additional examples with applications to water-limited agriculture are given in Chapter 13. Optimization is an implicit goal in selection of partitioning patterns but plant scientists have done little in this area using crop simulation models and they lack training in optimization theory and in alternative techniques such as linear programming. Givnish's (1986) monograph and a review by Bloom *et al.* (1985) provide helpful introductions into biological cost–benefit analyses of adaptive traits in terms of C, N, and water. The example of optimal distribution of N within a foliage canopy discussed above, and analyses by Cowan (1986) and Gutschick (1987) of optimal strategies for dealing with limited water, offer insights into complex ideotype concepts.

11.6   Review of key concepts

Respiration

- Respiration is metabolism of (some) photosynthesis products to provide energy to maintain other metabolic activity, including construction of complex chemical constituents of plant structure and function. These processes are central to production ecology and agricultural productivity. Although there is just one overall process, it is convenient to describe these two aspects of respiration, separately, as maintenance respiration (R_m) and growth respiration (R_g).

- R_m depends upon plant size and temperature with a Q_{10} of around 2, i.e., a doubling with each $10°C$ increase in temperature. R_g is independent of rate of growth, and hence of temperature. The amount of substrate required to support chemical syntheses depends upon the nature of material being constructed. Proteins and lipids, for example, are costly to synthesize compared to cellulosic material and non-structural carbohydrates.
- Costs of construction (R_g) can be calculated from the sequence of individual chemical steps of synthesis but more directly from chemical analysis of the product.

Growth

- Carbon that is not expended in respiration becomes part of crop structural biomass or accumulates in storage materials. Even in vigorous crops, 40 to 50% of gross photosynthesis is expended in respiration related to maintenance (R_m) and growth (R_g).
- Comparison of productivity between crop types must consider chemical composition and not just mass. Relative primary production (glucose equivalent) required to produce unit mass of cereal (wheat), protein (soybean), and oil (canola) grains is in the sequence 1.63:2.64:2.71. Without N fixation, and utilizing nitrate, the ratio for soybean would be 2.01.

Partitioning

- Capacity for growth is distributed throughout plants in meristematic tissues. Meristems are relatively independent of each other with the result that partitioning of assimilate is dynamic and highly plastic in response to the environment and strengths of competing sinks. Inheritance of many aspects of partitioning, e.g., functional equilibria between roots and shoots, growth–differentiation balances, and compensation among yield components, are highly polygenic.
- Few features of partitioning have been altered during domestication of crop species: gigantism of useful organs and harvest index are the main ones; improvements in photosynthesis and respiration efficiency have been elusive. Natural selection operates through reproductive success of *individuals*, and it remains to be seen how far we can go in identifying and selecting for rules that will optimize *community* performance.
- Exciting questions emerge about whether we can define new crop plant ideotypes possessing ideal physiological and morphological traits for particular cropping systems, yield enhancement, and stress tolerance. Thus far, progress has been mainly with simple aspects of form and dry matter partitioning. Critical questions about optimal patterns of carbon and N use, and how to manage them, remain relatively unexplored.
- Our limited understanding of how partitioning is controlled can sometimes be formalized in operational rules. These rules can be incorporated into simulation models

that integrate cellular and molecular-level information for predictions of whole-plant behavior.

Terms to remember: apical dominance; functional equilibrium; glucose requirement (GR), equivalent (GE), and value (GV); growth yield, apparent and true; ideotype; nutritional control; production value (PV); respiration, growth and maintenance; respiratory quotient; sink capacity; sources and sinks.

Part IV

Resource management

Plant biology, soil science, and the impact of the environment on crop and soil interactions are central issues in crop ecology. Placing this knowledge into practical farming systems is the purpose of agronomy. In Part IV we give attention to how farm management integrates ecological principles with technology options in the design and management of cropping practices. Given the possibility to select a portion of the annual cycle for production, to choose appropriate cultivars and planting patterns, and to modify the environment by tillage, drainage, nutrient input, weed control, pesticides, correction of soil pH, and other means, the range of management options is extremely broad. While ecological factors determine what may be grown where, human decisions about labor supply, economics, and available technology are equally important in management.

Bridging the gap between environmental constraints and management to overcome them is considered for soil management (Chapter 12), how cropping may be adapted to supplies of water in rainfed systems (Chapter 13), and how water supply and productivity may be enhanced through irrigation in Chapter 14. Dependence of farming on energy and labor, the capacity of agriculture to provide motive energy as well as dietary energy (food), and whether energy supplies may be adequate in the future are presented in Chapter 15.

12 Soil management

Interactions among soil properties, soil processes, and management outcomes are the subject of this chapter. Soil properties constrain, through supplies of water and nutrients, the type of farming that may be practiced. In turn, soils are altered by farming in ways that affect their long-term value for agriculture. As a result, proper management of soil resources is the key to sustaining agriculture. Here we consider how management affects soil characteristics important for short-term productivity and longer term sustainability with a focus on ability to supply essential plant nutrients and water, and ease of tillage.

12.1 Spatial variability

Soil management aims at creating favorable and reasonably uniform conditions for plant growth in all parts of individual fields. Spatial variability of landscapes limits attainment of these goals, and soil profiles are seldom uniform. For optimum management, places with different textures, profile depths, slopes, drainage, and native fertility would be farmed differently. Although potential for within-field variation in soil properties increases with field size, small fields introduce another set of problems related to access roads, fencing, unused headlands, and excessive turning space in mechanized systems. Efficient farming requires compromises between field size and degree of heterogeneity although the tendency has been towards larger field size in mechanized cropping systems. Typical field size in the US Corn Belt, for example, is 30 to 60 ha; in Mato Grosso, Brazil, fields in soybean-based systems are often > 200 ha.

Farmers are aware of the larger variations in their fields, such as spots of low fertility or poor drainage, and sandy areas that may be droughty. Uniform tillage and crop management help reduce heterogeneity in fertility and physical properties over time except when caused by patches with markedly different texture. Other practices including drainage to control wet areas, and amendments of manure, fertilizer, and lime can be varied for different portions of fields. Leveling and terracing reduce variations in elevation, but variations in soil nutrient supply may be introduced by movements of topsoil. Some of these practices are also employed with pastures. Grazing commonly contributes to another source of variability through non-uniform distribution of dung and urine, which can be reduced through strategic location of water, salt licks, and other supplements.

Recent advances in information technology, software, sensors, and machinery provide new tools for dealing with soil variability through **site-specific management** (SSM). Geographic positioning devices and geographic information software are widely available at reasonable cost. Harvesters and tractors can be equipped to produce yield maps and apply fertilizers, soil amendments, and seed at variable rates across a field. Making variable rate applications, congruent with differences in soil properties or crop status within a field, is key to achieving an economic return to justify investment in SSM. Site-specific approaches fall into three categories: those based on spatial variation in soil properties, those based on spatial variation of plant status in the growing season, and a combination of both.

12.1.1   Site-specific management based on soil properties

Soil properties that have a large influence on crop response to management include SOM content, water-holding capacity, pH, salinity, texture, resistance to penetration (impedance), and supply capacity for essential nutrients. Of these, SOM, water-holding capacity, pH, and texture are relatively slow (recalcitrant) to change and thus investment in mapping them within a field provides useful information for many years. Other properties that are more difficult to measure, for example hydraulic properties, can be estimated from relatively stable primary soil properties through suitable **pedotransfer functions** (van Alphen *et al.* 2001). Spatial variability in supply of nutrients that are relatively immobile in soil (e.g., P, Ca, Mg, and to a lesser extent K) is also relatively stable over time. In contrast, spatial patterns of soil NO_3^- and soil biological characteristics (microbial numbers and species composition, enzyme activities, etc.) are highly variable from year to year and during each growing season. While data for some of these recalcitrant properties are available from the public domain, such as the Soil Survey Geographic Database from the US Natural Resources Conservation Service (NRCS) (http://soils.usda.gov/survey/geography/ssurgo/), they are provided at a scale too coarse to guide SSM at the field level. Instead, initial attempts to implement soil-based SSM have utilized soil maps derived from measurement of one or more soil properties by grid or similar regular sampling pattern across a field and geostatistics based on **kriging** techniques. Kriging interpolates values of a given soil property at any point within a field based on its measured value at nearby locations. The cost of such approaches is high because a large number of samples must be collected and analyzed by standardized laboratory tests.

Soil-based SSM has greatest potential for fields that contain distinct zones with large differences in recalcitrant soil properties. Thus, although soil NO_3^- at planting can be a sensitive parameter for estimating economically optimum N fertilizer rate in maize, it is too ephemeral, varying greatly from year to year at individual locations within fields, and so would require annual measurement and mapping for SSM of N fertilizer requirement. In contrast, soil pH and plant-available P (and buffering capacities for each of these parameters) change little over time without large applications of lime or P. Within a single field, these properties are often closely associated with SOM content and soil texture, which in turn are closely related to soil color. Thus, bare-soil photographs taken from an airplane or high resolution image from a satellite can sometimes provide detailed spatial data using soil color as a proxy for delineating management zones for variable rate application of lime or P fertilizer. **Soil sensors** pulled by a tractor or small vehicle for low-cost mapping of electrical conductivity, pH, SOM, texture, and compaction are under investigation for mapping NO_3^- and available P and K (Adamchuk *et al.* 2004b).

12.1.2 Site-specific management based on plant status

The most promising SSM approach for in-season N fertilizer management involves remote sensing of plant N status as the basis for timing of top-dressings. Leaf [N] is well correlated with chlorophyll content that absorbs radiation in blue and red spectral regions while reflecting light at visible green and near infrared wavelengths (Section 6.1.1). Photosensors estimate canopy "greenness" as a proxy for [N] and use this information to determine which portions of fields require N top-dressing. Sensors are mounted on tractors or mechanized boom sprayers that apply liquid N fertilizer on the go. Several types of sensors are now commercially available and have been used on wheat (Raun *et al.* 2002, Section 16.7.3), maize, cotton, and other crops.

Canopy temperature is closely associated with water stress because temperature rises when stomates close in response to a water deficit (Section 9.6). Remote sensing of plant water stress by thermal sensing with small drone aircraft is currently under investigation for timing of irrigation and to identify localized malfunction of drip irrigation systems in large commercial orchards (Berni *et al.* 2009). Some geneticists envision a future when transgenic crops can "report" degrees of stress from macronutrient deficiency or major disease following insertion of a gene coding for a protein in crop canopies that can be sensed remotely. The sensor-based application of herbicides and other crop protection chemicals represents another potential application of SSM (Timmermann *et al.* 2003).

For both soil- and plant-based SSM, current capacity and precision of measuring and mapping spatial variation exceeds our ability to use it in a cost-effective manner. While the approach is theoretically robust, costs have not consistently justified expenses. This may change, however, if fertilizer prices continue to rise in tandem with higher energy prices, and if public concerns about the negative environmental impacts of nutrient losses lead to regulations on fertilizer use, or incentives for improving fertilizer efficiency. Continued improvements in SSM technologies and access to lower cost data on soil and plant spatial variation also will promote adoption. Commercial satellites such as Quickbird (www.digitalglobe.com) or IKONOS (www.geoeye.com/CorpSite/) are now available with spectral sensing capabilities at spatial resolutions of 2.4 or 4 m for multi-spectral and 0.6 or 1 m panchromatic wavelengths. Yield maps themselves can be useful to guide soil management as described in Box 12.1.

12.2 Plant nutrition

Plants require mineral nutrients for growth so crop productivity is tightly linked to nutrient supply. Deficiency of any required nutrient reduces yield and is often

Box 12.1 Mapping yield for improved crop management

Harvesters fitted at relatively low cost with a geographic position system (GPS) provide useful information on spatial distribution of yield within large fields. Sensors measure grain flow rate and moisture content passing through the harvester, which is converted to yield after correction for speed and harvest width using a **Geographic information system** (GIS) software converts these data into yield maps, as shown below for 3 successive years from an irrigated maize field in Nebraska (from Adamchuk *et al.* 2004a). Inspection of these maps illustrates two points. First, there is considerable year-to-year variation in yield at individual locations. Second, the spatial pattern of variation is consistent with clearly defined zones of higher or lower yields. Soil sampling may explain such variation by differences in SOM, P or K levels, pH, or other soil traits.

| 1999 | 2000 | 2001 |

Yield level (t ha^{-1})

| 3–6 | 6–10 | 10–13 | 13–16 |

In the above maps, spatial distribution of yield is shown in four equal classes by shading.

Site-specific management (SSM) to reduce within-field yield variation and increase input use efficiency is most likely to be profitable where yield variation is large, spatial patterns are consistent over time, and the cause of variation can be eliminated by management. Adamchuk *et al.* (2004a) recommend the decision tree below to determine whether investment in SSM is a promising option. Consistent spatial patterns of yield variation are less common in rainfed systems, especially on sloping land: highest yields occur on upper slopes in wet years but are lower in dry years when low-lying areas experience a more favorable moisture regime.

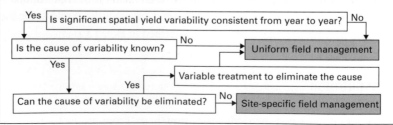

Table 12.1 Elemental composition (% dry matter) of aboveground biomass of several crops and forages and maize and soybean grains

All of the elements shown, except Si, are essential for growth of these species. Essential micronutrients (Fe, Zn, Cu, Mn, B, and Cl) are not shown.

Material	C	O	H	N	S	P	K	Mg	Ca	Si
Sorghum biomass[a]	43.0	44.0	5.7	2.0	0.62	0.14	1.80	0.32	0.26	1.2
Maize biomass[b]	43.4	44.5	6.2	1.5	0.17	0.20	0.92	0.18	0.23	1.2
Maize grain[b]	44.5	45.3	7.0	1.7	0.12	0.29	0.37	0.14	0.03	na
Alfalfa[c]	47	40	6	2.7	0.26	0.22	1.54	0.28	1.27	na
Bluegrass[c]	46	41	6	2.7	0.29	0.34	1.98	0.17	0.33	na
Soybean seed[c]	54	29	7	6.8	0.24	0.65	1.82	0.29	0.27	na
Wheat grain[d]	46	45	6	2.3	0.15	0.43	0.49	0.13	0.05	na

Notes: na: not available.

[a] At anthesis; data from Lafitte and Loomis (1988a).

[b] At maturity; all data from Latshaw and Miller (1924) except that nutrient contents of grain are from Tables 1 and 2 of National Research Council (NRC 1982).

[c] Approximate data developed from Tables 1 and 2 of NRC (1982): N from Table 1; C, O, and H from proportions of proximate and NDF fractions of NRC Table 1 using elemental compositions for those fractions from McDermitt and Loomis (1981); and minerals from NRC Table 2. Alfalfa and bluegrass biomasses were both at the early stages of flowering.

[d] Hard red winter wheat; N content varies from near 1.8% for soft wheats to over 2.6% for hard wheats (NRC 1982); data developed as per footnote *c*.

associated with visual foliar symptoms that are diagnostic. Here we learn about plant nutrient requirements and how crops respond to nutrient supply.

12.2.1   Essential nutrients

A small number of chemical elements are known to be **essential nutrients** for plant growth. Plant mass is composed mainly of C, acquired from atmospheric CO_2, and O, from CO_2 and H_2O. As shown in Table 12.1, C and O normally make up more than 85% of plant dry mass and H is usually near 6%. The remainder is composed of N and mineral elements acquired from soil, and is the basis of widespread occurrence of nutrient deficiencies in agriculture.

The presence of an element in plant tissues does not mean that it is essential for growth. It may be a surface contaminant or a passive companion of water uptake. Appreciable silicon (as silica, SiO_2) is found in biomass of sorghum (Table 12.1) and of rice, but Si is not yet known to be an essential nutrient for any crop. Essentiality is established by showing that a plant grows poorly, or not at all, in the absence of an element and that it cannot be replaced by another element. The essentiality of C, H, O, N, S and several metallic elements is further confirmed by their presence as structural components of specific metabolic compounds.

Essential nutrients can be divided roughly into two groups, **macronutrients** and **micronutrients**, depending upon the amounts required for plant growth. In agriculture, deficiencies of the macronutrients N, P, and K are most common. Requirements for some micronutrients are so small that it was exceedingly difficult to establish whether or not they are essential for growth, and the roles of other trace elements remain uncertain. A small requirement also means, however, that chance of deficiency in agricultural soils is also small. Among micronutrients, Zn deficiency is the most common. A classic case of micronutrient deficiency occurs in Australia where legumes respond over large areas to very small additions of molybdenum (Mo) that is essential for N fixation by rhizobia.

Global stocks of most nutrients are essentially infinite relative to the needs of agriculture. Nitrogen gas, for example, makes up 78% (v/v) of the atmosphere, and nearly 50 times that much is present in the igneous rock of the Earth's mantle. By contrast, the amount in soils and crops is only 0.006% of that in the atmosphere. Potassium is abundant in oceans and salt deposits. Concern exists about relative scarcity of high-grade P ores suitable for mining. Low-grade sources of P are abundant but they require considerable treatment for use in agriculture. In future, it may become practical to recycle P from sewage or from the ocean floor where it accumulates. The problem for agriculture is not limiting supplies of nutrient elements, rather that work must be expended to concentrate them into forms suitable for use as fertilizer.

Elements are sometimes present in amounts too great for healthy growth. The most widespread toxicities arise from Al^{3+} and Mn^{2+} in acid soils and from BO_3^{3-} and Na^+ in saline and alkaline soils. In arid regions, Se levels in forages and drainage waters are sometimes toxic to animals. Special problems occur with soils derived from serpentine minerals high in Mg and low in Ca. Not only do plants grow poorly under low Ca/Mg ratios, but they also encounter toxic levels of heavy metals such as Ni and Cr. Fe^{2+} can reach toxic levels in rice grown on flooded soils with high Fe content and strongly reduced conditions. Under these anaerobic conditions, relatively insoluble Fe^{3+} becomes an electron acceptor and is reduced to more soluble Fe^{2+}, which then reaches high concentrations in the soil solution. Symptoms of Fe toxicity include reddish-colored leaves, which can have [Fe] levels 10 to 100 times normal, and roots with a reddish-brown coating of precipitated iron oxide from oxidation of Fe^{2+} to Fe^{3+} in the rhizosphere. Movement of atmospheric O_2 through parenchymatous tissue in rice stems to roots keeps the rhizosphere oxidized, even when bulk soil is highly reduced.

12.2.2   Crop responses to nutrient level

Crop response to nutrient supply is seen most clearly in fertilizer experiments. The diminishing-return relationship between maize grain yield and increasing supplies of fertilizer N illustrated in Fig. 12.1a is the general pattern for all nutrients. Responses of this sort are usually clear evidence that the supply or availability of that nutrient was inadequate. Yield obtained without added N depends entirely on uptake from that supplied by soil, i.e., **indigenous soil N supply**. As in medieval farming (Section 8.7.1), the soil in Fig. 12.1 was extremely deficient in N and supplied enough N (31 kg N ha^{-1}) for only 900 kg grain without fertilizer. The highest grain yield was obtained with 224 kg N ha^{-1}. The slope of yield response to added N provides a measure of apparent fertilizer-use

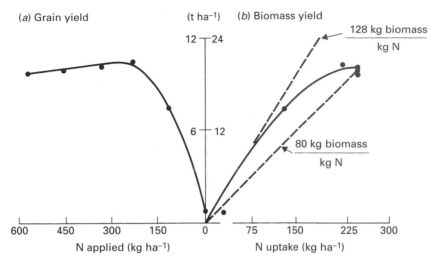

Fig. 12.1 Maize response to nitrogen fertilizer. The experiment was conducted in California with ^{15}N-depleted ammonium sulfate by Broadbent and Carlton (1978). The graphical method used here and in Figs. 12.4 and 12.5 follows van Keulen (1982). (*a*) Grain yield as a function of increasing applications of the limiting nutrient, nitrogen. (*b*) Biomass yields as a function of nitrogen actually taken up by the crop. The dashed lines indicate the maximum and minimum amounts of biomass produced per kg N.

efficiency. Efficiency is large when nutrients are strongly deficient (rising portion of the curve). In this example, 58 kg of grain yield were obtained per kg N applied over the range of 0 to 112 kg N ha^{-1}. The ^{15}N-depleted fertilizer used in this experiment allowed researchers to estimate the proportions of crop N uptake derived from fertilizer or indigenous soil sources. Applied N resulted in a greater apparent uptake of indigenous N to 50 to 80 kg N ha^{-1}, due to either promotion of greater N mineralization (priming effect) or to stimulation of root proliferation and exploration of the soil profile. About 64% of fertilizer N was captured by aboveground portions of the crop with application of 112 kg N ha^{-1} or 224 kg N ha^{-1}. With less than 224 kg N ha^{-1} applied, little inorganic N remained in the soil after harvest but that increased sharply at application levels that exceeded the requirement for the highest yield. The field was kept moist and between 15 and 22% of applied N seems to have been denitrified; the balance of applied N accumulated in soil organic matter.

The transition between deficiency and adequate supply (the breaking region of the curve in Fig. 12.1*a*) is rather sharp with N, and particularly so with maize. The economic optimum for adding N as fertilizer, and maximum efficiency in use of external energy (Fig. 15.2) typically corresponds to a yield level that is 90 to 95% of maximum, depending on the price ratio of grain to fertilizer. In Fig. 12.1*a*, the optimum would be somewhat less than 224 kg fertilizer N ha^{-1}. Where the transition zone is broad, i.e., a strongly diminishing return, or the nutrient source is expensive, economic optimum is found at lower yield levels on the deficient side of maximum yield. The plateau region of N response is quite broad for most species indicating a wide tolerance of excess supply. In

Fig. 12.1a, yield declines slightly beyond 224 kg N ha^{-1} owing to a salt effect in this irrigated field rather than to N toxicity. For some crops prone to lodging, such as wheat and rice, even a relatively small excess in N supply can lead to a marked decline in yield. Excess N that leads to lush vegetative growth can result in greater susceptibility to foliar diseases within a more conducive canopy environment. As a rule of thumb, N fertilizer efficiency is highest and N losses smallest when N fertilizer level is at or below the economic optimum.

In Fig. 12.1b, biomass yields (calculated from grain yield assuming HI of 0.5) are plotted as a function of N uptake. This curve also follows a diminishing return but stops short of a plateau. The slope of the curve with units of kg dry matter kg^{-1} N applied defines nutrient input efficiency (NIE, Table 1.2), i.e., how well the crop converted an additional kg of N uptake into biomass.

Two important concepts relating to the range of nutrient content in plant tissues are revealed in Fig. 12.1b. The first is seen in the slopes of the dashed straight lines. These define maximum (128 kg) and minimum (80 kg) amounts of aboveground biomass that maize can produce per kg N taken up. Reciprocals of these numbers correspond, respectively, to the minimum (8 g N kg^{-1} biomass) and maximum (12 g N kg^{-1} biomass) possible N contents of maize at maturity. The range is rather small, being determined by the compositions of wall material (very little N) and protein content (high in N) (Table 2.2). The second arises because the crop ceased taking up N when it reached a maximum possible N content. In contrast to ideas about "luxury consumption" as a general phenomenon, overfertilized maize crops in this example (e.g., 560 kg N applied ha^{-1}) contained only 12% more N than the crop receiving an optimal amount of fertilizer (224 kg N ha^{-1}). Some of that increase would be found in the crop as nitrate serving a useful role as osmoticum.

12.3   Management of soil fertility

Soil fertility is defined in large part by its ability to supply essential nutrients to support crop growth. Indigenous nutrient supply can be estimated by the nutrient content of aboveground biomass at maturity when a crop is grown without applied nutrients and without yield loss from pests and diseases. Without nutrient application, indigenous nutrient supply is depleted over time because removals by leaching and crop harvests generally exceed natural inputs. Here we discuss the factors governing soil fertility and the steps that must be taken to maintain fertility to sustain productivity.

12.3.1   Nutrient depletion

An example of yield decline that occurs over years with exhaustion farming is illustrated in Fig. 12.2. Yields of rye grain did not decline to zero because that crop is not demanding of nutrients and small supplies of nutrients come in rainfall, through free-living N fixation (Section 8.6), and through solubilization of soil minerals. In this example, yield stabilized

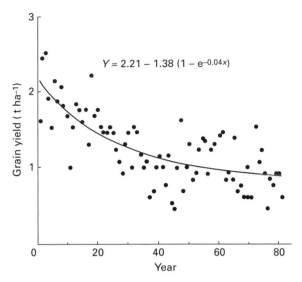

Fig. 12.2 Yields of rye grain over the period 1880 to 1960 in a German experiment without fertilizer inputs. Yields declined in a negative exponential fashion to an equilibrium level near 900 kg ha^{-1}. (Adapted from Schouten 1986.)

near 900 kg ha^{-1}. Estimated average annual removals in 900 kg rye grain ha^{-1} are 20 kg N, 4 kg K, 3 kg P, and 1 kg or less of both Ca and Mg.

Similar patterns are evident in long-term experiments with maize on the Morrow Plots in Illinois (Fig. 7.5) and with wheat in "old-field" experiments at Rothamsted in England. As with rye above, those experiments recreate situations found in low-input agriculture with yields near 800 to 1200 kg grain ha^{-1}. Such yields were characteristic of European agriculture through the medieval period (Section 8.7.1), and they remain the current situation for subsistence grain crops like sorghum, millet, and maize in many parts of sub-Saharan Africa. The reason that agricultural fields under exhaustion farming sustain production at a low level is that soils receive a small steady input of nutrients from several sources. We identified sources for significant inputs of N. Weathering of parent material, rainfall and irrigation water, and dust are the principal sources of the mineral nutrients K, Ca, Mg, and S. For cereals, the limiting nutrient is usually N. Legumes, on the other hand, supply their own N but deplete P, K, Ca, and Mg strongly and those elements come to determine legume productivity.

12.3.2   Diagnosing nutrient needs

Nutrient deficiencies can be diagnosed from visual symptoms, from soil or plant analyses, and from responses to added nutrients. Marginal deficiencies, not evident except by reduced growth, can sometimes be diagnosed by chemical analyses of plant material, or predicted from soil analyses. The most practical evidence is an observation of increased plant growth when nutrient supply is increased. A combination of methods is needed in some cases. Crops grown on soils deficient in both N and S, for example, may respond well to $(NH_4)_2SO_4$ but poorly or not at all to N fertilizer without S. And improved growth with application of sulfate (SO_4^{2-}) may actually be a response to P displaced from soil fixation complexes by sulfate.

Visible symptoms When the rate of nutrient uptake is less than the current demand for growth, plants develop characteristic symptoms that vary according to the mineral element and plant species involved. In some cases, new growth is affected most, in others, older tissues show symptoms. Shortage of N, for example, generally leads to yellowing and senescence of older leaves as N is mobilized from those tissues and utilized in new growth. New leaves may have normal green color but they are generally smaller and fewer in number. In addition to a smaller leaf area, photosynthetic ability per unit leaf area also is less in N-deficient crops. In combination, these two effects can lead to sharply smaller growth rates and yields. Recognition of such symptoms is important and is aided for many crops by picture atlases.

Soil analyses Chemical analyses of soil samples are used to predict if the supply of nutrients is adequate to avoid deficiencies. Soil analysis requires support by chemical laboratories and establishment of **critical soil nutrient concentrations** (CNC_s) above which nutrient supply will be adequate for crop growth and below which it will be deficient. Analyses on a single sample can include a range of nutrients as well as measures of salinity and pH (and thus the need for lime). Locally established CNC_s values are needed because soils differ and because a supply that is adequate for one species or level of production may not suffice for others. Difficulties arise in defining "plant-available" supply and because there is no way of knowing whether a crop will actually access the nutrients. Plant-available P, for example, has been defined variously as the water-soluble, acid-soluble, or bicarbonate-soluble fraction.

Assessment of plant-available N is especially complex. The available mineral pool (NH_4^+ and NO_3^-) is dynamic, reflecting variable rates of mineralization and immobilization. In some methods, soil samples are assayed twice for mineral N, immediately after sampling, and again after moist incubation for several weeks, to estimate the amount of organic N that may be mineralized during a growing season. Blackmer *et al.* (1989) have simplified that to a single analysis for nitrate in soil samples collected in late spring after crop emergence. Nitrate concentration at that time reflects the net of fallow-season mineralization, immobilization due to residues incorporated during primary tillage, and fertilizer additions. With their system, similar critical nitrate levels for maize production were found for a large number of soils within a climatic region. Spatial variation in soil [NO_3-N] is often a challenge to crop management when using soil-based diagnosis of nutrient needs, as discussed in Section 12.1.

Plant analysis Concentrations of nutrients in plant tissues reflect the balance between uptake and dilution by growth, i.e., past success in acquiring nutrients relative to subsequent growth rate.

Plant nutrient status is assessed from concentration of a nutrient in total biomass or just in selected organs. The actual concentration in plant biomass relative to the two lines in Fig. 12.1*b*, for example, can be related to the degree of deficiency. Individual organs are easier to sample than biomass, however, and nutrient levels in young tissues tend to correlate best with recent nutrition of crops. For most elements, analyses are

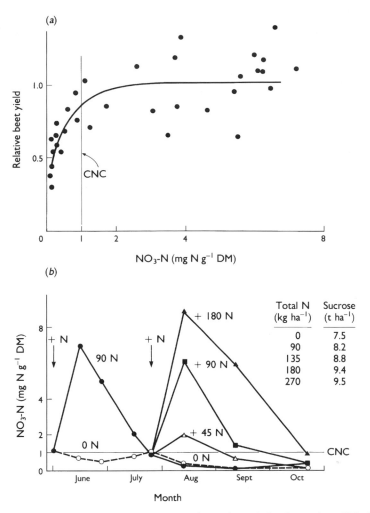

Fig. 12.3 (*a*) CNC_p for nitrogen status of sugarbeet, defined near 1 mg NO_3-N g^{-1} dry tissue from petioles of recently matured leaves. The relationship was established with plants grown in a field experiment with different supplies of nitrogen. (*b*) Time course of petiole NO_3-N levels for sugarbeet crops supplied with different amounts of N split into two applications. Also shown are the CNC_p levels established in (*a*). Yields ranged from 7.5 t sucrose ha^{-1} with 0 N to 9.4 t with optimum nitrogen supply (90 + 90 kg). (Redrawn from Ulrich *et al.* 1959.)

done for total content but with N, greater sensitivity to uptake–assimilation balances is sometimes obtained by analyzing for unassimilated [NO_3-N] rather than [total N]. Similarly, [SO_4-S] is a more sensitive indicator of crop status than [total S]. By selecting the form of nutrient associated with uptake and a sensitive **index tissue**, a reasonably sharp **critical plant nutrient concentration** (CNC_p) can be defined as is illustrated in Fig. 12.3*a*. With sugarbeet, [NO_3-N] in petioles of recently matured leaves is the most sensitive index.

An application of periodic "petiole analysis" to nutrient management in sugarbeet production is presented in Fig. 12.3b. CNC_p established in the experiment shown in 12.3a is drawn as a horizontal line in Fig. 12.3b, where levels and trends of [NO_3-N] define crop N status. [NO_3-N] in petioles increased following fertilization and declined when crop utilization of nitrate through reduction to NH_4^+ and incorporation in organic N molecules, such as amino acids and nucleic acids, exceeded its ability to acquire additional N.

Unlike soil analysis, plant analysis cannot carry a prescription for the amounts of nutrients that are required. Values less than CNC_p indicate that a crop is deficient, but not why that occurred or how much nutrient is required to correct it. A decline in plant nutrient content may be due to soil exhaustion, or it may be that disease or soil dryness has sharply reduced uptake. Conversely, deficiency of one nutrient slows rates at which other nutrients are diluted by growth, and their concentrations then tend to increase.

Local field experience with a crop is needed to define the amount of nutrients to apply, and that is closely linked to the average yield level. In general, plant tissue testing is not routinely used on field crops because of expense, rather only to confirm whether visual symptoms are due to nutrient deficiency or some other factor. Instead, most routine tissue testing is performed on high-value fruit, nut, and vegetable crops where monitoring of tissue nutrient level is critical for yield and quality.

12.4   Fertilizer practices

Given a diagnosis that supply of particular nutrients is inadequate, decisions are required regarding the form, amount, timing, and method of applying fertilizer. Use of organic amendments was examined in Section 8.8. Here we consider mineral fertilizers.

Managing nutrient supply Regardless of the method or material used, the objective of fertilization is to satisfy crop needs, not the soil's. Efficient use of fertilizers is important for both economic return and to minimize losses that cause negative environmental impact. The first step is identification of deficient nutrients; the second is accurate estimation of amounts needed to correct these deficiencies. The final step is to utilize the most appropriate fertilizer formulation, timing, and placement to maximize uptake efficiency of applied nutrients. Identification of all deficient nutrients is sometimes difficult because less severe deficiencies may not be expressed until the most severe deficiency is alleviated. For example, indigenous soil K supply may be sufficient at low yields when a crop is severely deficient in N but not when N supply is increased and higher yields are obtained. In this case, foliar symptoms of K deficiency can appear and yields are then limited by K supply. If, however, the need for K is evaluated without increased N supply, there will be no response to applied K and foliar analysis may not indicate deficient tissue levels. In a similar manner, the need for micronutrients is often masked by more severe macronutrient deficiencies.

In high yield systems there is no way to precisely predict the amounts of nutrients that must be supplied so that crop growth is not limited during a growing season, yet at the same time not be in surplus. While yield forecasting can provide an estimate of crop demand as described in Box 12.2, mineralization–immobilization balances are

Box 12.2 Crop nutrient requirements based on yield forecasts

Plant tissues require a minimum concentration of all essential nutrients to maintain active function and growth. For N, the challenge of predicting actual requirements is more difficult than for other nutrients because of potential for N losses when inorganic soil N exceeds short-term uptake. Achieving synchrony between supply and demand requires several N applications during the growing season rather than one large application at planting. While sensors that estimate plant N status (Section 12.1.2) can identify crop N deficiency, they do not estimate how much N to apply. Crop simulation models, coupled with historical weather data of at least 20 y, provide a means to estimate probability distributions for final yield that narrow as the growing season progresses. The figure below shows weekly yield forecasts of the Hybrid-Maize Simulation Model during 2003 growing season for a rainfed field in Mead, Nebraska (Yang *et al.* 2006). Each data point represents a simulation forecast using actual weather to that day and alternative sequences of historical weather data until harvest. The range of likely yields is bounded by alternatives that give the best or worst final yields (top and bottom yield forecast lines), and years that give the 75, 50 (median) and 25th percentile yield. Long-term median yield is shown as a dashed horizontal line and final measured yield as a solid line. Yield predictions gradually converge and the model is shown to explain the lower than average final yield (the solid line) when complete weather data became available. In this season, probability of below-average yield was evident before silking in mid-July. While this is late for adjusting N rate for maize, a forecast at a similar phenological stage for wheat would provide useful guidance for N top-dressing to ensure grain protein levels meet bread wheat quality standards.

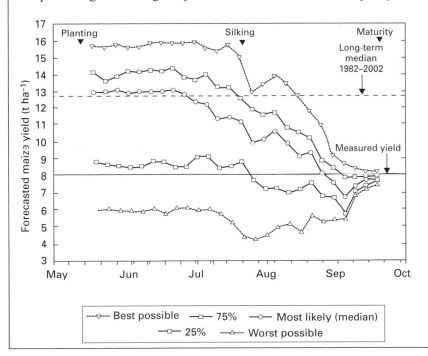

always uncertain; additional variations in supply arise from effects of weather variation, insects, and disease on nutrient use. Uncertainties about availability and plant access to applied nutrients are controlled to some extent with proper decisions about form, timing, and placement of mineral fertilizers, but farmers must rely heavily on field histories and past experience for expectations about yields and nutrient requirements. Good record keeping is essential. Soil and plant analyses and simple experiments are also useful. Strip trials with a normal rate of a nutrient and ± 10 to 25% of normal provide farmers and crop consultants with a means to fine tune nutrient management. Geographic positioning system (GPS) units, variable rate applicators, and crop yield monitors facilitate the establishment and analysis of such trials in mechanized systems. In small-scale agriculture, small test plots can be used in a similar manner.

Except for N, most surplus nutrients remain in the soil and can be taken as credit towards future crops. The cost is one year's interest on the investment and most farmers accept that in preference to risks of serious nutrient deficiencies. For N, a large portion of residual $NO_3{}^-$ is lost during fallow, which forces management to strive for as little excess N as possible. In fact, N management represents a unique challenge because too little results in deficiency and yield reduction, and too much increases risk for losses and associated environmental concerns. The key is to achieve a tight coincidence between crop demand at each point during the growing season and N supply from soil and applied N inputs.

The importance of livestock manure for meeting nutrient requirements of crop agriculture was discussed in Section 8.8 with information about handling, nutrient content, and application rates and methods. Suffice it to say here that it is more difficult to meet all nutrient needs of a crop with manure, or other organic sources, because nutrient contents are generally not well balanced with the stoichiometric requirements of crops. For example, the P:N ratio in manure is much greater than in plant biomass. Thus, excessive P builds up in soil when manure is the primary source of N, and P losses to surface waters in runoff can become a serious problem.

Uptake efficiency The response to added N illustrated in Fig. 12.1 can be extended with relationships between N uptake and N applied as shown in Fig. 12.4c. Uptake is generally a linear function of fertilizer supply until biomass yield begins to plateau (van Keulen 1982; van Keulen & van Heemst 1982). The slope of that line defines fertilizer recovery efficiency. It varies from experiment to experiment due to variations in weather that affect plant growth rates, form of nutrient applied, method of application, soil properties, and other factors. Indeed, control of those differences is among the reasons why mineral fertilizers can be managed more precisely and with less environmental concern than is the case for manure.

Commercial fertilizers The importance of chemical form is especially clear with phosphorus fertilizers. Apatite ores ("rock phosphate", $Ca_{10}Z_2(PO_4)_6 \cdot CaCO_3$, where Z is generally F or Cl, and Ca and P are variously substituted) are the principal sources of phosphorus. Even when finely divided, these are only slowly soluble and fertilizer-uptake efficiency is small compared with more soluble materials such as "superphosphate" and "concentrated superphosphate" fertilizers. Superphosphate [$Ca(H_2PO_4)_2 \cdot H_2O + CaSO_4$] is obtained by treating rock phosphate with sulfuric acid; treatment with

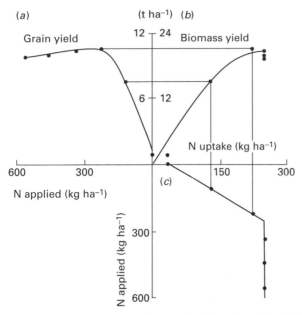

Fig. 12.4 Maize response to nitrogen fertilizer: (*a*) grain yield and (*b*) biomass yield are the same as in Fig. 12.1. (*c*) Relation between nitrogen applied to the soil and nitrogen uptake by the crop. The slope of the rising portion of the curve defines fertilizer-uptake efficiency.

phosphoric acid yields concentrated superphosphate [$Ca(H_2PO_4)_2 \cdot H_2O$]. Distillation of apatite ore concentrates phosphorus even more to polyphosphoric acid, $(H_3PO_4)_n$. A wide array of NH_4^+- and K-substituted materials are now made from phosphoric acid. Experience in a large number of field trials in Western Australia (Bolland & Gilkes 1990) illustrated the general principle that crop response to finely divided rock phosphate sources was only 5 to 20% that of superphosphate based on the amount of applied P. In those experiments, rock sources were unable to support as great a yield as superphosphate and their residual effectiveness in subsequent years was much less.

Potassium chloride (KCl) is the most common K fertilizer. Compared with potassium sulfate (K_2SO_4), which is less common, there are vast reserves in underground mines on several continents that are easily extracted with hot water due to high solubility. Either form is an equally effective source of K in most situations.

The behavior of N compounds depends upon their solubility and whether NO_3^- or NH_4^+ ions are the main source of N. In soils, ammonium ions tend to be adsorbed quickly by the cation exchange complex whereas NO_3^- ions move more freely in soil water. Exchanged ions are available only to nearby roots or, in the case of NH_4^+, after they are nitrified by bacteria and move to roots. Ammonia gas and aqueous solutions of NH_3 should be injected into the soil; applied in bands, these materials sterilize a small zone delaying nitrification while bacteria invade the area. Ammonium sulfate and urea are soluble. They are dispersed by water through a larger soil volume than ammonia and consequently nitrify more rapidly. Ammonium nitrate is highly soluble and is almost as quickly available to plants as calcium nitrate. Order of response times is: nitrate

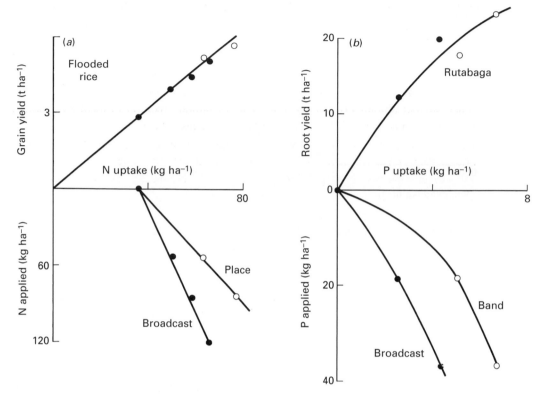

Fig. 12.5 Economic yields for flooded rice (*a*) and rutabaga (*b*) as a function of nutrient uptake (redrawn from van Keulen & van Heemst 1982). (*a*) Rice grain yield responses in Indonesia to urea fertilizer broadcast to the surface (27% recovered) or placed 0.1 m deep in the reducing zone (56% recovered). (*b*) Fresh root yields of rutabaga obtained in England in response to phosphorus fertilizer broadcast and incorporated or placed in a band near the seed.

sources > urea > ammonium sulfate > ammonia (Loomis *et al.* 1960). Urea and NH_3 (gas or in aqueous solution) are currently the cheapest and most widely used N fertilizers although concern about safety in handling ammonia is reducing its use. In the USA, liquid N fertilizers with various combinations of ammonium nitrate, urea, and calcium nitrate are being used more widely because they are safer than NH_3 and allow greater flexibility in timing and placement, especially for top-dressing.

Placement of fertilizers Broadcasting of dry materials onto soil surfaces is the simplest method for applying fertilizers, and it is the only suitable method for pastures. Surface applications are not efficient, however, without subsequent incorporation with soil by tillage, or being carried down from the surface by rainfall or irrigation. Ammoniacal materials, for example, are particularly vulnerable to losses through volatilization when they remain on the surface.

Fertilizer uptake efficiency depends on placement. Denitrification loss from flooded rice fields was described in Section 8.4.2; Fig. 12.5*a* compares rice yields obtained when ammonium N was applied to the soil surface or placed into the reducing zone, where it is protected from denitrification. With row crops, knifing fertilizer in bands parallel to rows allows plants to reach it quickly. Banding provides for more interception by crop roots than broadcasting (Fig. 12.5*b*). Another reason that banded applications are more efficient is that less nutrient is exposed to fixation and immobilization. As much as 300 to 500 kg P ha^{-1} may be needed with a broadcast application to saturate the P-fixing abilities of soils high in iron and aluminum oxides (Section 7.3.2). Bands containing much smaller amounts of soluble P are able to saturate fixation in local zones and a larger portion of applied P remains available to plants. Phosphorus dissolving slowly from rock phosphate, by contrast, may be captured mainly by soil fixation processes. Soil fixation can also be a problem for K in soils with illite and vermiculitic minerals. In some cases more than 1000 kg K ha^{-1} is required to overcome depletion where K-fixing soils have been cropped for long periods without adequate K fertilizer inputs (Cassman *et al.* 1989).

Nutrient distribution within soil profiles can also affect crop performance, especially for deeply tap-rooted crops like cotton as shown in Box 12.3. Over time, crop root systems extract nutrients from the entire active root zone, which may extend well below 1 m depth for most crops. In contrast, nutrients are returned to the surface soil layer in crop residues or with applications of fertilizer or manure. In soils with fixation properties for P and K, and for most cations, which are strongly adsorbed to the exchange complex, movement of these nutrients through leaching or diffusion is very slow. As a result, subsoil layers below a shallow 20-cm tillage layer, or the top 5 to 10 cm in no-till systems, become relatively enriched in nutrients while deeper layers are mined to deficiency. To correct this imbalance some researchers have tried periodic tillage in no-till systems or deep placement of nutrients to achieve a more uniform distribution in the profile.

Soil nutrient fixation can be avoided by foliar applications of nutrients. Application from airplanes is useful when the ground is too wet for traffic or when ground travel would damage crops. Only small rates of fertilizer can be tolerated by foliage, however, because leaves suffer "burning" from high concentrations of soluble nutrients. Where irrigation is practiced, nutrients are sometimes run with the irrigation water ("fertigation", Section 14.6.2). Uniformity of nutrient delivery is generally poor in flood or furrow systems, but it increases substantially with pivot and drip irrigation. Fertigation avoids the need for trafficking fields, and cost of application is small.

Time of application Plants have their greatest need for nutrients during the vegetative phase to support leaf growth. To make full use of growing seasons, some nutrients must be available early so that full cover and maximum growth rate are attained quickly. The easiest procedure is to apply all fertilizer at or before planting although some loss may occur during seedling growth when nutrient uptake is small. Losses can be reduced for row crops by splitting total application between a "starter" amount, applied in bands near the seed, and a subsequent "side-dressing" made just before canopy closure. The amount of the second application can be adjusted according to intervening weather and crop growth. The capability to deliver nutrients in irrigation water allows more frequent applications to synchronize nutrient supply with crop demand.

In small-scale cropping systems, farmers typically apply fertilizer by broadcasting. Because field size is usually less than 1 ha, fertilizer application does not require much

Box 12.3 Nutrient uptake and root systems

Root system architecture varies considerably among crops. The greatest differences are found between fibrous-rooted cereal crops and tap-rooted crops such as cotton and legumes. Root proliferation also responds to localized supply of N and P. Thus, greater root density is found near banded fertilizer applications. Some crops have a greater capacity to respond to non-uniform nutrient distribution than others, especially when non-uniformity involves vertical stratification. For example, the fibrous root system of barley is found dominantly in topsoil regardless of soil depth, while the taproot system of cotton is more evenly distributed. These contrasting patterns are seen in the figure below from an experiment by Gulick *et al.* (1989). The authors constructed soil profiles with varying depths of topsoil from the San Joaquin Valley in California where cotton is the only annual field crop to experience K deficiency and require K fertilizer. While topsoils in these areas have adequate available K, subsoils are severely K deficient. In the figure, soil profile treatment designations are shown at the top of each column and range from D1 (entirely subsoil) to D6 (topsoil to 36 cm depth). The black line is the transition from topsoil to subsoil.

Cotton is sensitive to K deficiency in these soils because its vertical root distribution does not coincide with that of available K. In contrast, barley root distribution is well coordinated with nutrient distribution. As a result, K uptake by barley per unit depth of topsoil was six-fold greater than for cotton. Cotton roots do not explore topsoil due to a relatively high soil moisture threshold for root growth. This appears to be an adaptation to survival in its center of origin on littoral beach zones of Central America where its survival depends on deep tap roots that reach fresh water lenses at depth. Severe top-soil drying also occurs between irrigation events, which limits cotton root growth in uppermost soil layers. Barley root growth has a much lower soil moisture threshold than cotton, which allows root proliferation to respond to greater nutrient supply in topsoil.

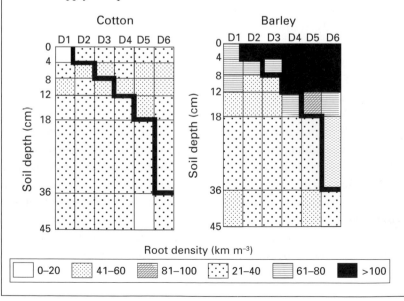

time or effort, which makes it possible for several in-season applications. This capability is especially advantageous in small-scale irrigated rice systems of Asia to improve fertilizer efficiency while maintaining or increasing yield. Based on the close relationship between chlorophyll content and leaf [N] as discussed in Section 12.1.2, Peng *et al.* (1996) found a hand-held chlorophyll meter can be used to determine the best timing of N fertilizer top-dressings on rice for large increase in recovery efficiency. Because most small-scale rice farmers cannot afford such meters, a low-cost green-leaf color chart was developed as a proxy. Application of the technique has substantially improved N fertilizer efficiency on a "field specific" basis in on-farm demonstrations in several major rice producing countries (Dobermann *et al.* 2002).

Applications of N late in the vegetative growth phase, or after flowering, favor nutrient accumulation in reproductive organs. Responses are species specific. Wheat, for example, can absorb considerable N after flowering and responds with increased protein content in grain. Late application of N has an unfavorable effect on sugarbeet since it stimulates vegetative growth rather than sucrose accumulation and leads to a higher content of non-sucrose materials in the beet. Sucrose yield then is less and extraction is more difficult. In general, late applications of N increase the risk of greater carry-over of mineral N and thus of greater losses during the fallow period.

The solubility of phosphate minerals is strongly dependent on temperature (Fig. 7.10). Plantings in cool soils sometimes have very large responses from starter bands of soluble phosphorus where crops show no response to added phosphorus during the warm season.

Replacement fertilization Some agriculturalists have advocated the application of N, P, and K in amounts removed by crops as a way to educate farmers about the extractive nature of farming. It is the wrong approach to crop nutrition, however, because a soil may be able to supply one nutrient indefinitely while being only marginally deficient in another, and woefully deficient in a third. Replacement fertilization emerged among environmentalists in an amended form: farmers should not be allowed to apply more nutrients than they remove. That concept ignores the inevitable losses of some nutrients and the fixation of others. The approach would result, in most instances, in a downward spiral of yields and fertilization, with accompanying inefficiencies in the use of energy, labor, capital, radiation, and water, and shortages of food supply.

12.5   Tillage systems

Movement and mixing of soil through tillage accomplishes several important functions. The most important are the management of residues and weed control. Tillage also is used to control flow of water, distribute fertilizers, pesticides, and amendments, and to create favorable environments for sowing, stand establishment, and root development. Each soil and each farming system present unique tillage problems and solutions, which fosters tremendous innovation by farmers and industry in developing tillage implements and approaches that best fit given soil–climate–cropping system combinations.

Recently, **conservation tillage** and **no-till** systems, that leave significant amounts of residues on the surface, have been developed to protect soils from erosion. Several trade-offs are involved in the choice of tillage systems. Conventional tillage deals most effectively with heavy stubbles and residues. It also provides the most uniform seed bed, best assurance of a crop stand, and best weed control. Yields tend to be greater with conventional tillage than by other methods but risk of erosion during cropping is also greater. Dependence on herbicides for weed control increases with each reduction in tillage.

Here we examine the principles that determine the most appropriate tillage system for a given situation.

12.5.1   Conventional tillage systems

Primary tillage Use of a moldboard plow, usually preceded and followed by a disk, is termed **conventional tillage**. Tillage is used to incorporate crop residues, uproot weeds, and to crumble soil for better planting conditions. Incorporation of residues promotes decay and efficient cycling of nutrients, and can also help control stem and foliar diseases, and insects. Coarse residues (e.g., maize) and heavy yields of fine straws (e.g., wheat) may have to be cut with a shredder or disk prior to plowing. Without that, seedbeds may be too loose and rough for control of planting depth and good moisture contact with seed.

Nitrogen content of residues influences methods of disposal. Residues with C/N ratios greater than about 45 (N content <1 % of dry matter) decompose slowly while immobilizing available N from soil. When surface residues are burned prior to tillage, N and C are lost to the atmosphere but foliar diseases are controlled, subsequent tillage is simplified, and complications from N immobilization are avoided. Where livestock are included in farming systems, stubbles can be grazed or straw can be harvested, used as bedding, and then returned to fields as composted manure. Complete incorporation or removal of residues is not always the best solution since surface residues are important for erosion control and for maintaining SOM. On sloping land, however, residue cover is critical for avoiding severe erosion and no-till systems are especially effective in this regard (Section 12.4.2).

Tillage implements The moldboard plow has been the basic tool for tillage in Western cultures since its invention in the 1700s. Its great contribution is to provide a way of cutting and inverting topsoil such that crop residues are well distributed throughout the plow layer. Usual depths of tillage are 20 to 25 cm; with a standard moldboard, only about 10% of residues remain on the surface. Fields are left in a rough state resistant to water erosion. Moldboards are the preferred plow for heavy soils, heavy stands of weeds, and/or heavy residues.

Disk plows and heavy disk harrows also achieve inversion with less incorporation of residues (c. 50%). Disk plows employ a gang of several large (up to 1 m diameter), angled disks that till soil to a depth of 12 to 15 cm. Heavy disk harrows use a larger number of smaller disks (up to 0.6 m), usually mounted in "tandem" fashion on two axles. These are set to a smaller angle of attack and less depth (8 to 13 cm). Disks are useful for primary tillage with moderate residues (e.g., for small grains); scalloped disks are sometimes used to aid in cutting residues.

Chisel plows consist of a gang of rigidly mounted chisels fitted with small sweep points. Set to 15 to 20 cm depth, they create a churning action and leave soil in a rough condition with more than 50% of residues on the surface. Infiltration rates are high and resistance of the plowed surface to wind and water erosion is generally excellent. Chisel plows with twisted shanks and/or larger points produce greater churning action and residue incorporation. Field cultivators, essentially light chisel plows equipped with wide (30 cm or more) sweep points, also create a churning action. Run at about 15 cm depth, field cultivators are more effective against perennial weeds than are chisels and they leave as much as 90% of the residues on the surface. They are used for primary tillage with light residues.

Chisel plows require only about 0.6 of the drawbar power per meter width as mold-boards and, given a large tractor, they can be pulled in wide spans at higher speed dou-bling labor productivity of conventional moldboards. Disk plows also use less energy than moldboards. The energy requirements of various tillage implements and methods are presented in Table 15.4b.

Secondary tillage focuses on seed bed preparation and post-sowing weed control. The looseness of freshly plowed ground makes a poor seed bed because planting depth and seed contact with soil moisture are not assured. Several tools find use in smoothing and firming plowed ground. Field cultivators (set to a shallow depth), light disks, and spring- and spike-toothed harrows are the most common. Toothed harrows present arrays of curved springs or short, rigid spikes. In semi-arid regions where low-organic soils often are tilled dry, ring rollers (closely spaced, toothed rings that rotate loosely on a common axle) are useful for breaking clods and firming soils.

Incorporation of fertilizer, herbicides, and amendments such as lime are sometimes integrated with secondary tillage operations. Broadcast fertilizer, for example, can be incorporated with a field cultivator. In some systems, planting is done on ridges or beds for the control of erosion, drainage, or soil temperature, or to facilitate irrigation. Ridges and beds are formed with a wide variety of furrowing and bed-shaping devices. Fertilizer and herbicides broadcast prior to furrowing end in a band through the middle of these beds or ridges.

12.5.2   Conservation tillage systems

Conservation tillage systems take many forms, from reduced tillage, usually with chisels, disks, or field cultivators, to no-till systems in which there is no soil disturbance. Any system that reduces soil disturbance and leaves more surface residues conserves energy and reduces the potential for erosion compared to moldboard plowing and secondary tillage. After devasting erosion during the 1930s "Dust Bowl" years in the US Great Plains, the Soil Conservation Service (now the Natural Resource Conservation Service) was established to identify ways to protect against soil loss. In common with most semi-arid regions, rainfall in the Great Plains is not sufficient to support continuous cropping. Instead, farmers followed a two-year wheat–fallow system that stores soil moisture during fallow for use in the cropped year (see also Section 13.8). European tillage methods with a moldboard plow and disk were used to incorporate residues and prepare seed beds, leaving soil vulnerable to wind erosion.

While researchers recognized the value of residues on the soil surface during the fallow year, weed control was difficult without primary tillage. A **stubble-mulch** system was developed that used a broad sweep plow blade to slip under the soil at 4 to 8 cm depth and sever roots without turning soil. This operation left wheat straw and weeds standing and retained snow that increased the capture of soil moisture. With the development of herbicides in the 1950s and 1960s, it became possible to eliminate tillage altogether. Today, no-till systems are entirely dependent upon pre- and post-emergence herbicides, and rotation for weed control. Recent development of herbicide-tolerant crops has further improved the reliability of weed control in no-till because use of the broad spectrum glyphosate herbicide allows greater flexibility in timing of application.

Greater water retention has allowed intensified cropping with three- or four-year rotations of wheat–maize or sorghum–fallow, or wheat–maize–sorghum–fallow in parts of the Great Plains. Innovations in planting equipment improved the reliability of sowing into large amounts of surface residues. As a result, no-till systems moved eastward into the Corn Belt where they reduced sheet and rill erosion on sloping land and improved soil tilth and organic matter in topsoil. Reduced fossil fuel use in tillage operations also favored adoption as energy prices increased (Table 15.4b). Recent survey of tillage practices shows that about 20% of total US maize, small grain, and soybean area is under no-till, while another 20% is under various forms of **mulch till**. Mulch till covers a wide range of conservation tillage systems that utilize chisel plows, sweeps, and field cultivators to partially incorporate residues while leaving at least 30% of the soil surface covered by plant residues. No-till systems are now widely adopted on 106 Mha in many parts of the world, including USA (27 Mha), Brazil (26 Mha), Argentina (20 Mha), Canada (14 Mha), and Australia (12 Mha).

Strip-till is a form of conservation tillage in which only a narrow strip of about 15 cm is tilled but provides most of the benefits of no-till. Strip-till allows more uniform seed placement than no-till after crops that produce large quantities of residue, such as maize. Soil in uncovered strips is warmer than under residue due to absorption of solar radiation, and soil warming improves the rate of germination and emergence.

Disadvantages of no-till in more humid climates include delays in planting due to slower soil drying in spring than with conventional tillage, less flexibility in the method of fertilizer application, and greater incidence of disease and insect pests – especially in continuous maize systems. In semi-arid areas surface protection with residues promotes the penetration of rainfall and less runoff. Adoption of no-till in unmechanized systems of developing countries is rare because, on the one hand, manual planting into large amounts of surface residues is very difficult, and on the other, where surface resides are not large, they are often used for animal fodder. Even systems that use small-scale mechanized equipment are not generally amenable to no-till.

12.5.3   Planting and cultivation

In conventional systems, most planting operations involve some tillage. Lister planters used for sorghum production in the US Great Plains represent an extreme example: a small plow clears a symmetrical furrow ahead of the planter so that seed is placed into moist soil at the bottom of the furrow. Furrows are closed during later cultivations, which also provide weed control and promote adventitious rooting from stem bases. Knife openers are common on planters where surfaces are free of residues, whereas various types of coulters or single- and double-disk openers are needed in systems with surface residues. In no-till systems with heavy surface residues, sweeps are added to soil openers to clear a planting line. The last components of most planters are a covering device and packing wheels to firm the soil–seed contact. Recent advances in sowing equipment allow precision planting with greater uniformity in intra-row spacing and seed depth and higher speeds, which improves timeliness of planting, especially in large-scale cropping systems.

Weed control can be accomplished with several types of tillage. Wide spans of tine weeders (a dense array of small spring tines) and rotary hoes (closely spaced spiked wheels) that can be pulled at high speed ($c.$ 18 km h^{-1}) are useful for eliminating small weeds and crusts during early stages of crop growth. Light ring rollers are also popular as crust breakers. A basic tool for inter-row cultivation is an array of sweep-pointed shanks arranged to till spaces between rows. These operate slowly (6 km h^{-1}) in narrow spans so labor and fuel costs are high. Control of weeds within rows is achieved with sweeps that move soil to form a slight ridge within the row.

12.5.4   Modification of soil temperature

Tillage can be directed towards avoiding or modifying unfavorable soil temperatures. In extreme cases, fields can be converted into hot beds by shaping soil to create a non-uniform distribution of the limiting resource, short-wave radiation. The sloping beds illustrated in Fig. 12.6 represent a sophisticated way of forcing vegetables. Such practices are obviously labor intensive and expensive, and thus suited only to high-value crops. They are less expensive, however, than glasshouse production or transplanting seedlings from hotbeds. As long-distance aerial transport has increased, forcing of fruits and vegetables in local areas has been replaced by conventional production in warmer regions and in the opposite hemisphere. Although intuitively this would seem to be energy intensive, transport represents only a very small fraction of total energy use in high-value fruit and vegetable production.

The strongly competitive markets faced by field crops have never justified horticultural forcing. Nevertheless, attention to land form, residues, and timing of cultivations and irrigation can be rewarding. For example, earliness can be achieved by ridging soil in fall and planting into warm ridges in spring. Ridge-till systems are a form of conservation tillage in which residue is swept into furrows so that only ridge tops are uncovered. Where soil temperatures are too high for plant survival, effective practices are planting

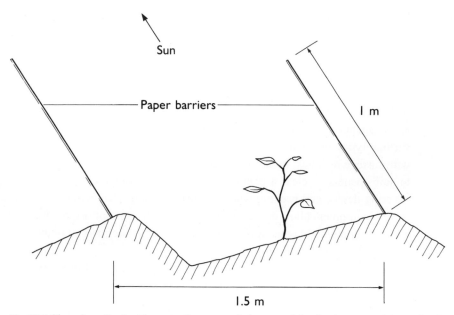

Fig. 12.6 Sloped seedbed with protective paper shelters used for forcing vegetable production in a cool climate. The arrangement creates non-uniform soil temperatures (cold behind the barrier, warm towards the sun) while protecting plants from wind and, at night, from radiation frost.

into deep furrows (lister planting), under stubbles or residue mulches, or cooling soil by irrigation.

Surface residues affect soil temperature by elevating the plane of R_n exchange and by protecting soil from turbulent transfer. In temperate regions, retention of stubble through the winter traps a layer of snow, insulating crops such as winter wheat, and extending their range into colder climates. Moldboard plowing in the fall protects fallow ground against deep freezing because increased pore space reduces soil thermal conductivity. In spring, dark, absorbing, surfaces of tilled soil warm and dry sooner so spring operations of seed-bed preparation can be accomplished earlier than with surface mulches.

12.6   Drainage

Drainage practices aim at relieving soil profiles of excess water and preventing flooding or seepage onto farmlands. Annual evapotranspiration is less for cultivated fields and heavily grazed pastures than for natural vegetation because the duration and/or degree of green cover are less. As a result, runoff and drainage terms are also larger (see Eq. 9.1, the hydrologic balance). Consequently, levees, channels, and tile drains are essential to divert water and to drain the root zone in areas subject to seepage or flooding.

Few crops, with the exception of rice, can withstand anaerobic conditions of water-logged soil for more than a few hours or days. The benefits of drainage extend well beyond improved survival of crops, however. In temperate climates, wet soils are slow to warm in spring. That and/or a high water table limit the diversity of species that can be grown and the depth and extent of their root systems. Decay and mineralization are slowed in wet soil while denitrification is accelerated and problems with metal toxicities and root diseases are more severe. Wet soils will not support machines or animals, sharply restricting timeliness of operations. Tillage of wet soil degrades its structure, and wet areas of pastures are destroyed by animal traffic. The principal effect of drainage is a marked increase in yield. Stability is also improved because a major source of spatial and temporal variability is eliminated.

Many seasonally flooded or boggy upland areas, both regional and on individual farms, have been converted to productive land by drainage. In addition, significant areas of floodplains and wetlands receive drainage. The earliest known examples of drainage were made by Roman and Chinese farmers over 2000 y ago. Between AD 1100 and 1300, a significant expansion of farmland and colonization in eastern Europe was based on drainage. Today, nearly 70% of the farm lands in Britain and the Netherlands are under drainage compared with about 25% in the USA. Nearly all irrigated lowland rice throughout the world requires drainage infrastructure for control of floodwater depth.

The first requisite for drainage is an outfall sufficiently low in altitude, or provided with pumps, to rapidly remove excess water. Individual farmers then connect their own drains to these channels. Outfalls are usually provided by a regional authority through construction of principal drainage canals or by straightening the course of meandering rivers. Channel straightening steepens a river's gradient increasing its flow rate for channel scouring. Slow-moving streams tend to fill with sediment causing seasonal flooding from the overflow as well as the cutting of new channels.

Simple systems of field drainage engage only excess surface water through channels constructed between topographical depressions and an outfall, or along the base of hills to prevent flooding of the foot plain by runoff or seepage. Saturated soil profiles require more elaborate works. The oldest approach, still used in many parts of the world, depended on the construction of mounds or raised beds of land separated by drainage channels. A cambered surface to beds provides runoff while height above the channel provides subsurface drainage. Scottish farmers revolutionized drainage practice through experiments in the 1820s with permanent subsurface drains. In their first designs, a trench was dug, partly filled with gravel, then back-filled. Spectacular benefits and the high cost of gravel led quickly to the development of clay tile, first as simple inverted U-channels and later as cylinders, and of tile-making machines. Scotsman John Johnston of Geneva, New York, introduced stone and then tile drains to his farm, beginning in 1835. His experience and writings spread the practice in USA. By the mid-1900s, mechanized trenchers had greatly eased the work of laying drainage lines.

Modern subsurface drains are constructed with plastic pipe that is perforated at intervals to admit water. These are laid with a trencher or pulled in with a "mole" plow while maintaining a constant gradient towards the outfall. The depth and separation of parallel drain lines are established with engineering formulae depending upon ground water conditions and soil texture. Lines must be spaced more closely in clayey soils having low hydraulic conductivity than in light-textured soils.

Drainage to remove salts is a critical component of all irrigated systems as discussed in Section 14.3.4.

12.7   Erosion

The Earth's surface is highly active with uplift and formation of new land mass countered through continual erosion by wind and water. Erosion over geologic time by glaciers, water, and wind is given perspective by examining the 5000 m of sediments laid open in a transect of N. America's Grand Canyon, Vermilion Cliffs, Zion Canyon, and Cedar Breaks in Utah. Natural erosion has slowed since gymnosperms evolved and flowering plants provided protective cover to the land during the past 100 million years. Regions such as California and New Zealand, with highly active landscapes due to tectonic plate movement and volcanic activity, continue to have very high natural rates of erosion. Rapid erosion can also occur with agriculture and threaten its sustainability through loss of soil. Here we learn about erosion processes and their management as a critical tool for sustaining the productivity of agriculture.

12.7.1   Loss and replacement of soil

Erosion reduces potential productivity of soil through removal of nutrient-rich surface layers, reduction of the profile depth capable of holding moisture, and by altering landforms. Off-site damage through air and water pollution and silting of river channels and reservoirs may accompany erosion. That humans have long recognized these problems is evidenced by ancient terracing and river works and windbreaks in many parts of the world. Serious mistakes continue to be made, however. In this century, the Dust Bowl crisis of the US Great Plains in the 1930s, failures in the Soviet New Lands projects in the 1960s, and continuing high rates of erosion in many areas signal a need for greater understanding and effort directed at erosion control. Indeed, topsoil is best valued by its contribution to crop productivity, which can be estimated in a region by the relationship between crop yield and depth of topsoil.

Some erosion during cropping cycles is inescapable but all sites also have some **tolerance to erosion**. A strict tolerance when soil loss rate is equal to or less than soil formation rate would allow production to continue indefinitely. Estimates of formation rates at undisturbed sites are generally near 0.1 to 0.2 mm depth y^{-1} (i.e., about 1.3–2.6 t ha^{-1} y^{-1}). Rates increase with rainfall. These values are found by dividing the depth of profile by time since the last great disturbance, for example, the last glaciation, but ten-fold greater rates (1–2 mm y^{-1}, 13–26 t ha^{-1} y^{-1}) are observed during the first century or two following such disturbances. Formation rates for eroding agricultural soils probably fall into the category of greater rates because subsoils are less insulated from tillage, percolation, leaching, and weathering as topsoil is removed through erosion (Morgan & Davidson 1986).

Soil conservation workers view tolerance in terms of the time for measurable yield decline of, say, 10%, or in relation to the depth of profile available for formation.

"Acceptable" rates of erosion generally have time spans of 25 to 50 y for measureable yield decline. That is a short period, however, relative to views of sustainability. In practice, yield declines are measurable only with extreme erosion; in most cases, they have been masked by technological advances in farming practices and yield. In addition, yield level can usually be restored quickly by some alternative management, for example by fertilization or by rotation to pasture.

Erosion tolerances of deep loess soils are generally large: 10 t ha^{-1} y^{-1} or more corresponding to a depth of about 0.8 mm y^{-1}. In China, some loess soils have sustained even greater rates for thousands of years but China also has numerous examples of extensive areas ruined by gullying. In contrast to loess, shallow soils with underlying agglomerations or bedrock may have little or no tolerance to erosion.

In future, it is likely that tolerance to soil loss will be judged not just on sustainability but also on more demanding criteria associated with levels of atmospheric pollution, sediment and chemical fluxes, and turbidity of surface waters. In other words, the public may take an increasingly strong position about off-site effects of erosion.

12.7.2   Erosion processes

Water erosion   Rainfall and snowmelt cause several types of erosion. Uniform erosion over a soil surface is termed "sheet" erosion. A film of water, agitated in some cases by rainfall, can detach and move soil particles along slopes. With long slopes, transport accelerates and small channels, or "rills", form. A continuing stream of snowmelt can cause considerable rill erosion in soils loosened by freeze–thaw cycles. Erosive action within rills can be rather powerful. Long-term, down-slope, movement by sheet and rill erosion causes gradual downwearing of hilltops, backwearing of slopes, and accumulation of alluvial footplains at the base of hills.

Gullies develop when large torrents repeatedly follow the same tracks. Whereas small rills are easily disrupted by tillage or vegetation, gullies destroy land for cultivation. Gully formation can start with rills, from subsurface tunneling by drainage water, or, most commonly, by the scouring action of water passing into small depressions on hillsides. Gullying is sometimes spectacular but can be addressed easily if caught early, or avoided completely with simple engineering works and management practices to divert down-slope water movement.

An impressive body of physics and engineering knowledge exists relating to the explanation, prediction, and prevention of water erosion. A summary of the main factors affecting water erosion is presented in Table 12.2. Sheet and rill erosion are greatest when surface soils are saturated by snowmelt or intense rainstorms. Impacts of raindrops loosen and suspend particles from saturated soils. The ability of rainfall to dislodge soil depends upon its intensity (mm h^{-1}) and the kinetic energy of raindrops (J m^{-2} land mm^{-1} rainfall). The relationship varies with drop size and thus with type of rainfall. The example in Fig. 12.7 illustrates that maximum energy was achieved at that site with rainfall intensity greater than 25 mm h^{-1}. Only a small fraction of storms have sufficient intensity and duration to cause significant erosion. As a result, a single storm per year,

Table 12.2 Factors influencing soil movement by water

Letter codes (*R, K, LS, C,* and *P*) are those used in the Universal Soil Loss Equation.

Factor	Influence on soil movement
R, rainfall	Erosion increases with intensity (mm h^{-1}) as well as amount of rain (mm).
K, soil erodibility	Fine sands and silts are moved more easily than coarse sand and clay. Soil particles are detached more easily from poor structure with poorly aggregated clays due to low Ca or humus, or high Na content.
LS, gradient and length of slope	Erosion rates increase rapidly as the gradient increases above 4% and as the length of slope increases.
C, surface protection	Erosion is greatest with bare ground not covered by plants or residues. Standing crops break the force of raindrops. Stony surfaces and residue mulches absorb the impact and prevent movement.
P, management	Erosion is greater with up and down rows than with contour plantings. Terracing and drainage lessen erosion.

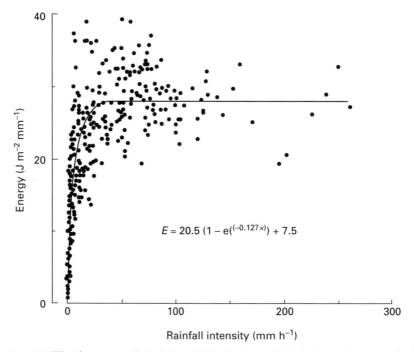

$$E = 20.5\,(1 - e^{((-0.127x)}) + 7.5$$

Fig. 12.7 Kinetic energy of rainfalls at Holly Springs, Mississippi, as a function of rainfall intensity. (Redrawn from McGregor & Mutchler 1977.)

or just one in 5 y, may account for nearly all long-term average erosion in a field or watershed.

Coarse sand and soil aggregates are large enough to resist movement by water; very fine sand and silt are most susceptible to runoff when rainfall exceeds infiltration capacity. Loess soils, lacking a sand component that might protect their surfaces, are particularly susceptible to erosion. Infiltration through clay can be as little as 2 to 5 mm h^{-1} compared to rainfall that commonly falls at >20 mm h^{-1}. In contrast, infiltration into a drained sand will accommodate almost any rainfall.

Wind erosion Wind erosion is a hazard mainly in semi-arid and arid regions because dry soil will detach with wind whereas wet soil does not. In addition, strong winds are more common where sensible heat flux dominates the energy balance. The processes of wind erosion are rather different to those with water. Aggregated silt and clay are not easily detached by wind, and coarse sand is too heavy for its cross-sectional area to move. Erosion begins with the movement of easily detached, fine sand (0.1 to 0.25 mm). Blown sand bounds along the surface ("saltates") with a highly abrasive action. Seedling plants are particularly vulnerable to abrasion because they offer little drag (Section 6.8) to slow movement of air and sand. On striking dry soil, sand detaches aggregated clay resulting in further erosion and dust storms. Saltation increases with length of fetch, which makes wind erosion most severe from broad, flat areas.

Prediction equations have been developed for wind erosion (e.g., Woodruff & Siddoway 1965). These are generally more complex, but also better based on mechanical principles, than those for water erosion.

Maintenance of vegetative cover is the basic technique for control of wind erosion. Vegetation displaces the zero plane (Section 6.7) so that wind velocity near ground is too small to be erosive. Residue mulches on fallow ground are helpful, particularly if they remain anchored to soil, a key feature of stubble-mulch systems with subsurface tillage. Fallow ground can be protected to some extent by ridge tillage or by rough, cloddy surfaces created by plowing moist soil. Tillage operations such as plowing or furrowing are used in emergency situations to control active wind erosion. Where wind erosion is a common hazard, permanent grass is the best means for control.

Windbreaks ("shelterbelts") of trees and shrubs planted in rows at right angles to prevailing winds are an ancient solution to wind erosion. They are still employed in many regions. Hedgerows, with about 50% porosity to wind, control erosion by creating a layer of slow-moving air near the ground. Tallness is important; as a general rule, a permeable windbreak will reduce relative wind velocity near the ground by 50% for a distance downwind of 10 times its height. In this way, a 10 m tall windbreak protects about 100 m of cropland but the break itself will occupy perhaps a 10-m wide strip.

Faced with strong onshore winds, early farmers along the California coast planted rows of tall-growing eucalypt trees in shelterbelts. These trees usurp moisture and light from a wide band of land, however, and a current trend is to replace them with rows of maize distributed at intervals across fields. A special problem of wind erosion of peaty soils in the California Delta region was solved in a similar way. Culture of white asparagus required continued tillage to keep the asparagus covered as it elongated, but that led to storms of black peat dust. Control was achieved with sprinkler irrigation on windy days and by interplanting with rows of barley.

12.7.3   Estimating erosion loses

The Universal Soil Loss Equation A number of quantitative tools have been developed to predict erosion (Morgan & Davidson 1986). The Universal Soil Loss Equation (USLE), for example, was developed by the USDA Soil Conservation Service (Wischmeier & Smith 1978) from statistical analyses of more than 10 000 experimental plot–years covering diverse soils, crops, and weather in humid areas of the USA. The original USLE attempted to predict average annual soil 'loss', A (ton acre^{-1} or t ha^{-1}):

$$A = R_i K_i LSCP \tag{12.1}$$

R_i is a rainfall-runoff index relating to the power of storms in the region; K_i is the erodibility index of local soil; LS combines length and gradient of slope; C represents crop and residue cover; and P is the beneficial effect of preventive management practices such as terracing. The parameters LS, C, and P express erosion relative to rates that would occur with bare soil, of specific slope and length, and cultivated along the slope. Length (L) and steepness (S) of hills generally dominate the equation as is illustrated with LS values drawn from tables in Wischmeier & Smith (1978):

Length		Slope:	
	2%	*8%*	*16%*
30 m	0.20	0.99	2.84
90 m	0.28	1.72	4.92

Cover provided by crops and residues protect soils from raindrop impact and slow runoff. The C value for bare soil is 1.0; dramatically smaller values are found when soils are provided cover by crops or residues, as they are in no-till systems:

Type of cover		Percent cover:	
	0%	*50%*	*100%*
1 m tall crop	1.0	0.65	0.30
Residues	1.0	0.30	0.03

Good meadow crops (grasses and legumes grown for hay) and grass pasture have C values less than 0.01. Inclusion of a meadow phase in rotation sequences sharply reduces the probability of soil movement.

The term "loss" in the USLE name is inaccurate because movement within a field rather than export from the region is predicted. Annual flux of sediment in the Mississippi River, for example, comes more from gullying and bank cutting than from field erosion. Total movement of sediments and dissolved minerals in that river is only about 10% of

Fig. 12.8 Terraced cropland in the Loess Plateau, DingXi County, Gansu Province, China. The crops grown are mainly maize, rice and soybean, and various vegetables.

soil movement predicted in its watershed by the USLE. The equation either overpredicts erosion or most eroded material is deposited on lowlands, or is captured in reservoirs.

Empirical equations such as the USLE require an enormous amount of experimental work to establish appropriate parameter values for local soils and practices, and to be adjusted to changing technology. In practice, rarity of erosive storms creates uncertainties in R_i, and K_i sometimes must be arrived at by guesswork. Use of the USLE in the USA is supported by detailed soil maps, extensive tables, and nomographs providing quantitative values for equation parameters, and by local experience. It is a poor predictor of actual erosion, however, and research continues to upgrade or replace it. Process-based models are less site specific and seem to offer more promise (Rose 1985). Despite the limitations of the USLE, its parameter values reflect the relative effects of alternative farming practices on erosion sufficiently well to be used in the USA as a basis for national and local conservation programs. While there are more recent updates of the USLE (http://fargo.nserl.purdue.edu/rusle2_dataweb/userguide/RUSLE2-2-3-03. pdf), the structure remains similar.

12.7.4 Managing erosion

Terraces and contours In addition to no-till and other conservation tillage practices (Section 12.5.2), other land management options help minimize erosion. Farming along elevation contours reduces erosion to about half that observed with up–down cultivation. Strip contouring and terracing give further benefit by shortening the length of slopes. With short, rapidly changing gradients, however, contours can only approximate the general trend of the land.

The remarkable effectiveness of terraces is seen in level benches with stone retaining walls that have continued in use for centuries in Asia, South America, and Mediterranean regions. The photograph of Fig. 12.8 displays extensive terraces currently maintained on the highly erodible hills of the Loess Plateau in central China. Unfortunately, great cost and human effort now limits their maintenance (and expansion) even for high-value crops. Simpler terraces with "narrow" (grassed) or "broad" (farmed-over) bases shaped with earth-moving equipment are widely used for field crops. These are designed to slow runoff and to either convey it off the land (humid climates) or to conserve it through infiltration (semi-arid climates).

Seasonal watercourses Those formed by the intersection of two slopes present special problems to farming. Such courses collect seepage from adjacent hills and may be too wet to support traffic. In addition, heavy flows of water can turn the course into a gully. One solution involves planting grasses along shaped courses to channel water while avoiding rill erosion. It also leaves grassed areas sufficiently dry for crossing traffic.

Farm boundaries and erosion Erosion and drainage are watershed problems whereas farm boundaries typically are based on different principles and rarely follow contours. Small farms generally control only a portion of a problem hill or water course and, for them, allocation of land to permanently grassed headlands and water courses has a greater relative effect on annual income than it does for large farms. Large farms tend to do a better job of setting aside or correcting problem areas, in part because they can more easily afford the capital costs of conservation and changes in tillage practice. Present worth of future benefits from projects such as terracing, however, is invariably less than the cost of conservation. As a consequence, most Western nations have government programs that partially subsidize the costs of soil conservation projects such as terracing.

12.8   Land value and capability

The value of land for farming is reflected in its current market price, an index that integrates economic factors such as commodity prices, distance to market, and discount rates as well as traits such as water supply and production potential. We also need means for assessing the merit of land independently of current economics. In the following sections we examine different approaches to land value and its capability to support agriculture.

12.8.1   Land capability classes

Two methods have emerged to evaluate land suitability for agriculture. It is useful, for example, to evaluate land from its potential for cultivation, giving weight to ease of management and to hazards that may accompany farming, rather than to production potential. Factors such as wetness, erosion hazards, ease of tillage, spatial variability, and depth of profile can be included in the evaluation. The Land Capability classification system of the USDA Natural Resource Conservation Service (NRCS) is an example of this approach. Alternatively, we can consider production potential expressed in terms of attainable yields of principal crops. Regional and local planners make considerable use of both approaches.

Table 12.3 USDA Land Capability classes

Class I. Few limitations to cultivation. Soils are level, with deep, well-drained profiles, never subject to flooding; inherently productive or responsive to fertilization.

Class II. Moderate limitations restrict the range of crops that may be grown or require some special management or conservation practice. Subclasses e (erosion), w (wetness), s (soil morphology problem), and c (climate) indicate the nature of the limitation.

Classes III–IV. Severe and very severe limitations, respectively. Subclasses are the same as for Class II.

Class V. Lands without an erosion hazard but unsuited to cultivation owing to wetness, shallow soil, or other factor.

Classes VI–VIII. Severe and very severe limitations make these lands unsuited for agriculture except, in some cases, grazing.

Source: Klingebiel and Montgomery (1961).

The USDA Land Capability system places each unit of land into one of eight classes based on ease of farming. The scheme, summarized in Table 12.3, provides a basis for map overlays characterizing possible agricultural uses. By emphasizing erosion and drainage problems, the approach relates well to sustainability. Within regions, classes are further divided into subclasses and "units" denoted by Roman numerals. Such land suitability indices can also be used for land-use planning by government agencies to avoid development on prime farmland.

Similar land capability systems are now used in other countries. Portugal's system follows the same general pattern as the USDA's. In England and Wales, Class V is omitted and land gradients are weighted more to their suitability for machines than to erosion. Canada also omits Class V while retaining an emphasis on erosion.

Land capabilities may change as investment in infrastructure and new technology allows successful farming of soils with severe constraints. For example, investment in regional drainage systems or a system of protection dykes can make areas prone to chronic flooding available for farming. Improved understanding of soil chemical and fertility constraints can lead to innovations in soil management to overcome them. Such is the case in the Brazillian cerrado where acid-infertile soils were brought into production through the use of lime and large initial inputs of P fertilizer to overcome P fixation capacity and erosion controlled by conservation tillage (Section 17.4.2).

12.8.2   Production indices

While land capability classes provide qualitative statements about ease of farming, production indices make quantitative statements about success in farming and are more easily interpreted by farmers. The USDA-NRCS recently produced a National Commodity Crop Productivity Index (NCCPI) for corn, soybean, small grains, or cotton based on soil properties and climate to estimate a soil's suitability for each of these crops under non-irrigated conditions in the USA (Dobos *et al.* 2008). Although the NCCPI does not estimate crop yields per se, higher values are assumed to be associated with greater yield potential, at least on a regional basis. A more explicit emphasis on yield is the basis of land capability evaluations in many European countries. For example, Dutch scientists make considerable use of potential yield models dependent upon radiation. As is discussed in Chapter 18, Buringh and van Heemst (1979) used de Wit's model of potential production at various latitudes along with local data on soils, rainfall, and length of growing season to construct estimates of agricultural production for each region of the globe. Estimates were expressed in wheat equivalents as a standard unit. It is likely that improved quality and access to geo-referenced soil maps and weather databases will facilitate the development of improved land productivity assessments based on robust crop models that provide more accurate estimates of potential crop yields.

12.9   Review of key concepts

Soil as the basic resource

- Soil quality, largely the capacity of soil to provide water and nutrients for uptake by roots, is central to the sustainability of agriculture. Cropping depletes soils of their nutrient stocks, and tillage exposes them to increased erosion. Adequate drainage is an important attribute of soil quality. Good soil management must account for significant spatial variability, even in seemingly uniform fields.
- Management is facilitated by adjustments in size and shape of fields and of crop sequences applied. Tillage, soil amendments, and drainage can be varied according to spatial patterns in individual fields. Geo-positioning guidance equipment and geographic information software are increasingly applied in mechanized agriculture to achieve these objectives.

Plant nutrition and nutrient management

- In addition to C, O, and H, at least 15 elements are essential for plant growth. Six are considered macronutrients (N, P, S, K, Ca, and Mg) because they are required in larger amounts than micronutrients (Fe, Zn, Cu, Mn, B, Cl, Co, and Mo). Diagnosis of deficiencies can be made through soil and/or plant analysis based on critical sufficiency concentrations.
- Nutrients are recycled in residues and manure but those sources are inadequate to replace losses that occur through runoff, animal feeding, and the human food chain. Diagnosis of nutrient deficiencies and proper attention to fertilization are important elements of farm management. Deficiencies of N, P, and K are most common because their supply in soils is small and crop requirements are relatively large.
- Fertilizer-use efficiency depends upon the degree that a nutrient is deficient (response functions are typically diminishing-return in nature), on timing and placement of fertilizer, on form of the nutrient applied, and on events such as immobilization and fixation within soil. Management of N supply is the most difficult.

Tillage

- A wide range of primary tillage systems is in use today. Conventional moldboard plowing provides best incorporation and recycling of residues, followed by disk plows and harrows. It is also most time consuming and costly and, under some circumstances, subjects fields to erosion due to lack of residue cover.
- No-till and conservation tillage systems, although dependent upon herbicides, are less costly and more effective in erosion control. These systems result in strong vertical gradients of nutrients and pH in soils. In temperate climates, slow warming and drying of soil under no-till can be a disadvantage to early crop establishment in wet years. In drier regions, no-till conserves water.

Conserving soil

- The sustainability of agriculture depends upon maintaining nutrient levels and controlling erosion to below the rate of soil formation. Soil formation is more rapid in tilled than in untilled soils but information about both formation and erosion is sketchy. Experiments have revealed large differences in erosion with different farming practices and their relative effect can be embodied in erosion-prediction equations. Cover provided by residues and growing crops, especially perennials, is effective in minimizing erosion by both water and wind. Rotation with sod crops and strip cropping (on contours for water or normal to prevailing wind) are also helpful in conserving soil in erosion-prone areas.

Terms to remember: geographic information system (GIS), position system (GPS); index tissue; kriging; land capability classes; pedotransfer functions; plant nutrients, critical nutrient concentration (CNC_p), essential macro- and micronutrients, and uptake capacity; primary tillage; replacement fertilization; site-specific management (SSM); soil nutrients, critical concentration (CNC_s) and indigenous supply; tillage systems, conservation, conventional, mulch, and no (zero)-till; terraces and contours; tolerance to erosion; Universal Soil Loss Equation (USLE).

13 Strategies and tactics for rainfed agriculture

Most agriculture is practiced under rainfed conditions with varying supplies of water. Farming strategies and tactics divide roughly into those dealing with too much water and those for coping with too little. A discussion of the large diversity of methods that has been developed for rainfed conditions reveals critical relations between production and water supply. We begin with comments on the management of farming in wet regions where hazards and management practices to counter them are rather different from those in dry regions. Chapter 14 examines the principles of irrigation as a separate topic.

13.1　Agriculture in wet regions

Rainfed agriculture in humid regions would seem blessed with a free good in its generous water supply. But that supply is seldom ideal, varying from excess to transient deficiency. Excess supply leading to surface flooding and saturated soils is a major problem that generally requires drainage works (Section 12.6) but new biological solutions are being found to some problems. A successful example of combating flooding damage of rice, a major problem in many areas of southeast Asia, is presented in Box 13.1. Water erosion (Section 12.7) and nutrient loss are also greater concerns with abundant rainfall than in drier regions. Leaching of N was considered in Section 8.4.3. Because vegetative cover serves to control erosion and nutrient losses, many sloping sites are maintained in pasture. Many low lying sites also remain in pasture because cold, wet soil and poor drainage combine to make them unsuited to cropping. Special techniques such as ridge culture are sometimes employed to overcome cold and wet conditions.

Despite high rainfall, drought is an occasional hazard in most humid zones and chosen crops may be poorly adapted to drought. They also may lack opportunity for acclimation, for example by development of appropriate root:leaf ratios that provide resistance to sudden dry spells (Section 9.6.2). As a result, the effects of drought can be more severe than in the same crops in dry regions. Weeds are a general problem in wet areas and opportunities for mechanical cultivation can disappear quickly. The problem is not easily solved with herbicides. Topical applications may encounter the same weather limitations as cultivation, and pre-emergence herbicides incorporated into soil must be chosen with the correct balance between solubility and persistence. An effective solution for some crops is now provided by cultivars that are resistant to specific herbicides. The most widespread examples are found with glyphosate-resistant soybean, maize, cotton,

---

Box 13.1 Flooding tolerance in rice

Rice is the only crop plant adapted to waterlogged environments. Root systems do not suffer **anoxia** under those conditions because well-developed **aerenchymatous** tissues facilitate oxygen diffusion through continuous air spaces from foliage. Flooding that causes submergence prevents oxygen flow and is detrimental to plant growth, survival, and yield. Most cultivars cannot withstand submergence of more than one week, a frequent event in the rainfed lowlands of SE Asia, home to 140 million poor farmers, and so are unfavorably affected by transient flooding.

Flooding tolerance has been a long-standing goal of rice breeding (Mohanty *et al.* 2000) and recent research at IRRI (Xu *et al.* 2006) has identified and cloned a gene locus (Sub-1A) for flooding tolerance in the *O. sativa* spp. *indica* landrace (FR13A). Presence of the locus enhances activity of alcohol dehydrogenase in roots thus providing protection from anoxia. A widely used cultivar, Swarna-Sub1, transformed with the locus using marker-assisted breeding, is now grown over 5 Mha in India and Bangladesh. It displays a flooding tolerance of two to three weeks duration and resultant yield benefits of 1 to 2 t/ha without loss of grain quality.

The gene has another potential benefit in rice production systems. The ability to withstand flooding opens opportunities to use controlled submergence to kill weeds and so reduce dependence on herbicides. This technique is being evaluated with direct-seeded rice.

---

and canola now used in many parts of the world. The glyphosate herbicide inhibits amino acid formation in all green plants except those specifically provided with genetic resistance. Glyphosate is used as an overall post-emergence spray on resistant crops and provides a longer period for effective weed control (Sections 4.3.4 and 12.5.2).

The most persistent problem is the disruptive effect of frequent rainfall on farming operations. Delays in tillage and sowing due to wet soil tend to have a cascading effect, exposing crops to greater attack by insects and disease, to early or late frosts, and delaying harvest. Opportunities for critical operations are sometimes very short because tillage of wet soils can seriously damage soil structure. Farmers can counter that problem with no-till methods using herbicides, or tillage during preceding non-cropped fallow seasons, and with oversized machinery so that critical tasks can be completed through round-the-clock efforts whenever soil dries. Fields can also be prepared, when operations are delayed, to substitute a different cultivar (or even crop) having a shorter growing season. Because rain also disrupts production and damages hay, ensilage is a common solution to rescue late or damaged crops, and for fodder conservation in humid areas. Wet weather after grain and oilseed crops reach maturity can delay dry down, reduce grain quality, and, for some crops such as wheat, lead to sprouting, which can result in a total loss of value. In contrast, dry weather during the final phase of grain filling and during the dry-down period supports good grain quality and also high-quality planting seed.

Finally, it bears emphasis that climates of humid and dry regions differ in more than just rainfall. Cloudiness and a humid atmosphere result in significantly less solar radiation and smaller diurnal amplitude in temperature than is found in dry regions. The major effect is to reduce production potential.

13.2   Principles for efficient use of water

Water supply available to crops consists of stored soil moisture plus rainfall during the growing season. Efficient use of that water relies upon the application of relatively few principles (Loomis 1983). Given many possible interactions between water supply, water use, and crop growth, a wide range of management options can, however, be employed to achieve it. To increase crop yield under water-limited conditions, management must be directed in sequence toward:

- maximizing total water available for each crop;
- maximizing transpiration (E_p) by minimizing soil evaporation (E_s); and
- selecting and managing crops with high transpiration efficiency (TE_Y) in production of economic yield.

Options aimed at maximizing ET include control of water supply by varying fallow duration, promoting deep root systems, reducing runoff, increasing infiltration, selecting soils with a high capacity for water storage, and sowing long-season cultivars. Those that seek to maximize E_p (minimize E_s) include weed control, surface mulching such as in no-till systems, increasing plant populations, sowing early, and selecting cultivars with rapid early growth. Other options seek to maximize TE_Y by avoiding periods of high evaporative demand through the use of cultivars of suitable length of growing season and sowing them at optimal times, by controlling pests and diseases, and by use of fertilizers (e.g., Cooper *et al.* 1987).

A general principle for effective use of a scarce resource is to concentrate it through a **non-uniform distribution in time and/or space** (Section 8.7.1). Examples include fallow–crop sequences that concentrate water supply in time, thus increasing and stabilizing crop yield. Wide-spaced rows and lower plant density extend the period when crop roots are extending into soil at field capacity and thus delay the occurrence of stress. They also provide a larger catchment for rainfall per unit LAI. In both cases, sufficient water to complete the crop cycle is a critical goal of management that also requires attention to weed and disease control and appropriate levels of fertility.

13.3   Patterns of water shortage and crop types

Success in rainfed farming depends upon the selection of appropriate crops and their cultivars. Crops will be most successful when their developmental cycle (Chapter 5) escapes, avoids, or tolerates periods of water shortage and makes best use of the pattern of water supply in formation of yield. A common characteristic of semi-arid and arid zones is large year-to-year variability of rainfall. It is variability rather than low rainfall that is the major challenge to management of productive agricultural systems in these climates.

Determinate crops are best suited to regions where length of growing season is relatively constant from year to year. In practice, however, determinate crops, particularly cereals, are the dominant food crops of semi-arid agriculture and are particularly sensitive to terminal drought that shortens the period for yield formation. Where terminal drought is expected, as in Mediterranean climates, terminal stress can be minimized by timely sowing of short-season cultivars at low density. The pitfall of this approach is that such crops make poor use of better years and may, over the long term, yield less than longer season cultivars. This variability in yearly performance opens the way for tactical variation by adapting crops to anticipated weather conditions as discussed later with various examples in Section 13.9.

Indeterminate flowering seed crops are well suited to regions where growing seasons are punctuated by unpredictable periods of drought. In crops with total life cycles of 100 to 150 days, an indeterminate flowering period of, say, 50 days allows replacement of yield organs after a short drought whereas a determinate-flowering crop might lose a substantial part of its attainable yield.

Root and tuber crops provide an interesting option in drought-prone areas because yield accumulation is indeterminate and short bursts of growth can be added to yield organs with each rainfall. Cassava is an excellent example of a drought-resistant root crop. It is perennial and can survive extended drought periods (>3 months) while maintaining a high internal water status by stomatal closure, restricted leaf expansion, establishment of a favorable LAI/fibrous root-length ratio, and ultimately by leaf fall (Connor *et al.* 1981). Recovery of leaf area following rain is rapid, probably involving mobilization of starch from storage roots. Over one or more years, cassava can amass considerable yields (>20 t fresh weight ha^{-1}, *c.* 35–40% dry matter).

Pasture and forage crops are useful in regions with highly variable water supplies where low probability of sustaining reproduction of seed crops generally precludes successful yield. Pastures of annual species require seed to regenerate each year, so soil seed banks (Section 5.4.4) serve to overcome years of low seed production.

13.4   Optimum patterns of water use

There are two components to the determination of water-use efficiency of crops. First, is the basic relationship between biomass production (B) and ET (WUE$_B$ $= B/$ET) that varies with environmental conditions affecting both growth and transpiration. The second concerns the relationship between seasonal patterns of growth and formation of economic yield. At that level, overall performance of crops is expressed as WUE$_Y$ ($= Y/$ET). With forage plants, for which yield is seasonal accumulation of growth, no distinction need be made between WUE$_B$ and WUE$_Y$.

Details of relationships between crop water use and growth were introduced in Section 9.7, where it was emphasized that transpiration (E_p) is the part of ET related

Table 13.1 Growth and water-use efficiencies of winter- and spring-sown sunflower at Cordoba, Spain

	Performance at CGR_{max}				Seasonal performance		
Sowing date	CGR (kg ha^{-1} d^{-1})	Duration (d)	ET (mm d^{-1})	WUE$_B$ (kg ha^{-1} mm^{-1})	biomass^a (t ha^{-1})	ET (mm)	WUE$_B$ (kg ha^{-1} mm^{-1})
December 15	300	27	5.0	60	12.2	481	25
March 15	244	16	6.4	38	5.3	372	14

Note: ^{a} harvest index of 0.25 was unaffected by sowing date.
Source: from Gimeno *et al.* (1989).

to growth. The term E_p is proportional to saturation vapor pressure deficit ($\Delta e = e^* - e_a$) and therefore $B/E_p = k/\Delta e$, where k is a crop-specific efficiency factor (Tanner & Sinclair 1983). As a result, efficiency of water use in production of biomass by any crop:

$$WUE_B = B/ET = k(1 - E_s/ET)/\Delta e \qquad (13.1)$$

is greatest when soil evaporation (E_s) is minimized to maximize E_p and when crops are grown while Δe is small and radiation and temperature are suitable for growth.

The influence of seasonal conditions on WUE$_B$ is illustrated in Table 13.1 for sunflower sown on two occasions, December and March, at Cordoba, Spain (latitude 38° N). December sowing is very early by local standards; crops sown at that time establish slowly and are less competitive with weeds than those sown in March. Early-sown crops develop more slowly but despite a longer total crop cycle are exposed less to summer drought. In consequence, these crops have higher maximum growth rates, maintain them longer, and have higher WUE$_B$. In this case, there was no additional effect of sowing date on HI ($= 0.25$) and therefore WUE$_Y$ responded proportionately to growth. Soriano *et al.* (2004) have extended these field observations and, using a simulation model, have demonstrated the wider applicability of early sowing to the region provided that weeds are controlled effectively.

13.4.1   Pre- and post-anthesis water use

How to optimize growth duration in relation to water supply, and how to optimize allocation of that supply between pre- and post-anthesis periods of growth are important questions with annual crops. If water supply is small it must be carefully divided between both periods. On the one hand, development of excess numbers of floral primordia by anthesis uses water that would have been better saved for post-anthesis growth. On the other hand, any crop that fills grain and has water remaining at maturity would have yielded more by development of additional grains.

Fischer (1979) analyzed the problem of efficient use of water in wheat production using data from southern Australia (Fig. 13.1). Increasing pre-anthesis growth (abscissa) leaves progressively less water for source activity during grain filling and sets a "water-limited" boundary to grain yield. The other boundary shown in the diagram is potential sink capacity, an estimate of maximum crop yield calculated as the product of grain

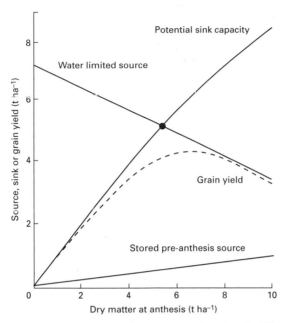

Fig. 13.1 The relationship between growth and grain filling in wheat under post-anthesis drought. Development of increased sink capacity during the pre-anthesis period leaves less water remaining to support source activity after anthesis (after Fischer 1979.)

number at anthesis and potential grain size. This analysis shows that source activity after anthesis becomes water limited at high biomass. If 10% of biomass at anthesis is stored as assimilates that can be mobilized to grain, then the optimal allocation of water to pre-anthesis growth is achieved, in this example, at a biomass at anthesis of 6.2 t ha^{-1}, for which grain yield is 4.2 t ha^{-1} and HI is 0.43. This flowering and growth strategy would maximize yield for the cultivar represented here under average conditions in this environment.

Those concepts are further developed as a general model of determination of HI in determinate crops (Box 13.2) that fitted experimental data for wheat and sunflower with $R^2 = 0.75$ and 0.81, respectively (Sadras & Connor 1991). Harvest index adjusted for production value (PV) of yield organs (HI$_{pv}$) (Section 11.2) depends upon the proportion of transpiration (p) during post-anthesis growth normalized against vapor pressure deficit (Δe), and two cultivar parameters (a, b) related to the contribution of pre-anthesis growth to reproductive yield. Optimal distribution of transpiration occurred at values (a/b) of 0.35 for wheat and 0.49 to 0.68 for three cultivars of sunflower. The model was able to distinguish between the responses of cultivars and so offers assistance to cultivar selection and improvement. In this case "potential" HI$_{pv}$ ranked Beauty > Cannon > Dwarf II. Beauty and Cannon have similar phenological patterns, but Dwarf II, which partitions a smaller proportion of biomass to stems, also has a shorter post-anthesis phase (34 vs. 41% of total). Intra-specific differences quantified by a and b were not detected when non-normalized HI was fitted to non-normalized p.

Box 13.2 Determination of HI in determinate crops

The relationship between crop yield (Y) and growth (B), the harvest index ($HI = Y/B$), depends upon the seasonal distribution of crop growth. The HI of determinate crops can be related to transpiration before and after anthesis and crop product type as portrayed by

$$HI_{pv} = p\,[1 - (a - bp)]^{-1} \tag{13.2}$$

where, HI_{pv} is HI corrected for production value (PV) of harvested product relative to cellulosic biomass (Section 11.2) to account for large differences that exist between mass yields of cereal, legume, and oilseed crops, relative to assimilation, and

$$p = (E_p^{a-m}/\Delta e^{a-m})/(E_p^{e-m}/\Delta e^{e-m}) \tag{13.3}$$

is the proportion of crop transpiration after anthesis modulated against evaporative demand (Δe) during emergence-to-maturity (e–m) and anthesis-to-maturity (a–m).

The parameter "a" is "potential" contribution of pre-anthesis assimilate to yield, i.e., the limit of $(a - b\,p)]$ as p tends to zero. The limit of HI when p tends to unity is $(1 - a + b)^{-1}$ and represents "potential (sink-limited)" HI. Yield is source limited when $p < a/b$ but is sink limited when $p > a/b$. The optimal pattern of crop water use occurs when pre-anthesis growth produces a sink that can just be filled by post-anthesis growth plus mobilized assimilate. Any other pattern of water use produces sink capacity that is either too large or too small for growth capacity after anthesis.

13.4.2   Water-use efficiency of rainfed wheat

A relationship between the yield of rainfed wheat and seasonal ET (WUE_Y) from many individual measurements, grouped for four regions of the world, is summarized in Fig. 13.2 (Sadras & Angus 2006). Mean and maximum values reveal substantial gaps from **attainable yields** in all regions, ranging from 43% in China to 59% in the North American Plains. Part, but not all, of that could be reduced by management and better adapted cultivars. The explanation for differences between individual crops and regions is found in TE_Y rather than WUE_Y.

Figure 13.2 identifies the maximum yield achieved at each level of ET by a line of gradient 22 kg ha^{-1} mm^{-1} and intercept of 60 mm ET. The line can be taken to describe the performance of crops for which seasonal distribution of rainfall minimizes E_s (mean value = 60 mm) and optimally supports the development and expression of grain yield in pre- and post-anthesis phases, i.e., that maximizes TE_Y (= 22 kg ha^{-1} mm^{-1}). Maximum WUE_Y is also found along the fitted line but varies with ET according to $WUE_Y = 0.022(ET - 60)/ET$ (= 17.6 kg ha^{-1} mm^{-1} at 300 mm ET). The lowest yielding crops at each level of ET might reflect limitations by factors such as sowing date, disease, nutrient supply, or competition from weeds. Management can more easily

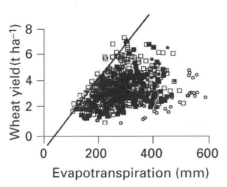

Fig. 13.2 Scatter plot of grain yield and seasonal evapotranspiration (ET) in four world regions. The line has an intercept of 60 mm and a slope of 22 kg grain ha^{-1} mm^{-1} (after Sadras & Angus 2006).

reduce the part of a yield gap caused by those factors than by non-optimal distribution of rainfall to which the greater part of the yield variation can be attributed in these generally well managed crops.

The authors also report ancillary data that reveal TE$_Y$ of Australian wheat crops has increased at 0.08 kg ha^{-1} mm^{-1} y^{-1} since 1860 to a present value of around 22 kg ha^{-1} mm^{-1}. This, they conclude, has been achieved by better adapted, shorter season, disease-resistant cultivars and improved nutrient management.

13.5   Cultivars and sowing time

The selection and timely sowing of adapted cultivars are basic steps for efficient use of water in rainfed agriculture. Appropriate cultivars are those that use efficiently (high WUE$_Y$) all water that rainfall and management can make available. In the absence of other limitations to growth, efficiency depends upon avoiding or tolerating transient water shortages during the crop cycle and, as explained earlier, distributing seasonal water use optimally between the vegetative and reproductive phases.

Crop development depends upon thermal, photothermal, and vernalization responses (Chapter 5). All crop species offer a wide range of developmental patterns that can be manipulated in breeding programs. Examples of optimum developmental patterns for rainfed environments are found in wheat cultivars grown in Mediterranean climates. These commonly are spring types that flower without vernalization. Introduction of a small vernalization requirement, however, controls flowering time so that crops sown early in autumn, that otherwise might develop too quickly, will flower in spring when the danger of frost has passed. This offers growers a valuable tactical alternative in years when rains begin early in autumn.

A question arises whether such cultivars use additional water for vegetative growth that would be better used for grain filling. This was studied in experiments with spring and winter wheat (cvs Banks and Quarrion), of similar genetic background, sown on four

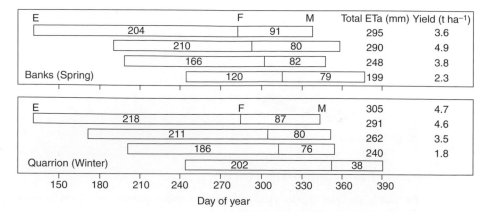

Fig. 13.3 Development and yield of spring (Banks) and winter (Quarrion) wheat cultivars at four times of sowing (May, June, July, and September) at Werribee, Victoria, Australia. The numbers within the histograms record mm ET during the respective phenophases (after Connor *et al.* 1992).

dates, near Melbourne, in southern Australia (Connor *et al.* 1992). Total ET decreased as sowing was delayed from fall to spring (Fig. 13.3). Although late-sown crops matured under more evaporative conditions, crop cycles were significantly shortened and total ET reduced. The winter cultivar had a longer crop cycle and used more water (higher total ET) than the spring cultivar, particularly at the last sowing. The winter culti-var performed better at May and worse at August sowing, but behaved comparably in development, water use, and yield when sown in July. A major imbalance of water use between the vegetative and reproductive phases was observed with the winter cultivar sown in September. Of a total ET of 240 mm, only 38 mm (16%) was used during grain filling and grain yield was restricted to only 1.8 t ha^{-1} compared with the high-est yield of 4.9 t ha^{-1} achieved in the same year from the spring cultivar sown in June.

13.5.1   The C3 versus C4 option

Comparisons of crop species presented in Table 9.4 reveal that C4 types have greater TE$_B$ than C3. This may lead to the erroneous generalization that C4 species will perform better than C3 species in all water-limited environments. In practice this is not the case because C3 species are more productive at lower temperatures and therefore at lower ET$_0$ than C4 species. In water-limited environments, there is often more to be gained from restricting crop production to cooler, less evaporative seasons than concentrating on C4 crops that only flourish under higher temperatures and more evaporative conditions during summer. This point was made earlier (Section 13.4) with the performance of early-sown sunflower in Spain.

13.6   Crop rotations and fertilizer

Strategies to maximize water-use efficiency will be most successful when no other factor limits yield. Soil-borne diseases and root-feeding insects and animals reduce length and density of roots thereby decreasing their effectiveness in extracting water. Weeds use water that might otherwise remain available to crops. Severe deficiencies of N and P and soil acidity adversely affect root growth and the capacity of roots to recover water from a soil profile. Hardpans, formed by accumulations of minerals in the subsoil or through compaction by implements, physically restrain the depth of penetration of roots and restrict infiltration of water into soil profiles. Farmers can avoid or reduce some of those problems through rotations. They can also select crop species to match water availability by tactical sequencing of crops that have different patterns and amounts of water use. In that way cropping systems can be adjusted to rainfall variations over years.

13.6.1   Weeds

Weeds use water that might have otherwise been available to support crop growth and yield. Without weed control, it is not possible to establish the optimum patterns of water use required to maximize yield under water-short conditions. Weeds can be controlled by cultural operations and by competition with crops established and managed in rotational farming. In addition, there is an increasing range of general and specific herbicides that can be used at sowing or during cropping to deal with individual problems. Within rotations, fallows must be maintained weed free if they are to conserve water and mobilize nitrate for subsequent crops (Section 13.8).

There is need for compromise when weeds of crops are also desirable components of pasture phases of rotations. This is the case in the wheat–sheep zone of southern Australia (Chapter 16), where annual ryegrass is a productive component of pasture phases but a competitive weed in wheat crops as well as an alternative host for certain root-borne diseases of cereals.

13.6.2   Diseases

Diseases divert assimilate from crop growth and yield, thereby lowering crop water-use efficiency. Any disease can reduce yield in this way, but **soil-borne diseases** exert an additional effect. They damage root systems and so restrict both the water-absorbing capacity of crops and their ability for thorough exploitation of soil water. If diseases are not controlled, there is little chance that other management tactics aimed at improving crop water-use efficiency can succeed.

Table 13.2 provides an example of the importance of disease and pest control in the yield of wheat in southern Australia. There, *Rhizoctonia* root rot and a soil-living parasite, cereal cyst nematode, combine to curtail the application of strategies developed to manage water and N supply. The table records wheat yields, water use, and WUE$_y$ for a range of rotations. The lowest yield (0.87 t ha^{-1}) was achieved in continuous wheat

Table 13.2 Crop water use, yield, and WUE_y of wheat under several rotations at Walpeup, Victoria, Australia, under a Mediterranean pattern of rainfall

Annual rotation[a]	Crop water use (mm)	Yield (t ha^{-1})	WUE$_y$ (kg ha^{-1} mm^{-1})
W–M–F–W	262	2.13	8.1
W–W–W–W	247	0.87	3.5
L–W–P–W	272	1.74	6.4
R–B–L–W	257	1.79	7.0
W–M–L–W	256	1.84	7.2
M–W–M–W	234	1.62	6.9
LSD (p < 0.05)	ns	0.31	

Note: [a] W = wheat; B = barley; M = medic; F = fallow; L = lupin; P = pea; R = rape.
Source: after Griffiths and Walsgott (1987).

and the highest (2.13 t ha^{-1}) immediately following a nine-month fallow period. In four other rotations there was no fallow and wheat followed a leguminous crop or pasture. All crops used equal amounts of water (mean 255 mm) so it is concluded that the major advantage of fallow here is in control of grassy weeds and associated wheat diseases rather than supply of additional water or N.

13.6.3 Soil fertility and nutrient management

Crops that are not limited by nutrient supply grow best and will have highest short-term water-use efficiency. During individual seasons, however, growth and water use at certain times can lead to water shortages at others. For example, early growth depletes soil water reserves and effects on yield are exacerbated when greater leaf area increases the subsequent demand for water. One result is that optimum nutrient supply for yield is less than for biomass production.

That response applies to any nutrient but, in practice, adjustments to N nutrition are usually most important. Wheat crops grown at high fertility in Mediterranean climates are known to senesce prematurely ("hay-off") in years of severe terminal drought (Russell 1967; López-Bellido *et al.* 1996; van Herwaarden *et al.* 1998). The cause is water stress during increasingly evaporative conditions of grain filling, exacerbated by low soil moisture content depleted by vigorous early growth. By contrast, low-N crops develop foliage cover slowly so E_p is small under the same weather conditions and crops continue through to maturity, although at low yield. With low fertility it is possible to match water supply to demand per plant, distribute water use more favorably between vegetative and reproductive periods, and maximize grain yield relative to water supply. Low N availability leads, however, to stable but unnecessarily low yields, and, as is the case with short-season cultivars, to low WUE$_Y$.

Under the variable conditions characteristic of rainfed agriculture, management of N is most easily accomplished with fertilizer because it is possible to control the availability in relation to crop requirements and weather. The alternative, reliance on legumes in crop sequences, provides uncertain additions related to variability of crop growth in response to weather. That introduces additional difficulty in matching crop N requirements to water-limited attainable yield in a given season.

A great challenge in semi-arid regions is to develop cropping strategies that operate at highest appropriate levels of N for long-term yield and allow for tactical responses to weather (and price) based on manipulations of N supply with fertilizers. In favorable seasons, N supply can be increased by topical applications in midseason. Examples are provided in Section 13.9 and in Chapter 16.

13.6.4   Grazing management

In zones where rainfall is too low or too erratic for crop production, grazing animals are often the only way to harvest fodder produced by grasslands. Proper management depends on choosing timing, distribution, and intensity of grazing that will maintain pasture condition for continuing productivity. Given the erratic nature of plant production this is a difficult task except in extensive systems where stocking rates are kept deliberately low. Excessive grazing first displaces palatable species but ultimately removes plant cover, permitting increased erosion. In practice, effective utilization of many low-rainfall pastoral areas is limited more by the provision of drinking water for stock than by the low or erratic nature of rainfall for plant production. Without well-distributed watering points, stock are not able to graze extensively during dry periods but instead concentrate near water, with resultant overgrazing and damage.

13.7   Density and planting arrangement

In water-short environments, crops are planted at low density, at the expense of increasing the fraction of ET lost as E_s, to maximize water available to each plant in the crop. The principle here is to provide as much soil volume (stored water) and catchment area (later rains) as possible per plant or per unit LAI and thus extend the duration of growth.

13.7.1   Density

Plants grown at low density experience greater radiation per LAI than do high-density plants and aerodynamic roughness of crops is greater, i.e., they are more closely "coupled" with the atmosphere (Section 9.6.3). In consequence, photosynthesis and evaporative demand per LAI are greater. This explains in part why relations between radiation absorption and E_p and LAI are non-linear (Fig. 9.14). Despite this disadvantage, low density is the only feasible strategy other than fallow that allows plants of long-season cultivars sufficient water supply to complete their life cycle in low rainfall environments where $E_p \ll ET_0$. In practice, low density is usually the first option employed and fallow is added in progression to areas of lower rainfall.

13.7.2 Planting pattern

Crop density can be manipulated by varying row width or row density. Densities can be maintained but development of cover and pattern of crop water use are altered by sowing plants more densely in wider rows. These effects operate mostly through the interaction between root systems within the soil volume available for root exploration.

The use of stored moisture by annual crops depends in part on how quickly roots expand into the available soil volume. Possibilities exist, through variations in planting pattern, to control when crops reach moisture. A uniform spacing (e.g., in a hexagonal pattern) allows roots to reach the perimeter of an allotted space most quickly. In contrast, crowding plants within wide rows, a further example of the **non-uniformity principle**, leads to early and intense competition for water within rows, thus restricting water use and early growth. Water use is distributed over longer periods before the root zone is exhausted.

Under some circumstances, planting patterns can serve purposes of both crop production and fallow because bare areas can accumulate and store moisture from rains early in the crop cycle. In general, choice of uniform or non-uniform arrangements depends upon soil depth and water-holding capacity as well as the relative amount and distribution of rainfall during crop and preceding fallow. If crops rely upon stored water, then non-uniform arrangements are appropriate because E_s will be small and stored water can be accessed slowly. Under these conditions, density would be adjusted to storage (soil type) and growth duration. On the other hand, if crops rely upon rainfall during the growth cycle then a uniform arrangement with appropriate spacing will enable each plant to extract water as it becomes available while minimizing unproductive loss by E_s and perhaps by drainage.

Myers and Foale (1981) have analyzed the results of a number of experiments to test the importance of row spacing in sorghum under dryland conditions. They compared yields obtained at row spacing between 0.25 and 1.00 m. At sites of low rainfall where yield at high density was less than 2.0 t ha^{-1}, the widest spacing gave the highest yield. At sites where the yield of narrow spacing exceeded 3.7 t ha^{-1}, narrow spacing was the most productive. In the range of 2.0 to 3.7 t ha^{-1}, differences in performance were small with advantage gradually shifting from wide to narrow spacing. The analysis was extended to weed control. In a comparison between yields of crops in rows 0.30 and 1.00 m apart they showed, not surprisingly, that weeds exerted their greatest effect at wide spacing and at low rainfall sites where yield at 0.30 m spacing was low. Only at yield < 0.4 t ha^{-1} was it better to use widely spaced (1 m) rows for weedy crops. Calvino *et al.* (2004) add further explanation in their analysis of sunflower response to row spacing as influenced by year, cultivar, soil depth, and sowing date at many sites in Argentina. Narrow spacing had a yield advantage in wetter conditions resulting from greater seed number, associated with greater radiation interception and crop growth. In dry years, advantage was with wide rows where crops achieved larger seeds, reflecting more water to finish grain filling relative to environmental demand.

Spacing arrangements serve as a practical means for managing the performance of root systems. An alternative, proposed by some workers (e.g., Passioura 1983), is to select cultivars with differing rooting patterns or resistance to water uptake. Selection for factors that determine resistance of root systems – root length density, root conductance, and timing of root growth during the crop cycle – could complement techniques to collect and conserve water in soil during a period before cropping (Section 9.6).

13.8   Fallow

In rainfed agriculture, extended weed-free fallows are commonly included in rotations to accumulate annual, and sometimes biennial, rainfall to improve the probability of a successful crop. In environments where water supply is relatively assured during part of the year (Drought types 1 and 2 of Fig. 9.10), storage of moisture in fallow is an effective way to extend crop growth duration. In others (type 3, Fig. 9.10), fallows may be essential to success but their value requires careful analysis of rainfall probabilities and soil water-storage capacity.

Fallows do not, however, only accumulate water. They also allow time for mineralization of soil N and reduction in weed load and soil-borne diseases, all of which can also contribute to yield responses.

13.8.1   Fallow management

The principles of fallow management derive directly from the water-holding characteristics of soils and evaporation (E_s) from exposed soil surfaces. Before rainfall can penetrate a surface layer and become less vulnerable to loss by evaporation, it must rewet that layer to field capacity, usually from an air-dry state. For sand, loam, and clay soils this requires up to the first 6, 10, or 12 mm of each rainfall event. As a result, the proportion of rainfall lost by evaporation increases with the frequency of showers and is greatest on clay soils into which infiltration is slowest. Although rainfall penetrates deeply into sandy soils, their total water-storage capacity is low, e.g., 50 mm m^{-1} of profile depth compared with 150 for loam and 250 for clay. Therefore, loss by drainage below the potential root zone of an ensuing crop is most likely to occur on sandy soils.

The management of fallows is aimed at maximizing the penetration of rainfall and minimizing losses by evaporation and drainage. Farmers can improve efficiency of fallows by careful management of residues and surface water. Even in semi-arid areas, rainfall can be sufficiently intense to exceed infiltration rate. Where rainfall intensity is high, residue retention, construction of terraces, contour banks, ridges, and other surface irregularities increases surface storage, thereby reducing runoff and risk of erosion. Terraces and ridging can also be used to concentrate water spatially so that infiltration occurs where plants will be grown. In some cases, ridging is even useful on relatively flat land. Where runoff is a problem, farmers must direct efforts toward maintaining infiltration capacity of soil. Infiltration is such an important component of the management of surface water that minimum tillage, deep ripping, organic matter incorporation, or application of amendments, such as gypsum, may be justified to improve water storage capacity of poorly structured soils.

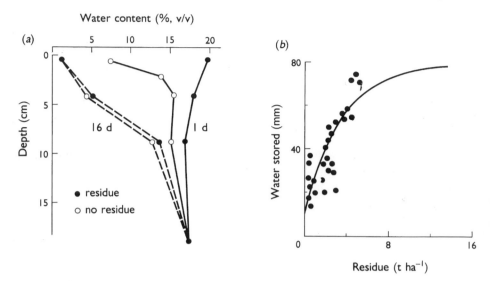

Fig. 13.4 Effect of the presence or absence of crop residue on soil water content of fallow. (*a*) Surface soil water content 1 and 16 d after rain. (*b*) Effect of residue level on soil water content after a short fallow (after Cornish 1987).

13.8.2   Minimum tillage

Previously, the only way to maintain weed-free fallows was by tillage after each significant rainfall. At some locations, during the early days of cereal cultivation in Australia and North America, fallow ground was tilled as many as 14 times with horse-drawn implements at great expense to soil organic matter and soil structure as well as risk of erosion. Farmers and researchers believed that tillage served to break soil capillaries through which water was lost by evaporation to the atmosphere. In reality, the main value of tillage lay in the control of weeds to prevent transpiration of subsurface water. Understanding the dynamics of fallows allows us to manage them without tillage. Evidence suggests that fallows maintained with herbicides ("chemical" fallows) are more efficient than cultivated ones, although clearly they have many different properties. A comparison of fallow techniques in the Great Plains, USA, showed that stubble-mulched fallows (partial incorporation of residues) without herbicides conserved 32% of rainfall whereas those maintained by herbicides conserved 42% (Smika 1970). Australian experience is similar and is reviewed in Cornish and Pratley (1987). Greater detail about tillage effects on soil properties is presented in Chapter 7, while the benefits of no-till are covered in Section 12.5.2.

Residue mulch protects wet surfaces from direct evaporation, especially during the initial energy-dependent stage of evaporation from wet soils. Because surfaces stay wet longer (Fig. 13.4*a*), there is more chance that subsequent showers will penetrate the drying surface layer. This helps explain why fallows with surface residues conserve more water than those where residues are removed or incorporated (Fig. 13.4*b*). Also, because surface soil remains moist longer after each rainfall, farmers have more opportunities to sow and establish crops at optimal times: a further critical component of efficient

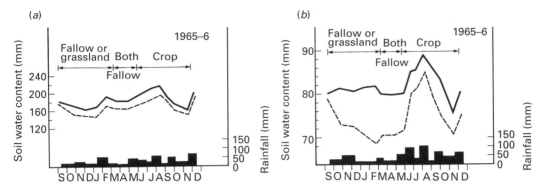

Fig. 13.5 Soil water contents and rainfall between recording dates on two soils during 1965 to 1966 in South Australia; (*a*) sand at Pinery; and (*b*) clay at Northfield. (Adapted from Schultz 1971.)

management with limited water. In contrast, in no-till systems in wetter climates, surface residues can delay planting because the soil is slower to dry compared to conventional tillage.

13.8.3   Fallow efficiency

Fallows are open to inevitable losses by soil evaporation, transpiration by weeds, and drainage below the root zone of a following crop. Consequently, their efficiency, i.e., the proportion of total rainfall that contributes to production during subsequent cropping, is low and variable from year to year. Fallow efficiency depends upon soil type, rainfall amount and distribution, and ET_0.

Figure 13.5 compares the performance of spring–summer fallow on two contrasting soil types (sand and clay) in South Australia. In this area of winter rainfall, the major effect of weed-free summer fallow was to prevent transpiration of soil water accumulated during the previous winter. Storage did not increase significantly during the dry season. The difference in storage of water between those two soil types was marked, reflecting their different water-holding characteristics.

The USDA (1974) compared fallow performance at nine locations in the US Great Plains (Table 13.3). In that region, rainfall is monsoonal with peak rainfall in summer. Annual cropping to wheat permits short (six-month) fallows during summer and early fall when 65 mm of 200 mm rainfall were conserved, an average efficiency of 32%. Cropped only in alternate years (fallow–crop system), longer (18-month) fallows received greater rainfall, 577 mm, and stored more, 107 mm, but with a lower efficiency of only 19%. In those long fallows, 50 to 70% of storage was achieved by the end of the first winter and 84% in the 12 months to midsummer. Only small additional gains were made in the second winter.

13.8.4   Yield stability

Fallow not only increases yield but also makes yield more reliable. In the Great Plains, a 30-year comparison revealed that long-term mean yields per crop were 0.73 (CV $= 0.70$) for continuous wheat and 2.48 (CV $= 0.30$) t ha^{-1} for fallow wheat (Smika 1970; USDA

Table 13.3 Precipitation and storage efficiency during uncropped periods of two cropping systems for spring wheat in the northern Great Plains

Data for individual stations cover periods of 20 to 47 y.

Station	Annual cropping		Alternate fallow–crop	
	Precip. (mm)	Storage efficiency (%)	Precip. (mm)	Storage efficiency (%)
Havre, Montana	159	30	454	24
Mandan, North Dakota	178	35	582	17
Dickinson, North Dakota	197	29	618	19
Huntley, Montana	211	30	546	18
Ardmore, South Dakota	211	37	615	18
Newell, South Dakota	214	28	618	17
Sheridan, Wyoming	250	40	637	25
mean	200	32	577	19

Source: after USDA (1974).

1974). Including the costs of production, 37% of continuous-wheat crops were economic failures, with yields below 0.4 t ha^{-1}, whereas break-even yield under fallow–wheat was 1.2 t ha^{-1} (Fig. 13.6). Yield per year for fallow–wheat was 2.48/2 = 1.24 t ha^{-1}, still significantly greater than 0.73 t ha^{-1} for continuous wheat. Thus, fallow–wheat systems are more efficient of seed, fertilizer, labor, and support energy.

13.9   Simulation models and analyses of cropping strategies

Simulation models that portray the responses of crops to environment, site, and management variables offer an alternative to seemingly endless experimentation for investigating crop–management interactions under variable environmental conditions. A range of examples presented here covers the use of models to assist in analysis of cultivar characteristics, design of cropping systems, and in tactical management. A recurring feature is the analysis of yield variability over long runs of seasons.

13.9.1   Crop duration, root depth, osmotic adjustment, and yield in sorghum

Genetic traits, such as those concerning disease resistance and product quality, perform consistently so assessment of their value to improved cultivars is relatively straightforward. Assessment of traits concerned with drought resistance and identification of appropriate **ideotypes** (Section 11.5) is, however, complex. There are many traits that might improve yield under drought but it is difficult to predict their performance in all possible combinations. Further, because values of trait combinations vary from season

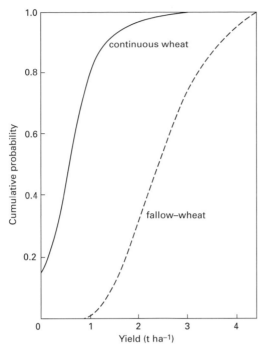

Fig. 13.6 Cumulative yield distribution for fallow–wheat and continuous wheat during 1940 to 1970 at North Platte, Nebraska (adapted from USDA 1974).

to season due to variations in water supply, they must be evaluated over a number of years.

Jordan *et al.* (1983) used a model of sorghum growth to evaluate drought-resistance attributes of grain sorghum at three sites in the US Great Plains. Cultivar characteristics of phenological development, root exploration, and osmotic adjustment were varied. Earliness to flower, at either 15- or 17-leaf stage, controlled the balance of water use between pre- and post-anthesis periods. Osmotic adjustment allowed crops to extract more water from soil. Further, increased root depth (2.8 versus 2.0 m) allowed crops access to additional stored water whenever rainfall was sufficient to recharge the profile to depth. Analysis of crop performance over 30 y weather data, with soil water balance carried forward each year, allowed assessments in a variable climate. Advantage of the long-season cultivar (Fig. 13.7) was substantial at Manhattan, Kansas, the wettest site, but it was also superior at Lubbock, Texas, the driest site, although advantage there was slight. Crop performance at Temple, Texas, demonstrated that the best cultivar performs differently from year to year. Over 30 y, the long-season cultivar outyielded the short-season cultivar. But, in 33% of years, when yields were less than long-term average, the short-season cultivar performed better. If it were possible to predict weather patterns, one could select the best cultivar for each year. Given variability in weather, however, growers must choose between maximizing long-term return and minimizing short-term

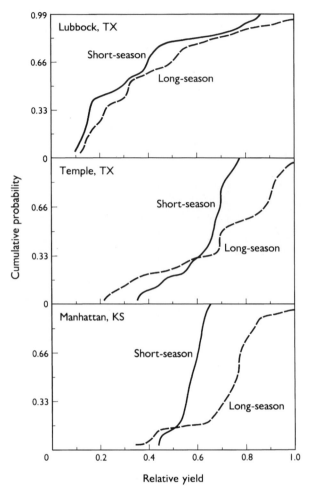

Fig. 13.7 Cumulative probability distribution of grain sorghum yield at Lubbock and Temple, Texas, and Manhattan, Kansas, for early and late maturity genotypes (Jordan *et al.* 1983).

risk. This is always the case when maximum use of a highly variable, limiting resource is attempted.

Jordan *et al.* (1983) found that osmotic adjustment offered no advantage through increasing soil-water availability at any site, either alone or in combination with deep rooting (Table 13.4). Because the model did not make allowance for the metabolic cost of **osmoregulation** (Section 9.6.3), that strategy may actually decrease productivity. Deep rooting increased yield by 50% in 20% of years at Temple and Manhattan. The benefit of deep rooting depends upon continuing replenishment of subsoil moisture and was least at the driest site. It would also depend upon the large carbon costs of growing and maintaining additional roots (Section 11.1). As was the case for osmotic adjustment, Jordan *et al.* (1983) did not consider these costs, so the real benefits will be less.

Table 13.4 Benefits from specific genetic modifications of sorghum and wheat cultivars grown at Temple, Texas (TEM), Manhattan, Kansas (MAN), and Lubbock, Texas (LUB) as predicted by crop models SORGF and TAMW

| | | Percentage of years that yield of modified crop exceeded normal crop by at least | | | | | |
| | | 20% | | | 50% | | |
Crop	Trait	TEM	MAN	LUB	TEM	MAN	LUB
Sorghum	Osmoregulation	3	0	10	0	0	0
	Deep rooting	38	29	20	21	20	7
	Osmoregulation plus deep rooting	38	29	27	24	20	10
Wheat	Deep rooting	21	30	28	14	20	17

Source: from Jordan *et al.* (1983).

Table 13.5 Climate and soil characteristics of two sites in the Northwest Victorian wheat belt

| | Site | |
	Dooen	Walpeup
Latitude (°)	36°40′ S	35°07′ S
Elevation (m)	155	110
Annual rainfall (mm)	414	331
Observed range (mm)	190–635	110–694
Annual ET_0 (mm)	1400	1800
Soil type	chromic vertisol	calcic xerosol
Texture	clay	sandy loam
Rooting depth (m)	2.0	1.5
Max. available water (mm)	367	121

13.9.2   Cropping strategies in low and variable rainfall environments

O'Leary and Connor (1998) evaluated the response of wheat to rainfall, N nutrition, and stubble management at two sites in Northwest Victoria, Australia. The location of sites and key environmental and soil characteristics are presented in Table 13.5 (see also Fig. 16.1). A weather generator was used to extend observed weather data so that simulations could be run for 200 y to include 100 fallow–crop sequences.

Although stubble retention is known to increase soil water content by reducing evaporation during fallow and early crop growth (Section 12.5.2), the effect in this study was variable depending upon soil type and amount and seasonal distribution of rainfall.

Effect of stubble retention Frequency distributions of yield are presented in Fig. 13.8 for fallow–wheat sequences with or without stubble retention versus removal

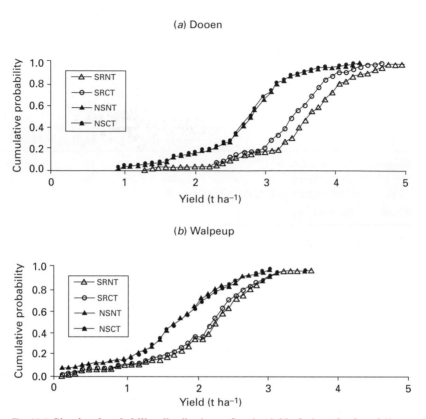

Fig. 13.8 Simulated probability distributions of grain yield of wheat for four fallow–crop management strategies at two sites in northwest Victoria, Australia. (*a*) Dooen and (*b*) Walpeup. The treatment codes are, stubble retained or not (SR, NS) and conventional tillage or not (CT, NT). For NT weeds were controlled chemically (adapted from O'Leary & Connor 1998).

(4 t ha^{-1} at Dooen and 2 t ha^{-1} at Walpeup) both with or without tillage during fallow. Data reveal a wide range in crop yield for all treatments, characteristic of these sites of low and variable rainfall. Stubble retention increased yield, especially at Dooen that benefited from greater water storage in wetter years. There, stubble retention has a median yield benefit of 0.9 t ha^{-1} compared to tilled fallow. Yields obtained and benefits of stubble retention are less at Walpeup where corresponding median yield benefit is 0.5 t ha^{-1}. The effect of tillage was small relative to that of stubble retention. A small advantage of zero tillage was evident in stubble-retained treatments at Dooen but not in either treatment at Walpeup where rainfall, soil water-holding capacity, and stubble quantity are smaller.

Effect of N fertilizer Treatments with stubble-retention–zero tillage and residue-removed tillage were compared with and without N fertilizer (40 kg N ha^{-1} at or near sowing), a typical amount and timing used by farmers. Nitrogen fertilizer produced a small increase (median 0.2 t ha^{-1}) in yield at Dooen, increasing to about 0.4 t ha^{-1} in wetter years when the stubble was retained and the fallows maintained without tillage. Without stubble retention, the yield response to N was about half (0.1 t ha^{-1}).

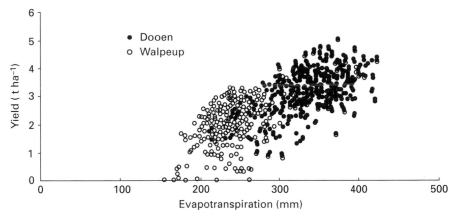

Fig. 13.9 Relationships between wheat yield and ET from the simulations presented in Fig. 13.8.

Similarly at Walpeup, the yield increase to N was greater with stubble retention (median < 0.1 t ha^{-1}) than with tillage (undetectable).

13.9.3 Rainfall distribution, soil evaporation, and crop yield

It is important to assess what natural variation can occur in field-determined yield–ET relationships (e.g., Fig. 13.2) owing to rainfall amount and distribution alone. This sets the yield framework within which manipulations to improve WUE$_Y$ must function. In Fig. 13.9, crop yields from the simulations of the previous section are plotted against ET.

The model predicts considerable year-to-year variation in yield at each level of ET. This variation reflects variable losses to E_s and drainage, which depend upon the amount and distribution of rainfall relative to crop cover and ET* during the crop cycle. It is important that farmers and farm advisors understand this variation. The management of fallow, selection of cultivar, sowing time and density, nutrition, and weed control can all help to increase WUE$_Y$, but optimal rainfall distribution is needed to achieve maximum values.

13.9.4 Tactical variations in fallow management

Performance of fallow varies from year to year depending upon weather, residue management and tillage, weed growth, as well as nutrient and disease status. In regions where fallowing is a valuable part of a long-term strategy, there will be years when the cost of fallow does not pay for its yield benefit. For example, tillage experiments during periods of continuous wheat production in southern New South Wales, Australia, revealed that in some years a few cultivations (VF, a variable length fallow) commencing when weeds first appeared after harvest (November–December) conserved more water up to sowing than did an unmanaged fallow (NoF). The question is whether VF is a good long-term strategy and what might be the value of certain tactical adjustments

Table 13.6 Simulated advantages of several fallow management tactics relative to no fallow

The analysis relates to wheat after wheat over the period 1943 to 1983 at Wagga Wagga, NSW, Australia.

Condition for initiation of cultivation	Years with fallow	Average extra AW[a] May 1 (mm)	Accumulated net yield benefit (kg ha^{-1})
First appearance of weeds after harvest (VF)	39(33)[b]	28	6010
On February 15 if weeds present	22	29	5480
After harvest if AW > 50 mm	25(13)[b]	35	6420
After harvest if AW > 50 mm and on Febrary 15 if weeds present	29(13)[b]	33	7120

Notes: [a] AW, available soil water content; [b] years when fallow began before February 15.
Source: after Fischer and Armstrong (1987).

to it. Fischer and Armstrong (1987) used a wheat simulation model (Stapper 1984) to analyze fallow management tactics over 41 y (1943–83) for which climatic data were available. Analyses (Table 13.6) revealed that NoF carried a substantial penalty in terms of available water at sowing. Under the VF strategy, available soil water for a fall sowing (on May 1) was increased on average by 28 mm, probability of a successful fall sowing was increased from 56% to 75%, and the best sowing date advanced by one week.

The model was also used to assess the alternative tactics by which fallowing could be initiated if soil moisture and/or weed conditions appeared favorable. After subtracting realistic costs for tillage, the authors expressed accumulated benefits of additional stored water as wheat grain. Tactics based on the presence of 50 mm soil water at harvest and appearance of weeds during summer were better than VF because they avoided or delayed fallowing in many years when yield advantage did not cover the cost of additional tillage.

13.9.5   Response farming

In regions of variable rainfall, farmers sometimes adjust their overall cropping strategy to suit the current weather conditions. In Mediterranean climates, for example, where cereals are the primary crop, farmers adjust first by changing to short-season cultivars of the intended crop as opening rains are progressively delayed, then by changing to a shorter season crop, such as from wheat to barley, and finally by deciding not to sow at all when the chance of a successful crop is unacceptably low. Each of these options requires access to seed of the different crops and cultivars.

These are examples of **tactical** responses to year-to-year variability of weather in which crops are changed in an attempt to improve a long-term "fixed" **strategy**. A further possibility is to vary the management of already-established crops once the likely outcome of a season can be identified. Such "in-season" possibilities widen in situations where seasonal outcomes can be identified early.

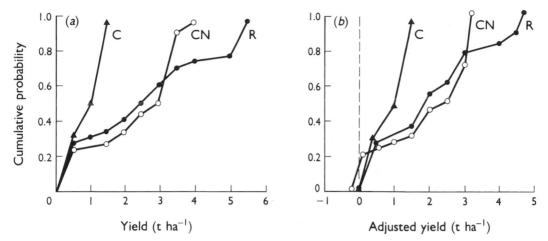

Fig. 13.10 Cumulative probability of a given yield of maize under alternative management systems during the "long" rains (March to June) at Katumani, Kenya. Conventional: C = 20 000 plants ha^{-1} with no N fertilizer; CN = 30 000 ha^{-1} and 30 kg N ha^{-1} at sowing. In response farming, R, the plant population was thinned from 50 000 to 20 000 ha^{-1} and N was applied at 20 to 60 kg N ha^{-1} depending upon seasonal conditions. (*a*) Grain yield. (*b*) Grain yield reduced for value of N fertilizer using a price ratio of 10:1 for N:grain (adapted from Wafula 1995).

An example of in-season tactics can be found in the **response farming** proposed by Stewart and Hash (1982) (see also Stewart 1988) following analyses of rainfall patterns in semi-arid Kenya. There, annual rainfall is bimodally distributed providing two distinct but variable growing seasons each year. The longer rainfall period ("long" rains) occurs from March to July. For that period, analyses identified workable correlations between total seasonal rainfall and timing and nature of opening rains. Response farming proposes that maize should be established with opening rains and then managed according to early-season rainfall to which follow-up rains are correlated. It is recommended that N fertilizer be applied to crops 40 to 50 days after sowing when the probability of greater-than-normal rainfall, and hence likelihood of yield response, is high. When the outlook is for less than normal rainfall, crop thinning is recommended at the same stage of growth to maximize yield.

Wafula (1995) used the CERES-maize model (Jones & Kiniry 1986) to assess response farming tactics. As a preliminary step, the model was tuned to Kenyan cultivars and soil characteristics and shown to handle response farming manipulations of fertilization and thinning. Analysis presented in Fig. 13.10 compares yield distributions for various management strategies and tactics simulated over a 29-y historical sequence of rainfall for cropping during March to June at Katumani, Kenya. Conventional management used plant population of 20 000 ha^{-1} without N fertilizer. Response farming included a number of options depending upon sowing time and early season rains. The range covers final plant populations of 20 000 to 50 000 ha^{-1} with 20 to 60 kg N ha^{-1}. CN is a variant of C with a fixed population of 30 000 ha^{-1} and 30 kg N ha^{-1} applied at sowing.

Figure 13.10*a* compares yield distributions of three management systems. Response farming provided consistently larger yields than conventional management, but performance of CN indicates that advantage is largely related to N fertilizer. This simpler system actually performed better than response farming except during seasons that produced the highest yields. Then, response treatment with greater population and more N made better use of wetter seasons.

Nitrogen fertilizer is not generally applied by Kenyan subsistence farmers because it requires substantial cash outlay. To investigate the financial return from fertilizer, a set of distributions of adjusted yield are presented in Fig. 13.10b. These were derived directly from Fig. 13.10a by reducing the yield of N-applied treatments by 10 kg for each kg N applied. In the yield range 0 to 1 t ha^{-1}, there was little advantage to expenditure on fertilizer although response farming did perform better than CN, which had smaller populations and less N. Response farming also performed better than conventional management, but advantage lay in N application and simpler approaches might suffice.

13.10   Review of key concepts

Efficient use of water

- In water-short environments, yield is most closely related to crop transpiration (E_p). Therefore, efficient use of water requires correction of other limiting factors, maximizing E_p and distributing it optimally between vegetative and reproductive periods. The latter is well defined in determinate crops. Water use before flowering should establish the maximum sink for yield (grain number) that can be achieved by water remaining for growth afterwards. Any other distribution of water use will result in either a smaller sink- or source-limited yield.
- The challenge to management is that many options exist to minimize losses to E_p by drainage, runoff, soil evaporation (E_s), and transpiration by weeds. Producers must select the most cost- or labor-effective options. These include choices about tillage, fallowing, crop and cultivar selection, sowing time and density, fertilizer practice, weed and pest control.
- In final analysis, inevitable losses to E_s, and less commonly to drainage, explain the wide year-to-year variation seen in water-use efficiency (WUE$_y$) of crops in individual locations. Conservation tillage practices contribute to reducing E_s by maintaining surface mulch, thereby increasing E_p and yield.

Fallowing

- In modern agriculture, fallows are usually periods when land is maintained free of vegetation by tillage or herbicide prior to crop establishment. Such fallows store water but efficiency is low (maximum $c.$ 20%). The value of additional water comes with associated benefits on disease control, especially of soil-borne pathogens and predators, and N mineralization, benefits that can be achieved in other ways. This explains why fallow–crop sequences, once widely used in semi-arid cropping regions, are now being replaced with more intensive cropping using crop sequences, fertilizers, and herbicides to give higher total productivity.

Strategies and tactics

- Farmers select crop sequences from a wide range of options to best meet objectives for their specific combination of climate, soil, and markets. Such cropping strategies optimize productivity and profitability at acceptable risk. Cropping strategies are not strict rotations, however, and can be modified with tactical variations to take advantage of favorable weather, high prices, or to reduce input costs. Tactics include changing cultivar or crop species as well as skipping a crop entirely when unfavorable weather conditions or price changes suggest that returns are unlikely to meet marginal costs.

System models

- Strategies and tactical variations for cropping in variable rainfed environments can only be established by analysis of performance over many seasons, preferably on many farms. Establishing tactical variations requires closer analysis of variability of crop performance in response to environment than is needed to establish the overall cropping strategy itself.
- Crop simulation models offer a powerful means to explore and explain crop performance under different management schemes and environmental scenarios and hence to design strategies and tactics (including new crop ideotypes). Continuing attention is required to improve and validate crop models for analyses of complex decision support.

Terms to remember: anoxia; aerenchymatous tissues; fallow, efficiency and management; non-uniformity principle; optimum pattern of water use; osmoregulation; potential sink capacity; response farming; simulation model; soil-borne diseases; soil seed banks; strategies and tactics; stubble retention.

14 Water management in irrigated agriculture

Irrigation avoids the constraints that inadequate water supply places on crop production in most parts of the world. The most spectacular irrigation schemes are those in arid regions of mid-latitudes where fully irrigated crops can achieve high yields made possible by high insolation. But irrigation practice covers a wide range of locations, environments, methods of water application, and production targets. Increasingly, as competition for water resources intensifies, so is the practice of deficit irrigation that is designed to increase the efficiency of limited water supply by removing only the most serious limitations of water shortage to crop growth. At the same time it is evident that excess irrigation in some places is not only wasting water but also reducing productivity and causing environmental damage, especially through salinization. This chapter deals with the principles of environmentally sound and productive use of water in irrigation as a key issue in crop ecology and natural resource management.

14.1 Irrigation and world food supply

Irrigated agriculture produces an estimated 40% of world crop production from 280 Mha, or just 18% of the total cropped area (FAOSTAT). For this it uses 69% of all withdrawals from streams and ground water compared with 21% for industry and 10% for domestic use. As population increases, irrigated agriculture is facing increasing competition for water from other uses at a time when the expansion, not restriction, of irrigated area is an evident way to increase food production (Wallace 2000).

There is much water on the Earth, but fresh water comprises less than 1% of it; the major part is found in oceans and polar ice. Current withdrawals of existing fresh water for all purposes vary greatly from place to place, ranging from 53% in the Near East and North Africa, through 36% in South Asia, to 2% and 1%, respectively, in sub-Saharan Africa and Latin America. There is insufficient fresh water to irrigate all suitable lands but extending irrigation onto some of the 82% of cropland that is rainfed is an obvious way to better feed an expanding world population on existing farmland. Given increasing pressure for other uses of water, expansion of irrigated land is slowing at the same time that some currently irrigated land is being lost to production due to deterioration and urban expansion. Prospects for large net expansion under current conditions are not promising. Only a cheaper source of energy to allow expanded desalination, now increasingly used to supply water for industrial, domestic, and agricultural purposes, would markedly change that situation.

In the immediate future, a significant practical challenge is to improve irrigation practice to achieve greater productivity per unit water used, and to extend the irrigated area in the developing world where food demand is great and current water withdrawals are small. Without greater productivity, irrigation will not be able to maintain the current relative contribution of irrigated agriculture to food supply. Greater productivity from irrigated agriculture would provide a valuable buffer for the more variable productivity of rainfed agriculture. For this, improved irrigation techniques are required that are appropriate to local water resources available, food demands, economic returns, and needs to conserve rivers and riparian habitat, groundwater, and wetlands. A range of techniques is available to improve present efficiency of water use in irrigated agriculture, but first, some general considerations.

14.2   Water and salt – an inescapable combination

Rainfall carries considerable quantities of salts added by dust and spray during ocean storms. Irrigation water has potential to add much more. All surface and ground waters contain dissolved salts. In arid regions Na^+ and Cl^- are common together with Ca^{2+}, Mg^{2+}, K^+, HCO_3^-, CO_3^{2-}, NO_3^-, SO_4^{2-} and BO_3^{3-}. Chemical composition of water reflects its origin and determines its quality for irrigation. Direct surface runoff from vegetated catchments is usually "cleaner" than water from rivers, which includes water that percolated through soil and underlying parent material. Ground water accumulates by deep percolation, so it usually has a higher concentration of dissolved salts. Quality of runoff, stream flow, and ground water varies substantially from place to place depending upon surface characteristics, geology, leaching history, and present hydrological balance.

The practice of irrigation, therefore, requires management of both water and salt. The global hydrological balance has already concentrated most salt in the oceans and that process is continuing. Large-scale irrigation changes the hydrological and salt balances of entire regions by diverting water from natural drainage systems that mostly lead to the sea. The design of irrigation systems must allow for the dispersal of large amounts of salt that evapotranspiration leaves behind in the soil. But it is not always possible to reroute salt to an ocean.

Water and salt management in irrigation usually requires action at regional levels as well as on individual farms. At regional levels there are two considerations:

- selection of areas for irrigation schemes; and
- management of overall water supply and associated drainage.

On farms, there are four components to good irrigation practice:

- irrigation scheduling for individual crops;
- optimal strategy for allocation of available water between alternative cropping activities;

- accurate and timely application and uniform distribution of water to crops; and
- provision for, and management of, drainage.

When water comes from surface runoff, management includes its collection, storage, and distribution to individual farms. When water is pumped by individual irrigators from aquifers, regional management may be needed to match withdrawal and replenishment. In both cases, management of ground waters into which excess water drains is critical because of the dangers that rising water tables pose to cropping and that ground water pollution poses to livestock and humans when water supply is drawn from them.

14.3   Salinity and alkalinity

Soils with a high content of neutral salt (e.g., Na^+ with Cl^- or SO_4^{2-}) are said to be **saline**. Sodium is generally an important constituent and terms such as sodic and natric are used in soil descriptions. Sodium ions (Na^+) plus weak anions $\left(HCO_3^-, CO_3^{2-}\right)$ give rise to **alkaline soil** with pH > 8.5. Alkalinity coupled with a high salt content results in a **saline–alkaline** condition.

14.3.1   Salt accumulation

Soluble and insoluble salts accumulate in soils formed under semi-arid and arid conditions. That occurs when salts released by weathering of parent material and those added by rainfall are not leached away through the profile. With shallow leaching, salts tend to accumulate in B horizons. Concretion nodules and "hardpan" layers of $CaCO_3$ and cemented clay are common at leaching depths.

Soluble salts such as NaCl are more mobile. Where hardpans and low rainfall prevent deep leaching, such soluble salts may be carried by internal and lateral drainage, and by surface runoff, to oceans or to topographically lower areas where they become concentrated by evaporation. There, sodium-saturated clays remain dispersed, effectively sealing low-lying areas from further drainage. In Chapter 16, a problem with that type of saline seepage is examined in relation to dryland farming.

14.3.2   Water quality

The quality of water for irrigation is primarily determined by total salt content, summarized as **total dissolved solids** (TDS) in mg L^{-1} but better described by its particular chemical composition. Concentration measured indirectly as **electrical conductivity** (EC) in deciSiemens (dS) m^{-1}, takes some account of the range and proportion of ions present. In soils, the ratio $Na^+:Ca^{2+}$ is in the range 2:1 to 5:1 (cf. 40:1 for sea water). Given an average ionic composition of salt found in arid regions, it is possible to make the following practical correspondences: EC $= 1$ dS m^{-1}; TDS $= 640$ mg L^{-1}; and molar concentration $= 11$ mM. For comparison, the EC of sea water ranges from 50 to 60 dS m^{-1}.

Table 14.1 Guidelines for the interpretation of water quality for irrigation

Irrigation problem	Degree of problem		
	nil	increasing	severe
Salinity (affects water availability)			
EC (dS m^{-1})	<0.75	0.75–3.0	>3.0
Permeability (affects infiltration rate)			
EC (dS m^{-1})	>0.5	0.5–0.2	<0.2
SAR			
montmorillonite	<6	6–9	>9
illite–vermiculite	<8	8–16	>16
koalinite–sesquioxides	<16	16–24	>24
Specific ion toxicity			
sodium (SAR)	<3	3–9	>9
chloride (meq L^{-1})	<4	4–10	>10
boron (mg L^{-1})	<0.75	0.75–2.0	>2.0
Miscellaneous effects (susceptible crops)			
NO$_3$-N or NH$_4$-N (mg L^{-1})	<5	5–30	>30
HCO$_3^-$ (meq L^{-1})^a	<1.5	1.5–8.5	>8.5
pH (normal range)		[6.5–8.4]	

Note: ^{a} by overhead sprinkling.
Source: after Ayers and Westcot (1985).

Salt content and **sodium adsorption ratio** (SAR; Section 7.4.2) determine the suitability of water for irrigation. The SAR is of particular importance in the irrigation of clay soils because they have a high exchange capacity and form structures that are susceptible to destruction by dispersion (called slaking) when irrigated with water of high SAR. Table 14.1 provides guidelines for the safe use of irrigation water according to its salt and sodium content.

The amounts of salt added by irrigation depend upon the quantity of irrigation water as well as its salt content. Salt additions to crops are significant even with high-quality water. For example a 500 mm (5 ML ha^{-1}) irrigation season with good quality water (EC = 0.5 dS m^{-1}; TDS = 320 mg L^{-1}) will add 1.6 t ha^{-1}. Without leaching, this salt may accumulate on and near the surface and ultimately reach levels detrimental to the productivity of crops.

Table 14.2 summarizes the electrical conductivity of irrigation water from various locations around the world. The list separates river and well waters because they form distinct classes of quality with respect to suitability for irrigation. Comparison cannot be strict, however, because the locations are few and individual measurements do not capture substantial spatial or temporal variation that can occur at individual sites. The data illustrate, however, that river waters are generally less saline than ground waters and that the two classes overlap. Overall, the range is great and shows that some waters have the potential to cause moderate to severe salinity problems. Ayers and Westcot (1985) have compiled an extensive analysis of the salinity and chemical composition of irrigation waters.

Table 14.2 Electrical conductivities (EC, dS m^{-1}) of irrigation water from rivers and wells around the world

Country	River	EC	Well	EC
Afghanistan	Kunduz at Seh Dorak	0.60	Kunduz (D-92)	4.50
Australia	Goulburn at Nagambie	0.10	Shepparton	3.40
Colombia	Cauca	0.87	Hda. Marsella	0.38
Egypt	Nile at Cairo	0.40	Nakel	2.20
India	Ganges at Patna	0.31	Rohtak	1.98
Mexico	Colorado in Mexico	1.35	Mexicali Valley	1.27
Malawi	Tangazi	0.15	Tambordera	0.37
Philippines	Jalaur at Iloilo	0.31	Laguna (P-18)	0.48
Spain	Guadalquivir at E. de Mengibar	0.89	Bardenas Alto	2.70
USA	Feather at Nicolaus	0.09	Denver	0.63
USA	Rio Grande at El Paso	1.32	Riverdale	0.97

Source: from Ayers and Westcot (1985).

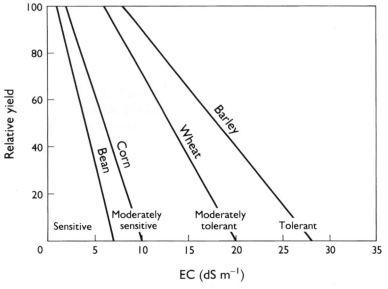

Fig. 14.1 Sensitivity of some agricultural species to soil salinity. EC is electrical conductivity of the saturation extract. In relation to Eq. 14.1, b is the slope of the response line and t is the intercept at 100% relative yield. (After Maas & Hoffman 1977.)

14.3.3   Salinity tolerance

The sensitivity of crops to salinity can be summarized by linear relationships (Fig. 14.1) between relative (%) yield (Y/Y$_0$) and EC of the soil saturation extract as:

$$Y/Y_0 = 100 - b(EC - t), \ EC > t \qquad (14.1)$$

Table 14.3 Salt tolerance of agricultural crops

Crop	Threshold $(t)^a$ (dS m^{-1})	Sensitivity $(b)^a$ (% reduction per dS m^{-1})
Sensitive		
Bean	1.0	17
Carrot	1.0	14
Strawberry	1.0	33
Onion	1.2	16
Orange	1.7	16
Peach	1.7	21
Moderately sensitive		
Radish	1.2	13
Lettuce	1.3	13
Clover, white	1.5	12
Broadbean	1.6	10
Maize	1.7	12
Potato	1.7	12
Sugarcane	1.7	6
Alfalfa	2.0	7
Tomato	2.5	10
Vetch	3.0	11
Rice	3.0	12
Moderately tolerant		
Sudangrass	2.8	4
Wheatgrass	3.5	4
Zucchini	4.7	9
Soybean	5.0	20
Ryegrass	5.6	8
Wheat	6.0	7
Sorghum	6.8	16
Tolerant		
Date palm	4.0	4
Sugarbeet	7.0	6
Wheatgrass	7.5	4
Cotton	7.7	5
Barley	8.0	5

Note: ^{a} See Eq. 14.1.
Source: after Maas (1984).

t is the threshold EC (dS m^{-1}) below which there is no effect of salinity on productivity (i.e., $Y = Y_0$), and b is salt sensitivity, expressed as relative yield reduction to salinity (% per dS m^{-1}) for EC $> t$. Table 14.3 records the effects of salinity on various crop species. Sensitive crops are affected by EC < 2 dS m^{-1}. Some field crops such as barley, cotton, and sugarbeet perform reasonably well up to 8 to 10 dS m^{-1}, but little production can be achieved beyond 15 to 20 dS m^{-1} with any crop. Toxicity to Na$^+$ and reduced

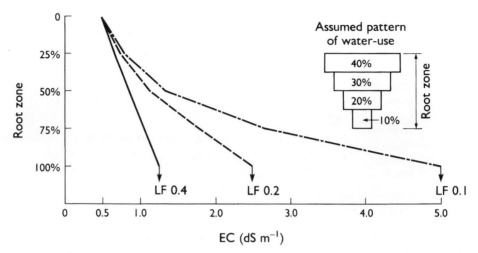

Fig. 14.2 Salinity profiles expected to develop after long-term use of water of $EC = 1.0$ dS m^{-1} at leaching fractions (LF) = 0.1, 0.2, and 0.4 (after Ayers & Westcot 1985).

availability of soil water are the main reasons. As a group, grasses are significantly more tolerant than legumes.

14.3.4   Leaching requirement

Continued farming in the face of rising salinity depends on removing excess salt through drainage. The annual hydrological balance (Eq. 9.1) of cropping systems must therefore allow for drainage (*D*). Because storage and runoff are small relative to other terms, drainage will occur when $(P + I) >$ ET. That is accomplished by reducing ET (e.g., shortening the duration of cropping) or increasing *I*. The additional amount of *I*, termed **leaching requirement**, can be calculated by assuming a steady state salt balance (salt$_{in}$ = salt$_{out}$). Most salt input comes from irrigation water with a conductivity of EC$_i$, so the necessary salt balance is:

$$EC_i I - EC_d D = 0 \qquad (14.2)$$

With the EC$_d$ of drainage water set to tolerance of the crop and *I* to ET $- P + D$, then, after rearrangement:

$$D = EC_i(ET - P)/(EC_d - EC_i) \qquad (14.3)$$

In this equation EC$_i$ and *P* are known. ET is also known or can be estimated from relationships as presented in Table 9.1; EC$_d$ is set according to sensitivity of the crop. Thus if EC$_i = 0.5$ dS m^{-1} and ET $- P = 500$ mm, then for EC$_d = 2$ dS m^{-1}, $D = 167$ mm. If EC$_d$ can be relaxed to 4 dS m^{-1}, then *D* falls to 71 mm.

Salinity profiles that are established in soils depend upon the leaching fraction and the pattern of water uptake from the root zone. Figure 14.2 shows the salinity profiles expected to develop after long-term use of water of 1.0 dS m^{-1} at three leaching fractions,

0.1, 0.2, and 0.4. In this case, 40% of transpiration requirement is extracted from the upper layer, declining to 10% in the lowest of four layers of root zone.

In practice, where farmers are restricted by the salinity of irrigation water, EC_d is often set to the value corresponding to 50% reduction in crop yield. If increase in EC is linear from soil surface to drainage depth (cf. LF $= 0.4$ in Fig. 14.2), average salinity in the root zone under this strategy will reduce yield by 5 to 10% for many crops (Reeve & Fireman 1967).

There must also be a place for drainage water to go. In the early years of irrigation projects, it is common to simply move salt down below root zones. This carries the danger of adding salt to ground waters, which may be an irrigation source for others, or of filling the profile with saline water. **Perched** water tables due to impervious zones in soils represent another problem. With a water table near the surface, **capillary rise** tends to return salt to the root zone and it may form crystalline deposits on the surface. Permanent solutions require the installation of subsurface drains, now commonly perforated plastic tubes laid 1 to 1.5 m deep and emptying to an outlet drain. Finding an appropriate outlet for drainage is an increasingly worrisome problem in irrigated areas. Disposal of drainage water, and methods for reclamation, are discussed further in Section 14.8.

It might seem that the use of salt-tolerant plants (e.g., barley) would provide a solution to continuation of agriculture under conditions of increasing salinity. While such crops offer an economically viable option for farmers, gain is likely to be transitory because without adequate drainage soil salinity will continue to increase beyond EC $= t$ (Eq. 14.1) as the system equilibrates to new, higher levels. Unless adequate drainage is built into irrigation systems, most will eventually succumb to salt accumulation.

14.4 Efficiencies of water use in irrigation

Subsequent sections deal mostly with **crop water-use efficiency** (WUE_B or WUE_Y), relationships between crop productivity (biomass B, or yield Y) and crop consumptive use of water (ET_a). It is valuable, however, to acknowledge that a broader range of efficiencies is used to evaluate performance of irrigation systems (Hsiao *et al.* 2007; Jensen 2007; Mateos 2008). The compilation in Box 14.1 identifies losses and returns of water during delivery to farms, application to crops, in their productive response, and in the control of salinization.

Important messages are, first, that losses can be avoided and efficiency gains made in irrigation systems other than on farms. And second, that water applied in excess of consumptive use (ET_a) may be necessary to manage salt or may not be lost to entire irrigation systems if it returns to rivers or ground water for subsequent use.

14.5 Water use and productivity

The principle has been developed (Section 9.7) that maximum efficiency of water use in the production of biomass (WUE_B) will be achieved when direct loss by soil evaporation

Box 14.1 Some irrigation efficiencies and performance parameters

Term	Meaning	Comment
Irrigation delivery efficiency	Water reaching crops as a proportion of water entering a delivery system	Accounts for losses due to evaporation and seepage, in canals and other operational losses
Irrigation water application efficiency	Water retained by soil in root zone as a proportion of water applied to crop	Accounts for losses due to runoff and drainage as determined by soil characteristics, irrigation amount, and application method
Irrigation efficiency	Irrigation water used beneficially as a proportion of water applied to a crop	Beneficial use includes ET_a and leaching fraction. Does not include return flow of irrigation water
Net irrigation efficiency	Irrigation water used beneficially as a proportion of irrigation water applied to a crop less return flow	This calculation discounts the part of return flow that is used beneficially
Irrigation consumptive use efficiency	Proportion of applied irrigation water that contributes to ET_a	Beneficial use is limited to ET_a (cf. irrigation efficiency above)
Crop irrigation-use efficiency	Ratio of crop yield to water applied to crop	Expresses yield response to applied water ignoring concurrent rainfall
Crop water-use efficiency (= water productivity) (WUE_Y)	Ratio of crop yield to ET_a	Relationship between crop yield and ET_a supported by rainfall and irrigation

is minimized and when cropping occurs while ET^* is lowest consistent with crop temperature and radiation requirements. Crop reproductive yield will be maximized when the pattern of water availability least restricts crop development, growth, and expression of yield.

These ideas of crop yield (Y) response to seasonal pattern of water supply can be condensed (Jones 1983) as:

$$\max(Y/Y_0) = \prod_{i=n}^{i=1} (ET_a/ET^*)^{\lambda i}$$

(14.4)

Table 14.4 Susceptibilities of some irrigated crops to water shortage during various stages of growth

The sensitivity parameter λ, for final yield, is defined in Eq. 14.4.

Crop	Growth stage	Sensitivity (λ)
Maize	vegetative	0.25
	silking	0.50
	tasselling to soft dough	0.50
	after soft dough	0.21
Cotton	prior to flowering	0.00
	early flowering	0.21
	peak flowering	0.32
	late flowering	0.20
Rice	vegetative	0.17
	reproductive and ripening	0.30
Pea	vegetative	0.13
	flowering to early pod	0.46
	pod growth to maturity	0.43
Soybean	vegetative	0.12
	early to peak flowering	0.24
	late flowering to early pod	0.35
	pod growth to maturity	0.13

Source: after Hiler *et al.* (1974).

Growth during each of *n* successive phenophases of a crop is actually proportional to relative transpiration E_p/ET^*, but for these purposes can be approximated by ET_a/ET^*. The effect of growth restrictions on final yield (Y) relative to attainable yield (Y_0) depends upon timing. The variable λ_i ($i = 1$ to n) describes the sensitivity of final yield to stress in each phenophase. In this model, the cumulative effect of stress during successive phenophases is multiplicative, as indicated by the symbol Π (compare with Σ for summation). Examples of such effects on a range of crops are presented in Table 14.4.

Equation 14.4 provides a practical basis for the management of water supply in irrigated agriculture. When water is insufficient to meet total seasonal demand, crop yield is greatest when water is supplied in a pattern that avoids serious effects of stress on yield, i.e., when Eq. 14.4 is maximized. This requires that water be managed to ensure that crops suffer water shortage least during sensitive stages (large λ in Table 14.4). This approach is now being superseded (see Section 14.7) by the use of crop models with more general responses to the environment than just water supply.

14.6   Irrigation methods

A range of methods by which water can be applied to crops is summarized in Table 14.5 according to the degree of control they afford over the amount and placement of water

Table 14.5 Irrigation methods arranged according to degree of control over application rate and distribution

Capital costs increase in the same order.

Surface	basin	
	border-check	↓
	furrow	increasing
Sprinkler	hand move	control
	wheel line	over
	center pivot	quantity
	linear move	and
	permanent set	placement
Micro-irrigation	drip emitters	↓
	micro sprays	

to be applied. The costs of installation and maintenance of irrigation systems generally increase with the degree of control. The choice of irrigation method also depends upon other costs involved in producing crops, potential income, and topography, soils, salinity, and the need for frost protection. Not included here is **subirrigation** by which water is applied to crops by capillary rise from a water table. This technique finds use in naturally wet areas where rising water tables can be controlled by drainage pipes that, when connected to a water source, can also counteract falling water tables to maintain crop water supply during dry periods.

14.6.1   Surface irrigation

In surface irrigation, water flows onto and over land where it is managed to control infiltration into soil. Soil physical properties determine infiltration rate and slope determines the duration that water remains on the surface for infiltration. The most primitive form is **wild flooding**, seen where crops are sown as river floods recede, or **water spreading** where banks are used to divert and hold flood waters more widely. Most forms, however, require channels to control the delivery of water and land grading to control the evenness and rate of flow of water onto individual cropped areas. In **basin irrigation**, individual level crop areas are surrounded by levees so that water can be ponded for required periods. These areas are flat and often small, their size depending upon the slopes on which they are constructed. **Border-check** and **furrow irrigation** require flatter land and careful grading so that water can be introduced to flow down slopes at rates appropriate for infiltration. Border-check systems are commonly used for pasture areas established between two parallel levees. Furrow systems are more common for crops. When areas between furrows contain more than one crop row, the term **bed** planting is often applied. Flat land is most suitable for surface irrigation because the cost of field development is least. It is, however, possible to reform almost any land surface into a series of irrigable areas as witnessed by ambitious terracing projects of ancient peoples in South America and those that still operate in Asia, e.g., in the Philippines, but with enormous labor costs for construction and maintenance.

Surface irrigation is not suitable for soils of high infiltration rate (sands and sandy loams) because it is not possible to run water down suitable slopes rapidly enough to avoid excessive drainage through the profile without causing serious erosion. On soils with low permeability it is possible to maintain standing water at low and spatially even rates of infiltration. This technique is used in basin irrigation of paddy rice, where already low permeability soils are "**puddled**" to further reduce infiltration (Section 17.3). On other soils, the size and slope of individual managed units and the rate of water supply are matched to the infiltration capacity of soil. Uniform wetting of each bay or furrow is difficult to achieve leading to addition of excess water and its subsequent recovery as tail-water flowing off each bay or furrow. Depending upon the irrigation layout on a farm, this tail-water may be cycled back to irrigation supply channels or used for lower-lying fields (see Box 14.1).

Surface irrigation subjects crops to a sequence of conditions of water supply. During irrigation and shortly afterwards, the upper soil root zone may be waterlogged. That is followed by a period of optimum supply of water and oxygen as upper layers drain. Towards the end of each inter-irrigation period, transpiration may extract all readily available water (RAW, Section 7.5.3) reducing soil water content below the level that sustains maximum growth. Despite these limitations, heavy soils perform best under irrigation because they store the largest amounts of water and because transitions in Ψ_{soil} during extraction near WP are more gradual (Fig. 7.9a). An exception occurs on soils with shallow topsoil over a slowly permeable subsoil. Such subsoils can only be recharged by prolonged irrigation and hence waterlogging of surface soil. Irrigators face this problem on many soils of medium to heavy texture, for example on the Duplex soils of northern Victoria, Australia (Cockroft & Mason 1987). There, when soil is watered with 70 mm of (ET* − P), corresponding to a ten-day supply at a daily rate of 7 mm, the period of optimum water supply may persist for only four to five days, which is less than 50% of the total growing period.

In **basin** and **border-check** irrigation, the penetration of irrigation water into soil is even, depending upon spatial homogeneity. Consequently when saline water is applied, salt moves evenly down profiles except for minor concentration by evaporation from border levees. With furrow irrigation, however, salt is concentrated by evaporation in each irrigated row, especially when crop cover is incomplete and E_s is high. This can be especially hazardous to seedlings, which are often more sensitive to salt than mature plants and have shallow root systems with little access to cleaner water below. Only rainfall can leach salt to depth so, in severe cases, a change of irrigation practice may be necessary to bring a field back into production (see Section 14.8).

In surface irrigation, the placement and infiltration of water is vastly improved by precise land forming. With laser-controlled equipment it is now possible to set precise grades, often as small as 1:1000, and to do that over large fields and long furrows. The outcome has been improved productivity, more efficient use of water, reduced drainage, and less expense of labor. Reduction of total area given over to control levees, and a change from irregular-shaped contour basins, have themselves lifted productivity

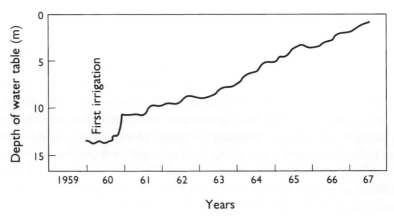

Fig. 14.3 Rising water tables under irrigation in Solano County, California (after Houston 1967).

on well-managed irrigated farms of the Central Valley of California by 7 to 10%. Laser leveling has been especially important to improved irrigation efficiency and crop management in irrigated rice where uniformity of floodwater depth is critical for seedling establishment, minimizing losses of N fertilizer, and weed control.

The measurement of water applied by surface irrigation requires attention to both input and loss. Weirs and wheels (e.g., Dethridge wheel) can measure flow in ditches but it is unlikely that many irrigators can know within 20% how much water has entered soil. The usual reaction to this uncertainty, especially to satisfy minimum leaching requirements, is to overirrigate. Over an irrigation season of 500 mm this may add an additional 100 mm to ground water or drainage. Added to a clay soil that holds 400 mm between saturation and wilting point (Section 7.5), the water table would rise 0.25 m y^{-1}. If soil were initially wet, its water-holding capacity less, or if annual accessions were augmented by rainfall on recently irrigated land, then the water table would rise more quickly. This has been a relatively common occurrence in irrigation schemes around the world (e.g., Fig. 14.3), leading to **waterlogging** and reduced crop yields.

14.6.2   Sprinkler irrigation

A range of **high-pressure** irrigation equipment has been developed to apply water to crops. They are especially suited to light soils and rolling land. Application rates can be adjusted to suit topography and infiltration characteristics of soils. With care, an even wetting pattern can be achieved.

Most systems are mobile, using equipment **(simple lines, side roll, water cannons)** that farmers move periodically while in others, movement is automatic and is mostly slowly continuous **(wheel line, hose drag, water cannons)**. Automation is most highly developed in **linear move** systems. With a width up to 400 m and supplied by a channel of 1000 m length, they command an area of 40 ha. A more widely used mobile sprinkler system is the **center pivot**, which rotates around its central point of supply. With an arm of up to 500 m in length and one complete revolution in one day or so, such systems can irrigate an area of 80 ha. During a period of 20 years up to 1975, center pivot systems were used to develop 4 Mha of irrigated cropland in the US Great Plains and they are replacing surface irrigation systems elsewhere because of improved irrigation efficiency. More recently there has since been a major shift from overhead sprinklers on linear move and center pivot systems to low-pressure downward-facing nozzles on hanging lines that deliver water closer to the soil surface. This innovation reduces cost of operation and improves irrigation efficiency substantially. Rather than spraying water into air high above the canopy, water is delivered much closer to the ground, which avoids the disruptive effects of wind and reduces direct loss by evaporation.

Permanent systems are used mainly with high-value horticultural crops. They are usually **set sprinklers** fed by a network of underground pipes with a riser to each outlet nozzle. In recent years many permanent systems have been replaced by micro-irrigation.

In sprinkler irrigation supply water is confined to pipes so measurement of water applied is straightforward and accurate. It is possible to account for losses by evaporation from foliage and soil surface and to estimate the water that enters soil with an accuracy of around 5%. Unlike surface irrigation, these systems do, however, require moderately clean water so that sprinkler nozzles are neither blocked nor damaged by suspended sediment. The use of sprinklers, linear move, and pivots also allow for application of nutrients ("fertigation"), which can be applied in several split applications to increase fertilizer use efficiency (Section 12.4).

Sprinkler irrigation can provide even wetting and hence even movement of salt vertically through soil profiles. If, as is common in horticultural crops, sprinklers are used to wet rows or root zones of individual plants (trees), then salt will migrate horizontally to the margins of the wetting front.

14.6.3   Micro-irrigation

Various forms of micro-irrigation are a relatively recent development in irrigation technology. Pioneered in Israel in the 1960s, these techniques have spread rapidly worldwide. Micro-irrigation offers precise control of the amount and placement of irrigation water and opens new possibilities for control of crop productivity while reducing water use (see Section 14.7). Micro-irrigation delivers water at low pressure through low-output emitters, sometimes to all plants individually. Emitters may be **sprays** or **drippers**, the latter now commonly an integral part of supply lines, that can be arranged to control the shape and volume of soil and root zone to be wetted. Restriction of E_s^* to a small wetted part of an orchard surface contributes, despite increase by micro-advection from immediately surrounding dry soil, to high water-use efficiency compared with surface and sprinkler irrigation. The advantage is most marked with drip irrigation because the wetted area is smaller and is frequently shaded by the crop canopy.

Bonachela *et al.* (2001) used micro-lysimeters and a soil evaporation model (as presented in Section 9.2) to estimate E_s^* from wet, and E_s from dry, areas in a drip-irrigated olive orchard in Córdoba, Spain (45% cover, 10% wetted area). They established that E_s would contribute 105 mm (19%) to orchard ET of 541 mm during a rainless irrigation season, days 155 to 277, with $ET_0 = 760$ mm. For comparison, E_s from a spray-irrigated

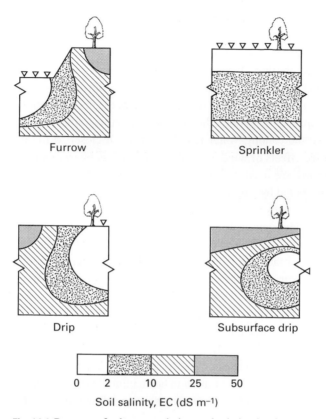

Fig. 14.4 Patterns of salt accumulation under irrigation by a range of techniques in an orchard (after Oster *et al.* 1984).

orchard (ten-day irrigation interval) would comprise 224 mm of ET of 660 mm in the same environment. By drip-irrigating individual trees, the inter-row space remains dry, weed free, and trafficable for orchard management.

Although ideally suited to high-value perennial crops, these systems are now extending into annual horticultural crops where removal or replacement of supply lines and emitters is necessary to allow cultivation and sowing as each new crop is planted. An interesting development, extending also into perennial crops, is **subsurface drip** irrigation. In this case, supply lines with in-line emitters are below the cultivation layer at 0.2 to 0.4 m depth but otherwise operate in a comparable manner. This technique has even higher water-use efficiency than surface drip irrigation because it avoids losses of applied irrigation water by E_s^*. In the orchard study referred to above, conversion to subsurface drip could save a further 58 mm irrigation. The model of Testi *et al.* (2006) includes spatial patterns of E_s for calculation of ET in irrigated orchards.

Complex patterns of wetting that are characteristic of drip irrigation produce equally complex distributions of salt in soil profiles (Fig. 14.4). Salt becomes concentrated at the peripheries of wetted volumes, away from emitters located close to plants. This can lead to difficulties during a crop cycle if infrequent heavy rain redistributes the salt back

Table 14.6 Trends in irrigation method, California, 1970 to 2000

Method	% Irrigated area			
	1970	1980	1990	2000
Surface[a]	80	78	67	50
Sprinkler	19	20	17	16
Micro-irrigation	1	1	15	33
Other	–	1	1	2

Note: [a] the survey excluded rice growing that has not changed from permanent flood.
Source: data from Orang *et al.* (2008).

into the root zone of seedling plants. With completion of each crop cycle, especially perennial ones, attention must be paid to eliminating pockets of high salt concentration to enable success of subsequent plantings.

As with sprinkler irrigation, water supply in micro-irrigation is confined to pipes and accurate measurement of flow is possible. Control of volume of water entering soil is possible within 1%. The systems do, however, require clean, filtered water. Unless due care is taken, emitters can become blocked with either sediment or salt. Fertigation and application of biocides against insect pests and disease are also possible with micro-irrigation systems.

14.6.4   Trends in irrigation methods

The worldwide trend is toward more controlled forms of irrigation to increase water-use efficiency and to reduce deleterious effects on the environment. A major example of change within surface irrigation methods is the current expansion of aerobic, at the expense of flooded, rice production (Box 14.2). More generally, data from California (Orang *et al.* 2008) document more extensive changes that have occurred in irrigation practice over the past 30 y (Table 14.6). There, a dramatic shift from surface to micro-irrigation has also been accompanied by a change in cropping pattern. There are now fewer field crops and more high-value horticultural crops. Many field crops are still irrigated by surface (furrow and bed) methods but increasingly by linear-move and pivot sprinkler systems. Horticultural crops, including vines, have increased in area and are now mostly irrigated by micro-methods rather than sprinkler. The same trend is recorded elsewhere. During the period 1970 to 2008, for example, micro-irrigation in Spain increased from 17 to 43% of total irrigated area (3.5 Mha). Recent developments of low-cost drip ("bucket") systems are extending the application of micro-irrigation to small-scale farmers in developing countries (Postel *et al.* 2001).

Changes in cropping practice reflect changing economics of alternative irrigation methods, largely attributable to product value and energy use. When the cost of delivery of water is low there is considerable energy saving with surface irrigation. However, when water is pumped to fields at great cost, e.g., 5.9 GJ ML^{-1} in the San Diego

---

Box 14.2 Changing irrigation methods to reduce water use: opportunities for aerobic rice

Rice is the world's most important staple crop. Annual production of 568 Mt is achieved from 156 Mha (Table 4.1), mostly in Asia, but with significant production around the world. Rice culture originated in naturally flooded lowland areas of Asia where rice could withstand anaerobic conditions but most weeds could not, where yield was maintained by annual additions of nutrients in floodwater, and soil organic matter was maintained by slow decomposition (Section 17.3). Such systems remained productive for many centuries and supported relatively high population densities.

From that beginning, irrigated rice production has expanded widely through the development of levees and water conveyance infrastructure to maintain flooded basins to which current rice cultivars are well adapted. But increasing competition for water from non-agricultural sectors threatens the sustainability of this form of rice production, especially on soils that are permeable and not well sealed by **puddling**.

When rice is grown in aerated soils competition is greater from a wider range of weed species that thrive under aerated soil conditions. Nutrient supply and seasonal patterns of availability are modified substantially, especially for nitrogen and phosphorus. Without the buffering effects of floodwater, diurnal temperature and humidity in crop canopies fluctuate more widely in aerobic systems, with marked effects on disease and insect incidence and severity. Cracking soils undergoing wetting and drying cycles disrupt rooting in surface soil. New rice cultivars must be developed to withstand these changed conditions.

Initial efforts have produced new cultivars with some tolerance of aerobic growth conditions in furrow, bed, or sprinkler irrigated systems as used for major crops such as maize and wheat. However, rice yields in these irrigated aerobic systems are smaller than those grown under permanent flood. Xue *et al.* (2008) report yields of 6.0 to 6.8 t ha^{-1} in the North China Plain, which is lower than attainable yield with permanent flood, but with less water and so higher irrigation water-use efficiency. Currently 80 000 ha of aerobic rice are grown during summer on the North China Plain where water limitations threaten the viability of irrigated agriculture. Likewise, rice production under pivot irrigation is expanding rapidly in Missouri, USA. Continued efforts to improve productivity and management of aerobic irrigated rice will become increasingly important to meet the global demand for rice in a world with limited water resources and a population projected to reach 9.2 billion by 2050.

River area, California, additional cost of pressurization, 0.6 GJ ML^{-1}, is more than offset by 20% saving in water use (Fig. 14.5). That explains why the energy cost of sprinkler irrigation is less than surface irrigation at that site. The relative costs of various irrigation techniques will vary as components of cost change. It is unlikely, though, that there will be major reversals (surface < sprinkler ≪ micro-irrigation) because a law of economics, if not of ecology also, is that the cost of biological effort (here, labor) and its replacement (here, electricity) are functionally related (Chapter 15). Likewise if values

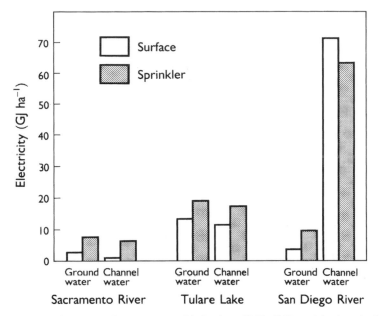

Fig. 14.5 The seasonal energy cost of irrigating alfalfa (950 mm) in three hydrologic basins in California. For each basin, a comparison is made between surface and sprinkler irrigation with water obtained from channels or from ground water. The calculations assume sprinklers use 20% less water than surface methods (after Knutson *et al.* 1977).

of crop commodities rise, as they have recently, it may become cost-effective to invest in higher cost irrigation systems for some field crops. For example, some maize producers in Nebraska, USA have switched from furrow to a subsurface drip irrigation on large fields.

14.7　Irrigation scheduling

A decision of when next to irrigate a crop and how much water to apply will depend upon many factors. It requires an estimate of ET*, knowledge of soil water-holding capacity and depth of root zone, an understanding of the effects of water deficits on yield, an estimate of amount of the irrigation water available to complete the crop cycle, and the probability of rainfall during that time. It also requires information on salt content of irrigation water and crop sensitivity to salinity.

14.7.1　Scheduling to save water

Scheduling water application to maximize yield with minimum water or to minimize yield loss when water supply is limited can be accomplished with crop water budgets calculated with models of the type presented in Box 9.2 and crop-specific stress factors as presented in Table 14.5. This is still done in many parts of the world but more advanced

---

Box 14.3 Irrigation scheduling: crop models and expert systems – an example with cotton

Cotton is grown in many parts of the world, most successfully in areas of low rainfall and high insolation where disease and insect pressures are least, but where irrigation is essential for productivity and fiber quality. The response of cotton to irrigation has been captured in a series of crop simulation models (Baker *et al.* 1983; Hearn 1994; Hearn & da Rosa 1985; Jackson *et al.* 1988). These models are able to explain, on physiological principles, the performance of crops in response to sowing time, density, row spacing, nutrition, water stress, and irrigation (frequency and amount) under prevailing weather conditions. They now form components of expert systems that have been developed to assist growers decide in real time when next to irrigate and how much water to apply (McKinion *et al.* 1989; Plant *et al.* 1992; Richards *et al.* 2008).

The decision of when to irrigate, and how much water to apply, is complex. Successful farmers (experts) would evaluate outcomes of alternative options based on long-term experience of local weather and performance of crops they have grown. An expert system seeks to emulate that process and extend the knowledge and decision skill to more farms and a wider region. It does this by combining site information (soils, historical weather, known crop responses) with analyses (weather probabilities, output from crop simulation models) to evaluate outcomes of optional irrigation tactics. Users can interact with the system, e.g., adjust models to the current status of their crops and soil, ask questions, and restrict options, etc. Results appear as optional actions, each assigned a probability of success for the complete crop cycle. These probabilities assist growers in forming decisions.

Importantly, expert systems are gradually improved by use. New information and improved analyses and models can be included without change of structure. Further, measured outcomes of individual crops evaluated by the system increase the empirical database of crop responses and allow analysis and review of individual in-crop decisions. As short-term weather forecasting improves, these expert systems can incorporate information about the probability of rainfall within the next two to three days into the decision process.

---

approaches are now available that use **decision-support systems** (DSS). These are built around simulation models of crop water balance and productivity but include other considerations of environment and management that determine crop productivity and profitability. An example of irrigation scheduling of cotton is presented in Box 14.3. On some farms, scheduling is assisted by measurements of crop condition and soil water content. The latter is now made routinely possible using measurements with neutron probes and other devices. Observation of crop water status can also be used. Direct measurements of leaf water potential, leaf turgor, leaf conductance, stem diameter, and sap flow are possible but require specialized equipment and skill and are less commonly used in practice. Indirect measurements are more easily applied. Visual detection of color change, movement, and wilting of leaves can indicate when irrigation is required. If such observations are made on parts of a field with soils of smaller water-holding capacity known to experience water stress first, a form of local calibration, this technique is highly successful. A range of infrared thermometers is also available to measure the onset of stress from temperature of representative areas of crop canopies. Canopy temperature rises above ambient when stomata close and transpiration falls (Section 9.6). The technique is widely applied and works best on windless days when hot non-transpiring leaves are least cooled by air and when the temperature differential between the canopy and air is greatest.

Adoption of water scheduling has been most rapid where the cost of water and/or consequences of over-irrigation are most expensive to producers. In areas of low rainfall, scheduling irrigation is relatively easy because low probability of rainfall simplifies the decision of how much of immediate water requirement should be added as irrigation. In such areas, irrigation may be effectively scheduled chronologically and, in community schemes, water is often delivered to farms on this basis.

Regrettably, emphasis can too easily be placed on frequent irrigation to maintain high soil water, low soil salt, and high productivity. Instead, irrigation practice should allow for maximum utilization of rainfall and hence for minimization of runoff and control of drainage. Fully charged profiles cannot accept additional rainfall that must contribute to runoff and/or drainage. Runoff can be spectacular and may cause erosion, sedimentation, and eutrophication of streams and lakes. Excessive drainage, in contrast, is unseen but is often insidiously more dangerous. Therefore, less frequent irrigation so that soil has storage capacity for unexpected rainfall during most of the growing season can reduce irrigation requirement and reduce runoff and leaching.

14.7.2   Scheduling to improve crop performance

Yield is, generally, maximized when $ET_a \rightarrow ET^*$ during all growth stages (Eq. 14.4) but there are some crops whose yield is increased by limited (**deficit**) irrigation. Cotton is an excellent example of a field crop in which favorable conditions promote vegetative growth at the expense of reproductive yield. Research in California (Grimes & El-Zik 1982) led to scheduling irrigation so that Ψ_1 declined as crops advanced from vegetative growth to "squaring" and at peak bloom. While biomass is certainly reduced by this watering regime, economic yield is not. In the San Joaquin Valley of California maximum yields of cotton are achieved with irrigation totaling 400 to 500 mm during a growing season in which ET^* approaches 800 mm.

Perennial horticultural crops also have reproductive relations that can be manipulated by irrigation to improve yield, including fruit size, color, and flavor, and to reduce irrigation input. Two schemes, **regulated deficit irrigation (RDI)** and **partial root zone drying (PRD)** (Box 14.4) are widely applied in horticulture to a range of stone and pome fruits and wine grapes (Kreidemann & Goodwin 2003). Micro-irrigation provides temporal and spatial control of irrigation water that the methods require. Substantial savings of irrigation water are reported compared with full-season micro-irrigation, 20 to 30% for RDI and 30 to 50% for PRD, depending on rainfall, ET_0, soil type,

---

> **Box 14.4** Regulated deficit irrigation (RDI) and partial root zone drying (PRD): two irrigation strategies to save water, improve yield, and increase water-use efficiency
>
> Regulated deficit irrigation was developed in peach production. In that crop, fruit growth proceeds in double sigmoidal fashion with a pause coinciding with the development of seed and associated tissues. If water is withheld during this stage, vegetative growth is reduced and on rewatering the growth of fruit may be enhanced. This technique is referred to as regulated deficit irrigation (Chalmers *et al.* 1981) and offers advantages in reduced irrigation input and amount of summer pruning needed to expose fruit to light that improves color and quality. Regulated deficit irrigation has been shown to be effective in many tree fruits (Fereres & Soriano 2007).
>
> Partial root zone drying has its origin in wine grape production (Dry & Loveys 1999; Dry *et al.* 2000) and involves alternating irrigation between sides of the root zone using dual irrigation lines. The goal is to promote root signals (abscisic acid and kinetins) to close stomates, control growth, and reduce overall water demand, without exposing plants to severe water stress. These responses have been mostly recorded in container-grown plants. In practice, the ability to achieve alternate wetting and drying depends upon rainfall and importantly on soil water-holding capacity, being most effective in soils of light texture. It now appears, however, that irrigation water deficits imposed by PRD have similar effects on growth, water relations, and productivity as those produced by RDI when the quantity of water applied and soil surface area wetted by emitters is the same (Fereres *et al.* 2003; Sadras 2009). The benefits from double irrigation lines may reside in a larger wetted zone and a more extensive root system in arid areas with marginal soils.

and any reduction in yield that may accompany improvement in quality. The methods, discussed further in Box 14.4, are based on distinct physiological responses of plants to water deficit but share similar dangers of reduced orchard longevity and limited control over salinization.

14.8   Management of water supply and drainage

Good irrigation practice depends upon water being available when it is needed. In each irrigation district, patterns of crop production and hence demand for water are generally similar from farm to farm. This presents great difficulties in provision of water, depending upon the size of major storage facilities and delivery canals. In practice, irrigators do their best work when they have their own supply, i.e., they pump individually from rivers, lakes, or ground water, or can store delivered water on-farm. On-farm storage can accept water when it is available even if rate of delivery is less than that required for efficient, direct irrigation. On-farm storage also assists recycling of tail-water from

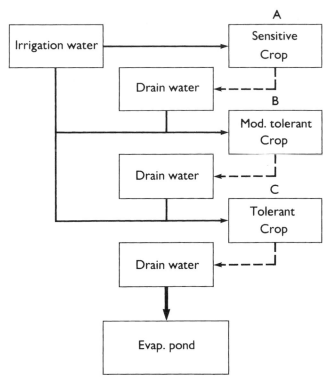

Fig. 14.6 A scheme for use of drainage water in irrigated agriculture. At each step, drainage water is mixed with high-quality water from the irrigation source. Crop types A, B, and C have increasing salinity tolerance.

surface irrigation and introduces options for incorporating drainage water into irrigation reserves on farm.

At regional levels, the need to maintain separate drainage and supply systems depends upon water quality. Drainage water can be added to supply systems provided the chemical composition of both is known and, if necessary, greater leaching requirement is provided for irrigation with saline water. Figure 14.6 illustrates an on-farm scheme for the utilization of drainage water. Crops can be chosen (Table 14.3) to accommodate the quality of the available water. This scheme also applies to entire irrigation districts because few have completely separate supply and drainage systems. For this reason it is characteristic that the quality of irrigation water gradually decreases as it moves through a district. This is seen dramatically in Fig. 14.7, which records salinity levels along the Murray River of southern Australia. As the river passes through irrigation districts it picks up additional drainage so that its quality for agriculture, industry, livestock, and human consumption gradually declines as it flows towards the sea.

Figure 14.8 presents a long-term analysis (1942–1973) of salinity of water from the lower Colorado River as it passes the Imperial Valley irrigation district of southern California. The steady increase over this period, 1 to 1.5 dS m^{-1} reflects the gradual

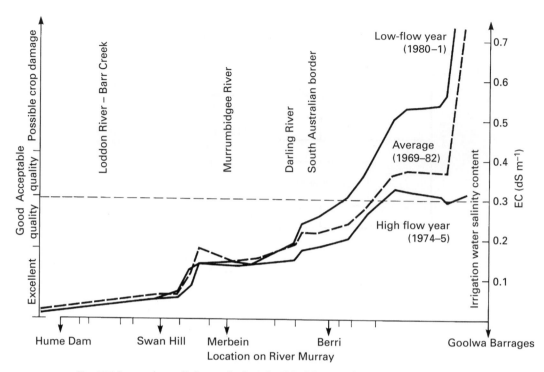

Fig. 14.7 Increasing salinity up the length of the Murray River, southern Australia. Salinity was recorded as it passed from the Hume Dam on its upper reaches through the irrigation regions of New South Wales and Victoria to South Australia. The locations of three tributaries that enter from the irrigation areas are also shown (after Anon. 1988).

development of other irrigation activities upstream in Colorado and Utah. For completeness, the graph also shows salinity of drainage from the area as it enters the inland Salton Sea. Increased salinity of effluent reflects, in part, reclamation of soils having greater initial salinity (see next section), and increasingly conservative use of water resulting from smaller licences for drainage allowances.

Reservoirs usually also serve purposes other than the provision of irrigation water including generation of electricity, flood control, and recreation. These objectives sometimes conflict. For example, secure irrigation requires maximum storage at the start of irrigation seasons whereas flood control requires that reservoirs should reserve capacity to handle sudden runoff from catchments. Recreation in reservoirs is usually best when they are full, but electricity generation and downstream recreation require continuous release of water.

14.8.1   Reclamation of alkaline and saline soils

Saline soils (EC > 3 dS m^{-1}; TDS = 2000 mg L^{-1}) will reduce growth of moderately tolerant crops and if EC exceeds 10 dS m^{-1} replacement of Na^{+} (Section 7.4.2) will

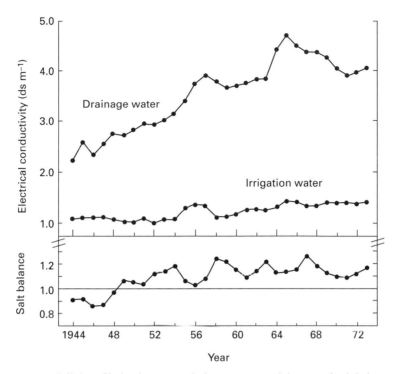

Fig. 14.8 Salinity of irrigation water, drainage water, and the annual salt balance (output/input) of the Imperial Valley irrigation area of southern California, USA, over the period 1944 to 1973 (after Kaddah & Rhoades 1976).

improve productivity of any crop (Fig. 14.1). Reclamation of salt-affected land requires that farmers coordinate their irrigation and drainage.

With neutral salinity (Section 14.3.1), medium and light-textured soils can be improved quickly by leaching. Theoretical analyses (Gardner & Brooks 1957) showed that 1.5 to 2 pore volume changes are needed to reduce salinity to below 30% of initial value. In practice this can be achieved by leaching profiles with an equivalent depth of water (1 m water to leach the top 1 m of the profile). Heavy-textured saline soils are difficult to reclaim by irrigation because that causes them to disperse, sealing them against further leaching. Calcium ions (Ca^{2+}) must be supplied to replace excess Na^+ on the exchange complex. Cation exchange proceeds slowly when the common ameliorant, gypsum ($CaSO_4$), is used because its solubility is low. In practice it is possible to dissolve 1 t of gypsum in 1 ML of water with mechanical mixing. Repeated treatments over several years are usually required to reclaim soil to 1 m depth. Direct application to soil and physical incorporation by plowing may improve its effectiveness, but this requires more gypsum. Intermittent flooding and sprinkler irrigation are more effective than continuous flooding because they cause more water flow through small pores leading to more efficient leaching.

Alkaline and saline–alkaline soils present a different problem. Often, the best way to lower pH is to incorporate sulfur. This is readily oxidized by microbes in most soils to SO_4^{2-}, releasing H^+ from water for Na^+ displacement.

It has been proposed from time to time that crops could be used to reclaim saline land by harvesting them to remove their salt content. This has been shown to be unrealistic, however, because most plants limit their uptake of salt. Lyerly and Longeneker (1957) measured average salt removal of 200 kg ha^{-1}, of which 60% was NaCl. To grow those crops, 900 mm of irrigation with a salt content of 6000 kg ha^{-1} was applied so the crops actually removed less than 5% of extra salt that was added. Even halophytes, which may accumulate 50% of their small biomass as salt, are unlikely to remove more than 200 kg ha^{-1}, and their growth would also require irrigation.

14.9   Selection of areas for irrigation schemes

Most irrigation schemes are located in arid and semi-arid regions. The impact of irrigation is greatest there despite the difficulties and disadvantage of low water-use efficiency (WUE, Section 9.7) due to low humidity (high vapor pressure deficit). As was explained earlier, soils and subsoils of arid and semi-arid regions often have high salt concentrations because they have experienced little leaching. That salt is mobilized in water. Ideally one would choose non-saline areas and best-quality water to minimize difficulties and costs of amelioration. Whereas all irrigation areas require drainage, those in saline areas require it larger, better, and sooner! There are many cases where realization of the need for adequate drainage has come too late.

In the Goulburn Valley of Victoria, southern Australia, 500 000 ha are irrigated in a semi-arid region. The natural drainage is the Murray River that provides water for irrigation, industry, and for human consumption down the length of its course (Fig. 14.7). It is not generally acceptable to use the river as a drain. Alternatives are to divert drainage water to inland depressions or to pipe it south to the ocean with a lift of 500 m over the coastal range. An inland drain is not acceptable because its consequences are unknown, and the cost of a pipeline to the ocean is prohibitive. The remarkable social issue is that the necessity of managing salt balance for continuing productivity of irrigation should come as such a surprise to so many.

In arid and semi-arid regions, ET^* is high and efficiency of water use in irrigation is small. Given comparable temperature and radiation regimes more production can be achieved from the same quantity of water in more humid areas. There, rainfall would provide the major proportion of water required and, with less transpiration, greater efficiency would be achieved from irrigation. Irrigation management would encounter fewer problems with salinity but more variable climates would add more disease pressure and complicate irrigation scheduling. If crop commodity prices rise, it is likely that the irrigated area will expand into sub-humid rainfed areas where water resources are available to support it.

14.10   Review of key concepts

Water supply and use in irrigation

- A major part (69%) of all water diverted from rivers or drawn from underground aquifers is used for irrigation and increasingly competes with need for other uses, forcing societal demands for more efficient use.
- Global irrigated area is around 280 Mha, just 18% of arable area but contributing 40% of world food supply. While there are regions where further expansion is possible (Africa, Latin America), loss of irrigated land is occurring in regions where ground water is being overextracted and where reservoir systems are being reduced by siltation. The prospects for significant net expansion of irrigated area are not promising. If irrigated agriculture is to maintain its relative contribution to world food supply by 2050, efficiency of water use must increase by 75%.

Water and salt – an inescapable combination

- The quality of water reflects its origin and its suitability for irrigation. Rivers, fed mostly by surface runoff, are low in dissolved salts compared to underground water that has passed through soil and rock. But all irrigation waters contain salt that can accumulate in surface soils as water is lost by evapotranspiration. 500 mm of irrigation with relatively clean water of 0.5 dS m^{-1} adds 1.2 t NaCl ha^{-1} salt equivalent.
- Irrigation practice requires a minimum leaching requirement, the leaching fraction (LF), to move added salt below the root zone. The leaching fraction increases with salt concentration of irrigation water and must be disposed of in drainage water.
- Most crop species are sensitive to low levels of salinity. Changing crops or breeding more salt-tolerant cultivars can, however, only offer temporary advantage without attention to salt management. The consequence of attending to salt tolerance rather than to drainage is that salinity will continue to rise, further degrading soil productive potential.

Irrigation methods

- A range of methods is available to apply water to crops, ranging from wild flood (water spreading, flood recession cropping), to basin and border-check methods of flood irrigation, to sprinkler and spray from fixed (set sprays) and mobile systems (center pivot, linear move), and to micro-irrigation by micro-spray or drip.
- Micro-irrigation offers the greatest control over the amount and timing of water application compared with other application methods. Micro-irrigation is increasingly applied to horticultural crops and is the key method to apply "deficit irrigation" (RDI and PRD) to reduce water use in irrigated agriculture.

Water-use efficiency

- Significant opportunities exist to increase irrigation water-use efficiency at all stages in delivery, application, use, and also in yield per unit of crop evapotranspiration (ET).
- Relationships between productivity and ET, and consequences of stresses at critical growth stages during development, are well understood. The challenge is to apply these principles to take into account evaporative demand, irrigation technique, soil water-holding capacity, crop response to stress, and expected rainfall. Crop models and expert systems are increasingly available and used in commercial production.
- Control of irrigation cannot be perfect because rainfall disrupts irrigation schedules and can damage recently irrigated land by waterlogging and runoff. Optimal management is most easily established in arid and semi-arid areas where infrequent rainfall least interferes with irrigation practice, but is achieved at low water-use efficiency due to high evaporative demand.

Terms to remember: alkaline and saline soils; capillary rise; decision-support systems; deficit irrigation, regulated defict (RDI) and partial root zone drying (PRD); electrical conductivity (EC); irrigation scheduling; irrigation systems, basin, border-check, furrow, micro-, spray (center pivot and linear move), subsurface drip, and surface; leaching requirement; puddled soils; sodium absorption ratio (SAR); subirrigation; total dissolved salts (TDS); water spreading and wild flooding; waterlogging.

15 Energy and labor

All human activity requires energy. The inescapable minimum is dietary energy to maintain the population. In earlier times, if each hunter-gatherer could collect around 33 MJ every day for a family unit (man, woman, and two children), then survival was possible. In practice, additional organic materials, mostly non-dietary, were needed for shelter, clothing, and combustion (cooking and warmth).

Agriculture provided a way to secure that supply, and more, with less environmental hazard and less competition from other organisms. The development and maintenance of industrialized cultures is based upon the substitution of energy for labor in mandatory activities of food provision. By success in raising and stabilizing yields, agriculture has supported an increasing population and released an increasing proportion from labor in food production. Greater participation in cultural, leisure, recreational, and scientific activities improves well-being for all and advances human civilization.

The purpose of this chapter is to explain the extent, pattern, and significance of energy use in agriculture so that we might understand how agriculture at various stages of development can respond to changes in the supply and cost of energy and labor.

15.1 Sources and utilization of energy

Earth systems capture energy that originates on Earth and beyond. Earth energy comprises a small geothermal heat flux and the essentially "limitless" nuclear energy of matter. Energy captured from outside is dominantly the flux of radiant energy originating in nuclear fusion reactions in the Sun (Section 6.1), and supported by kinetic energy in ocean currents and tides caused by gravitational forces of planetary motion. Solar energy is collected by the Earth's physical and biological systems. Temperature, wind, rain, and flowing water are all maintained by solar energy. In addition, chemical bond energy of current organic matter originates in solar energy collected by green plants in photosynthesis (Section 10.1), as does that in fossil fuels preserved from earlier geological eras.

15.1.1 Renewable and non-renewable

A further distinction between energy types is useful. The Sun can be considered permanently available, so current solar flux and recent solar energy contained in wind, water,

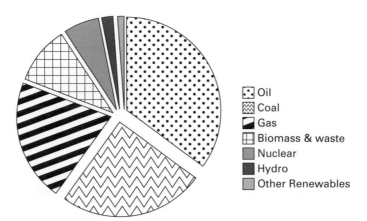

Fig. 15.1 The contribution of each energy source to current world annual energy use of 400 EJ (from FAO 2008).

and the standing biomass of vegetation can, together with geothermal energy, be taken as **renewable** forms of energy. Nuclear energy produced by fusion could be included here, but the process is not currently technically feasible. In contrast, accumulated solar energy fixed by previous vegetation in **fossil fuels** (petroleum, coal, natural gas, shales, and tar sands), and nuclear energy in radioactive compounds, comprise Earth's **non-renewable** energy resource.

15.1.2   Utilization

Current world energy use of c. 450 EJ y^{-1} is obtained from a variety of sources identified in Figure 15.1. Non-renewable fossil fuels, petroleum, coal, and natural gas provide most energy (81%). Hydroelectricity (2%), and other renewables (wind, solar, etc., 1%) are relatively small contributors. Biomass, largely by combustion, is currently the major renewable energy source, now providing 10% global energy. The average in developed countries is 3 to 4% but it is 25% in Latin America, 34% in Asia, and 60% in Africa (EMBRAPA 2006). At current usage rates (c. 360 EJ y^{-1}), fossil fuel reserves are estimated to be sufficient for 40 y (IEA 2008). In contrast, the potential of nuclear energy based on uranium and breeder reactor technology is estimated to be hundreds of times larger than the original supply of fossil fuels.

Serious limitations exist to continuing dependence on both fossil fuels and the expansion of nuclear energy. Society has become concerned over the contribution of fossil fuels to global warming (Section 6.10.4), risk of accidents in nuclear power stations, and safe disposal of nuclear wastes. While expansion of renewable energies is certain, the challenge is to find a mix that can provide required energy supplies at an acceptable cost to society and the environment. It is anticipated that world annual energy demand will increase, perhaps to 1000 EJ, by 2050. There are many options to fill this need. The most promising emerging energy technologies are wind turbines, photovoltaic cells, geothermal energy, and H_2 gas as a cleanly combustible fuel. Direct conversion of solar energy to electricity by photovoltaic cells is improving rapidly. Presently available cells have efficiencies of 15 to 24%, which exceeds the maximum conversion efficiency of solar energy to biomass (5%, Section 10.1). Designs are also improving for electricity generators driven by wind and tidal power. Energy from crops is also receiving much attention and appropriately receives special treatment in Section 15.5.

Electricity is valued for ease of distribution and versatility of application. Renewable sources fit well with electrical technology. Electricity can be added to distribution grids by hydro, geothermal, solar, tidal, wind, biomass, and nuclear sources. Those sources will grow in importance as the availability of fossil fuel declines. Electricity is not readily stored, however, so the key to efficient application is synchronization of generation with utilization, or breakthroughs in storage technologies. Strategies such as pump storage of water for subsequent hydroelectric generation, or compressed air for turbines, are under investigation for energy collected from the sun or wind.

15.2   Energy in food production

Energy used on farms is only part of the total used in food production. To complete the chain ("field to fork" or "paddock to plate") we must consider energy use in the entire production life cycle. A **life cycle assessment** includes energy to manufacture machinery and agrichemicals used on farms, energy used on farm for all field operations, energy used in transporting farm products to points of sale, energy used in food processing and storage, and finally energy used beyond points of sale in transportation to homes and in preparation for consumption. We need an appreciation of energy use in all those steps to interpret the significance of that portion to total life-cycle energy used in food production up to and on-farm, i.e., to the "farm gate".

15.2.1   The work of farming

Farming requires expenditure of considerable effort during relatively short periods distributed unevenly through the annual cycle. Without assistance of draft animals or powered machines, operations of tillage, sowing, fertilizing, irrigating, removing weeds, controlling pests, harvesting, and storage of products involve much human muscular effort: digging, hoeing, lifting, bending, and carrying.

Data presented in Table 15.1 emphasize the extent and irregular seasonal distribution of effort for lowland paddy rice production, as it was practiced in the 1960s in Taiwan, before mechanization. There, an average of around 850 h ha^{-1} was performed by teams of up to seven over a total of 40 d. A large part (14 d) was devoted to preparing the paddy for transplanting. Such labor-intensive rice systems remain the norm in parts of Bangladesh, Vietnam, China, India, Indonesia, Myanmar, and Vietnam.

Human work is slow because our bodies have poor energetic performance. The human body can convert around 25% of dietary energy to mechanical energy as physical work and most humans can continuously sustain a maximum power output of 75 W. Many laboring tasks in agriculture, such as digging, lifting, and carrying, require work rates

Table 15.1 Field operation schedules for lowland paddy rice production in Taiwan using hand tools, 1963

Operation	No. in team	Days	h ha^{-1}	% Sample using practice
First plowing	1	4.4	44	100
Application of manure	2	2.2	44	20
Harrowing and puddling	1	14	140	100
Transplanting	7	1.7	119	100
First additional fertilizer	1	1.3	13	
& cultivation	5	1.7	85	52
Second additional fertilizer	1	1.3	13	
& cultivation	4	2	80	63
First disease control	1	1.2	12	91
Third additional fertilizer	1	1.3	13	
& cultivation	4	2.1	84	54
Removing barnyard grass	2	1.5	30	91
Second disease control	1	1.3	13	83
Harvesting	8	2.9	232	100
Cleaning, drying, transport	2	5	100	100
Average for farm			855	

Source: adapted from Chang (1963).

of 300 to 600 W (Stout *et al.* 1979) and so must be interspersed with substantial rest periods.

In most tasks, rates of effective work are considerably less than 75 W. In harvesting, for example, 75% of muscular effort is expended in lifting our bodies, tools, and containers, and just 25% in getting the harvest into storage. When manual harvest operations are expended over a 10 h work day that includes rest periods, work rate is only 22 W. Animals are superior to humans because they are bigger and stronger, not because they are intrinsically more efficient. A horse, for example, can sustain 750 W (1 HP) and over a 10 h day can deliver 27 MJ, but once again only around 6% of that is converted to useful work. In contrast, moderate-sized tractors can continuously deliver 35 to 50 kW, converting (non-dietary) fossil fuel to useful work at efficiencies around 35% at the crankshaft, and 25% at the drawbar under average soil conditions.

Farming operations can be performed more expeditiously, and with less human effort, with draft animals and powered machinery. For example, inclusion of one draft animal per worker reduces the time needed in paddy rice production from 850 to around 500 h ha^{-1}. In mechanized systems total work required is 25 h ha^{-1}. Comparison of duration of operations using human, animal, and mechanical power in cereal production is compiled in Table 15.2. Again, without machines and fuel, many workers are required to maintain the schedule that successful crop production demands. When a large proportion of a society toils in fields to provide staple foods, it is not possible for that society to develop a high standard of living.

Table 15.2 Time requirement for component field operations (h ha^{-1}) in cereal production at various levels of intensity of human labor

	Production system		
Operation	Manual	Draft animal	Power equipment
Land preparation			
Spade	500–1000		
Plow		16–27	2
Harrow		3–5	0.5
Sowing			
Broadcast	2–4		
Machine	10	1–6	0.5
Fertilization	2	1–2	0.3
Weed control			
Hoeing	65	3–5	1
Herbicide application	4–5	2	0.5
Harvesting[a]			
Scythe	1000		
Machine steps		500	
Combine			2
Total	1183–1686	526–547	6.8

Note: [a] based on a yield of 6 t ha^{-1}.
Source: adapted from Unger (1984) and van Heemst *et al.* (1981).

In fact, the price of human labor has to be very low to compete with internal combustion engines. The human body and mind are more suited to control independently powered machinery than to perform strenuous and monotonous physical tasks. Machines can perform tasks better, more quickly, and more efficiently than humans. Hence, human ingenuity has devised efficient ways to replace physical activities in non-mechanical ways. Examples include tractors pulling tillage implements and seeders, harvesters, biocides that supplement physical effort (human or machine) of hoeing weeds or squashing insects, and N fertilizers that replace effort of growing green manure (legume) crops in rotation with grain crops for food.

15.2.2   To the farm gate

Complete analysis of on-farm energy use must include all energy expended directly and indirectly in production. Together they form the **embodied energy** of products (Box 15.1). Such an energy budget is presented in Table 15.3 for maize production in Iowa. In this example, total energy used to produce 10.8 t ha^{-1} of maize grain (14% moisture) was 16.6 GJ. The major component of that **support energy** was N fertilizer (43%), followed by liquified petroleum gas (LPG) for grain drying (21%), herbicides (12%), diesel fuel for tillage, planting, spraying, and harvest (9%), and P and K fertilizers (7%).

Box 15.1 Embodied energy in food production

Direct energy consumption occurs on farms as food for human and animal labor and as fuel for machines. Indirect expenditure includes energy cost of manufacturing machines, structures, tools amortized for their useful life, and agrichemicals applied to support crop and animal production. Energy costs of those goods and services are the sum of human effort and (usually) fossil energy that was expended in their manufacture. That sum is termed **embodied energy**. A tractor, for example, has embodied energy reflecting human labor and fossil energy used in its manufacture, i.e., mining, smelting, fabrication, distribution, sales, and maintenance.

Further, proper comparison between the energetics of human labor and machines requires details of appropriate lifetime share of embodied energy of all inputs to humans. The real (i.e., embodied) energy cost of human labor is determined more by lifestyle and standard of living than by dietary energy for survival. The major components are provision and heating of homes, clothing, meal preparation, medical services, transport, care and education of replacement workers, and care of retired workers. An analogy to a dairy herd is helpful. In addition to the dietary needs of lactating cows, dairy farmers must also budget feed for dry cows, replacement heifers, and bulls, as well as energy expended in provision of shelters, feed storage, and milking facilities. In the USA, about 40% of on-farm consumption of fossil fuel is in maintenance of the farm household. Compared with a dietary cost of 12.5 MJ d^{-1}, embodied energy of an average American or Dutch farm worker, i.e., the employment-generated energy requirement, is around 600 MJ d^{-1} (Fluck 1981), comparable with that of urban counterparts. Viewed this way, human labor is seen to have a high energy cost, especially in countries with high standards of living.

Human labor, amortized energy to manufacture farm equipment and machinery, and seed each represent 2% or less of total energy use to the farm gate. In this energy-intensive system, proper accounting of embodied energy of human labor makes little difference to total energy use of 16.6 GJ ha^{-1}. It becomes an increasingly important issue, however, as labor intensity of farming systems increases and yield decreases. Low-input, low-yield farms and subsistence agriculture serve as examples where energy cost of labor is a major component.

The maize system in Table 15.3 produced 180 GJ ha^{-1} of dietary energy from the solar flux and had a net energy ratio (NER) of 10.9. That crop actually fixed about twice as much solar energy into chemical bonds but the rest, in stover (stalks), has little dietary value except to soil organisms and cattle, and is not included in this analysis. Solar radiation also is not included because it is a renewable source of energy that is lost to food production if not captured by agriculture (see also Jones 1989). Analysis of potential productivity of biological systems (Section 3.2.5), showed that potential efficiency of solar energy capture is 4 to 5%, while 2.5% is a reasonable value achieved under favorable conditions of nutrient and water supply.

Table 15.3 Energy budget for maize production in Iowa, 2005[a]

Input category	Input type	Quantity per ha	MJ ha^{-1}
Labor[b]		3.8 h	280
Machinery	145 ha farm	1 ha share	320
Fuel and energy[c]	diesel	43 L	1557
	gasoline	11 L	356
Grain drying[c]	LPG	135 L	3498
	electricity	18 kWh	167
Fertilizer[c]	nitrogen	158 kg	7110
	phosphorus	23 kg	380
	potassium	56 kg	763
Seeds[c]		22 kg	213
Herbicides[c]		5.4 kg a.i	1922
Total energy input			16 566
Maize grain yield		10 800	180 160
Output/input ratio		10.9	

Notes:
[a] Data from USDA-NASS for 2005 yield; USDA-ERS (2005) for 2005 input levels.
[b] Daily proportion embodies energy of labor of 600 MJ d^{-1}.
[c] *Sources*: Liska *et al.* (2009) for energy inputs for field operations, fertilizer, herbicides, seeds, and for energy content of fossil fuels and electricity (including transmission efficiency); Wilcke (2008) for LPG and electricity for grain drying.

Data useful in the construction of energy budgets in agriculture are presented in Tables 15.4*a* and *b*. Table 15.4*a* presents energy contents and energy costs of a range of materials and inputs relevant to crop production; Table 15.4*b* compares energy use by various field operations.

15.2.3   From farm gate to table

Farming produces commodities, and except for some types that can be eaten raw (e.g., some nuts, fruits, vegetables), additional labor and energy are required to provide and prepare finished products for consumers. For individual products, this requires varying amounts of transport, storage, processing, refrigeration, merchandizing, and cooking. In developed countries, energy used in this second part of food production systems far exceeds that used that used in crop and livestock production to the farm gate.

This pattern of energy expenditure in provision of food is evident even in staple foods such as wheaten bread that is produced and delivered comparatively efficiently to points of consumption. In a Californian study, Avlani and Chancellor (1977) estimated that only 40% of embodied energy of bread at the table was used up to the farm gate. Further, more than half of total energy was expended after bread left the bakery or, for less efficient home baking, after flour left the mill. That more energy is consumed in toasting a slice of bread than in growing the wheat it contains is perhaps the most startling point of this analysis.

Table 15.4 (*a*) Energy values for analysis of energetics of cropping system

Material	Energy content	Material	Energy content
Heats of combustion		*Gaseous fuels*	MJ m^{-3}
Plant material	MJ kg^{-1}	Natural gas	35
Carbohydrate	14–16	Propane	86
Protein and lignin	25	Coal gas	18
Fatty oils and fats	38–40	Producer gas	6
Waxes	45		
Terpene hydrocarbons	45	*Energy costs of production*	MJ kg^{-1}
Typical biomass	17	*Machinery*	
		Tractors	87
Liquid fuels	MJ L^{-1}	Implements	70
Gasoline	38		
Diesel	39	*Fertilizers*	
Ethanol	22	Nitrogen (via NH$_3$)	45
Biodiesel	33	Phosphorus ⎫	16
LPG	26	Potassium ⎭ (mine and refine)	14
		Herbicides[a]	356
Solid fuels	MJ kg^{-1}	Pesticides[a]	358
Coal	17–30		
Wood	18–23		
Bagasse (30% moisture)	15		

Note: [a] calculated for active ingredient (a.i.).

Table 15.4 (*b*) Fuel requirements of various field operations from McLaughlin *et al.* (2008)

Operation	Depth (cm)	Energy requirement (MJ ha^{-1})
Tillage (clay loam soil)		
Deep ripper	34	616
Mold board plow	19	782
Disk ripper	19	427
Chisel sweep	18	659
Chisel plow	17	503
Strip till	17	235
Disk harrow	6	264
Fluted coulter	5	141
Other (US maize system)		
Planting, spraying, cultivation, harvest (combined total)		111

Foods other than staple grains and root crops generally have much larger embodied energy per unit dietary energy. Frozen, canned, and prepared foods, and fruits and vegetables, particularly those grown in controlled environments or transported by airplane from the opposite hemisphere, substantially increase energy cost of the food production system. For the USA, energy expended after the farm gate is five times that expended up to the farm gate. Comparable values apply to other developed countries; in Australia the ratio is 3:1 (Gifford & Millington 1975).

15.2.4   National energy use

If farming and its inputs of machinery, fuel, and fertilizer are the smaller part of energy expended in the food production system, how does it compare with total national energy use? This becomes an important issue in developing policies to promote greater energy-use efficiency. Industrialized societies use 10 to 15% of total energy on their food production system, 3 to 5% to the farm gate, and the balance on processing and distribution. Given the magnitude of national energy use, it is surprising how little, not how much, energy is used in the essential activity of farming. It shows that most opportunities to adjust to changing costs of energy exist not just outside farming, but also outside the entire food production system. Within agriculture itself, however, energy use is important because it represents a significant cost of production that can be reduced with improved efficiency.

15.3   Improving efficiency of energy use

Energy budgets such as that presented in Table 15.3 identify component energy costs of agriculture. They identify where energy savings can have most impact as agriculture, along with other societal activities, adjusts to the changing costs of energy.

It has been proposed that society will soon be unable to afford food because of increasing energy costs. Given the small proportion of energy allocated to food production, consumers will have plenty of other adjustments to make before they have to return to the fields to grow their daily bread. A bigger problem exists in those societies where price of food is kept so low that farmers are not able to purchase necessary inputs of fertilizers, fuel, and agrochemicals that would sustain their productivity and profitability.

15.3.1   Mechanization

The mechanization of crop production substitutes for human and animal muscular effort and removes the drudgery of heavy physical work. It enables human labor to be more effective, efficient, and timely in performing the tasks required in crop production. Many improvements to crop production practice cannot be applied without mechanization. Thus crop cultivars with maturity that finely matches phenological development to climate must be sown quickly if they are to realize that advantage. Tillage, or application of biocides to control weeds, pests, and diseases, must be accomplished expeditiously, especially when they are a component of integrated pest management (IPM). As crops mature, farmers rely on machines to harvest quickly with minimum loss. Looked at another way, mechanization enables crops to be grown successfully in areas where hand labor would have insufficient time for critical operations.

Farms and farm machinery are gradually increasing in size in response to rising costs of human labor. Operations cannot be performed as quickly over larger areas simply by driving implements faster. In tillage, for example, increased ground speed requires specially designed implements to avoid damage to soil structure. Big tractors cover ground faster by drawing wide implements at appropriate speeds. In addition they are more economical of fuel and labor (Hunt 1983) and offer security that operations can be carried out quickly while conditions remain favorable. This contribution to risk avoidance is an important aspect in the selection of machines of appropriate work capacity (e.g., Whitson *et al.* 1981).

Geographic information systems, information technologies, and remote sensing are introducing new possibilities to mechanization and thereby improving energetic efficiency of crop production (Sections 12.1; 16.7.3). For example, seed drills can be programmed to adjust seed and fertilizer rate to variation in soil type within a single field. Such decisions can be based upon prior surveys of microtopography, soil maps, and/or direct measurements of soil chemical and physical properties. Advanced imaging can direct weed treatment to where weeds occur rather than as a uniform treatment over entire fields. In the future, when tillage, sowing pattern, fertilizer level, and weed control can be adjusted according to currently sensed conditions, perhaps cross-checked against spatially referenced data collected in previous years, e.g., yield recorded during the harvest of the last crop, then farmers really will have "smart" machines in their service.

15.3.2   Tillage

Previously, tillage was the only effective way to control vegetation for establishment of new crops and pastures. Fire and grazing were valuable adjuncts, but tillage was the key to success. Under those conditions, tillage, not fertilizer (Table 15.3), was the major component of energy use. Now herbicides offer possibilities to control vegetation at significantly less cost and energy without tillage.

Among conventional tillage methods, mold board plowing requires greatest energy and a disk ripper much less (Table 15.4*b*). Secondary tillage by disk harrow or fluted coulter is also relatively efficient. Among conservation tillage methods, a chisel sweep requires about twice the energy of strip till, while no-till systems eliminate the need for any form of tillage (Section 12.5). Another advantage of no-till and strip-till systems is maintenance of good soil structure because conventional tillage operations often contribute to compaction. Where compaction occurs, additional energy can be required to alleviate the problem by deep ripping. Total energy requirements for planting, spraying, and cultivation, are relatively small compared to requirements for most types of tillage operations, but not so, for harvesting high yielding crops. For example, energy to harvest a maize crop of 15 t ha^{-1} can be twice that of disking for land preparation.

Herbicides are now replacing tillage in mechanized cropping systems in many parts of the world. Once commenced, this trend is further supported by development of transgenic herbicide-tolerant crops, new herbicide products, better application equipment, and an improved understanding of how to use them (Box 4.2).

15.3.3   Crop nutrition

Fertilizers In the Middle Ages, cereals in Europe yielded up to 1 t ha^{-1}, compared with present average yields of around 7 t ha^{-1}. That increase results from higher fertility, achieved mainly by the use of fertilizers. Without that and improved husbandry, modern cultivars could not reach their attainable yield. Fertilizers have raised productivity of agricultural systems, lifting efficiency of utilization of short-wave solar radiation in photosynthesis from < 1 to 5% (Section 3.2). But fertilizers require energy for their manufacture (N) or for mining and processing, and the amount of energy is substantial. For the maize system in Table 15.3, fertilizers account for 8.3 GJ ha^{-1}, which is 50% of total energy used to the farm gate. That fertilizers represent such a large proportion of total energy use in on-farm production is common to high-yield cropping systems, even though total energy use to produce fertilizers accounts for < 2% of global energy use. Fertilizers can, however, be viewed as a substitution of energy for land because without higher productivity, more land would be required to achieve equal production.

Nitrogen is the most expensive nutrient because it is required in large amounts as a constituent of biomass (1.5–3%) and because considerable energy (45 MJ kg^{-1}) is required to produce it (Table 15.4a). For this reason most emphasis on energetics of fertilizer use focuses on N. It is an ironic feature of plant evolution that N$_2$, abundant in the atmosphere, is unavailable to all except leguminous plants. Most crops must be supplied directly with N fertilizer or indirectly by rotation with legumes. Compared with energy cost of N, that required to mine and refine other macronutrients (16 MJ kg^{-1} P; 14 MJ kg^{-1} K) is small (Table 15.4a). Because N fertilizer represents such a large fraction of total energy use in most commercial cropping systems, improving energy efficiency through better N fertilizer management can reduce N losses and the associated negative environmental impact.

Leguminous rotations As the price of N fertilizer increases relative to that of products, farmers reassess their mix of legume and fertilizer N within the economics of alternative production strategies. In general, economically viable rotations will be found where markets are available for products from all crops within them. Nitrogen fertilizer is most likely to remain profitable in systems well supplied with water and other nutrients because leguminous rotations are unable to supply sufficient N for attainable yield (Section 8.8). For example, a maize crop with grain yield of 11 t ha^{-1} accumulates 246 kg N ha^{-1} in biomass at maturity (Section 17.2.5). Of this, 140 kg N ha^{-1} is removed with grain harvest. In most cases, leguminous rotations cannot supply this quantity of N to an ensuing crop, and certainly not each year.

Heichel (1978) has analyzed savings of energy achieved by growing maize within leguminous rotations including soybean and alfalfa. Yields of individual maize crops were maintained equal in all systems with supplementary N fertilizer. In rotations including legumes, energy savings relative to continuous maize (25–50%) relate mostly to less N fertilizer and those rotations with alfalfa to less tillage. Since most N content (60–70%) of leguminous seed crops is removed by harvest, soybean contributes little N to ensuing maize. In contrast, leguminous forages may contribute significantly. In this case, residual contribution of fixed N by alfalfa after three years of hay production was

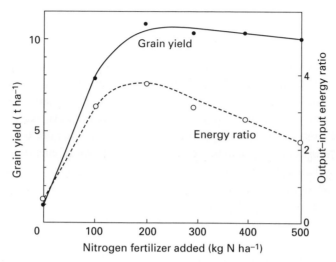

Fig. 15.2 Response of maize grain yield to fertilizer N and the overall input–output ratio for each system of production (from unpublished data of P. R. Stout, UC, Davis).

estimated to be 150 kg N ha^{-1}. But while energy efficiency was increased with legume rotations, mean annual harvested yield (grain in the case of maize and soybean, and forage for alfalfa) decreased compared to continuous maize.

So what is the effect of N fertilization on the energetics of crop production? Figure 15.2 adds the output–input energy ratio to response of maize yield to N fertilizer previously presented in Fig. 12.1. Without N fertilizer, yield is extremely low and energy ratio for the entire operation is about unity. Yield and energy ratio increase with N application up to a maximum yield of 11 t ha^{-1} achieved with around 200 kg N ha^{-1}. There is no suggestion in these data that high energy-use efficiency is achieved at suboptimal levels of N supply, a widely promulgated conclusion of an early analysis of energetics of N fertilization (Pimentel *et al.* 1973).

15.3.4   Biocides

The management of water and nutrients frequently comes to nothing in the face of competition from weeds, disease, or attack by insects. In those situations, small amounts of energy, judiciously applied as biocides, provide highly leveraged improvements in efficiency of production.

Herbicides The use of non-selective herbicides in minimum and zero tillage was introduced earlier (Section 15.3.2). In addition, selective herbicides find application in control of weeds in crops and in the management of species composition in pastures. In all cases their relatively low energy cost (including application) replaces substantial quantities of mechanical, animal, and human labor. In maize production, for example, a single well-timed spray that eliminates or reduces weed competition might easily prevent a yield loss in excess of 20% (2 t ha t^{-1} containing 34 GJ of combustible energy). Including around 1.9 GJ ha^{-1} in chemicals and a much smaller additional energy input for application in a total energy expenditure of 16.6 GJ ha^{-1} (Table 15.3) is a relatively small energy cost that provides a significant marginal economic benefit.

Efficiency of herbicide use is improved by biological studies on target weeds that identify the nature and timing of their competitive effects on crops, modes of dispersion, and the most vulnerable points in their life cycles. Such information provides the basis for IPM and increases management options, not only leading to less herbicide use but also to biological alternatives of weed control.

Pesticides The use of pesticides (fungicides, insecticides) does not represent a major component of energy use in the production of most field crops. Many crops, including cereals, are often grown without them, although others, such as sunflower and soybean in temperate latitudes and many crops of the tropics, are often protected with one or more sprays of insecticide and fungicide during each growing season (Chapter 17). The situation is different in fruit and vegetable production, and in crops such as cotton. Those crops incur more problems with insects, and concern about physical appearance and quality leads to more intensive use of pesticides.

The development of resistant cultivars reduces the need for biocides but steady deterioration of resistance means that biocides will often be needed to stabilize yield. In practice, development and management of resistant cultivars and better biocides should proceed together to ensure a clean, safe environment and to reduce the rate at which cultivar resistance breaks down. Ideal biocides are highly selective of their target species and decompose rapidly into non-toxic by-products. Research on the development of such chemicals should have high priority because of the efficiencies that biocides afford crop production. Recent development of transgenic crops with resistance to specific insect pests can greatly reduce the amount of pesticide used in crop production, although deployment of such transgenic cultivars must be combined with a broader IPM framework (Box 4.2). For example, insecticide use on US maize is very small because the majority of cultivars are Bt hybrids with resistance to a number of major insect pests. This is evident in the energy balance for maize in Iowa presented in Fig. 15.2.

15.4   Low-input farming

Farmers have been criticized (Odum 1967) for "subverting nature" and creating systems that "convert oil rather than sunlight into food." The charge is untrue; the reality is that greater energy use in food production has enormously increased the amount and efficiency of solar radiation fixed in farming. Odum's energy fundamentalism has become a tenet of proponents of some types of "alternative agriculture" who consider it is somehow wrong to use fossil energy in food growing, and others who see a return to low-input farming as the answer to "inevitable" depletion of fossil fuels and the energy shortages that would ensue.

Underlying these views is the assumption that energetic efficiency of farming will be improved by reducing energy-expensive inputs of machines and fertilizers. This assumption seems to arise from several misinterpretations: (i) high efficiency ratios reported for low-yield subsistence agriculture that ignore embodied energy cost of labor; (ii) failure to properly construct and understand energy budgets for modern agriculture; and particularly, (iii) failure to recognize sources and magnitudes of nutrient requirements for crop production. In addition, those views have ignored the relationship between population, productivity, and availability of arable land. Low input, low productivity systems are adequate for small populations with access to a large amount of arable land, which allows fallowing of sufficient duration to restore fertility. Such systems cannot, however, meet food demand for a large population. Low-input systems are not efficient convertors of sunlight, nor are they efficient in the use of scarce resources of water, nutrients, and land.

Johnson *et al.* (1977) made a detailed analysis of one type of low-input agriculture by comparing productivity and energy use of conventional farmers with those skilled practitioners of the art of organic agriculture, the Amish farmers of Pennsylvania, Illinois, and Wisconsin. In each comparison, the two groups produced the same crops and animals under identical climatic conditions within the same market constraints as the wider community. Analyses show how withdrawal of energy from conventional farms might affect productivity and energy use.

The Amish used considerably less energy in production and lifestyle than their conventional neighbors. The percentage energy use compared to conventional farms was 48% (Pennsylvania, dairy), 49% (Illinois, maize and swine) and 13% and 20% for two groups of the strictest Amish, both dairying without fertilizer, and using only stationary engines, in Wisconsin and Pennsylvania, respectively. Effect on yields varied from place to place. In Pennsylvania, yields were comparable despite 50% reduction in energy use. In Illinois, on flat land and better soils, higher energy use of conventional farmers was rewarded by a corresponding increase in yield so input–output relationships were similar. For the strictest adherents in Pennsylvania and Wisconsin, yields were less, but not in proportion to the substantial reductions in energy use.

A more striking difference between the Amish and their conventional neighbors lies outside farm energy use and productivity but is relevant to issues of energy and lifestyle. In Pennsylvania, conventional farm families used energy outside farming at 1800 MJ d^{-1}, comparable to that in the wider community, and in stark contrast to Amish families, who consumed a mere 175 MJ d^{-1}. The Amish achieve that by their preference for horse transport, their rejection of networked electricity, and their limited use of fossil fuel (bottled gas) for cooking and lighting. It is their frugal lifestyle, not agricultural productivity, that enables them to purchase more land so that their children may achieve what is to them the most fulfilling of occupations: farming.

Some of those who propose that the assured future of agriculture is in small, labor-intensive, low-input farms have attempted to farm in that way. Not surprisingly, an increasing number of these "small farmers" generally compete poorly with large-scale mechanized agriculture. The many social issues involved are beyond the scope of this book but the comparative energetics of small and large farms are not. Quite apart from demonstrable economies of scale in farming, support of small farms in terms of personal transport, roads, homesteads, and power networks greatly increase energy cost per unit farm product. On energetic grounds, it would be preferable to have fewer large farms for production of food or fuel. In the case of small farms, it might be preferable to concentrate farm families in villages where services can be supplied more efficiently. This is how the manorial system operated in the Middle Ages and how small-scale farming is still practiced in many parts of Europe and Asia.

In practice, possibilities for low-input farming are greatly improved when, as now, it is carried out on a limited scale within a region that is dominated by conventional farmers. Under these conditions, low-input farmers are, like the Amish, able to buy supplementary feed from conventional farms for their animals and accumulate its nutrient content on their farms. For others, use of pesticides and herbicides by conventional neighbors reduces invasion by weeds and pests. Organic farmers often rely for financial success on premium prices that some consumers will pay for their products. This is a subsidy that society pays for their form of stewardship of the land. Of course that premium price for organic products is only possible when conventional farmers produce a cheaper product. Ironically, the biological subsidy, the transfer of nutrients in feed and manure from conventional to "organic" farms or fields, does not infringe legal requirements that secure this premium. Scientists argue that nutrients are the same from any source, but organic regulations require one cycle through plants to cleanse some "artificial" plant nutrients.

The major problem facing humanity is the provision of adequate food for all. To achieve this while sustaining land and water resources to secure future agricultural potential will require care of soils and landscapes. The better solution is more likely to involve more intensive management of crops for high yields on good quality soils and protection of more marginal soils in extensive agriculture, forestry, and conservation zones. In both cases, crop yields will not meet world food requirements without adequate nutrients. Fertilizers enable greater production from each unit of land and the opportunity to protect remaining natural ecosystems from conversion to agriculture.

15.5   Crops for energy

Agriculture and forestry now supply 10% of world energy (Fig. 15.1), largely by direct combustion for heat and electricity generation, but society is expecting a bigger contribution in future. A convergence of issues has turned attention to agriculture as a future significant source of liquid biofuel to power internal combustion engines, currently essential for transport. Liquid forms of fuel are important because they have high energy density (MJ L^{-1}) and, unlike natural gas or hydrogen (H_2), can be stored in unpressurized tanks.

Biofuel for internal combustion engines is not a recent innovation. The diesel engine was designed to run on vegetable (peanut) oil and the first automobiles in the USA ran on bioethanol. More recently, biofuel production expanded significantly during the petroleum price shock of the 1970s. Now, additional concerns over the inevitable exhaustion of petroleum, its contribution to climate change by release of CO_2 upon combustion (Section 6.10.4), and questions of national energy security have led many governments to mandate a required usage of biofuel and provide subsidies for production and use.

15.5.1   Biofuel types

Currently, two biofuels, biodiesel and bioethanol, from a small range of crops, provide essentially all renewable liquid transport fuels. Other liquid fuels, such as synthetic gasoline and diesel, play minor roles although their use might increase if cost-effective thermo-chemical conversion technologies from cellulose are developed.

Biodiesel (*c.* 33 MJ L^{-1}) is formed chemically by trans-esterification of vegetable oils obtained by physical and/or chemical separation from oilseed crops. The process reduces long branched molecules, less suitable as fuel, to short straight-chained fatty acid methyl esters of lower viscosity and higher cetane number (more easily combustible). Trans-esterification uses methanol (or ethanol) and also produces a useful co-product, glycerine.

Bioethanol (21 MJ L^{-1}) is produced by the biological fermentation of glucose and fructose that are easily obtained from sucrose crops, such as sugarcane or sugarbeet, less readily by milling and hydrolysis of starches [(C12)n] from a wider range of crops, including grains of many cereals and tuber crops such as potato and cassava. Fermentation is followed by several energy-demanding steps of distillation and dehydration to produce fuel-grade alcohol. Some of that energy can be supplied by burning biomass residues or by-products, as commonly done in sugarcane refineries. Fermentation produces organic co-products that find use as animal feed.

Bioethanol can also be made from cellulose, also [(C12)n] but with different chemical bonding than starch, by two routes (Badger 2002). The first produces ethanol by fermentation as above, following depolymerization of cellulose by various physical, chemical, and enzymatic treatments. In principle any plant material can be used, but unlike sugar or starch, cellulosic material is variable in chemical content, especially in woody plants that contain large quantities of lignin compounds. This variation complicates commercial production. The second is a set of processes that convert biomass to liquid fuels, directly as bio-oil via pyrolysis, or indirectly in various forms, including methanol, ethanol, and dimethyl ether, from **synthesis gas** (H_2, CO, and CO_2) following gasification (Brown & Wright 2009).

15.5.2   Current crops and productivity

Maize (USA), sugarcane (Brazil), and cereals and sugarbeet (EU), provide the bulk of feedstock for bioethanol production at 1.9 EJ y^{-1} (91 GL) in 2008. Other crops, barley, cassava, potato, and rice are used in various countries. Main crops for biodiesel 0.4 EJ (12.5 GL) are oil palm (Malaysia and Indonesia), soybean (USA and Brazil), rapeseed/canola (EU), and sunflower (Eastern Europe). Peanut, cotton, sesame, and coconut are also used. The production level, but not the range of crops, is changing rapidly. Most countries, those of the EU are an exception, use locally produced crops.

Table 15.5 Average energy yield of the best two producers of ethanol and biodiesel in various countries (adapted from Liska & Cassman 2008)

Crop	Country	Yield (t ha^{-1})	Product	Biofuel (L ha^{-1})	Energy[e] (GJ ha^{-1})
Oil palm[a]	Malaysia	20.6	biodiesel	4736	155.8
	Indonesia	17.8		4092	134.6
Sugarcane[b]	Brazil	73.5	bioethanol	5475	115.5
	India	60.7		4522	95.4
Sugarbeet[c]	France	80.0	bioethanol	6080	128.3
	USA	60.0		4560	96.2
Maize[d]	USA	9.4	bioethanol	3751	79.1
	China	5.0		1995	42.1
Cassava[c]	Brazil	13.6	bioethanol	1863	39.3
	Nigeria	10.8		1480	31.2
Rapeseed[d]	China	1.7	biodiesel	726	23.9
	Canada	1.5		641	21.1
Soybean[d]	USA	2.7	biodiesel	552	18.2
	Brazil	2.4		491	16.1

Notes: [a] fruits harvested fresh; [b] aboveground material harvested fresh; [c] roots harvested fresh; [d] grain harvested at low water content; [e] energy contents are bioethanol 21 MJ L^{-1}, biodiesel 33 MJ L^{-1}.

Productivity of various biofuel crops, presented in Table 15.5, illustrates large differences between crops, biofuel types, and environment. Oil palm and sugarcane perform best because of their year-round growing environments. Maize and sugarbeet also perform well in warm and cool climates, respectively. Except for perennial oil palm, the annual oil crops, soybean, sunflower, and rapeseed, produce small yields.

Data in Table 15.5 also reveal important characteristics shared by two productive crops; C4 sugarcane for warm and C3 sugarbeet for cool growing conditions. Both are long season crops, both have vegetative storage organs (sugarcane the stem and sugarbeet the primary root), and both store sucrose. These characteristics have physiological implications for yield. First, vegetative organs are able to accept assimilate for storage over longer periods than grain crops that depend on flowering and successful fruit set to provide storage capacity. Vegetative storage reduces the opportunity for "feed-back" restriction to yield accumulation when grain set is reduced by environmental stress. Second, sucrose is the least transformed storage product of photosynthesis and therefore subject to smallest losses by subsequent metabolism (Chapter 11) or conversion to biofuel. A disadvantage of sugarbeet in the USA, however, is that the crop requires irrigation to achieve high yields.

The use of residues from existing crops such as maize or wheat to produce biofuel has been constrained by development of cost-effective technologies for harvest, transport, storage, and conversion via fermentation or thermo-chemical processes. Conversion technologies are challenged by the recalcitrant nature of cellulose and hemi-cellulose

Table 15.6 Ranges of net energy ratio (NER) reported for various biofuels (FAO 2008)

Crop	Biofuel	NER[a]
Sugarbeet	bioethanol	1.2–2.2
Sugarcane		2.2–8.4
Wheat		1.2–4.2
Maize		1.2–1.8
Oil palm	biodiesel	8.6–9.6
Soybean		1.4–3.4
Rapeseed		1.2–3.6

Note: [a] biofuel energy/support energy (production plus processing).

that comprise a large portion of these materials. Another issue is how much residue can be removed without decreasing soil organic carbon (SOC) that is high in cellulose and hemi-cellulose. For maize systems at current yield levels, the amount that can be removed without negative impact on soil is relatively small (Wilhelm *et al.* 2007).

15.5.3   Energy efficiency

Net energy gain in biofuel production is the energy content of the product less that used in two stages of its production. First, is support energy to the farm gate discussed earlier in Section 15.2; second, is energy cost of subsequent processing, including transport to the conversion facility. The second stage can include a credit for energy value of co-products. A comparison of efficiencies of various biofuel production systems is presented in Table 15.6 as **net energy ratio (NER)** – the ratio between biofuel energy and total energy used in production.

Two points stand out. First, NERs are small except for sugarbeet and sugarcane. Second, NER is equally variable among crops that produce bioethanol or biodiesel. The explanation for high values for sugarcane and oil palm is found in high productivity in year-round tropical environments, and low support energy in processing. Palm oil benefits from small energy cost of expressing oil from seed. Sugarcane fermentation benefits not only from small pre-treatment required to enter fermentation but also from energy obtained by burning residual biomass. Finally, comparison with NER = 10.9 of maize production to the farm gate (Table 15.3) reveals that more energy is used to process maize grain to bioethanol than to grow it.

15.5.4   Greenhouse gas balance

Greenhouse gases are emitted during crop production to produce feedstock and subsequently during conversion of feedstock to biofuel. At issue is whether emissions saved when biofuel replaces petroleum-derived motor fuel (i.e., gasoline or diesel) are greater than the total emissions from the biofuel life cycle. An important reference here is the "carbon intensity" of gasoline, i.e., the C equivalent of its discovery, extraction, refining, and combustion. The value is currently about 96 g CO_2 equivalents L^{-1} (Liska & Perrin 2009) but it is steadily rising as petroleum supply relies more heavily on new sources of oil, such as those extracted from tar sands or deep-water petroleum reserves. In contrast, the carbon intensity of biofuels is decreasing as farm yields increase, farmers adopt improved management practices to increase N and water use efficiency, and as biorefineries are engineered to be more efficient in conversion of crop feedstock to biofuel.

Recent evaluation of direct emissions from maize ethanol estimates a carbon intensity of about 54 g CO_2e L^{-1}, almost half that of gasoline (Liska *et al.* 2009). This value includes an energy credit for maize replaced by use of the co-product, "distillers grain", that is valued in cattle rations for its content of protein, lipids, and fiber. It comprises about 30% of initial grain mass. On the basis of total use, the credit is equivalent to 19 g CO_2e L^{-1}, which offsets about 27% of total emissions from the maize ethanol life cycle.

Sugarcane ethanol in Brazil has a smaller carbon intensity than maize ethanol for the same reasons that it has a larger NER. Recent studies estimate a carbon intensity of about 16 g CO_2e L^{-1} for sugarcane ethanol in Brazil (Macedo & Seabra 2008), which is well below that of maize ethanol.

The values of carbon intensity of maize and sugarcane bioethanol production assume a steady state for soil organic carbon (SOC). In general, direct emissions from biofuel production life cycles must also include changes in SOC content. If SOC increases under a cropping system that produces biofuel feedstock, an emissions credit is appropriate for the sequestered C. If SOC decreases, then a C cost should be added. A special case occurs where new land is cleared for crop production because diversion of food crops for biofuel production reduces supply and raises crop prices, called indirect land use change. In this case, the carbon debt is large, especially when the converted land is carbon-rich rainforest or wetlands (Section 17.4). In the case of ethanol produced from US maize, it has been estimated that more than 100 y of biofuel production would be required to overcome the indirect land use change penalty and achieve a positive GHG balance (Searchinger *et al.* 2008). The world and its energy requirements will be very different by then!

15.5.5 New options for biofuel crops

Concern to avoid criticism from diversion of food crops to biofuel has pushed dedicated energy crops towards the center of debate for future options. Many crops are proposed and four types, non-edible oil plants, short rotation trees, perennial grasses, and algae, are distinguishable (Table 15.7). Justifications advanced for developing **dedicated energy crops** that include less intensive production requirements, suitability to lower quality land, and intrinsically greater efficiency of crops developed for energy rather than food do not, however, withstand scrutiny.

The proposition that energy crops can be grown with less intensive production methods on land unsuitable for food crops is largely untrue. Short-rotation trees and grass crops may extract less nutrient per unit of biomass than food crops but that will be largely offset

Table 15.7 Some crops favored for investigation as dedicated energy crops[a]

Cellulose crops		Non-edible oil crops	
Short rotation trees and shrubs	Eucalypts (various) Poplar Willow Birch	Field crops	Castor oil Physic nut Oil radish Pongamia
Perennial grasses	Giant reed Reed canary grass Switch grass Elephant grass Johnson grass Sweet sorghum	Trees and shrubs	Souari nut Buruti palm Grugri palm Neem Various native (Brazilian spp.)
Aquatic plants			Various algae

Note: [a] data from various sources.

by harvesting a greater proportion of total biomass. Under all circumstances, efficient and continuing production will require substantial inputs of fertilizer, and irrigation if available, to justify effort and investment in manufacturing plants and transportation infrastructure for large-scale commercialization. From an agronomic perspective, the use of perennials to protect soils in areas that are not currently arable is commendable, but the definition of cropping land has changed in recent years with the introduction of no-till production methods.

Of a range of non-edible oil crops, *Jatropha* (Euphorbiaceae), a perennial frost-sensitive shrub, native to North East Brazil, is attracting considerable attention as a non-food biodiesel crop in tropical and subtropical zones. It also embodies the challenges that must be overcome to develop a new biofuel crop. The seed contains 30 to 35% of easily extracted oil that is suitable as a fuel even before trans-esterification to biodiesel. The plant is unpalatable to livestock and its seed, oil, and seed meal are toxic to humans. Its main use to date has been as hedgerow plants in dry areas and in other places for soil conservation and local production of soap and medicines. Large yields have been claimed, including in areas of low rainfall and on "wasteland" soils (Francis *et al.* 2005), but plans of major projects are more modest (WWF 2008). Yields for established (8 y-old) plantations on good soils in favorable rainfall environments (> 600 mm y^{-1}), some with supplemental irrigation, are expected to range between 5 and 10 t seed ha^{-1}, i.e., 1.8 to 3.5 t oil ha^{-1}. Available physiological information (Jongschaap *et al.* 2007) reveals no special productive capacity or water-use efficiency advantage compared to established oilseed crops like palm oil, and there is an absence of solid data on which to evaluate crop productivity and costs of production on marginal soils to which proponents claim the crop is adapted.

The current expansion rate of *Jatropha* is substantial (WWF 2008). Planted area is currently 0.9 Mha in Asia, Africa, and Latin America and is set to increase to 5 Mha by 2010. A further increase to 13 Mha is expected by 2015. Most activity is in Asia (65%) and in all regions this expansion is dominated by large plantations >1000 ha. Just 26%

Box 15.2 Biotechnology for more productive biofuel crops

Biotechnology is promising much and attracting the attention of policy makers and private investment (e.g., *Nature* 2006; DOE 2006; Economist 2008). Readers of such literature should ignore the phrases "smart breeding", "novel crops", "innovative cropping systems", "high energy grasses" and look for defensible quantitative results. They should also be aware that the same promises for major productivity gains in food crops have remained unfulfilled for decades. Valuable contributions from biotechnology have been in manipulations of end products, such as oil and starch chemistry, and in disease and herbicide resistance controlled by one or few genes. There has been no contribution to a greater efficiency of photosynthesis. Evolution has been working on that for millennia. Gains in crop yield have arisen mainly from changing partitioning patterns: less stem and more grain. Biotechnology will likely have success in biological transformation steps from biomass to biofuel but it is the sustainable production of biomass, controlled by many genes, that will restrict dedicated biofuel crops to a small and contentious role in world energy production.

Finally, it is important to stress that crop adaptation and breeding is a prolonged process as exemplified by successes with relatively few food crops that feed our world, and the continuing effort required to maintain them. Genetic potential for productivity must be matched with understanding of water and nutrient requirements, of rapidly evolving pests and diseases, and the development of production systems appropriate for mechanization. Special "bioengineered" energy crops are unlikely to appear within the next few decades and more slowly unless clear specifications for ideal energy crops emerge. Concentration on few crops without rejection of existing food crops would increase the chance of success.

---

of area is in plantations less than 5 ha. This rapid expansion will soon test the potential of *Jatropha* in many environments.

Biotechnology has plans to develop specialty crops for energy that are intrinsically more productive and more suited to transformation to biofuel than food crops (Box 15.2). The USDA (DOE 2008) favors switchgrass as a dominant component of their biomass mix and have identified a range of lines suited to regions within continental USA. Reported yields are reasonable but not large, 7 to 16 t ha^{-1} y^{-1} in the southeast, 5 to 6 t ha^{-1} y^{-1} in the western Corn Belt and 1 to 4 t ha^{-1} y^{-1} in North Dakota, but data on nutrient and other management inputs required for sustainable production are insufficient. European Union countries, in contrast, place reliance on a range of bio-engineered dedicated energy crops with predicted yield gains of 2.5% pa for the decade from 2020 (EEA 2007a). The hopes for high productivity of dedicated energy crops sit uncomfortably with a requirement for 30% low yield, "ecologically oriented agriculture" in the accompanying strategy for food crops.

A number of reports claim high yields of oil from algae, e.g., 12 000 L ha^{-1} y^{-1} biodiesel (e.g., *Nature* 2006), which is twice that achieved consistently by oil palm.

The production system is distinct but so is the infrastructure required and there was no information on support energy or overall NER.

Clearly, design for optimum biomass energy crops has yet to be formulated and their contribution yet to be established. We support a focus of effort on perennial grasses but without emphasis on cellulose production alone (Connor & Hernández 2009). That, we interpret as a misdirected result of avoidance of crops also suitable for food. Perennial grasses will likely best succeed as energy crops when the emphasis moves from cellulose production to that plus sugar content. If such analogs of sugarcane can be successful for cool environments, they would be potential food crops also.

15.5.6   Potential for expansion

Options (non-exclusive) to increase biofuel production can be found in expansion of crop area, greater crop yields including dedicated energy crops, use of crop residues as well as the sugar, starch, or lipid yield components, and more efficient transformation methods. Individual countries will find different solutions but at a global level, the requirement to increase global food production to feed an increasing world population (Chapter 1) will place a serious limit on land available for energy crops. Biofuel crops will always compete with food crops for land, nutrients, water, labor, and investment. That explains existing concern with the moral, food security, agronomic, and ecological issues associated with biofuel production (Thompson 2008). It seems to us that the appropriate approach to biofuel production is to focus on residues of agriculture and forestry, so long as it does not lead to a reduction in soil quality. The limited future of biofuel could well be found in highly productive systems able to feed the world, and divert some residues and surplus grain to integrated animal and biofuel production systems. Such systems produce crop residues in excess of that required to maintain SOC, and in some cases farmers currently burn this residue because it interferes with tillage and seeding operations. In the meantime, current biofuel production from food crops such as maize will help raise commodity prices and the interest of scientists and policy makers in agriculture in general and food security in particular. Such interest is critical to maintain investment in research and development of improved technologies to drive future growth of food production capacity.

The commercial reality of a residue option depends upon the economic viability of cellulosic ethanol production that is yet to be established. Economically successful transformation is the first challenge, second is sustainable supply in sufficient quantity without deleterious effects to environment or elsewhere. These are important questions that require analyses by agronomists, foresters, and other ecologists.

Without detailed analyses, it is difficult to estimate what contribution residues might make to biofuel production in terms of availability and competition from other energy extraction chains. Worldwide annual production of crop residues in the range 3 to 4 Gt, of which perhaps 30% is burnt annually (Smil 1999), is a considerable energy resource. Production of 3 Gt corresponds to 52 EJ y^{-1} (at 17 MJ kg^{-1}). A future scenario of crop production might increase that by 50%. If 30% of 4.5 Gt residues were available for removal that would correspond to 12.8 EJ (608 GL as ethanol at 400 L t^{-1}).

This is seven times the current world ethanol production of 91 GL y^{-1} from grain and sugar crops.

Some national data are also available. The USA has a biomass database on a county basis and has estimated that agriculture and forestry could supply 786 Mt residues on a sustainable basis (crop 428 Mt and forestry 358 Mt) (Perlack *et al.* 2005). An associated analysis estimates that maize, the most widespread crop, could contribute 100 Mt residues, around half (55%) of those produced, without effects that could not be remedied by fertilizer, provided no-till production methods were applied (Graham *et al.* 2007). The US biomass analysis also includes surplus grain (87 Mt), animal wastes (106 Mt), and a proposal for 377 Mt from dedicated energy crops. The estimate of 1.3 Gt y^{-1} biomass, that could be harvested sustainably, is sufficient, at 100 gal ethanol t^{-1}, to replace 30% of national liquid fuel requirement. This is, however, not a net gain because it does not account for the support energy in collection, transport, or processing. The plans for Europe (EEA 2007a) concentrate on reorganization of land use to make 20 Mha available for dedicated energy crops by 2030. That area, together with concentrated effort to improve productivity could, it is proposed, produce 350 Mt biomass that, together with 148 Mt from forestry (EEA 2007b), is smaller than the "billion ton vision" for the USA (Perlack *et al.* 2005). Without the sort of rapid gains promised by biotechnology, Europe has little scope to divert cropland to biofuel. It has been estimated that with current technology Europe would have to divert 70% of cropland to biofuel to contribute the mandated 10% of transport fuel by 2020.

Two centuries ago the world ran on biomass but residues can now contribute only a small part to our now energy-demanding world. Technologies for conversion are yet to be evaluated and economics of collection of residues and distribution of products are unknown, as are the long-term consequences of residue removal. The biological world would be a very different place if all "spare" biomass were returned directly to the atmosphere through exhausts of internal combustion engines and not left as food for a web of consumers and decomposers.

15.6   Review of key concepts

Energy demand and supply

- World annual energy use is now around 450 EJ and could increase to 1000 EJ by 2050 as population and average standard of living increase. Fossil fuel (non-renewable) energy provides 80% of current energy, while biomass at 10%, mostly in developing countries, provides more than nuclear and hydro combined. There is still much fossil fuel to be extracted but concerns over global warming seek to slow its use. The goal is to control atmospheric CO_2 by encouraging more efficient use of energy and hastening conversion to renewable sources of energy from wind, water, solar, and biomass.
- Given the enormity of solar flux and nuclear energy of matter, there is no shortage of energy but there is uncertainty of how, when, and at what cost and environmental impact it might be made available. Technologies to capture energy from sun, wind,

and tides are developing rapidly but compete poorly with the cost and energy density of fossil fuels. There has been no breakthrough in nuclear fusion research and public concern over the safe storage of radioactive wastes from nuclear fission restricts its deployment.

Energy use in agriculture

- Support energy in the form of motive power, fertilizers, improved cultivars, and agri-chemicals, together with knowledge, amplify and stabilize productivity of agricultural systems. In this way, productivity per hectare and per worker increases so that a small proportion of population working in agriculture can support the majority engaged in alternative pursuits.
- Agricultural systems account for a relatively small proportion of total energy use. Developed countries devote *c.* 10% of total energy to food production but only one fifth of that is used in production to the farm gate.
- In developed countries, there are more opportunities to increase energy-use efficiency outside than within agriculture. Greatly improved efficiency to the farm gate has already resulted from widespread adoption of reduced tillage and improved N fertilizer efficiency. Reducing dependence on N fertilizer through wider use of legumes would require expansion of crop land area to allow more diverse rotations with fewer grain crops.
- The future depends upon the availability of energy. If energy is available for a techno-logical society, agriculture could become concentrated on robust lands where all inputs are used most efficiently to preserve more fragile lands for their natural habitat and biodiversity. If energy is scarce, land will be less productive and it will be difficult to protect fragile environments from conversion to agriculture. In either case, the major factor limiting population and quality of life will be the capacity to find solutions to inevitable conflicts between food production and conservation of natural resources.

Agriculture as a supplier of energy

- There are startling conflicts in concern for future energy supplies. Some believe the end of the fossil fuel era signals a return to the past with a greater proportion of humanity in energy-frugal agrarian societies. Others look to agriculture to supply combustible fuels as well as food, fiber, and other organic products for an intensely technological society. In the middle are those concerned for the fate of natural ecosystems that are vulnerable to conversion to agriculture, and yet others, ourselves included, concerned to establish a rational view of opportunities to balance productivity of agricultural lands with preservation of natural systems.
- A small number of food crop species are now used to produce biofuel from sugar, starch, and oil. Such use will remain problematic as the world struggles to increase food production to better feed an increasing population that currently includes *c.* 1 billion who are severely underfed. Dedicated, non-food energy crops are not an effective way to avoid competition with food production because they too require

land, water, nutrients, and other inputs. There is no evidence that non-food crops can be grown efficiently for energy production on land that could not also grow crops for food.

- Residues from agriculture and forestry are important potential sources of energy. However, processes to transform this cellulosic material efficiently into liquid fuel are not yet established. Likewise, amounts of residues that could be sustainably utilized without degrading soils and/or ecosystem are unknown in most cases. Resolving issues of sustainable supply of residues is of equal importance to research and development of commercial transformation pathways.
- Greater production of food crops will require significant productivity gains because limited land is available for expansion of agriculture. Concentration of research and development on food production increases the chance to feed the world and provide residues for biofuel production.
- Evidence suggests biofuels can make a modest ($< 10\%$) contribution to national transportation fuel supply in countries with large land resources relative to population size. Few countries will likely be significant exporters of biofuels. Biofuels cannot be a major source of transportation fuel in a highly populated and energy demanding world.

Terms to remember: biofuels, biodiesel, bioethanol, cellulosic bioethanol, and synthesis gas; dedicated energy crops; embodied energy; energy, renewable, non-renewable, support; fossil fuels; life cycle assessment; net energy ratio (NER).

Part V

Farming past, present, and future

In the first edition of this book, published in 1992, we presented two case studies of important farming systems, wheat–sheep farming in southern Australia and maize–beef production in central USA, discussing how they operated based on current knowledge, technology, and economic conditions. The purpose was to integrate the knowledge and principles presented in earlier chapters and show how success in farming depends upon this integration. Economic survival required a sustainable system, which could be analyzed in terms of balances of water, nutrients, and capital. In addition, farmers had to be nimble in adapting to variable climatic and economic environments. We also hoped that those analyses would serve to encourage readers to make detailed analyses of other farming systems. Those examples remain available to readers on the internet and can continue to meet those goals. However, and unsurprisingly, these analyses are substantially out of date with regard to current cropping systems in those regions. Knowledge and technology have advanced rapidly, global population and demand for agricultural products have increased substantially, and economic conditions have changed dramatically.

In this edition we wish to emphasize that dynamic nature of agricultural and farming systems and stress how developments in technology have and will continue to provide farmers with the capacity to remain economically viable over the short term, and sustainable over the longer term. For this we present two new chapters. Chapter 16 examines the development of wheat production systems in southern Australia since inception of the industry in 1850. Chapter 17 is a shorter term analysis, since the 1950s, of technological change and productivity in two of the world's most important cropping systems, and a third that is rapidly matching their importance also. These systems, maize–soybean in the USA, rice-based systems of Asia, and soybean production in Brazil, now produce 40% of the world's food.

The new chapters describe how farming systems operate in relation to endowments of soil and climate, their strengths and weaknesses, and the underpinning science and technology critical to their survival. Because all farming systems will continue to evolve as world food demand continues to increase, until at least 2050, and as new technologies are developed and economic and environmental conditions continue to change, the dynamic adaptation of cropping systems is critical for global food security and world stability in coming decades.

Many in society express disillusionment with technology because of concerns about the environment; some attack the premises and methods of modern agriculture. Whether such concerns are justified, whether proposed alternatives will be sufficient for the future, and issues that require particular attention, are addressed. We conclude Part V and the book (Chapter 18) with a look to the future by considering how agriculture may respond to intensifying challenges of productivity gains and natural resource conservation in a world where human numbers are still increasing, and average income must rise to encourage smaller population growth and a stable population within the coming decades.

16 Evolution of wheat production systems in southern Australia

Wheat has been grown in Australia since European settlement, initially to feed colonists, but soon as an export crop. Although total national production, 19 Mt (five-year average 2003–2008), remains small by world standards, the high proportion (60%) that is exported ranks Australia fourth, after the USA, Canada, and the EU, among wheat exporting countries.

This chapter describes the continuing evolution of wheat-cropping systems in semi-arid southern Australia (annual rainfall 300–500 mm) using yield data for the State of Victoria from soon after inception of the industry c. 1800. The analysis reveals how a sequence of cropping systems has developed in response to **technological innovation**, **economic incentives**, and **societal pressures**. Economic pressure to compete on world markets has been, and will likely remain, a major driver of change in these cropping systems. Producers receive little subsidy to relieve competitive pressure. Among OECD countries, subsidies account for 25 and 40% of farm income in the USA and the EU, respectively, but only 6% in Australia. Driving forces for change may be further complicated by widely anticipated climate change.

Producers, agronomists, and researchers now have access to new tools to meet the increasingly complex objectives that must account for variability in climatic and economic environments, and also address societal interests. The principles and range of strategies and tactics available to combat crop response to low and variable rainfall have been presented in Chapters 9 and 13.

16.1   The wheat belt of Northwest Victoria

The location of wheat production in Victoria, Australia, is presented in Fig. 16.1a together with mean annual rainfall isohyets. Seasonal distribution of rainfall is presented for one station, Horsham, during ten successive years in Fig. 16.1b according to average phenological development (Chapter 5) of wheat crops in the zone. The summer period, January to May, is without crop, June to September is vegetative growth (sowing–anthesis), October to December is grain filling (anthesis–maturity). These data reveal characteristic variability in annual and seasonal rainfall. In this ten-year period, annual rainfall varied from less than 200 to almost 600 mm, while rainfall during grain filling ranged from 20 to 150 mm. The long-term mean for the site is 410 mm, while during the 30 y to 2008 it was 394 mm. Reference crop evapotranspiration (ET_0) exceeds rainfall

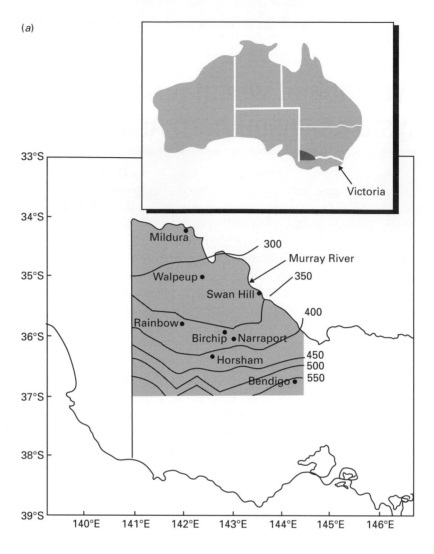

(a)

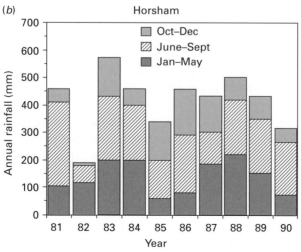

(b)

Fig. 16.1 (*a*) Wheat production zone in Northwest Victoria, Australia: location and annual rainfall isohyets (mm). (*b*) Seasonal distribution of rainfall during ten consecutive years at a representative site, Horsham.

across the entire zone, increasing, contrary to the trend in rainfall, from 1400 mm in the south to 1800 mm in the north.

The landscape is flat to undulating with a gradual decrease in altitude from 150 m in the south (Horsham) to 100 m at Mildura on the Murray River. Soils in the south are clayey, having developed on lacustrine sediments while in the north, light-textured soils have developed on overlying wind blown sands. The original vegetation types in the south were grasslands and savannah woodlands dominated by species of *Eucalyptus* and *Casuarina*. In the north, typical vegetation was shrublands of various eucalyptus species, comprising small trees with several slender stems to 2 m height arising from large basal lignotubers, called "mallee" by the aboriginal inhabitants.

16.2   Evolving systems

Rainfed cropping systems, despite their limitations in comparison with others where irrigation is available, provide many opportunities for management to increase productivity. The key challenge is to adopt strategies that make optimal use of the water available to each crop by concentrating it in space and time and by minimizing losses to runoff, drainage, and direct evaporation from soil (Section 13.2). These "unproductive" losses of water are substantial so that opportunities exist to improve the productive use of water. Sadras (2003) provides model-based estimates of the fate of annual rainfall in the wheat zone of southeast Australia. Runoff accounts for 1 to 6%, drainage 0 to 18%, soil evaporation 37 to 81%, while transpiration is always less than 53% (range 18–52%). Strategies to reduce the unproductive losses will also seek to balance nutrient and water supplies during the growth cycles of individual crops by sowing cultivars of suitable phenological development at times and densities that best allow expression of their intrinsic water-use efficiencies and drought resistance mechanisms.

The design and management of these cropping systems must importantly take account of variability in climatic and economic environments. In semi-arid zones, rainfed cropping usually extends towards a desert margin. Especially there, at the dry edge, the challenge for sustainable productivity is accentuated by extremely variable rainfall such that crops may express attainable yield in some years and yet fail in others. It is the relative frequency of these extreme years that determines the location of cropping boundaries as well as the optimum crop sequence strategy. In southern Australia, this boundary is long, so that gains in water productivity, from farming practice and genetic improvement, provide large gains in both crop area and production.

Strategies are made flexible by identifying tactical variations that seek to improve overall performance by responding to observed or anticipated changes in climatic or economic environments. These tactics include adjustments both to management of individual crops and to the cropping sequence itself. In general, best results are obtained when tactical variations are directed to maximize yield in seasons of high rainfall and to minimize costs in seasons of low rainfall (Connor & Loomis 1991).

An analysis of development of the wheat industry in Victoria is presented in Fig. 16.2. It displays mean crop yield for each year from 1840 to 2008. For convenience, several

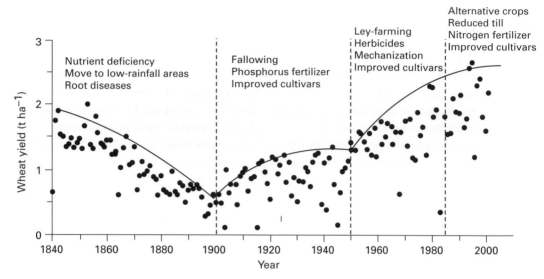

Fig. 16.2 Cropping strategies and average yield during development of the wheat industry in the State of Victoria, Australia (extended from Elliott 1987).

phases are identified, although actual transitions were never so sharp. In this case, lines passing through years of greatest yield display the long-term yield frontier. These years are interpreted to be the most favorable in this highly variable rainfall environment. This form of presentation offers two advantages to the present discussion. First, it shows how cultivar and management have progressed to raise yield, and thereby increase water-use efficiency, through almost 200 y. Second, it highlights the low yield stability that has confronted farmers during each stage of development, and continues to do so.

16.3   Initial development (1840 to 1900)

Crops were first grown near new settlements along the coast but soon land more suitable for cropping was chosen, and this was increasingly inland in areas of lower rainfall. There were constraints, such as effort and cost to clear land and access to transport. Development was most rapid where easy-to-clear woodland trees and low scrubby vegetation were found, in this case aided by an extending network of railways. By 1900, 0.8 Mha had been sown to wheat in Victoria.

In the mallee areas, land clearing and crop establishment displayed much innovation and invention. Initially, parallel teams of horses dragged large logs to knock down the scrubby vegetation, which when dry was burnt to remove aboveground debris and control regrowth. Important local inventions were a stump-jump plow and a stripper harvester. The first could be used without the need to remove lignotubers so that crops could be sown immediately upon clearing. The second allowed the mechanical harvest of grain while leaving straw. Burning the residue after harvest continued the control of regenerating eucalypts. Lignotubers were loosened at each plowing and could be sold as firewood to supplement income. With these two inventions, clearing land and cropping developed in unison.

During this time, wheat was mostly grown in continuous culture (WW) and initial yields, around $1.7\,t\,ha^{-1}$, declined steadily to $0.3\,t\,ha^{-1}$ by 1900. As the yield frontier fell so did year-to-year variability of yield (greater yield stability!) because factors other than rainfall variability came to dominate productivity. Loss of soil fertility due to **nutrient extraction** without replacement is likely the major cause of yield decline, but there were others. First, as wheat production expanded, it moved progressively, as explained above, to drier inland areas of smaller attainable yield, and second, it is almost certain there was a build-up of **soil-borne diseases** under continuous wheat cropping. The crisis of uneconomic yields, the economic depression in the 1890s, and a drought during 1897 to 1903 forced many farmers from their farms and encouraged the adoption of new cropping practices.

16.4   An early recovery (1900 to 1950)

Work at Rothamsted, UK, had recently established that application of P to crops and pastures increased yield. News of this technology was carried to Australia and, together with widespread introduction of a fallow year between crops, paved the way for recovery in productivity of the wheat industry. Fallowing had long been practiced in the UK to rest land by simply leaving it out of production. In Australia, however, fallows were kept more or less free of plant growth by a combination of tillage and grazing by sheep, a practice that became feasible once eucalypt regrowth was controlled. Especially in low rainfall areas, a long, tilled, up-to-18-month fallow period was found to make economic yield of wheat more certain (see Fig. 13.6). This fallow–wheat (FW) system came to dominate production, and re-establish a viable wheat industry. To explain success, much emphasis was placed on the benefits of conservation of water by fallow, *albeit* with low efficiency (Section 13.8.3), from the previous spring to the current winter–spring growing period.

There was plenty of land to meet the required doubling of cropped area and during this period area sown to wheat fluctuated widely from 1 Mha to a record of almost 2 Mha in 1929. There was a peak during World War I to provide additional grain for Britain, but a sharp decrease during World War II was a result of effective blockading of shipping by the German Navy. Yields recovered, however, but yield stability decreased as variable rainfall again began to exert its effect. Analysis of rainfall patterns at Horsham reveals that during 1937 to 1945, a majority of years were below long-term average, and this is reflected strongly in the frequent low yields recorded in Fig. 16.2. During this time there was improving understanding of soil science and measurement of soil physical properties and nutrient levels. Improved cultivars became available and timelier sowing was achieved through mechanization. Information and assistance became available to farmers. First, agricultural colleges, then university faculties, and then field stations of

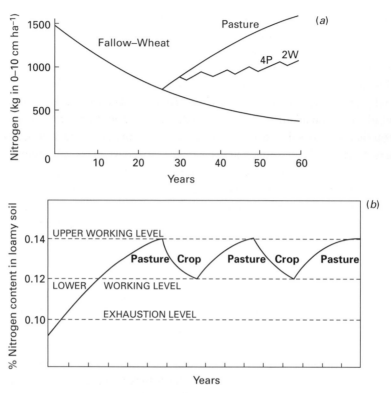

Fig. 16.3 (*a*) Nitrogen fertility trends in fallow–wheat, continuous legume pasture, and a pasture–wheat sequence in southern Australia. (*b*) Recovery of nitrogen fertility by pasture after continuous wheat and its maintenance with pasture–wheat sequences.

departments of agriculture devised and promoted new technologies. The beginnings of a knowledge-based system appeared.

For some time, FW looked like a sustainable system but new problems arose. First, maintenance of fallows to control weeds required frequent tillage, as many as 12 to 14 times during an 18-month period in some cases. The resultant loss of soil organic matter and structure increased susceptibility to soil erosion. Fallows lost soil to erosion by wind in dry years and, in wet years, to sheet and gully erosion on sloping land. Second, despite many contributions from improved cultivars, fertilizer amount and placement, and improved timing and precision of operations, yields did not exceed those obtained at the inception of the industry and were insufficient for economic survival under prevailing economic circumstances. Greater yields were required and these came from a revolution in cropping practice made possible by a dramatic rise in the profitability of wool production.

16.5　　Ley-farming (1950 to 1985)

From 1950 there began an important confluence of systems. Pasture improvement for low rainfall areas can rightly be claimed as an Australian innovation (Smith 2000). As early as

1900, subterranean ("sub") clover was recognized as a potentially important contributor to increasing pasture production in areas of short (annual) growing season. Soon, P, and later micronutrients (Mo, Zn), were routinely applied to pasture with spectacular results. Stocking rates of sheep could be increased from 2 to 10 ha^{-1}, and soil organic matter and N content rose dramatically. New strains of sub clover capable of persisting down to 300 mm average annual rainfall were developed. The same approach was used with annual medics on alkaline soils in the mallee areas of lower rainfall.

Adaptability of sub clover and wheat to the same zone made a new **ley-farming** system possible. Crops and pastures were no longer grown separately but were closely integrated in sequences on the same land so that fertility, accumulated during a pasture phase, could be utilized by subsequent crops (Rovira 1992, 1993). Figure 16.3a presents a comparison of soil N content under fallow–wheat, legume–pasture, and pasture–wheat management systems (Greenland 1971). In this example, a sequence of 4-y pasture followed by 2-y crop reveals slowly increasing fertility. In practice, optimum time spent in pasture (P) and crop (W) phases varied depending upon soil type and rainfall. The objective, as depicted in Fig. 16.3b, was to maintain fertility within a working range that was well above the exhaustion limit reached previously in FW cropping while, at the same time, avoiding excess fertility. High fertility can also cause yield loss when crops of high vegetative vigor, high early water use, and continuing high water demand are left with inadequate soil water during low rainfall conditions of grain filling (Section 13.4).

Fallow became less frequent, except in areas of lowest rainfall, as benefits of rotation from crops and pasture contributed previously unappreciated benefits of fallow, i.e., mineralization of N and control of root-borne diseases. Crop yields increased above those achieved at settlement and economic benefits of crop–animal enterprises even more so. Soil organic matter increased, frequently to higher levels than previously, and soils were more resistant to erosion. Farmers had to contend with less stable yields because no improvement to crop management could overcome the dominating effects of occasional severe drought. In contrast to the period 1937 to 1945, however, the two decades from 1950 to 1970 experienced greater than average rainfall, again evident in yield data (Fig. 16.2) when only one year recorded average yield below 1 t ha^{-1}. These were fortuitous climatic conditions in which to launch a new cropping system! As a result, this was a time of relative affluence in the wheat–sheep zone when high productivity and profit in favorable years could bridge years of low rainfall and income. Surely this is evidence of a sustainable system. Later, however, the second drought of this phase in 1982, average yield of 0.3 t ha^{-1}, emphasized the hazard of low yields to operating capital.

A now repetitive feature of agricultural development, transformations caused by the system to the operating environment itself, introduced new problems and a consequent need for change. In this case, a trigger for change came from external economic conditions. The world price for wheat had been in continual decline for decades and was approaching uneconomic levels. Expressed as Aus$2004 for comparison, the real value of wheat on the world market had fallen from $1000 in 1940, to $500 in 1975, and to $200 in 2004. Then towards the end of the 1980s, the price for the wool market indicator, fleece of 21 micron diameter, collapsed from Aus$10 to Aus$2 kg^{-1}.

Accompanying these dramatic trends, cropping faced two environmental challenges that had been building up for some time.

First, through a rise in soil organic matter and N content, legume-based pastures had gradually acidified soils in many areas to the detriment of legume activity, N fixation, and hence system productivity (Helyar *et al.* 1988). **Acidification** is a problem that varies from paddock to paddock. It depends upon soil type and pasture history and so is potentially under the control of individual producers. Acidification is a natural and inevitable process (Section 7.4.4) that can be reversed by liming, reduced by management, and temporarily, but blindly, avoided by the use of acid-tolerant species.

Second, replacement of native perennial vegetation with year-round ET by winter–spring active, annual crops and pastures, and inactive fallows, had changed the hydrological balance, increasing drainage through soil profiles. Salt, commonly present in these soils of a semi-arid zone, was leached more deeply and water tables that had gradually risen in lower parts of the topography brought salt to the surface in many areas (NLWRA 2000). Most of this salt had been brought inland by rain from the Southern Ocean and had accumulated in soil profiles because insufficient amounts were returned to the ocean in stream flow. The initial effects of this redistribution of salt in the landscape are seen in changing species composition of pastures and poor growth of crops. The final effect is bare, erosion-prone land and salinized streams. The nature of this (secondary) **salinization** problem is, therefore, distinct to that of acidification. It is a landscape and not a paddock-scale problem. Recharge areas, where increased drainage adds to ground water, can be at great distance from discharge areas that suffer salinization. Unlike acidification, there are few locations where an individual farmer can hope to reduce his salinity problem by the changing management of his own farm. Solutions are found in broad cooperation requiring regional planning and involvement of producers and consumers.

16.6   Intensification and diversification (1985 to present)

The response to the economic crisis was a dramatic reduction in sheep numbers, accelerated **intensification** of cropping, and a search for more profitable, alternative crops. Sheep numbers in the zone fell from 6 M in 1990 to 2.8 M in 2006. Many wheat–sheep farms with favorable soils removed sheep enterprises completely, sometimes including fencing to facilitate use of larger machinery. Cropping became more intensive and more diverse in zones where adapted cultivars of alternative crops, legumes and oilseeds, were available. In 2009, the century-old Horsham sheep show and ram sale was suspended because of low sheep numbers and lack of interest.

16.6.1   Crop diversification

Revival of a canola industry with cultivars resistant to the blackleg fungus (*Leptosphaeria maculans*) had an enormously beneficial impact on profitability of wheat-based systems. Not only is canola a profitable crop in its own right but cereal crops that follow it show greater yield as well. This response is caused by a reduction of soil-borne diseases

Table 16.1 Increasing crop diversity on a farm near Horsham, Victoria, Australia, expressed as percentage of area sown to various crops over the period 1982 to 2009

Crop	Year			
	1982	1992	2002	2009
Wheat	49	36	16	39
Fallow[a]	51		9	24
Barley		18	25	5
Field pea		8		
Faba bean		38		
Lentil			39	32
Canola			11	
Farmed area (ha)	1160	1547	2133	2400

Note: [a] fallow was maintained by tillage until 1982, then by herbicides.

by biofumigation and positive effects on nitrification bacteria (Kirkegaard *et al.* 1994, 1999). As some new opportunities were produced, so others were lost. Previously, field pea was the only legume suited to cropping sequences but new options, lentil and chickpea, became available. Chickpea expanded rapidly during the 1990s until it fell susceptible to *Ascochyta* blight. Resistant cultivars have been developed so growers can return to chickpea production without the use of expensive crop-protection chemicals.

An example of **crop diversification** is revealed in details of an individual farm presented in Table 16.1. In 1982, the farmers had already removed an earlier sheep enterprise and adopted a fallow–wheat rotation, each activity occupying 50% of farm area. Fallow was gradually reduced and omitted completely by 1992 to include a large proportion of legume crops (46%), mostly faba bean but also field pea. Cereal area increased slightly from 49 to 54% but with a significant proportion of barley (18%) and less wheat (36%). Ten years later (2002), barley increased (25%) at the expense of wheat (16%) and legume area, then lentil was reduced to 39% following the introduction of canola (11%). A small fallow area (9%) was maintained chemically rather than, as previously, by tillage. During the decade from 1999 most years experienced below average rainfall and yields were consequently low (see Fig. 16.2 for regional behavior). This explains the reappearance of fallow recorded in 2002. With that change of management, lentil remained a profitable choice (32%) and wheat was represented more strongly (39%) than barley (5%), but all on a smaller cropped area.

Newly developed semi-dwarf cultivars of wheat were adopted rapidly during the early years of this phase of diversification. They were accompanied later by minimum and zero tillage systems with consequent benefits to water and soil conservation (Section 12.5.2) and timing and efficiency of operations. Growers had then to adapt to new sowing and harvesting technologies, new crop sequences, disease control measures, herbicide types and sequences, and fertilizer strategies and tactics. Fertilizer management now includes N that was rarely used during the ley-farming era. Farmers discovered that legume crops, harvested for grain, did not leave adequate N in the soil for subsequent non-legume crops. They also discovered value in precise and timely supplementation with N fertilizer. How much N to apply, and when, have become decisions with great impact on crop yield and profitability.

The challenge to management is to match N application to rainfall and soil fertility for individual crops. Too little N, and crops cannot take advantage of seasons of high rainfall, too much and there is risk of yield depression in years of low rainfall. Success requires measurement of starting conditions of soil water and N content, and accurate "guesstimating" final water-limited yield early in the growing season at stem elongation so that an appropriate dose of N can be applied. Decision-support systems are being developed (e.g., Hochman *et al.* 2009) but, given uncertainty of rainfall, the range of yield prediction is inevitably wide so most farmers who adopt this management tactic prefer to rely on their own records and experience. As information accumulates, however, models do offer the possibility to narrow the range of predictions and extend the skill of the best farmers more widely (Box 12.2).

Although Australia was a leader in the introduction of transgenic cotton to assist pest control, public concern about food safety, together with possible loss of overseas markets, prevented the sowing of transgenic food crops until 2009. Transgenic canola with resistance to glyphosate herbicide can now be grown commercially in Victoria.

16.6.2   Understanding weather and climate

In established cropping areas, where long-term experience has helped refine cropping systems, farmers place a priority on good short-term weather forecasts so they can adapt their activities to favorable conditions. Suitable weather conditions are required, for example, for tillage, sowing, harvesting, and the application of herbicides. Such forecasts have been greatly improved during the evolution of wheat cropping in the zone and are now readily accessible to farmers by radio, TV, and the internet. Next week's weather is less a surprise than ever before and there is now more predictability with seasonal forecasts also, to help longer term decision making and in-season crop management.

Seasonal forecasts of break of season and later of in-season rainfall offer opportunities for major tactical variations in the management of wheat crops. Sowing cannot be delayed indefinitely so, as they wait for the break of season, farmers wonder if it would be better to sow dry and risk a false break that would jeopardize establishment, or not to sow at all. Farmers are also interested in seasonal forecasts to assist in longer term management decisions of established crops. For example, is rainfall likely to support a response to late N application, or at the other extreme would it be more economically efficient to graze immature crops before they dry off completely and lose value for either fodder or grain? Progress is being made on both issues. A recent historical analysis revealed the predominant meteorological conditions that are associated with breaks of season (Pook *et al.* 2009). Analyses based on the condition of the Southern Oscillation Index (SOI), and more recently on sea surface temperatures in Pacific and Indian Oceans (Ummenhofer *et al.* 2009), have improved skill in predicting seasonal rainfall and therefore crop yield for this zone (see Box 16.1). In the Victorian wheat belt, SOI analysis has little

---

Box 16.1 Forecasting Australian wheat yield with the Southern Oscillation Index (SOI)

The Southern Oscillation Index is a standardized pressure difference between Darwin in northern Australia and Tahiti in the mid Pacific Ocean. The long-term trend of SOI is strongly cyclical with frequent shifts from strongly negative to strongly positive values during January to April. Shifts during wheat-growing seasons are infrequent. Long-term values are available from the Australian Bureau of Meteorology (www. bom.au). Negative values correlate with periods of low, and positive values with periods of high, rainfall in Australian wheat production areas (McBride & Nicholls 1983).

Rimmington and Nicholls (1993) established correlations between Australian wheat yields and SOI. Using data for the period 1940 to 1990, they removed the time trend in yield attributable to better cultivars and management and sought correlations with monthly values of SOI. The analysis for Victoria is shown below.

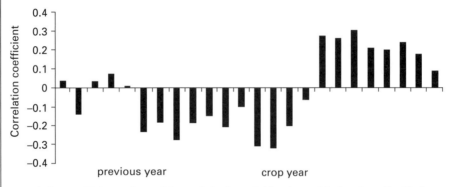

Correlation coefficient values of detrended wheat yield and monthly Southern Oscillation Index over the period 1948 to 1988 for Victoria, Australia. Correlations are shown for the crop year and the previous year.

Yields were positively related to SOI in the current year. Equally important, correlation analysis revealed the cyclical nature of the signal, often switching from negative to positive in succeeding years. This characteristic of SOI added improved prediction skill by combining values around sowing time with values measured earlier. The correlation coefficient for detrended wheat yield for the Victorian crop and the difference between average SOI for May, June and July of the current and previous years was −0.45. Correlations for the States of Queensland and New South Wales were higher, −0.50 and −0.56, respectively. The value for the national crop was −0.56. All values significant at $P < 0.01$.

---

predictive skill for autumn conditions but has greater value for later in-season management decisions.

An important example of association of yield with SOI is seen in the rainfall data of Fig. 16.1b and yield data of Fig. 16.2. Reversal of yield from 0.3 t ha^{-1} 1982 to the most productive crop to that time (2.5 t ha^{-1}) in 1983 was the first major switch from low to high rainfall in successive autumns that was clearly linked to changes in pattern of SOI (Nicholls 1991). Relationships developed from SOI are increasingly used in predictions of seasonal climate for decision-support systems to assist wheat growers with risk management.

16.6.3   Living with risk

The management of **risk** is a critical aspect of economic survival and farmers are understandably averse to risk in adoption of new technology (Pannell 1999). Many tend to be conservative (**risk averse**) at the expense of generating greater long-term income. Farmers with limited financial means are more appropriately advised to concentrate on reducing costs rather than maximizing yield. Those with sound financial conditions, often including off-farm investment, are able to take higher risk. Farmers that do adopt new technologies tend to do so slowly, initially testing new options on a portion of their holdings and gradually adapting their enterprise to change. An interesting example is seen in the adoption of canola. Production costs are about twice those of wheat so there is a greater chance of loss in years of low rainfall. Sadras *et al.* (2003) offer an analysis of a dynamic cropping strategy based on a putative association between start-of-season rains (April and May) and total seasonal rainfall (cf. response farming discussed previously in Section 13.9.5). They show the advantage to long-term income of switching from a cereal-only strategy in years of low rainfall to a more risky strategy of canola and cereal in years of higher rainfall. The study reveals the advantage of decision-support systems to assist farmers and their advisors to make complex decisions required to optimize cropping strategies in this highly variable environment.

16.6.4   Technology transfer

Two important activities lay behind these agronomic adjustments. First, adoption of new cultivars, crops, and practices was aided by continuing research and technology transfer by universities, government, private enterprise, and farmers' organizations. Farmers make large contributions as levies on production that fund half of all activities (Aus M\$120 in 2008) of the Australian Grains Research and Development Corporation. They also support Regional Cropping Groups that are active in adaptive research and technology transfer in various parts of the wheat belt. In Northwest Victoria, the cropping group at Birchip is nationally recognized for innovation. Second, there has been considerable structural adjustment, seen especially in the increase in farm size and search for economies of scale. The continuing story since the first farmers were settled on blocks of 320 acres is that some farmers survive at the expense of others. As an example of this change, the farm depicted in Table 16.1 that intensified and diversified its cropping activities also gradually increased in size from 1160 ha in 1982 to 2400 ha in 2009. Both aspects of change were instrumental in its survival as a productive enterprise.

16.6.5   Where to next?

The yield data in Fig. 16.2 reveal little gain to the yield frontier during the early years of the current phase. The more general pattern appears as a return to more frequent years of low yield, a feature more characteristic of the FW period, 1900 to 1945, rather than the ley-farming phase that continued from then until 1985. In consequence, Victorian data show no evidence of the continuing yield increase projected by Angus (2001), as a response to post ley-farming cropping strategies, in his analysis of the national wheat crop. That analysis used decadal yield averages that obscured the annual variability of yield shown here. It also could not foresee that the decade to 2010 would be a period of low rainfall equal in severity to the century-old drought of 1897 to 1903.

Only the passing of time will allow an assessment of the success of the present phase in the continuing evolution of wheat cropping systems of southern Australia. Farmers and scientists now grapple to construct farming systems that are ecologically as well as economically acceptable. Fortunately, this can be now approached with better appreciation of functioning of cropping systems and landscapes, and with the help of new technologies and tools.

16.7   Searching for new designs

Environmental issues bring society generally into the debate, and there are some strident views. One is that a revolution is needed to take land management away from the "European" farming systems to others that are more aligned with the "Australian" environment. Another view is that revolution is not needed because the history of development has been, and will remain, one of continuous and even dramatic evolution, as evidenced in Fig. 16.2. Within these positions, there are differences in views about the nature of "ecology", the role of native species, and links between diversity and sustainability (Connor 2001). There are clearly some important issues but also much distraction, increasing the risk that scarce research funds will be poorly allocated and solutions to important, pressing problems delayed.

For the immediate future, new designs can only be constructed with existing plants of economic value, i.e., in the absence of subsidies from government, those that either return a profit from production or that provide ecosystem services that increase the overall profitability of farming enterprises. We can expect new options in the future now that biotechnological methods have increased the scope and efficiency of plant breeding. It is unlikely, however, that new plants will solve the environmental problems directly. Rather severe caution must be expressed, for example, on the ecological consequences of inclusion of more acid- and salt-tolerant plants in these ecosystems. More tolerant crops would allow the systems to acidify or salinize further, to levels from which recovery would be more expensive or perhaps impossible. Breeding of such plants can at best provide a respite while management solutions are implemented to redress these problems. The emphasis must be on cropping systems that conserve the resource base and do not allow it to degrade. Indirectly, however, there can be realistic hope for benefits from plant improvement in the form of more productive crops and new crop products able to provide adequate economic return for the continuing investment essential to maintain the natural resource base.

Combinations of technologies that promise to redress the major problems of acidification and salinization will require careful scientific and economic evaluation.

16.7.1   Acidification

The main processes of **acidification** are removal of alkalinity in harvested products and leaching of nitrate, whether from legumes or fertilizer (Section 7.4.4). In the wheat belt, loss of alkalinity, 9 and 40 kg lime equivalent for each tonne of wheat or clover hay removed, respectively, has continued for 150 y, mostly without amelioration. Solutions are applied where the problem exists – in individual paddocks. Acidification can be slowed by recovering leached nitrate using deep-rooted perennial grasses (Ridley *et al.* 1999), but the more certain and long-established remedy is liming. Here, the questions are: how much to apply, how frequently, and how best to incorporate lime into soil? Cost of treatment and value of response are clearly linked for individual paddocks. Farmers will quickly adopt and adapt these solutions once liming becomes an economically feasible treatment. Scientists, in the meantime, evaluate and rightly worry if acidification processes are being allowed to progress too far for ready recovery. They have, however, also established good information on the extent and rates of acidification, on the amounts of alkalinity removed in various crop and animal products, and the buffering capacity of various soils. This has been used to derive liming strategies required to raise pH and then maintain it within limits consistent with productivity. Lime is widely available, and current use of lime in Australia is around 2 Mt y^{-1}. Although the rate of use is increasing, it remains small compared with the estimated 12 Mt y^{-1} required to maintain agricultural surface soils at pH 5.5.

16.7.2   Salinization

Much research is being undertaken into the nature, cause, and extent of secondary salinity (e.g., Rengasamy 2002; Petheram *et al.* 2002; Roberts *et al.* 2009). There is also much experimentation, mostly by farmers, that is providing options for system design. It is evident that a solution to **salinization** is to be found in reducing drainage by managing vegetation to create soil water "buffers" to hold more of the winter–spring rainfall in the soil zone explored by vegetation for subsequent use in transpiration. The questions are how, where, and how much buffer storage is required? What combinations of annual crops, perennial crops, trees, and nature reserves are required? What are rarely discussed are the alternatives that sacrifice some land as inland salt deposits, or promote the return of salt to the sea in drains constructed for the purpose. Is the timescale of sustainability in these discussions long enough? Would a vision for 1000 y change the set of solutions to be considered? It is interesting to ponder the effect of climate change on salinization. If rainfall were to increase, for example, salt would be more mobile than at present and would likely pose a greater problem, unless and until it were flushed back to sea. A long-term view of the salinity problem may favor disposing of salt rather than the current view of attempting to re-establish its place in the landscape and live with it!

Confusion in the search for solutions to salinization reflects the social and economic, rather than the scientific dimensions of the problem. Current experimentation by farmers and scientists is directed mostly towards reduction of leaching in order to keep salt below the root zone. The components are changes to fallowing practice and inclusion of summer-active species in cropping sequences. Matters to be resolved are the relative locations of the two areas, the role of fallow, and if economically attractive cropping systems can be generally available. Factors that favor deep drainage have been identified (Cook *et al.* 2001). They are coarse-textured soils, high rainfall, and large individual rainfall events. Such areas located close to main rivers will also have detrimental effects on water quality.

Elimination of fallow A major contribution to the control of salinization is already in place. During evolution of cropping systems, as previously described, a significant change has been toward crop diversification and the elimination of long fallow. This has occurred in other places also (Peterson *et al.* 1993). The close link between crop growth and water use provides assurance that systems without fallow do use more water, leaving less for drainage. Reduction in fallow is a major landscape change when compared with earlier periods of FW management. It changes landscape hydrology appropriately, but quantitative effects are uncertain and, in any event, just as salinity took a long time to develop, amelioration will occur after a lag phase of unknown duration. There have been relatively few complete measurements of leaching under fallow and the range of crops and tillage systems involved (O'Connell *et al.* 1995; O'Leary 1996). The challenge to experimenters is not trivial because drainage is infrequent and episodic, and occurs mostly in periods of high rainfall (O'Connell *et al.* 2003). The contrast between simulations (Keating *et al.* 2002) and measurements (Díaz-Ambrona *et al.* 2005; O'Connell *et al.* 2003) is marked, and will remain so until more experimentation improves understanding. Much work studying drainage in fallow–crop systems has been done on small plots; future work needs to address landscape responses at larger scales.

Summer-active herbaceous species Another option is to include summer-active perennials in cropping systems. The major herbaceous option is the perennial alfalfa that fits well into a wheat–sheep system, providing valuable summer feed as well as additional transpiration. Even at low densities required for survival in this semi-arid environment, alfalfa dries the soil to depth (Ridley *et al.* 2001; Tennant & Hall 2001) increasing the storage buffer by 50 to 150 mm. The drying effect is sufficient to incur a yield penalty, at least to the first wheat crop that follows (McCallum *et al.* 2001). Subsequent recovery of yield and re-establishment of earlier drainage patterns depend upon rainfall (Dunin *et al.* 2001; Ward *et al.* 2001). This system of several years of alfalfa followed by a sequence of annual crops requires tuning to soil characteristics and rainfall patterns of individual sites. The system has attracted much attention and a new name – "phase farming". In principle it is no more "phasic" than pasture–wheat rotations of ley-farming, but novelty has value in science also.

Woody perennials A range of options is being evaluated with woody perennials, grown in mixed systems. Experimentation is interesting because these mixed communities of perennials and annuals with distinct canopy and root geometries, and seasonal growth dynamics, offer a range of competitive and complementary interactions.

Deep-rooted, summer-active perennials dry soil to depth providing a large but horizontally discontinuous storage buffer (Lefroy & Stirzaker 1999). As with the elimination of fallow, there could well be a landscape solution here, provided commercially attractive tree species can be identified. This is not yet the case. Whatever existing products are contemplated, i.e., timber, cellulose, biomass energy, fruit, essential oils, etc., they must compete with production elsewhere, commonly in more favorable environments. Some people place hope in mixtures of novel native fruits and eucalypt oils but such solutions face enormous obstacles and challenges. An early step must be to identify a market to justify the extensive experimentation required to evaluate planting patterns and management to optimize such systems for productivity and drainage control.

16.7.3   New tools for new challenges

The challenge to design new crops and cropping systems that are economically and environmentally sound is enormous. It is unlikely that further major yield gains are possible with current plant types. Earlier gains were achieved by improvements to nutrition, disease control, and drought resistance that gradually brought the systems closer to water-limited productivity. That water supply now exerts primary control on yield is evident in the yield variability seen again during the last decade (Fig. 16.2). The best hope lays in improving drought resistance (Chapters 9 and 13) and the introduction of new crops, or new crop products with better market prosects. Some can hope for a genetically engineered change in the photosynthesis system that will increase transpiration efficiency (TE_Y through TE_B, Section 9.7), i.e., improve the exchange rate of carbon for water within the atmosphere. But in the meantime agronomists must get on with the task of designing cropping systems with plants that are currently available.

Important new tools for system design include those that allow rapid and cheap collection of large amounts of data and enable analysis of patterns and relationships. These are the information technology tools of remote sensing, geographical information systems, geo-spatial analysis, and simulation modeling of crops and cropping systems. Easy availability of global positioning systems (GPS) now presents farmers with a near saturation of information on the spatial nature of crop yield and its variance (Bramley 2009). New ways of analysing and interpreting this kind of data are needed.

New tools are finding application in the search for solutions to salinization. **Remote sensing** and GIS are providing maps of salt distribution and analyses of spread and relationships to underlying sediments. Geospatial analyses at paddock level are able to relate yield maps to subsoil salinity. Together these techniques are locating and quantifying problems, leading to suggestions of individual site-specific solutions (O'Leary *et al.* 2003). After three decades of development, the major focus of crop modeling remains on individual crops but with attention turning to long-term issues of cropping systems and spatial nature of response. The inherent complexity of cropping systems requires comprehensive models with socio-economic as well as physio-biological components for their solution.

Increased effort in modeling is a priority if agronomists are to contribute efficiently and effectively to the design of optimal cropping systems for the future. Guidelines that derive from understanding of dry land salinity, for example, have to be converted into answers to site-specific questions of what species, densities, planting patterns, and management to use, and all in the face of climate variability. Also to be stressed here is the need for more long-term field experiments, conducted at appropriate scales, and benchmark sites so that we can properly understand what sustainability really means with respect to crop and environment. At present we rely on few experimental sites established long ago. There is a need for new designs of new combinations and for new techniques of measurement and analysis.

16.7.4   Climate change

Designs for new cropping systems are being undertaken under the specter of climate change. Farmers appear less concerned with that prospect than they do with survival under present circumstances. They have better appreciation of climate variability than average citizens because previous dry and wet periods have impacted directly on their livelihoods. So why not again? None can personally remember the most severe recorded ("Federation") drought of 1897 to 1903 but a few can remember the "World War II" drought (1935–1949). And there have been other significant dry spells also, i.e., 1924 to 1931, 1965 to 1972, and 1999 to 2009. The question to be resolved is if an increase in mean temperature of $0.3°C$ reported for the zone since 1960 (Ummenhofer et al. 2009) will continue and will be associated with lower rainfall outside existing variability.

Projections from IPCC-AR4 (2007) are displayed on the CSIRO website (www.csiro. au/ozclim) to guide policy development and research into adaptation. Their range is sufficiently large, however, to emphasize inherent uncertainty. For the Victorian wheat zone, one can choose a new climate for 2030 with a temperature and rainfall change of between 0 and $2°C$ and -50 and $+25$ mm y^{-1}, depending upon the sensitivity of the climate model selected. The impact of those changes, and higher $[CO_2]$, the putative major cause of global warming, on productivity of individual crops can only be estimated with crop models (see also Box 5.3). Extensive comparisons compiled by IPCC-AR4 (2007) for wheat in mid-latitudes reveal that with adaptation, i.e., using cultivars with appropriate phenological development, yield increases from 20 to 30% are likely for mean increases in temperature to $3°C$ but falling to zero at $4°C$ and to -20% at $5°C$. Free atmosphere carbon dioxide elevation (FACE) experiments underway at two sites in the region have established the promotive effect of increased $[CO_2]$ from current 380 to 550 ppm at $+20\%$ for both biomass and grain yield under current climate (R. Norton, pers. comm.).

16.8   Role of society

There are some unfortunate aspects in the public debate about the relationships between agriculture and land management. Some discussions contain a culture of blame that perceives farmers as villains; in others, emphasis on "revolution" rather than "evolution" promotes the view that current farming systems are fixed and entirely wrong for Australian environments. In practice, however, all components proposed by revolutionaries: perennial pastures and forages, tree belts, agroforestry combinations, retiring land from agriculture, are already being tested by farmers. Farmers are visibly continuing the evolution (Fig. 16.2) that has always sought to adapt farming systems to changing economic and environmental conditions.

There is, however, a major challenge to bridge the gap between urban dwellers and farming with an education campaign of at least three objectives. First, to acknowledge that farmers do not operate independently but rather as providers of food and fiber to meet the demands of mostly urban markets, using a resource base that must be modified for use and yet also be conserved for future generations. Society as a whole, and not just farmers, must learn what constitutes sustainable agricultural production and that responsibility for change resides with consumers as well as farmers. Second, to explain that food is cheap in our society, even while an increasing proportion of profits are retained by retailing and distribution sectors. If consumers want farming done differently, then they should discriminate between products, not just on the basis of price but also on methods of production, supporting those that sustain farming and the natural resource base, while providing adequate food for all and retaining as much land as possible for nature. Third, to stress that economic pressure on farming drives farm amalgamations and that declines in rural population generally are also closely linked to conservation issues. Without adequate returns to farming it is not possible to make continuing investment necessary to maintain its natural resource base. Mechanisms to ensure continuing investment include price premiums for production methods, targeted subsidies, and regulations on land use. Farmers will respond to incentives by adopting those farming methods that meet the requirements of society.

Governments must be closely involved in the continuing transition to sustainable agriculture. There are many examples of national plans and guidelines for "good agricultural practice" in developed and developing countries. There are also international schemes (e.g., www.globalgap.org) that provide access to markets in developed countries for products of environmentally and socially acceptable agriculture. Governments can provide overall coordination and also targeted subsidies for appropriate land use and production practices. This is now clearly evident in the European Union, which is changing its production-based scheme of farm subsidies toward production methods with specific goals and financial support for rational use of water, control of erosion, improved use of native areas, and protection of biodiversity. The challenge to protect the agricultural environment in Australia is enormous. There is no large tax base to support agriculture and as long as most production is exported in unelaborated form, opportunities to share profits from processing and retailing will remain slight.

16.9   Review of key concepts

Continuing change is the norm

- Cereal production systems in semi-arid southern Australia have been in constant change since inception in 1800, applying technological innovation to increase productivity in response to economic and more recently environmental signals.

- Production systems have intensified and diversified, changing from continuous wheat production, to fallow–wheat sequences, to ley-farming, to current multispecies cropping systems. Farm size has increased and economic efficiency and survival (fewer farms) have been supported by a continuing sequence of improved cultivars, agronomic methods, mechanization, and agrichemicals.

Increased water-use efficiency

- Low rainfall is the ultimate limiting factor but technology and management, to control diseases and nutrient levels, have tripled yields and hence water-use efficiency since 1900. But yield stability has decreased, revealing that each system improvement is increasingly limited in years of low rainfall. It is uncertain how another step change in productivity might be achieved.

Environmental impacts and responses

- Conversion of evergreen perennial woodland and shrublands to annual crops and pastures has changed soil conditions and hydrological balance. Acidification and salinization of soil has gradually reduced productivity. Mobilized salt has reduced stream quality and threatened adjacent natural ecosystems. Acidification is a paddock-scale problem while salinization operates at a catchment scale across many farms.
- The challenge now is to design cropping systems that are economically as well as environmentally acceptable with control of salinity and acidification as major objectives. New systems under evaluation include less fallow and inclusion of summer-active herbaceous perennials, such as alfalfa, that are readily integrated into crop–livestock systems. Also of interest are agroforestry combinations of crops with various tree species.
- Design and appropriate distribution of new systems requires all available knowledge and technology. It is now aided by remote sensing techniques to locate areas of saline discharge and linkages to zones of ground water recharge, geographical information systems to arrange data for land system analysis, and computer models of crops and cropping systems.

Terms to remember: acidification; attainable yield; crop diversification and intensification; economic incentives; environmental impact; ley-farming; nutrient extraction; remote sensing; risk averse; salinization; Southern Oscillation Index (SOI); tactical application of nitrogen; technological innovation; technology transfer; yield forecasting; yield variability.

17 Technological change in high-yield crop agriculture

In the last half century, Earth's population has increased two-fold while land used to produce food, livestock feed, and fiber crops rose by only 13%. This considerable achievement has been called a **green revolution** made possible by a powerful combination of new technologies, including genetic improvement of major staple food crops, development and widespread use of mineral fertilizers and pesticides, and expansion of irrigated area. The result is that a few high-yield cropping systems now provide the major portion of human food from a relatively small area of arable land (Section 1.2). While these developments have had remarkable success in raising productivity and sparing natural ecosystems from conversion to agriculture, there are nonetheless growing concerns about negative environmental impacts. Here we learn how rapid technological advancement enabled conversion to high-yield systems and consider future challenges to sustaining them and their high productivity.

17.1 Common features of high-yield systems

High-yield cropping systems have evolved through intensification of traditional systems during the past 50 years. This process involved: (i) producing more crops per year per unit land area, i.e., temporal intensification; and (ii) more intensive use of inputs (fertilizers, manure, irrigation, pesticides) to alleviate yield-limiting abiotic and biotic stresses, i.e., input intensification. High-yield systems are not only found in developed countries where large-scale, mechanized agriculture predominates, but also in developing countries where small-scale, labor-intensive systems remain the norm. Examples of the latter include continuous rice, rice–wheat, and cotton and sugarcane systems in developing countries of south, southeast, and east Asia.

Intensive, high-yield systems share common features that distinguish them from the lower yielding agriculture they replaced. These include:

- individual crops grown as monocultures with relatively little diversification within crop sequences;
- use of modern crop cultivars, including hybrids, and rapid change in the most widely used cultivars;
- nutrient inputs to alleviate deficiencies of essential macro- and micronutrients;

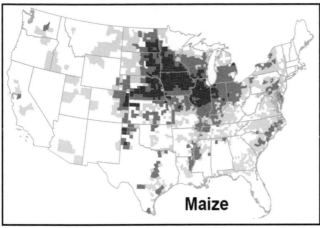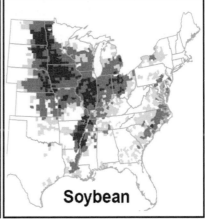

Fig. 17.1 Concentration of maize production in a "Corn Belt" that runs across north–central USA. Soybean has a similar distribution but also extends into the Mississippi Valley (bottom center of map) and along the Atlantic Coastal Plain. In total, 32 Mha maize and 30 Mha soybean were harvested in 2008. (*Source:* USDA-NASS website.)

- pest management using herbicides, insecticides, and fungicides along with crop sequences to keep pest populations below critical thresholds that cause economic yield loss;
- production focused on sale in local, regional, and international markets rather than on self-sufficiency;
- access to information and inputs to support the rapid adoption of new technologies.

Also common to these systems are concerns about sustainability of production and environmental impact, including long-term effects on water and soil quality, wildlife, and, more recently, greenhouse gas (GHG) emissions. Given the imperative to meet future food demand without large expansion of crop area, continuing yield gain is essential. The major immediate challenge is, therefore, to increase yield on existing cropland while minimizing negative environmental impact during further intensification. Understanding the structure and function of existing high-yield systems, and the technological advances that supported their evolution, provides a foundation to explore opportunities for future productivity gains through ecological intensification. In this chapter we draw conclusions regarding the future of high-yield cropping systems from three case studies: (1) maize–soybean in North America; (2) intensive irrigated lowland rice in Asia; and (3) soybean-based systems in Brazil.

17.2 Maize–soybean cropping systems in the North American Corn Belt

The USA produces 40 and 36% of global maize and soybean, respectively (2006–2007 average). Of this, more than 80% is grown in the north–central Midwest where a 2-y maize–soybean sequence is the predominant land use (Fig. 17.1). This region, known as the "Corn Belt", is one of the world's largest and most productive agricultural

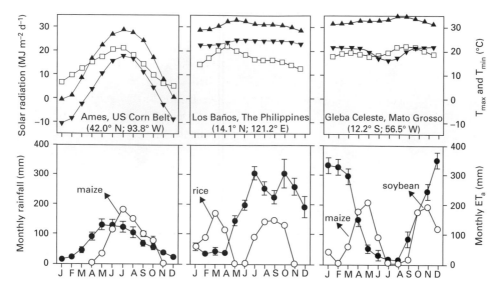

Fig. 17.2 Annual weather patterns in locations typical of USA maize–soybean, tropical rice in southeast Asia, and soybean production systems in Brazil. Solar radiation (□), T_{max} (▲), T_{min} (▼), rainfall (●), crop ET (○).

systems. It comprises 1 Mkm^2 (about 12% of continental US land area) and has a unique combination of deep, fertile soils and favorable climate for rainfed crop production.

17.2.1   Climate and soils

The climate of the Corn Belt is mid-continental and temperate with cold winters and a limited frost-free season of 140 to 170 days. Highest annual temperature, rainfall, and solar radiation coincide (Fig. 17.2). Summers are warm (July mean near 24°C with diurnal amplitude of ~8°C) and humid; 70% of moderate annual rainfall (700–1000 mm) is received during the growing season from May through September. Cloud cover limits daily total solar radiation to 20 to 23 MJ m^{-2} d^{-1} in midsummer. Crop ET$_a$ in mid-summer averages 5 to 6 mm d^{-1}, whereas rainfall averages only about 3.5 mm d^{-1} so most crops also depend upon soil moisture accumulated from spring rains and snowmelt. Severe drought, hail from severe storms, early frost in fall, and difficult conditions in spring (too dry, cold, or wet) are the principal environmental hazards. Drought is not uncommon and crops would sometimes benefit from supplemental irrigation but supplies of surface and ground waters are limited and capital costs of standby sprinkling equipment are rarely justified. Maize, with determinate development, is most susceptible to drought at silking, which typically occurs in mid-July. Soybean (indeterminate) can compensate for loss of early flowers and is therefore most sensitive to water deficit during seed filling in August.

The Rocky Mountain Range to the west produces a "rain shadow" that results in a strong rainfall gradient across the Corn Belt. The productivity of rainfed systems follows this gradient from west to east. Most irrigated maize is found in the western Corn Belt where exploitable ground water resources are available to supplement low rainfall for profitable maize production.

Glaciation was a dominant factor in landscape and soil formation in this region during the recent Quaternary Period. Soils are young and relatively unleached of their original nutrients. Early extensions of Canadian glaciers ("Nebraskan" and "Kansan" periods) deposited till 5 to 10 m deep over most of the region. During the "Wisconsin" glaciation (14 000 BP), ice lobes extended into central Iowa ("Des Moines lobe") and northern Illinois. Soils in those areas were formed on gently undulating till and surface drainage is poor so that marshes and lakes are common in central and eastern parts. South and west of those lobes, paleosoils of Kansan till (29 000–14 000 BP) were covered by as much as 15 m of silty loess blown from glacial outwash during the Wisconsin Period. In both regions, fire served to maintain disclimax grass prairie and oak-savannah, while protected areas supported climax vegetation of mixed hardwood forest (Section 3.1). In places, river valleys have cut through overmantles of loess and till, exposing earlier tills and paleosoils, and creating alluvial plains. Mollisols (FAO, Chernozems) dominate grassland sites, while Alfisols are found with woodland influence. Soils are prone to erosion when crops are grown on sloping land without conservation tillage (Section 12.5.2) and other measures to reduce runoff (Section 12.7.2).

17.2.2   Cropping systems and productivity trends

When farming began around 1830, tall grass prairies presented major challenges to farming. Traditional wooden plows were inadequate for breaking sod and shortages of wood for fencing and fuel were an added concern. Technological developments gradually provided solutions. John Deere's steel plow (1840) pulled by heavy oxen opened the land; invention of barbed wire fencing (1870) allowed development of mixed farming with livestock; and invention of grain storage elevators facilitated handling and shipping by rail and barge. Corn, wheat, and other small grains became the dominant cash crops because the climate is too cold, and growing season too short, for crops such as rice, cotton, sugarcane, and citrus. And while a number of fruits and vegetables can be grown, competing regions such as California and Florida produced those crops with higher quality, less risk, and at lower cost.

Rapid technological change remained a notable feature of Corn Belt farming. Hybrid maize resistant to stalk rot (1930), introduction of soybean (1935), low-cost ammonium fertilizer (1950), herbicides (1950), widespread use of maize hybrid cultivars, and genetic improvement of other crops such as wheat, alfalfa, oats, and barley, and mechanization of tillage, planting, and harvest operations led, by 1960, to a relatively diverse and highly productive agricultural system. Although dominated by maize, rotations with crops other than soybean were practiced on 37% of total crop area (Fig. 17.3). Since that time, however, Corn Belt systems have simplified such that 90% of crop area is now given to a maize–soybean two-year rotation.

A number of technological, social, and economic forces drove this simplification. Steady increases in yields of maize and soybean, greater availability of high protein soybean meal, and declining real grain prices gave economic advantages to large-scale

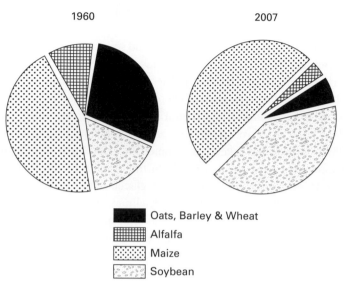

Fig. 17.3 Decreasing diversity in cropping systems in core Corn Belt states of Iowa, Illinois, Minnesota, and Nebraska (1960–2007). Total crop area increased by 17% from 27.5 to 32.3 Mha. Values shown are three-year means: 1959 to 1961 (1960), and 2006 to 2008 (2007).

livestock production facilities at the expense of small-scale enterprises on individual farms. Nitrogen fertilizer, herbicides, and conservation tillage, first with stubble-mulch sweeps or chisel plow and then with no-till, reduced erosion and gave greater returns to crop production over grazing on hillsides. Farmers, local machine shops, and equipment manufacturers played key roles in the development of new, more powerful, and larger capacity machines (e.g., four-wheel-drive tractors, no-till seeders, self-propelled pesticide applicators, larger tillage equipment, combines, and grain carts). As a result, farmers specialized in producing maize and soybean, farms consolidated, fences disappeared, and mixed farming with livestock became less common. **Specialization** also changed the nature of farming. While livestock require attention 365 days a year, farm families specializing in maize–soybean production find more opportunities for vacations, and cultural and recreational activities. As a result, farm number in Iowa, a central Corn Belt state, decreased by 50%, from 183 000 in 1960 to 89 000 in 2005, while average farm size increased from 77 to 145 ha.

Yield benefits to growers of hybrid maize are substantial and encourage the purchase of new seed required for each crop. In response, seed companies maintain investment and compete strongly in the research and development of improved hybrid cultivars. In contrast, there is less private investment for inbreeding crops, such as soybean and wheat. In these crops, yield benefits are smaller so farmers tend to sow their own grain and less frequently purchase new, more uniform seed of higher quality from specialist seed producers. Although selection of soybean cultivars also focuses on yield and yield stability, emphasis is also given to maintaining oil and protein content because soybean is processed into high protein meal and vegetable oil.

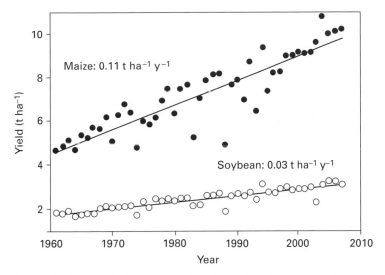

Fig. 17.4 Yield trends in US maize and soybean since 1960 (*source:* FAOSTAT).

Average farm yields of both maize and soybean have increased linearly over the past 50 y (Fig. 17.4). Contributions from genetic advance and improved agronomic management, although difficult to distinguish (Section 4.4.1), are taken to be approximately equal. Maize yields and rate of gain are always about three times those of soybean, explicable because a much higher concentration of protein and lipids in soybean seed requires about three times the photosynthate to support growth respiration in seed production (Section 11.2.3). Maize grain is dominantly carbohydrate (75%) with a small protein and lipid content.

Despite a 20% rise in atmospheric $[CO_2]$ since 1960 (Section 6.10.4), which should benefit C3 soybean more than C4 maize (Section 10.2.8), the relative rate of yield gain in soybean has remained relatively constant. This may reflect smaller total investment in soybean breeding by private sector seed companies, although average yields for both crops are relatively high by world standards. Sustaining these yield gains will depend on continuing investment in maintenance breeding and genetic improvement, as well as in continued improvement of crop and soil management practices.

17.2.3   Genetic advance

By the 1960s, demand for hybrid seed supported the establishment of numerous seed companies across the Corn Belt to serve local markets. New maize hybrids were evaluated according to yield and yield stability in field trials that increased in number as testing moved from initial crosses to evaluation of elite hybrids approaching commercial release. Geographic expansion of these evaluation programs allowed selection of broadly adapted hybrids and of fewer, larger seed companies also!

Despite significant consolidation, however, the seed business remains highly competitive. Seed companies distinguish themselves by providing additional services, such as

---

Box 17.1 Brute force breeding

Large seed companies use "brute force" selection based on numbers and scale (Duvick & Cassman 1999). Maize breeding programs of major seed companies involve continual development of new inbred lines and thousands of experimental hybrids from a large germplasm pool that is maintained to conserve wide genetic diversity. Experimental hybrids are grown in small-plot yield trials at multiple locations for several years from which no more than one hybrid per several thousand are advanced to more rigorous on-farm strip tests involving side-by-side comparisons of several promising hybrids. Hybrids that achieve the highest yield and stability in hundreds of trials over two to three years across the Corn Belt are considered for commercial release. The entire process typically requires five to seven years from initial cross to commercial release. Improved methods to evaluate hybrid performance based on weather variation and environmental classification (Loffler *et al.* 2005) and advances in marker-assisted selection (MAS) can facilitate this process.

---

agronomic advice about hybrid selection, sowing date, seeding rate, and seed treatment with fungicides and insecticides to protect against pests. But the deciding factors for success of commercial hybrids are yield and yield stability as described in Box 17.1. While other easily scored traits such as plant height, lodging, and disease or insect pest resistance are also considered, grain yield and stability are the primary selection criteria. Over time, this heavy focus on yield has also selected indirectly for a number of agronomic traits that contribute to yield (Fig. 17.5). From 1950 to 1990, grain protein content decreased by 1.5%, while grain carbohydrate increased by a similar amount. Because maize is used primarily as an energy source in livestock feed, more carbohydrate at the expense of protein does not diminish feeding value. The change did, however, reduce the amount of photosynthate required per unit grain produced (Section 11.2.3). Tassel weight decreased by 60% without loss in pollination efficiency, liberating more than 200 kg dry matter ha^{-1} for grain production. Leaf angle became more erect, in turn promoting greater total canopy photosynthesis (Section 10.5).

Although each of these traits contributes to greater yield, there is little evidence that potential yield of hybrid maize cultivars has actually increased significantly during the past 30 years. Instead, **brute-force selection** for yield and yield stability at increasingly higher plant populations, as used by farmers, has indirectly selected for resistance to multiple abiotic and biotic stresses, and this general stress resistance appears to be the major contributor to genetic advance (Duvick & Cassman 1999; Tollenaar & Lee 2002). Higher plant population reduces resource availability per plant and creates a general stress environment under all but the most favorable conditions. Delayed leaf senescence, or **stay-green**, that maintains canopy photosynthetic rates during late grain filling, appears to be associated with this general stress resistance (Lee & Tollenaar 2007). Analysis with a crop simulation model suggests that deeper rooting may have also contributed to genetic advance by allowing access to additional water from deeper soil layers (Hammer *et al.* 2009).

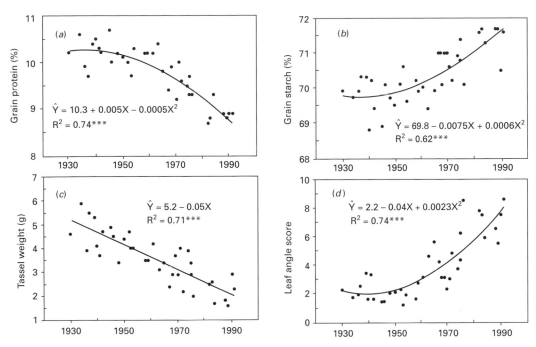

Fig. 17.5 Changes in (*a*) grain protein, (*b*) starch, (*c*) tassel dry weight, and (*d*) leaf angle (an increasing leaf angle score representing more erect leaf stature) in commercial maize hybrids and one open-pollinated cultivar released since the 1930s. Data are from field experiments in central Iowa, 1991 to 1994. (*Source:* Duvick & Cassman 1999.)

It is more difficult to identify specific traits associated with yield increase in soybean (Fig. 17.4). There is some evidence that greater lodging resistance, slower leaf senescence, and a general improvement in overall stress resistance, as expressed at high plant densities, have contributed to genetic advance (Specht *et al.* 1999).

Transgenic cultivars of both maize and soybean were developed and first commercialized in the 1990s. In 2009, transgenic soybean resistant to glyphosate herbicide was used on about 90% of US soybean area, while maize hybrids with Bt insect and/or glyphosate resistance are used on more than 60% of maize area.

17.2.4   Changes in management and resource-use efficiencies

Management practices In the temperate, sub-humid climates of the Corn Belt, growing season duration is limited by temperature (Fig. 17.2). Stored soil moisture and growing season rainfall support attainable rainfed maize yields of 12 to 15 t ha^{-1} in most years. These yields can be achieved if interception of solar radiation is maximized during the growing season. Contributing management practices include timely planting, adequate plant density with uniform intra-row spacing, early canopy closure, and appropriate cultivar maturity that optimizes the use of entire growing season, while also allowing adequate time for grain to dry before harvest. In practice, farmers are planting earlier and using later-maturing hybrids to utilize more of the available growing season. Larger tractors, planters, and combines facilitate these trends. Precision planters allow timely planting at higher seeding rates and greater speed while also improving uniformity of seeding depth and intra-row spacing.

In most years, crops experience transient water deficits during the growing season. No-till and other conservation tillage systems conserve water (see Section 12.5.2), especially important in the western Corn Belt. Compared to conservation tillage, plowing and disking reduce soil water supply in three ways: (i) incorporation of crop residue removes the protective effect on surface evaporation and reduces runoff; (ii) evaporation is greater from roughened (plowed) soil surfaces; and (iii) less snow capture reduces gain from melt in spring. A recent field study in the western Corn Belt revealed that conservation tillage reduced the irrigation requirement by about 75 mm (Grassini *et al.* 2011). The rapid adoption of transgenic glyphosate-resistant soybean and maize cultivars was motivated in large part by greater flexibility in timing of herbicide application. Glyphosate is a non-specific, wide-spectrum, post-emergence herbicide whose efficacy is less sensitive to growth stage of crop and weed than other products. Greater flexibility is especially useful in conservation tillage systems where seedling emergence through residue cover is more variable than with conventional tillage.

Efficient conversion of intercepted radiation to biomass and grain depends on adequate supply of nutrients, soil conditions that allow vigorous root growth and function, and minimizing yield loss from pests. Technological advances during the past 50 years have contributed to alleviating these constraints. The most important driver of productivity has been greater N input as fertilizer, which rose from an average of 45 kg ha^{-1} in 1964 to 157 kg ha^{-1} in 1981 to 1985 (www.ers.usda.gov/Data/FertilizerUse) but has remained relatively constant since. Most recent data report an application rate of 158 kg ha^{-1} in 2005. The development of laboratory tests to diagnose nutrient deficiencies in both plant tissues and soil has allowed farmers to identify fields where nutrients other than N give yield responses (Section 12.3.2). While most farmers apply N, P, and K to maize, soybean relies primarily on mineralization from SOM and symbiotic N fixation (Section 8.6) to meet N requirements, and residual fertilizer for P and K. Lime is applied to offset soil acidification caused by regular application of N fertilizers and N fixation (Section 7.4). Herbicides have replaced in-season tillage and hoeing as the primary means of weed management, to the advantage of reduced labor requirements, evaporative loss of water, and use of fossil fuel. In contrast, there is little use of insecticides. Cold winters kill or disrupt reproduction of most insect pests, and the recent widespread adoption of transgenic Bt maize hybrids has greatly reduced the need for control of stem and ear borers. Most insecticide use on maize is for seed treatment against root-feeding insects that become prevalent in continuous maize systems. The need for this may soon be reduced by release of new hybrids with Bt resistance to both borers and root worms.

Maize and soybean breeders select new hybrids with resistance to major diseases through brute-force selection (see Box 17.1). As a result, there is little fungicide use except in years when weather conditions are unusually conducive to disease. Alternate-year rotation between cereal and legume, each with different insect and disease pest complexes, helps keep pest pressures low, and most maize is rotated with soybean.

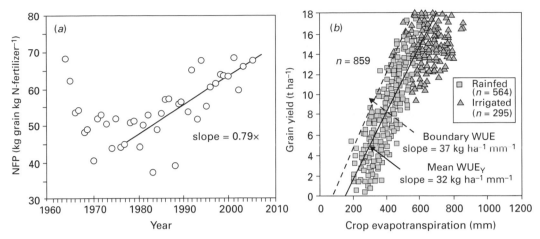

Fig. 17.6 (*a*) US nitrogen fertilizer productivity (NFP) based on national average yield and average N fertilizer applied to maize (source: USDA-NASS website for grain yield, USDA-ERS website for N fertilizer rate). (*b*) Transpiration efficiency of maize based on simulation of rainfed and irrigated maize yield over 20 y at 18 locations in the western Corn Belt (*source:* Grassini *et al.* 2009b).

Moreover, in response to concerns about overuse of pesticides, entomologists and plant pathologists developed **integrated pest management** (IPM) that relies on scouting to determine actual pest pressure such that pesticides are used only when insect numbers or disease incidence reach critical thresholds.

Resource-use efficiencies While crop yield responses to single production factors typically follow a diminishing return, i.e., yield increase per unit input decreases as input levels rise, synergistic effects of improving multiple production factors can result in a concomitant increase in both yields and input use efficiency (de Wit 1992). Increase in N fertilizer efficiency by US maize since the late 1970s is a case in point, as seen in Figure 17.6*a*. Soil nitrate tests were developed in the western Corn Belt to adjust N fertilizer rates to amounts of residual nitrate in root zones at planting. Farmers were advised to apply N in spring rather than fall and apply N in split applications – both to minimize loss by improving coincidence between N supply and crop demand. Improved maize hybrids with greater abiotic and biotic stress resistance gave more vigorous growth, especially at higher plant densities, in turn fostering rapid exploration of root zones and greater nutrient uptake. Application of nutrients other than N improved nutrition and contributed to higher yields without additional N. The combined impact of improved production factors increased N use efficiency, expressed as grain yield per unit applied N fertilizer, by 50% since the 1970s.

Boundary functions applied to yield vs. ET data are, as in Fig. 17.6*b*, interpreted as estimates of maximum transpiration efficiency (TE_Y, Sections 9.7, 13.4), i.e., estimates of best yield achieved per unit water transpired and achieved at all ET. The analysis in Fig. 17.6*b* establishes that C4 maize has a relatively high TE_Y compared to other C3 cereals or oilseed crops. In this case, TE_Y of 37 kg ha^{-1} mm^{-1} is nearly 70% greater than wheat (Sadras & Angus 2006; Fig. 13.2) and four-fold greater than soybean or sunflower (Specht *et al.* 1986; Grassini *et al.* 2009a). Actual TE_Y at the field level would vary about the regression line fit to these simulations of maize yield under optimal irrigated or rainfed growth conditions (32 kg ha^{-1} mm^{-1}) because year-to-year variation in solar radiation and temperature also affect yield in relation to water supply (Grassini *et al.* 2009b). Distribution of rainfall also contributes to variation about the regression line in rainfed systems. In years with favorable weather (i.e., moderate temperature and high solar radiation during grain filling), TE_Y would approach the boundary line but fall below it in years with unfavorable weather.

17.2.5   Environmental impact and sustainability

Negative impact on water quality is the greatest environmental concern regarding continuous maize and maize–soybean systems in the USA, and both are linked to crop N balance. Nitrate leaching from applied N has caused [NO_3^-] in ground water to exceed critical thresholds set for drinking water standards in some places. The N load carried by the Mississippi River is considered to be a main cause of **hypoxia** in the Gulf of Mexico (Turner *et al.* 2008). There, zones of low dissolved [O_2] due to vigorous growth of phytoplankton cause death of fish, crustaceans, and other life forms. Efforts to reduce hypoxic zones have focused on N losses from maize-based cropping systems because most N in the Mississippi River originates in the Corn Belt watersheds of the Ohio River Valley, Upper Mississippi, and Missouri-Platte Rivers.

A nitrogen balance of rainfed maize–soybean production in Iowa is presented in Table 17.1. Most striking is the quantity of N that cycles through these systems. Mineralization and N fixation represent 48% of total N supply, compared to only 21% from N fertilizer. Harvested grain and seed account for 86% of losses from the system while leaching, runoff, and gaseous losses account for the rest. Because soil C and N content are assumed to be at steady-state, there is no net gain or loss of N from this 2-y rotation. Most uncertain within this balance are N removal due to immobilization in SOM and losses from volatilization, denitrification, leaching, and runoff because there are few direct measurements from production-scale fields. If immobilization in SOM were less than shown, there would be a net loss of N. Greater immobilization than shown in Table 17.1 would indicate a net N gain as well as an associated gain in soil organic C content. Most comprehensive studies of C balance in maize–soybean systems show either no change or a small net loss in soil C over time (Verma *et al.* 2005; Baker & Griffis 2005; Dolan *et al.* 2006), which is consistent with the balance given in Table 17.1.

Fortunately there are large opportunities to reduce N losses from maize–soybean systems through improved management of soil and fertilizer (Section 12.4). Surveys report that adoption of current best N management practices is still relatively low. Site-specific N application and controlled-release fertilizers appear to be the most promising options to reduce losses while maintaining productivity.

Loss of C from SOM and N by denitrification (N_2O release) are causes of GHG emissions in maize–soybean systems. Management practices that can simultaneously increase yields, improve N fertilizer recovery efficiency (NRE), and reduce N losses have the greatest impact on reducing emissions per unit of grain produced. Higher grain yield

Table 17.1 Nitrogen balance (kg N ha^{-1} y^{-1}) of a two-year rainfed maize-soybean rotation in Iowa, USA, based on average yields and management practices

Source of N	Maize	Soybean	Two-year total
Nitrogen supply			
Deposition[a]	13	13	26
Soil mineralization[b]	133	133	266
Residual soil nitrate	33	25	58
Residue, previous crop[c]	84	105	189
Symbiotic N fixation[d]	0	175	175
Manure[e]	40	7	47
N fertilizer[e]	158	4	162
Total inputs	461	462	923
Removal from soil			
Plant uptake (include roots)	246	297	543
Soil immobilization[f]	133	133	266
Volatilization	9	2	11
Denitrification[g]	6	2	8
Runoff	6	2	8
Leaching	20	10	30
Total outputs	420	446	866
Balance			
Harvested grain[h]	−140	−213	−353
Losses[i]	−41	−16	−57
External inputs[j]	211	199	410
Net balance	**30**	**−30**	**0**

[a] Total of wet and dry N deposition (Holland *et al.* 2005).
[b] Estimated mineralization: turnover rate 0.025 y^{-1} of total soil organic N in a no-tilled surface soil layer (0.3 m) with 2.25% organic carbon, C-to-N ratio of 14, and bulk density of 1.1 g cm^{-3}.
[c] Aboveground residue (10.8 t ha^{-1} maize, 3.4 t ha^{-1} soybean, based on harvest index of 0.5 and 0.40, respectively) and roots (15% of total aboveground biomass); 0.75% N in aboveground maize residue and roots, 1.4% and 1.0% for soybean aboveground and root residue, respectively.
[d] N fixation by soybean at 59% of plant N uptake based on Salvagiotti *et al.* (2008).
[e] N fertilizer and animal manure based on 2005 USDA-ERS farm survey data.
[f] N Immobilization in soil organic matter (SOM) via humification process. Sources of N include crop residue, residual nitrate, and external N inputs. Assumes steady-state SOM content.
[g] Denitrification at somewhat higher levels than measured in well-managed maize fields in north and western U.S. Corn Belt (Venterea *et al.* 2005; Adviento-Borbe *et al.* 2007).
[h] Average Iowa maize and soybean grain yields (2005–2009) of 10.8 (1.30% N) and 3.4 t ha^{-1} (6.25% N).

also produces greater crop residue that helps maintain or increase SOM. Greenhouse gas emissions during production of N fertilizer by the Haber–Bosch process (Section 8.6.1) and emission of N_2O from applied fertilizer account for about 50% of total emissions from rainfed maize (Liska *et al.* 2009). Technological advances that increase crop yields and input use efficiency will further reduce GHG emissions from agriculture.

Phosphorus load is also an environmental concern in surface waters, streams, and rivers. Here the issue is not human health, rather the integrity of aquatic ecosystems.

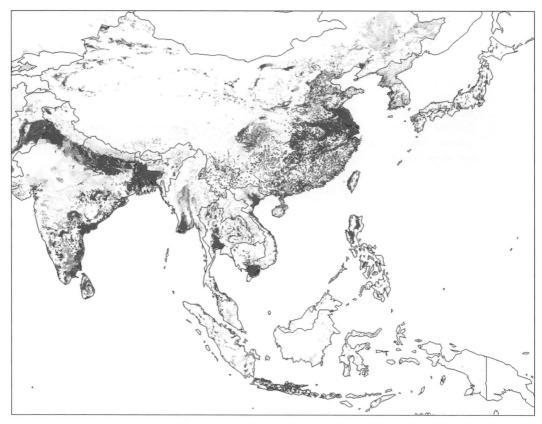

Fig. 17.7 Irrigated rice area in Asia. (Source: Portmann *et al.* 2010.)

Excessive growth of algae causes anoxic conditions, which in turn kill other life forms. Algal productivity is often limited more by P than N because a number of algal species can fix atmospheric N (Section 8.6.2). Preventing P movement from cropland to waterways helps avoid deterioration in water quality. Because P is relatively immobile in soil due to strong adsorption to soil mineral surfaces (Section 7.4.3), P losses to surface water are derived from runoff and erosion. Overall, soil management practices that reduce soil erosion, such as conservation tillage and contour bunds, and promote rapid canopy cover during early vegetative growth, also reduce P losses to surface water systems.

17.3   Intensive rice cropping systems of Asia

Rice provides more calories in human diets than any other crop. In 2008, the harvested area was 156 Mha with 90% of global production in Asia, as shown in Fig. 17.7. Although rice is also grown under rainfed conditions, 75% of total production is from irrigated systems (World Rice Statistics, www.irri.org) and about 40% is produced in intensive, continuous rice monoculture systems with two or three crops per year.

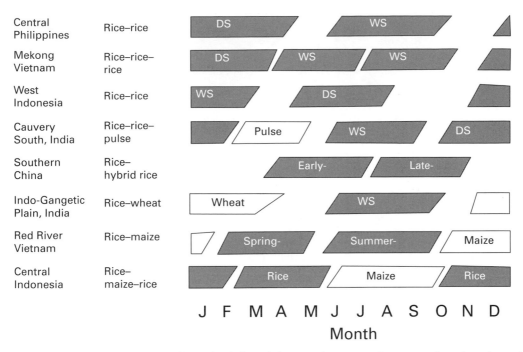

Fig. 17.8 Examples of intensive irrigated rice cropping systems in east, south, and southeast Asia (DS = dry season and WS = wet season rice crops). (*Source:* Dobermann & Cassman 2004.)

17.3.1   Climate, cropping systems, and soils

Intensive irrigated rice systems are the predominant land use in river valleys, flood-plains, and deltas of tropical and sub-tropical Asia. Rice requires warm temperatures and is especially sensitive to low temperature at microspore formation (i.e., spikelet differentiation phenostage) and anthesis. Spikelet infertility increases at mean daily temperature below 22°C (Yoshida 1981). A typical climate for tropical rice is found in the humid lowland tropics at Los Baños, the Philippines (Fig. 17.2). A dry season (DS) rice crop is grown from late December or January to the end of April and relies entirely on irrigation because ET_a then exceeds rainfall. Wet season (WS) rice is planted in July with the onset of rains that exceed ET_a by a large margin during the entire growing season. Triple-crop systems are possible here and elsewhere in the lowland tropics provided turnaround between harvest and planting is rapid. Similar patterns of continuous irrigated rice cropping are found throughout the lowland tropics and sub-tropics of Asia (Fig. 17.8). Sometimes rice is rotated with another crop. For example, a rice–wheat annual double-crop system occupies 17 Mha across south-central China and the Indo-Gangetic Plains of north India, Pakistan, and Nepal.

Intra-annual variation in temperature, daylength, and solar radiation is small in the tropics compared to temperate regions such as the US Corn Belt (Fig. 17.2). Peak solar radiation occurs during grain filling of DS crops, explaining their yield advantage over others grown during the cloudy WS.

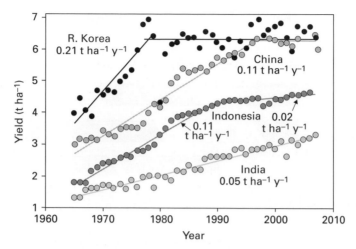

Fig. 17.9 Trends in rice yields, 1965 to 2008, China, Indonesia, India, and Republic of Korea. (*Source:* FAOSTAT.)

Irrigated lowland rice is grown in submerged soil with water depth maintained at about 5 to 10 cm by bunds 20 to 30 cm high around field margins. Rice is adapted to growth in flooded soil by aerenchymatous tissue that provides large pore spaces to facilitate the diffusion of oxygen from shoot to root (Box 13.1). Before planting, fields are flooded to saturate soil for "puddling", a tillage operation of several passes with a shallow plow that destroys soil structure, thereby reducing pore size and rate of water infiltration. Small downward percolation is needed to maintain permanent flood and explains why fine and medium textured soils are preferred for rice production (DeDatta 1981). Puddling also kills established weeds, while flooding during crop growth suppresses the germination and establishment of most weed species.

Irrigated lowland rice soils are often called "paddy soils", a term derived from the Malayan word meaning rice (padi). It is imprecise, however, because there is large variation in soil types used for lowland rice. Entisols and Inceptisols are common because they are found extensively in alluvial lowlands where natural flooding occurs (DeDatta 1981). Vertisols with high clay content and heavy texture are also used because percolation is slow when saturated. Alfisols and Ultisols are other common lowland soils used for rice production.

17.3.2   Productivity trends and genetic advance

The so-called green revolution began with the introduction of short-statured ("semi-dwarf"), stiff strawed rice and wheat cultivars that could respond to N fertilizer without falling over (lodging). Traditional cultivars were taller and lodged severely with even modest doses of N. Rice breeders Henry Beachell and Gurdev Khush developed the new rice cultivars at the International Rice Research Institute (IRRI) in the Philippines in the early 1960s. They were released quickly with immediate impact throughout Asia. By the mid-1960s, rice yields had increased rapidly in most countries (Fig. 17.9). Beachell and

Khush received the World Food Prize in 1996 to recognize their contribution to global food security (World Food Prize website).

Plants of short stature partition less dry matter to stems, allowing more assimilate for grain. The new cultivars also matured earlier than traditional cultivars, a change that enabled introduction of multiple cropping strategies seen in Fig. 17.8. Greater **harvest index**, ability to utilize applied N without lodging, shorter maturity, and multiple cropping supported rapid gains in total rice production that greatly exceeded the increased demand from population growth. From 1965 to 1985, for example, Asian rice production increased by 85% while population rose by 53%. Harvested rice area increased by 13% but actual land area devoted to rice probably remained constant, or decreased, because of multiple cropping. As a result, Asia enjoyed a period of food surplus that kept food prices low and fostered economic development while minimizing requirements for new arable land.

The rate and pattern of yield growth differs by country. Highest initial growth rates occurred in Korea and China where rice is grown almost entirely with irrigation (Fig. 17.9). In India, where less than 50% of total rice area is irrigated, the rate of yield gain was much slower. Rainfed production is lower yielding and more risky, so farmers use smaller amounts of nutrients and other inputs. In Indonesia, where 70% of rice is irrigated, yield growth was similar to that in China from 1961 to 1988 but then slowed considerably. Explanation is found partly due to logistical difficulties in providing adequate irrigation water from major reservoir systems during the DS when yield potential is highest. Data in Fig. 17.9 also reveal that rice yields are no longer increasing in other countries. Yields in the Republic of Korea reached an apparent ceiling of 6.3 t ha^{-1} in late 1970s, while yields in China have also been stalled at the same level since the mid-1990s.

Because rice is an inbreeding crop, cultivar development remained largely in the public sector at IRRI, national agricultural research institutes, and universities. After the release of the first modern rice cultivars in the mid-1960s, breeders focused on improving disease and insect resistance because pest problems became widespread with adoption of continuous annual double- and triple-crop systems. Grain quality was also high on the breeding agenda. As a result, there was little increase in attainable yield of inbred rice cultivars (Peng *et al.* 1999), and it became evident that yields were plateauing in several rice-producing countries (Fig. 17.9). A **yield plateau** occurs when average farm yields reach 70 to 85% of climate-adjusted attainable yield because it becomes technically more difficult and economically less rewarding to seek higher yields as the average approaches the attainable yield limit (Cassman *et al.* 2003; Lobell *et al.* 2009).

Increasing concerns about rice production provoked renewed interest to increase yield potential of rice and in 1990 IRRI initiated a major ideotype breeding program. While the objective to develop a new inbred plant type with higher attainable yield was not successful, a current, revised effort with hybrid rice has achieved some success (Peng *et al.* 2008). Hybrid rice was first developed in China in the 1970s and now accounts for more than 50% of rice area there. Although hybrids give a 10 to 15% yield advantage over inbred cultivars under favorable growing conditions and good management, cost and logistics of seed production have constrained widespread adoption in other countries.

Concerted efforts are now underway to expand hybrid rice production in India, Vietnam, and other countries of south and southeast Asia. The shift in emphasis to hybrids has brought investment by several private sector seed companies into rice breeding.

To date, transgenic rice cultivars have not been approved for use by farmers in any country despite the benefits of Bt insect resistance, and improved nutritional qualities. Recently, however, Chinese authorities gave safety approval for field testing of Bt rice and approval for widespread use is likely to follow. Once the world's largest rice producer allows production of transgenic rice, other major rice-producing countries are expected to begin a similar approval process.

17.3.3   Changes in management and resource-use efficiencies

Crop production in **flooded soil** presents unique challenges for efficient and profitable crop production and technologies have been adopted to address them. Because the average size of managed fields is relatively small in Asia, opportunities for mechanization with large tractors and harvesters are limited. Instead, mechanization has occurred at smaller scales with hand-held tractors for puddling and small stationary threshers following harvest by hand. At beginning of the green revolution in the 1960s and early 1970s, most rice was established by manually transplanting 15- to 25-day-old seedlings into saturated soil at equidistant spacing of 20 to 30 cm. This arduous operation was primarily performed by women and children, and required 10 to 20 person–days ha^{-1} (see also Table 15.1). As labor costs rose, broadcast seeding was developed and this method, using pre-germinated seed, is now widely adopted throughout tropical and sub-tropical Asia.

The ability to control water depth uniformly in a rice field is perhaps the single most important management factor because it affects the quality and efficiency of seedling establishment, weed control, N fertilizer efficiency, and crop growth. Fields are leveled by moving soil with a large plank behind a tractor or buffalo during puddling. Inadequate leveling results in uneven crop establishment, variable nutrient status, weedy patches, and reduced yield. At transplanting or sowing, soil should be saturated but not submerged because O_2 diffusion is too slow in water to support germination and early growth. Once plants are established, water depth is slowly increased with plant height to a maximum of about 10 cm. Fields remain flooded until one to two weeks before harvest, when they are drained to allow soils to dry for harvest.

Intensive rice systems require large inputs of N fertilizer for high yield because N uptake efficiency is typically less than 35%, a value notably smaller than for other major cereal crops (Cassman *et al.* 2002). Low N efficiency results from several factors. First, N transformations, i.e., mineralization, denitrification, and volatilization, are rapid in rice paddies due to warm temperatures so when soil conditions favor N losses, they proceed rapidly. Second, management of water level is critical to control denitrification that is slow in submerged soils but rapid if the soil surface becomes aerated, in patches due to poor leveling, or overall if water depth is too shallow (Buresh *et al.* 2008). Even with good control of floodwater depth, N fertilizer efficiency will be low, however, unless N is applied in several doses matched to crop demand. Recent advances in understanding temporal patterns of indigenous N supply from intensive irrigated rice systems (Cassman *et al.* 1998), and use of green color charts to determine crop N status, have helped to guide efficient **field-specific N fertilizer management** for small-scale farmers in tropical and subtropical Asia (Dobermann *et al.* 2002). On-farm trials in seven Asian countries have demonstrated both increased yields and decreased N fertilizer input from this approach.

17.3.4   Environmental impact and sustainability

Low diversity in cropping sequences, year-round warm temperatures, and humid conditions conducive to pest pressure have raised concerns about the frequency of disease and insect pest epidemics and over-use of pesticides. While breeders continually seek to maintain and expand resistance to major pests, continuous cropping of rice encourages the rapid evolution of disease and insect strains to overcome defences of current cultivars. The result is need for substantial investment in **maintenance breeding** to ensure adequate resistance to evolving pests and rapid turnover of elite cultivars because yielding ability is quickly lost when they are over-used. In the mid-1970s, for example, cultivar IR36 from IRRI was so successful that it was planted on more than 11 Mha in south and southeast Asia. Then, when it became sensitive to the brown leafhopper, yield losses were both large and widespread. While approval of transgenic Bt rice will likely help with stem borers and a few other insect pests, the challenge for control is great because there are a large number of other major pests and diseases that thrive in irrigated lowland rice paddies.

Maintenance breeding, cultivar replacement, and rotations are necessary but not sufficient to overcome pest pressure in intensive lowland rice systems. These practices must be complemented with IPM that was developed in the 1980s. Of particular concern then was the use of multiple insecticide sprays against foliar-feeding insects that also killed beneficial insects that fed on them. Research identified critical defoliation thresholds that proved to be much larger than most farmers realized. Early spraying was reduced and populations of beneficial organisms were established (Huan *et al.* 2004). Nitrogen fertilizer in excess of crop needs was found to create lush crop canopies conducive to development of blast and sheath blight diseases, while K deficiency predisposed rice to leaf spot. Thus, reducing pesticide use requires improved nutrient management, field scouting for foliar damage and critical thresholds in pest numbers, and continual improvement in cultivar resistance and deployment.

Because intensive systems with multiple continuous rice crops have only been practiced for 50 years, the long-term consequences of such cropping intensity are not known. A unique feature of these systems is the ability to maintain, or even increase, SOM and total N at relatively high levels, even when all rice straw is removed by harvest. Low [O_2] in submerged soil slows SOM decomposition (Section 7.2.4), and the soil–floodwater system also provides habitat for a wide range of heterotrophic and associative N-fixing organisms that contribute substantial N input (Section 8.6.2). Ability to conserve C and N can be seen in a field experiment comparing continuous annual double-crop rice versus a wet season (WS) rice–dry season (DS) maize annual sequence as shown in Table 17.2. Rice was grown under flooded soil conditions while maize was grown on raised beds with aerated soil; all aboveground crop residues were removed after grain

Table 17.2 Comparison of carbon balance in two annual double-crop systems, continuous rice versus rice–maize, maintained at two levels of N nutrition for a sequence of five crops each commencing with wet-season vice in 1993 and terminating with wet-season rice in 1995

SOC = soil organic carbon; DS = dry season; WS = wet season. (From Witt *et al.* 2000.)

Cropping system (DS–WS)	Rice–Rice		Maize–Rice	
Fertilizer-N (kg N ha^{-1}, DS/WS)	0/0	190/100	0/0	190/100
	(kg C ha^{-1})			
SOC, 1993 WS, initial	19 130	19 411	19 222	19 376
SOC, 1995 WS, harvest	20 973	22 147	19 005	19 833
Change in SOC, 1993–1995	+1843	+2736	−216	+457
C inputs from crop residues				
C-input ($\sum$1993–1995)	4930	7719	4153	7085
Mineralized C, 1993–1995	3087	4983	4369	6628
Mineralized C as % of C input	(63%)	(64%)	(105%)	(94%)

harvest. In continuous rice that received N fertilizer, the total annual grain yield was about 14 t ha^{-1} and soil C increased by 2.7 t C ha^{-1} over five consecutive rice crops. Without N fertilizer, rice yield was 8 t ha^{-1} and C increased by 1.8 t ha^{-1}. In maize–rice systems, soil C increased slightly (N-fertilized) or decreased (no applied N) due to more rapid mineralization of SOM under aerated soil conditions during maize cropping. In continuous rice without N fertilizer, N input from fixation was 50 kg N ha^{-1} per crop. Assuming that essential nutrients, other than N, are replaced by fertilizers and manures, intensive continuous rice systems are remarkably resilient in their ability to maintain soil fertility. Maize cropping in aerated soil accelerated mineralization of organic C resulting in little, if any, C sequestration.

17.4   Soybean-based cropping systems in Northern Mato Grosso, Brazil

Brazil produced little soybean before 1970 but it is now the world's second largest producer after the USA. The growing area has expanded to 22 Mha and total output to about 25% of global production. In contrast to the USA, where relatively little land is available for expansion, Brazil has large untapped land resources that could support a large increase in soybean area.

17.4.1   Climate, soils, and cropping systems

Soybean production is concentrated in the Cerrado Ecoregion, a vast savanna of 1.95 Mkm^2 of mixed grassland and shrubland that comprises 23% of national land area (Fig. 17.10). The State of Mato Grosso accounts for nearly 30% of production and there soybean is expanding north to the southern perimeter of the Amazon forest (Morton *et al.* 2006), a region of extraordinary diversity of plants, animals, and indigenous

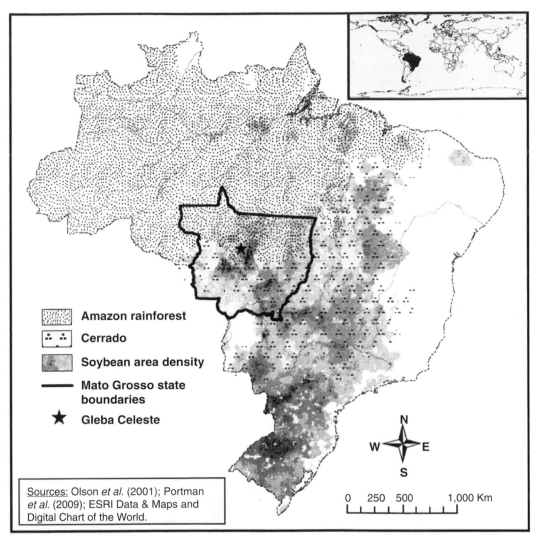

Fig. 17.10 Map of soybean area and outline of Mato Grosso and the Amazon Basin.

people. The Amazon region itself contains 4.2 Mkm^2 and accounts for nearly 50% of Brazil's land area.

Mato Grosso has a predominantly humid tropical climate with small intra-annual variation in temperature, similar to that in Los Baños, the Philippines (Fig. 17.2). The rainy season begins in September to October and continues through April to May. While rains follow a similar temporal pattern throughout the state, rainfall totals are greatest in the north and smallest in the south-central and southeast. Soybean is planted with the onset of rains and matures in February or early March. In the first years after soybean was introduced, many farmers grew a second crop of soybean in a continuous double-crop system. Mounting disease and insect problems encouraged more diverse rotations. Most farmers now plant corn, sorghum, or cotton immediately after soybean harvest. Corn and cotton are favored in central and northern parts of the state, and more drought hardy sorghum in the south. Although mean annual rainfall is relatively large, there is considerable year-to-year variation. Even in years with average rainfall, the rainy season ends well before maturity of the second crop, which experiences terminal drought such that grain filling relies on stored soil moisture, as shown in Figure 17.2.

Cerrado soils are mostly Oxisols, and Amazon soils are mostly Ultisols. Both types have very low cation exchange capacity (CEC) and high concentrations of Al and Fe oxides with a large capacity to fix P (Section 7.3.2). Typical soil pH under native vegetation is 4.0 to 5.0, a level at which Al is active and toxic to most crop plants, or at least inhibits root growth. Despite low fertility and high acidity, these soils have excellent physical properties that permit rapid water infiltration and root growth once soil acidity is alleviated with lime.

In Mato Grosso soybean is produced on a large scale. Fields are typically 150 to 200 ha and farms greater than 2000 ha, with some as large as 20 000 ha. Farming at this scale requires large equipment for planting, application of fertilizers and pesticides, and harvesting. Also essential are logistical skills to purchase and store inputs for timely application, to provide transport for harvest, and technical ability to keep machinery in full working capacity

17.4.2   Crop management and productivity trends

Soybean was initially introduced to the southern Brazilian States of Rio Grande do Sul, Parana, and Sao Paulo where native soil fertility is higher than in the Cerrado. Cerrado was considered marginal, unproductive land until research identified how to overcome the constraints to soil fertility. Solutions were complicated, requiring attention to multiple nutrient deficiencies, acidic pH, Al toxicity, and large P fixation capacity. In the late 1950s and 1960s, agronomist A. Colin McClung from North Carolina State University found the answers. Dolomitic lime (a mixture of Ca and Mg carbonates) was required to increase pH, reduce Al activity, and alleviate both Ca and Mg deficiencies. While more widely available, and at lower cost, calcitic lime (pure $CaCO_3$) raises pH but does not correct Mg deficiency. The management of P fertilizer also took time to refine because typical application rates, as used on soybean in the USA, did little to increase yields. Large amounts of fertilizer are needed to quench the fixation capacity of Al and Fe oxides in these soils before P becomes available for plant uptake.

Once **P fixation capacity** is quenched, subsequent P additions required to sustain yields are much smaller. In north-central Mato Grosso, initial P rate for soybean planted on virgin ground is 90 kg P ha^{-1}; in subsequent years applications are half that amount. Raising soil P levels is critical to support biological N fixation in soybean root nodules. Without adequate P supply, soybean does not fix much N and would depend on the application of N fertilizer. With adequate soil P, however, the soybean–rhizobium symbiosis can fully meet N requirements of productive crops as illustrated in Box 17.2.

This early work by McClung demonstrated the potential to increase crop yields three- to four-fold by improved soil management. Subsequent efforts by agronomist Edson Lobato of EMBRAPA (the Brazilian agricultural research organization) refined and extended this work to farmers in the 1970s, laying the foundation for rapid expansion

Box 17.2 Soil P and N fixation by soybean

Soybean is particularly sensitive to soil P deficiency because nodules represent a significant portion of total belowground biomass and are formed at the expense of root length development (Cassman *et al.* 1980). Because efficient P uptake from soil depends largely on root length density, the burden of investment in nodule biomass increases requirements for soil P availability. And while highly weathered tropical soils require a large initial dose of P to overcome fixation, adequate P supply supports N fixation at levels that eliminate the need for N fertilizer to achieve high yields. Such was the case in a field study conducted on an Ultisol (Cassman *et al.* 1993). Soybean was grown in four consecutive growing seasons. Seed yield averaged 1.7 t ha^{-1} without applied P versus 3.1 t ha^{-1} with about 75 kg P ha^{-1} applied in each crop cycle. Nitrogen fixation increased from 82 to 164 kg N ha^{-1} per crop cycle in these same treatments, which means these soybean crops fixed about 2.2 kg N per kg of applied P. Fixation of N per kg of P uptake, measured in aboveground biomass at maturity, was about 8:1 as shown in the figure below. With a total of 310 kg P ha^{-1} applied to four crops, net P balance was +235 kg P ha^{-1}. This surplus quenches some of the soil's P-fixation capacity so that a greater proportion of future P fertilizer input remains in plant-available forms.

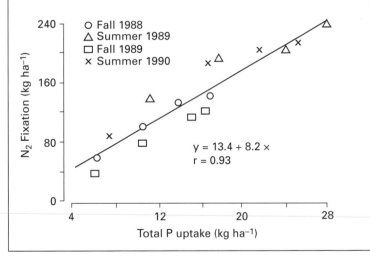

of soybean production in the Cerrado. In recognition of their contribution to converting marginal land into one of the world's most productive agricultural systems, McClung and Lobato, along with Alysson Paolinelli, Minister of Agriculture, Brazil, received the World Food Prize in 2006 (World Food Prize website).

New soybean cultivars were needed for the Cerrado because those from the USA and southern Brazil matured too quickly to take full advantage of the longer growing season (Box 5.1). Adapted cultivars mature later in the warmer climate. In contrast to US Corn Belt soils, where *Bradyrhizobium* populations are well established, soybean, as a new crop in the Cerrado, requires seed inoculation to ensure adequate nodulation for N fixation. Applications of K and micronutrients are also required to maintain high yields.

As with intensive irrigated rice in Asia, the warm, humid environment in the Cerrado and Amazon regions allows insect and disease pests to complete life cycles without major interruption. A number of native legumes that can serve as hosts for diseases and insect pests of soybean also foster persistence. As a result, soybean production requires multiple applications of insecticides and fungicides. Almost all soybean systems are no-till, which speeds management operations and reduces energy use. As in the USA, widespread use of transgenic herbicide-resistant soybean has facilitated weed control in no-till systems. More recent approval of transgenic insect-resistant Bt maize and cotton should help reduce insecticide use in these crops.

Today soybean production in Mato Grosso is an extremely high input and intensive system. While each technology currently used was developed previously for crops in other parts of the world, it is the magnitude of inputs, number of input applications, and scale of farming that distinguish Mato Grosso soybean systems. Although yields increased steadily from the late 1960s until 2000, yields have varied widely since then from a low of 2.2 t ha^{-1} in 2005 to record yields of 2.8 t ha^{-1} in 2003, 2007 and 2008. Much of this variation can be attributed to an epidemic of Asian rust disease (*Phakopsora pachyrhizi*). Average soybean yields in Mato Grosso are typically 10% larger than national yield levels.

17.4.3   Environmental impact and sustainability

The long-term impact of large inputs of pesticides and nutrients on biodiversity and water quality are the primary concern for soybean systems in Brazil's Cerrado and Amazon regions. A second concern derives from the success and productivity of these systems, namely, how much should agriculture expand into the Amazon forest?

Pesticide use on Mato Grosso soybean–maize systems is much greater than on maize–soybean in the US Corn Belt. While much of the difference can be attributed to the warmer tropical climate in Brazil, it may also reflect difficulties in implementing IPM based on scouting for threshold pest populations in the characteristically large fields. Recent epidemics of Asian rust have also increased reliance on chemical control because the disease established very quickly and caused severe yield loss (Yorinori *et al.* 2005). Current cultivars have little resistance to virulent new strains of the disease so farmers often seek control with up to three prophylactic fungicide applications.

A legal requirement to maintain native **vegetation buffer zones** along all major rivers and streams in Mato Grosso should, if strictly administered, greatly mitigate surface, although not underground, movement of nutrients and pesticide residues to waterways. The combination of high rainfall and water infiltration rates into the predominant Oxisol and Ultisol soils increases the potential for leaching and underground movement. Loss of P by leaching is unlikely but nitrate is a risk if N fertilizer is applied in excess of the uptake capacity of corn, sorghum, and cotton crops. Several currently used pesticides have been shown to be at risk for leaching in these soils (Laabs *et al.* 2000). Clearly IPM, including transgenic crops with insect and disease resistance, is essential to minimize pesticide load while maintaining yield.

Perhaps more difficult to answer is the question of further conversion of Amazon forest to soybean production systems. How much conversion should be allowed and who decides? Unsurprisingly, a recent study found a strong correlation between the price of soybean and the rate of deforestation in Mato Grosso where over 0.5 Mha were converted from forest to cropland during 2001 to 2004 (Morton *et al.* 2006). While there remains a vast reserve of Amazon forest to the north, much of it could be converted rapidly to pasture and cropland if it were profitable to do so. The key issue is to find an appropriate balance between economic development and conservation of natural resources and biodiversity for future generations.

17.5 The future of high-yield crop agriculture

The past 50 years was a period of unparalleled innovation and productivity growth in intensive agricultural systems that allowed food production to rise more quickly than population growth, except in Africa. Large investments in research and extension at universities and national and international research centers, and in technology development by private sector seed, fertilizer, and agricultural equipment companies have all contributed to this outcome. But development has been uneven. The greatest increases have been achieved in regions well endowed with good soils and reliable rainfall or with irrigation. Although much can be done in less favorable environments, e.g., semi-arid rainfed wheat production in Australia (Chapter 16), most agricultural lands have severe resource limitations that limit the capacity to markedly increase yields.

Some argue that high-yield, high-input cropping systems have inherent weaknesses that make them unsustainable. Concerns focus on water pollution from leakage of nutrients and pesticides, negative impacts on wildlife and biodiversity, and loss of soil quality to support future productivity. Each of these concerns is important and must be addressed through continuing research, development, and the adoption of new technologies and holistic management approaches. As was explained at the outset (Chapter 1), and stressed in subsequent chapters, agricultural production systems are designed and managed by humans for their benefit. There can be no continuing high productivity without management and inputs, and no advances without the application of science and technology. The great ecological benefit of high-input, high-yield systems is the relatively small area that is needed to supply society with food, fiber, and fuel, in turn leaving more land for nature and its other values.

It is unlikely that the basic structure of existing highly productive cropping systems will change greatly given population and income growth that is driving increased demand for food from limited arable land and water resources. Thus, the greatest environmental threat would likely come from the failure to sustain continued increases in crop yields in the major intensive cropping systems worldwide. Food shortages would demand a massive expansion of crop area at the expense of remaining tropical forests, wetlands, and savannah. The challenge, therefore, is to devise new approaches to high-yield farming that further increase yields while minimizing the negative environmental impacts. This will require new crop and soil management techniques, improved cultivars derived from both conventional and transgenic approaches, motivated young farmers, and scientists from a wide range of disciplines who have a functional understanding of crop ecology.

17.6   Review of key concepts

Characteristics and productivity of high-yield systems

- Intensive, high-yield cropping systems are relatively new yet produce the majority of world food supply. Common features include low diversity, substantial inputs of nutrients to support high productivity and to maintain soil fertility, and a constant effort to stay ahead of insect and disease organisms that rapidly evolve to overcome plant resistance.
- Improved agronomic management and genetic improvement contribute about equally to yield advance in high-yield systems. Sustaining further yield increase depends on fine-tuning all aspects of production to take full advantage of time, solar radiation, and water during the growing season. Technological advances in machinery, information technology, fertilizer technology, biotechnology, and conventional breeding will be required to sustain further yield gains.

Genetic improvement

- Yields are plateauing in some of the most productive high-yield cropping systems because average farm yields are approaching current attainable yield thresholds. Further yield gains in these systems will be small unless physiologists, geneticists, and breeders can identify traits that raise the attainable yield ceiling.
- Large investment in maintenance breeding is also required to provide a continuous stream of new cultivars with adequate resistance to evolving pests. Cultivar turnover is rapid. Integrated pest management is critical to minimize biocide use and to delay development of pest resistance, especially for transgenic crops with bioengineered insect, disease, or herbicide resistance.

Environmental impact

- Environmental concerns center on impacts from use of fertilizers and pesticides, and lack of diversity. Addressing these concerns requires explicit focus on ecological intensification that can sustain high yields while reducing, or eliminating, negative environmental outcomes. Holistic systems approaches are required that effectively integrate new technologies and management practices to avoid unwanted trade-offs.

- Because these intensive, high-yield systems are relatively new, the long-term impact on soil quality remains to be determined. Nitrogen and C balances suggest that intensive irrigated rice systems can maintain SOM while trends in US maize–soybean systems are less certain. Long-term field research is critical to determine the impact on soil quality to determine if present systems are sustainable.

Terms to remember: aluminium toxicity; Amazon soils (Ultisols); breeding, brute force and maintenance; Cerrado soils (Oxisols); deforestation; ecological intensification; environmental impact; field-specific N management; flooded soils; genetic advance; green revolution; harvest index; hypoxia, integrated pest management (IPM); multiple nutrient deficiences; P fixation capacity; puddling; resource recovery efficiency; semi-dwarf cultivars; simplification; soil acidity; specialization; stay-green; sustainability; technological advance; transgenic cultivars; vegetation buffer zones; yield plateau.

18 The future of agriculture

While the future is uncertain in terms of population, demand for food, energy supply, climate and weather, it is worthwhile to consider some scenarios, based on expected trends, from a crop ecology perspective. Critical to this exercise is whether agricultural production can be increased to meet food and fuel needs of an expanding population and whether that can be done safely with acceptable environmental impact. Success in achieving a sufficient agriculture will depend heavily on the rate of population growth, expected demand of individuals, and on decisions made about acceptable levels of natural resource conservation and energy use. Preceding chapters provide a basis for possible technological advances. This chapter reviews trends in population growth and food supply and considers the prospects for a food-secure world to 2050.

18.1   Population and need for food

World population of 6.8 billion in 2010 is continuing to grow but at a decreasing rate, recently predicted (in 2008) by the UN Population Division to reach 9.2 billion by 2050, and later decline toward the end of the century. The decline is good long-term news regarding overall food demand and environmental impact. But it also includes a crucial 40-year period when greater food production must be achieved in the face of demographic shifts that will have great impact on future agriculture. In fact, the time available for development of the required science and technology is much shorter than 40 y because it takes time for validation, adaptation, extension education, and adoption on farms.

18.1.1   Stabilization at last!

The intrinsic reproductive capacity of humans, near 40 births per 1000 population per year, is small compared to other animals but, unchecked, it is large enough to allow rapid increase in population size. Before the industrial revolution, starting late in the eighteenth century, birth rates were countered by high death rates due to disease and low and variable productivity of farming systems. The global population was around 1 billion and the growth rate was small. Then, with the creation of jobs outside agriculture and the development of new farming systems (private farms with legume rotations and recycled manure), global population rose quickly to 1.6 billion by 1900, 2.5 billion by 1950, and 6.4 billion by 2000. Continuing change in food production systems sustained this growth in human numbers, aided by increased longevity resulting from medical innovations in disease control. In sixteenth century Europe only 30% of the population survived beyond age 20 y. Now life expectancy at birth exceeds 80 y in 15 developed countries, 70 y in 100 other developed and developing countries, while only 15 countries, mostly African, have expectancies less than 50 y.

This historical background provides context for data presented in Fig. 18.1a that show the decadal increase in world population since 1750 and projected trends to 2050. Rapid increases in world population in the twentieth century, rising to 120 million y^{-1} around 2000, are projected to fall to replacement levels by mid-century. Population could also be reduced through wars, starvation, or pestilence although it is noteworthy that the population juggernaut continued unabated through two World Wars, numerous regional conflicts, and various epidemics in the past century.

18.1.2   Demographic shifts

Population increase to 2050 is associated with two important demographic shifts of significance to food demand and agricultural development.

First, population increase is not evenly distributed as shown by major region in Figure 18.1b. The largest increase, which occurred in Asia, contrasts with smaller increases in Africa, South America, and North America, and an actual decline in Europe. The result has been a shift towards heavy dominance of population in developing countries that is projected to continue as shown in Fig. 18.1c. In 1950, the proportion of the global population living in developing countries was 68%. It is projected to increase to 86% by 2050. These projections are closely linked to trends in fertility rate, i.e., average number of children per female. While fertility is decreasing generally, the trend is greater in developed than developing countries. According to the UN Population Division, fertility rate of countries that comprise 48% of world population was, in 2009, less than the 2.1 required for replacement. The proportion is projected to exceed 50% by 2015. The relationship between increasing wealth and decreasing fertility is strong. Recent 2009 analysis reveals that fertility rate, still 5 to 7 in some developing countries, falls to replacement level when annual GDP per person approaches US$4000 (www.gapminder.org).

Second, the world is about to pass a significant threshold of demographic distribution. For the first time, in 2010, rural and urban populations are approximately equal (Fig. 18.1d). The trend to urban living, well recognized in developed countries, where less than 10% of the population is involved in agriculture, is now extending to developing countries also. A larger urban population, that is projected to increase to 67% of global population by 2050, will require structural adjustment to agricultural systems to ensure sufficient and stable production.

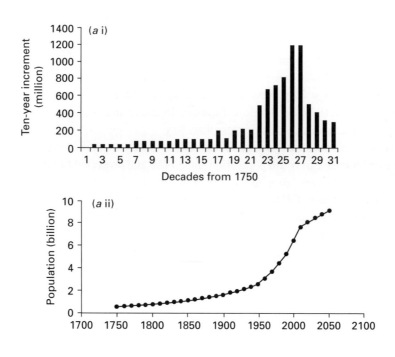

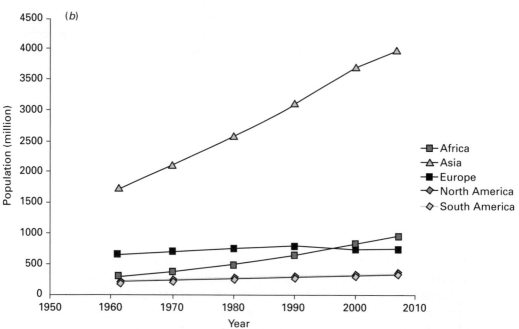

Fig. 18.1 Recent and projected population trends. (*a*) Decadal increase in global population, 1750 to 2050. (*b*) Population in major world regions, 1960 to 2008. (*c*) Population trends in developed and developing countries, 1960 to 2050. (*d*) Global trends in rural and urban population, 1950 to 2050. ((*a*) and (*d*) are from UN Population Division, (*b*) and (*c*) from FAOSTAT.)

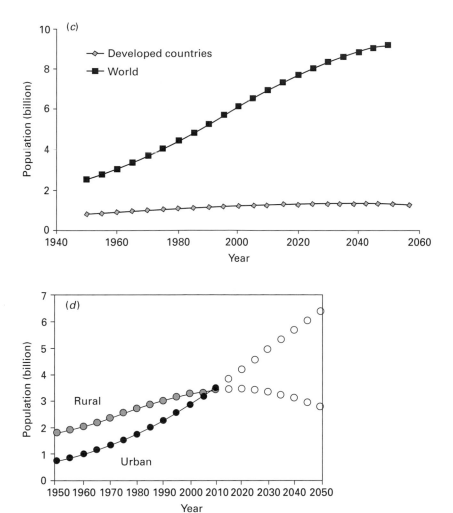

Fig. 18.1 (*cont.*)

18.1.3   Need for food

Production goals for agriculture are defined by size of the human population and its need for energy, protein, and other dietary factors. The Standard Nutritional Unit (SNU, 8.4 GJ cap^{-1} y^{-1}) (Section 2.4) provides a simple measure of food production needed per capita. It corresponds to 500 kg grain cap^{-1} y^{-1} (23 MJ cap^{-1} d^{-1}) at the farm gate, or its equivalent in energy and protein from other sources. After deducting allowances for planting stock, waste, storage losses, reserves, and alternative foods (animal products, fruits, and vegetables), this amount assures a food supply at the table similar to present supplies in developed countries (14 MJ cap^{-1} d^{-1}). World grain production in 2008 provides 371 kg cap^{-1} y^{-1} (Table 18.1). When production of other foods (sugar, potato, cassava, animal products, fish, fruits, and vegetables) is expressed in grain equivalents, the average approaches the 500 kg standard.

Table 18.1 Production of cereal grain from 1936 to 2008

Region	Area harvested (Mha)			Yield (t ha^{-1})			Production (Mt)		
	1936	1978	2008	1936	1978	2008	1936	1978	2008
World	551	715	713	1.16	2.21	3.54	640	1582	2525
Africa	38	637	1071	0.63	1.09	1.41	24	70	151
Asia	196	306	329	0.83	2.02	3.61	251	618	1188
Europe	185	194	127	1.14	2.39	3.96	210	464	504
N America	104	859	775	1.08	3.71	5.93	112	319	460
C America	–	12	13	–	1.83	3.19	–	22	41
S America	22	37	38	1.23	1.68	3.77	27	62	144
Oceania	6	15	20	0.83	1.71	1.69	5	26	35

Source: FAOSTAT.

Predicted increase in population to 2050 (35%) establishes a goal to increase 2008 food production by 70% because many people, estimated at 1 billion (FAO 2009), are undernourished and rising incomes will support more diverse diets and greater consumption of livestock products. Increasing food supplies may be complicated by limited availability of land (Section 7.3) and energy (Chapter 15) and also by climatic change (Section 6.10). Where and how agriculture can meet these challenges is not clear. The principal options are evident: increased yields and/or areas in production. A more vegetarian diet would assist (Chapter 2), while many current foods of Western societies may disappear or become more expensive: cold beer, alfalfa sprouts, and out-of-season fruits and vegetables, for example, embody high energy costs and/or inefficient use of land.

At issue is carrying capacity: areas of land that might be farmed, production levels that might be achieved, and choices about suitable technologies. Trends in areas of cropland by region since 1960 are presented in Fig. 18.2a. Except in Europe, where cropland was withdrawn from production ("set aside") in the 1990s to reduce surplus production from heavily subsidized agriculture, other areas show positive gains. Increases in Africa and South America were small compared to Asia, which experienced rapid expansion during 1980 to 2000, although have now slowed. Crop area in North America has remained relatively constant. As discussed in Section 7.3, areas of robust soils, globally, are limited and subject to severe competition from other uses. It is likely that the current area of 1500 Mha cannot be extended by more than 300 Mha. At the same time, loss of 120 Mha of existing cropland to other uses, mostly urban expansion, would limit net increase to 12%. Hence, to meet projected production goals, primary emphasis must be placed on greater productivity on existing cropland.

Demand for greater productivity will also vary among countries and regions depending on interactions between increasing cropland area and changes in population distribution. Analysis of recent data is presented in Fig. 18.2b. Cropland per inhabitant has decreased in all regions since 1960. In Asia, large and increasing cropland area still gives the smallest area per inhabitant with just 0.14 ha per capita in 2008 because of limited

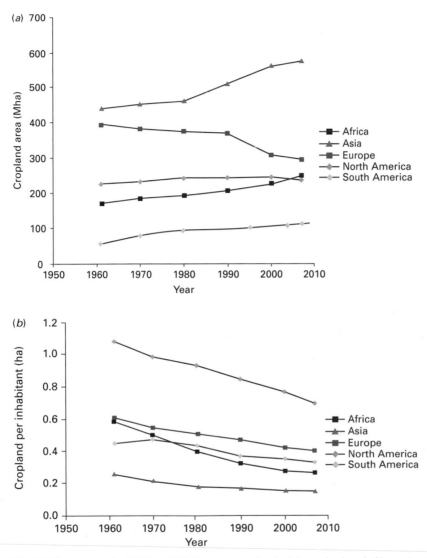

Fig. 18.2 Recent trends, 1960 to 2008, by major region in (*a*) cropland, and (*b*) cropland per inhabitant. (Data from FAOSTAT.)

supply relative to size of population. Africa, which had similar per capita area to Europe in 1960 (0.6 ha), has now slipped to 0.26 ha, well below Europe that, despite "set aside", maintains 0.50 ha per inhabitant. Gains in South America during 1960 to 1980 have reversed as the population has increased more rapidly than cropland area. North America commenced this time sequence with 1.1 ha and despite increases in population and adjustments to cropland area retains the greatest area per inhabitant at 0.70 ha. These are critical numbers in discussions of potential food supply and carrying capacity because they again stress the extreme importance of greater productivity to satisfy demand for food in the next 40 to 50 y.

Here, it is worthwhile referring again to Fig. 2.5 and the inverse relationship between yield and area required for a given amount of production. Relative land areas and resources needed to supply food are reduced sharply as yields are brought to higher levels through the relief of a constraint such as N deficiency. For example, a yield of 1 t grain ha^{-1} without added N, which is typical of maize in many parts of Africa, means that one hectare of land can supply adequate food for just two people (at 500 kg cap^{-1} y^{-1}). At that level, a country of 100 million people needs 50 Mha to support its population. With improved fertility producing a modest yield of 5 t grain ha^{-1} equivalent, ten people can be supported per hectare and the required area is reduced by 80% to only 10 Mha.

Under the above scenario, total nutrient extraction in harvested grain is the same so, by definition, water- and radiation-use efficiencies on cultivated land are increased five-fold. Labor and energy requirements for field operations are reduced by nearly as much (Chapter 15). While energy is then required for production of N fertilizer, it amounts to only 1.5 MJ d^{-1} for each of the eight additional people supported per hectare. This calculation is based on the 4.5 GJ ha^{-1} required to synthesize 100 kg fertilizer N to produce a yield of 4 t ha^{-1} (1.5% N) with allowance for losses. Other nutrients, particularly P and K, would then be depleted more rapidly, albeit from a smaller area, and replacement would require additional, but smaller, energy subsidies than for N fertilizer. Other benefits are also significant. In particular, soil erosion would be reduced because agriculture can be restricted to the best land with low erosion hazard. Also important is that 40 Mha can be left to nature reserves and other uses. The inescapable conclusion is that intensive farming systems are lower input and more sustainable per unit food supply than extensive ones when the goal is to provision a large population.

These concepts of carrying capacity are not new. They have long been evident in the organization of human societies but perhaps forgotten during the successful period of agricultural expansion in the second half of the twentieth century. The English economist Thomas Malthus (1798) was perhaps the first to attempt a scholarly analysis of links between population size and demands on food resources. He concluded that potential for food production is limited and, if human populations were not controlled by "prudence", they would inevitably be controlled by starvation, war, or pestilence. Malthus was not optimistic about preventive restraints and he, and others since, have been led to dire views of mankind's fate. Now, we are even closer to such a disaster should population growth continue beyond current projections. Fortunately indications of significant adjustment to population are hopeful, but danger of complacency remains to obstruct the search for sustainable global agriculture at 170% of current productivity by 2050.

18.2   Food production since 1940

Reviewing changes that have occurred in agriculture in recent times helps identify future options. In retrospect, it is rather amazing that agriculture was able to respond with improved human nutrition as world population tripled in size. Prior to 1940, when population was near 2.2 billion, farming was conducted throughout the world with organic methods and grain yields were only slightly greater than during medieval times. Food supplies were marginal, even for developed countries, and famine and undernourishment were common elsewhere. Remarkable increases in crop yields have been achieved in developed countries since then, followed by the much-publicized "green revolution" in developing countries. Wheat yield quadrupled in India, for example, and rice yields tripled in Indonesia. World grain production increased from 640 Mt y^{-1} in 1934 to 1938 to 1582 Mt in 1978 (Table 18.1) through increase in area under cultivation, improved cultivars and management practices (including irrigation, more effective pest and weed control, and mechanization), and, most importantly, improved plant nutrition.

Large differences in industrialization, energy use, and standards of living exist between developed and developing countries. At both ends of the spectrum there are concerns for the future of agriculture because production is either insufficient or is perceived to be unsustainable. In developing countries, concern is for access and affordability of nutrients and other inputs, and for new lands to develop for agriculture to meet increasing food demand or for export to generate income. In developed countries, particularly those with an exportable surplus of agricultural commodities, attention is turning to environmental costs of "excess" production. There, efforts are being made to remove marginal agricultural land from production, reduce nutrient losses to the environment, and retain as much land as possible in conservation reserves for wildlife and biodiversity, and to provide recreational enjoyment from nature.

18.2.1   Food increases in developed countries

Developed countries of Europe, North America, and Asia met the challenge of population growth through intensification. Fertilizer was the principal component of new technology with important contributions from plant breeding, irrigation, mechanization, and biocides for disease, insect, and weed management. Aggregate impacts of technology are illustrated in Box 4.4 with time-trends for global maize, wheat, and rice yields. Yields of all crops have increased linearly since 1965, at 63, 53, and 39 kg ha^{-1} y^{-1}, respectively. The data in Table 18.1 reveal similar trends in cereal yields in most regions.

Despite a doubling in population and, in some cases, a decline in cultivated area, people in developed countries are now well supplied with food (Table 18.2). Some countries, including the USA (85 Mt), Australia (20 Mt), France (16 Mt), and Germany (6 Mt), are net exporters of cereals (2006). Redistribution of population between rural and urban communities (Fig. 18.1d) has continued as people moved off farms to more profitable employment in industrial and service sectors. Mechanization substituted for labor such that only 7% of the workforce is now directly engaged in agricultural production in Europe, and fewer than 2% in the USA. Because agricultural yields have been brought closer to the limits imposed by climate and biology, potential for further increases in food production in these countries is now less than before. In most areas, however, actual yields are still less than what is attainable with more intensive methods indicating that significant reserve capacity for production remains.

Table 18.2 Recent (2008) status of agriculture and food supply

Attribute	Units	Africa	Asia	Europe	South America	North America
Population	million	987	4075	732	385	345
Population 2050	million	1998	4829	691	483	448
Agric. population	%	51	51	6	14	2
Cropland^a	ha cap^{-1}	0.25	0.14	0.40	0.33	0.65
Nitrogen fertilizer	kg N ha^{-1} crop	13	115	46	39	57
Grazing land	ha cap^{-1}	0.92	0.28	0.24	1.17	0.74
Cereal production	kg cap^{-1} y^{-1}	154	306	689	379	1331
Net cereal import	kg cap^{-1} y^{-1}	51	17	−18	−47	−329
Food energy	MJ cap^{-1} d^{-1}	6.3	6.5	8.2	7.0	9.0
Protein	g cap^{-1} d^{-1}	62	72	102	80	113
Animal protein	%	20	31	55	50	64

Note: ^{a} Includes annual and perennial crops.
Source: FAOSTAT.

18.2.2   Food increases in developing countries

In contrast to developed countries, birth rates remain high (Fig. 18.1c) in many developing countries. Two thirds of cities with more than 5 million inhabitants are now found in developing countries, and urbanization is continuing rapidly (Fig. 18.1d). Most people, however, remain in rural sectors with large proportions involved directly in agriculture, highlighting the importance of subsistence, rather than market economies. Subsistence systems place a greater burden on land because they must supply housing and fuel as well as food.

Continuing expansion of arable areas remains a feature of agricultural development in developing countries. This, together with increases in productivity, explains the significant increases in cereal production reported in Table 18.1. While those increases remain inadequate to feed the existing population, overall, they have greatly reduced the need for food import in populous Asia, but not in Africa. A major difference between these two regions is found in fertilizer use, and in turn reflected in trends of expansion in crop area (Fig. 18.2a). Africa accounts for just 1.5% of world N fertilizer use while Asia uses 63%. That amounts to an average rate of application to arable land plus permanent crops of 13 and 115 kg N ha^{-1} y^{-1}, respectively (Table 18.2). Data presented in Fig. 18.2a reveal that rate of agricultural expansion in Asia has slowed since 2000, but has increased in Africa. Because the most productive areas were farmed first, attainable yields from new lands are generally less such that disproportionately larger additional areas are required to satisfy food needs. In consequence, despite improved yields and greater areas under cultivation, food supply remains precarious in many developing countries as populations continue to expand rapidly and cropland area per inhabitant continues to fall (Fig. 18.2b). Under these conditions, even single years of below-average yield can cause famine in countries with small reserves of food or currency.

Interpretation of the time trend of land allocation to crop production is complicated by simultaneous increase in population and by multiple cropping in favored environments. The high rates of N application in Asia reflect the high frequency of multiple cropping there. The important question to ask is how long can expansion and intensification of cropping continue to increase production. Prospects appear much greater in Africa and South America than in Asia.

18.3 Immediate challenges

Given the constraints on land and water for agriculture, and continued growth in population and food demand, the inescapable challenges are how to move forward with agricultural development from a current situation of small food reserves to one in which a population, projected to reach 9.2 billion by 2050, can enjoy a sufficient and stable food supply. Further complicating this challenge is the need to preserve as much land as possible for nature and environmental services.

18.3.1 Global carrying capacity

Estimates of global carrying capacity require information on agricultural area, crop productivity, and dietary requirements. Excluding carbon-rich, biodiverse natural ecosystems such as rainforests and wetlands, new lands available for expansion of agriculture, perhaps a net 200 Mha (Section 7.3), are not evenly distributed around the world. In developing countries, where pressure to increase production is greatest, new lands are found predominantly in Brazil, e.g., 60 Mha of low productive pasture in the Cerrado, and grasslands with moderate rainfall and high evaporation in sub-Saharan Africa. Given competing pressures for water supply, there will be little opportunity to expand irrigated areas much beyond the current 280 Mha (Chapter 14).

Large differences in yield achieved with different levels of external inputs complicate estimates of productivity. Smil (2001) estimated that agricultural area required to maintain 10 billion people could vary from 800 to 3000 Mha depending upon production techniques and dietary energy and animal protein content. The range for production systems that provide reasonable diets of 11 MJ d^{-1} energy, with 30% supplied by animal protein, is between 1100 and 1500 Mha. In other words, with more restrained eating in some places, better eating in others, and reduction of waste, production of adequate food for 10 billion is possible on existing agricultural area. This is not a surprising conclusion. One can calculate, using the SNU of 500 kg cap^{-1}, that an average yield of 3.3 t ha^{-1} on 1500 Mha of current cropped area would feed a population of 10 billion. But the challenge is to achieve that result, bearing in mind the limitations to soils and rainfall, and competing demands for land and water.

Analyses of future production require estimates of productivity gain. We have explained earlier (Section 4.4) why linear rates of cereal yield increase, that were nearly 3% of average yield in the 1960s, have fallen to about 1% of average yield by 2010 as yield has increased. Even at a compound rate of 1%, production would increase by only 50% on the same agricultural area to 2050. And, without some major break-through that rate is most likely to decrease. We conclude, therefore, that some expansion of agricultural area will be necessary and the question is "How much?" Whatever the answer, it will be reduced by success in refinement of agronomic practice to achieve higher productivity through more efficient use of land, water, and nutrients, combined with effective pest and disease control. Success in those is the only way to minimize the need for additional land. And despite claims made for the promise of biotechnology, it is not realistic to rely on presently unforeseen "breakthroughs" to increase the intrinsic productivity of crops.

For comparison, various authors have also assessed potential productivity of "traditional agriculture" representing a high state of organic farming without external inputs of fertilizer or biocides. Buringh and van Heemst (1979) estimated that without energy for irrigation and drainage, the maximum extent of land that might be cropped was 2500 Mha or 1.6 times the present area. Estimates of grain production proved difficult because yields with limiting nutrients vary with lengths of fallow periods and with amounts of land assigned in rotation to forage legumes. Their conclusion was that best methods of organic farming (all wastes and manures recycled plus legume rotations) applied to all possible land could supply only 3.2 Gt grain, sufficient for 6.4 billion people. Applied to present farmlands, traditional agriculture of this kind would support less than 60% of present population. Despite recent claims to the contrary (Badgley *et al.* 2007), organic methods cannot feed the world (Smil 2001; Connor 2008).

Two satellite points emerge from studies of land use and land capability. First, present agriculture in most developed nations operates at less than achievable yield; few countries, including perhaps Japan and several in Europe, are truly intensive in their agriculture. Second, more than 60% of farmland well endowed with good soils and water supply from rainfall or irrigation is found in Asia and North and South America. Australia's proportion (3%) is small, although quite large in relation to its population. By contrast, Africa, where a population 40 times that of Australia is increasing at 0.028 y^{-1} ($t_d =$ 25 y), has only 6% of the world's good land but a large supply of soil with low fertility and/or limited water supply. This situation suggests that North and South America and Australia will continue to have a role during this century in meeting world food needs through exports. To continue as suppliers of food for developing countries, exporters must remain highly efficient in production and thus low in cost.

18.3.2   The specter of climate change

It is easy to be frustrated with current "debate" about climate change. That climate changes is a reality, with major consequences in the historical record. Recent small increases in terrestrial temperature motivated some scientists to evaluate the impact of increasing atmospheric concentrations of GHG on climate. Prediction of more rapid anthropogenic global warming (AGW) during the twenty-first century is the outcome (IPCC-AR4 2007). As a result, the media now persistently attribute unusually hot weather, even if it falls within recorded experience, as proof that unprecedented climate change is already with us, and having deleterious impacts. Politicians around the world also find convenience to attribute dwindling city water supplies to climate change even when greater demand on decades-old infrastructure is the root cause. Sensible debate is impossible unless variations, that comprise weather, are separated from long-term trends, that are climate change, and there is realization that both weather and climate can change independently of human influence.

There are many estimates of impacts of climate change on crop production based on models (crop simulation and global circulation) of varying complexity. Higher temperatures hasten phenological development and are more or less optimal for growth of current crops depending upon species and location. Greater $[CO_2]$, on the other hand, promotes growth and increases crop water-use efficiency generally. The result of regional assessments, that usually do not include effects of new pests and diseases, are unsurprisingly for either smaller or greater productivity, depending upon current temperature regime. To be useful, such predictions should be made for crops, cultivars, and cropping systems most adapted, according to current knowledge, to these "future" climates. That is not always the case as most studies evaluate current cropping systems in future climates, so unrealistic disasters are well represented in the literature and equally beneficial responses likely underestimated.

Modeling studies reported by IPCC-AR4 (2007) suggest that major cereals will gain in productivity (10%) in high to mid latitudes with temperature increase up to 1 to 3°C, but lose productivity in low latitude environments with increases of 1 to 2°C. Overall, predicted climate change to 2050 (less than 2°C) would not reduce global carrying capacity but would certainly change production systems, productive regions, and export opportunities. As a result, there would be winners and losers.

18.3.3   Increasing and relieving demand on production

It is important, from an ecological perspective, to consider new forces that might increase or decrease demand for food in the near future, and change the trajectory based on population increase alone. Two factors that might increase demand are production of biofuel and greater proportions of animal protein in developing country diets. Demand for both energy and livestock products increase rapidly as incomes rise, as illustrated in Fig. 18.3.

Biofuel Recent expansion of biofuel production is not yet reflected in the crop production data in Table 18.2, except that previous cereal grain surpluses in North America and Europe did encourage politicians to imagine that therein lay part of a solution to challenges of energy supply, energy independence, and climate change mitigation (Chapter 15). World production of biofuel, mostly in the USA, Brazil and Europe, has expanded rapidly since 2003, reaching 100 and 27 GL, respectively, of ethanol and biodiesel in 2010 and projected to increase to 135 and 40 GL in 2015 (Anon. 2010). In 2008, 83 Mt of US maize grain (27% of total production) was used in ethanol production and the current federal mandate for maize-ethanol will require 138 Mt by 2015. This increase is much larger than the rate of yield gain in maize (Fig. 17.4), which means less available for export. The USA accounts for 52% of global maize exports while imports to developing countries in Africa, South and Southeast Asia, Central America, and Mexico

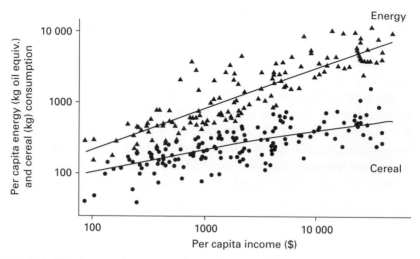

Fig. 18.3 Global per capita energy and cereal consumption versus per capita income, 2003 to 2004. Note that both axes have logarithmic scales. Each datum is for an individual country. From Naylor *et al.* (2007).

were 30 Mt in 2007. Most feedstock for biofuel used in Europe was imported edible oilseed crops producing similar effects on world food supply, although indirectly through third countries. While it is not possible to attribute the large increase in grain prices of 2008 to competition with biofuel production alone, it certainly played a significant role.

Brazil plans further expansion of biofuel production from sugarcane. The USA, in addition to expansion of maize-grain ethanol mentioned above, has mandated the production of a further 60 GL as "second generation" biofuel from cellulose and other non-food biomass as feedstock. In Europe there is more interest in importation of fuel or feedstocks. Sources are developing countries where, in response, major land conversion and agricultural development projects are underway. Current world area of oil palm is 13 Mha but Indonesia alone plans immediate expansion to 20 Mha. The new crop Jatropha was quickly planted on 1.5 Mha by 2008 with expectations for 13 Mha by 2010 (Section 15.5). While full realization of these plans depends on continued high prices for petroleum, these projections provide insight into potential competition for land between food and biofuel crops. And, despite frequent claims to the contrary, it is unlikely that land suitable for profitable production of non-food biofuel crops could not also be suitable for food crops. It will be interesting to see how the story unfolds. Although there are many dangers to food supply and environment, there are also significant opportunities for benefit in countries with large land area and small internal demand for motor fuels.

Dietary change Feasible suggestions to reduce demand are found in dietary control, including the expansion of vegetarianism. Significant reductions in intake are possible in many developed countries where food intake exceeds requirement for active lifestyles. Overconsumption has deleterious effects on health as seen in the rising frequency of obesity. Data in Section 2.3 provide a basis for discussion. While food requirements for individuals vary with body size and physical activity, a useful standard, taking into account the large size of European people, is 10.5 MJ cap^{-1} d^{-1}, in part provided by 50 g protein. Comparison with data in Table 18.2 reveals excess consumption, or considerable waste, over requirements in developed nations: +33% in Europe and +52% in North America.

Meat produced from grain-fed livestock requires a significant increase in equivalent grain production per person. Thus, decrease in meat consumption, and expansion of vegetarian diets, would reduce demand for grain in developed countries. The gain would, however, be less than is often quoted. While grain-to-meat conversion by mass for lot-fed beef is 8:1, cattle spend most of their life cycle grazing pasture or forage crops not suitable for humans. Their life-cycle grain-conversion efficiency of human-edible food is about the same as for poultry and swine (c. 3:1) (Section 2.3.3).

The situation in developing countries is distinct with low consumption of energy and animal protein, the latter insufficient, in many cases for adequate nutrition, especially for children (Demment et al. 2003). Both energy and meat consumption need to increase in developing countries, and will do so as incomes rise. Because major expansion of population will occur in developing countries (Fig. 18.1d), dietary change there, more food and more animal products, will greatly increase demand for crop production, regardless of dietary changes in developed countries.

18.4   The importance of a technological agriculture

Present technological levels of farming systems in developed countries are variously characterized as "conventional", "integrated", or pejoratively as "industrial". Developments important in modern agriculture are increased scientific and technological knowledge, replacement of human and animal power, principally by internal combustion engines, widespread use of fertilizers and biocides, and rapid replacement of cultivars, including some with transgenic traits. We prefer the term "integrated" to emphasize the combination of natural cycles and agrichemicals that, respectively, promote productivity and protect it. Replacement of human labor and animal power by machines changed the scale, speed, and timeliness of operations. It provided sharp increases in productivity and released large areas of land from production of fodder for work animals. Enormous increases in productivity and food supplies are the important benefits of modern agriculture.

Advances in scientific understanding, in particular of the chemistry and biology of soils, of nutrition, metabolism, and reproduction of plants and animals, and of pests and disease, have allowed significant refinements of many long-standing practices. Our understanding of weather and microclimate also has advanced although we still are unable to predict future weather with a high degree of certainty. In contrast to earlier times, agricultural practice has become more chemically based, not just in agrichemicals, but also in its entire approach to the management of biological processes central to productivity.

It is, however, a serious mistake to assume that modern agriculture can be identified as a single concept involving uniform application of knowledge, machines, and agrichemicals. Diversity of farming practices occurs because soils, climates, access to markets and technology, and cultural influences differ greatly between regions. Farmers are numerous and independent and seek to apply best practices for their land and resources. Similar diversity exists within research and extension professions. This diversity has been a priming force for continuous streams of innovation and change.

18.4.1   Problems with modern agriculture

Increased productivity has not been achieved without social or ecological costs. Change was essential for increased levels and efficiencies of food production and some disruptions followed as natural consequences of change; others occurred through poor choices and mistakes. The range of concerns raised in the minds of the public and agriculturalists is justifiably wide.

- Pollution of surface and subsurface waters by agrichemicals, their residues and contaminants, rendering water unsafe or unsuitable for other uses, and disturbing natural systems.
- Contamination of products with residues of agrichemicals rendering them unsafe or uncertain for consumption.
- Depopulation of rural areas, as farms are aggregated into larger more mechanized units able to produce food at lower prices and raise standards of living for fewer farmers.
- Economic strain in seasons of low yield or low returns because agrichemicals and purchased seed have become a significant part of production costs.
- Extension of farming into fragile land made possible by use of inputs, and thereby increasing competition with nature conservation and, when poorly managed, leading to problems of soil erosion and siltation.
- Salinization through changing hydrological balance of rainfed cropping in some semi-arid areas and failure to provide for adequate drainage in irrigation.
- Reliance on non-renewable petroleum as primary energy source for motive power and for production of inputs such as N fertilizer and pesticides.
- Uncertainty about the environmental impact and safety of transgenic crop cultivars with insect and/or herbicide resistance.

The question now is how to address these concerns while seeking to increase production. Regardless of directions taken, the basic strategy must be to supply sufficient food for humans through manipulation of environments and plant communities in ways that provide for efficient use of scarce resources. That requires protection from losses during production and afterwards during storage, processing, and distribution. Our view is that current problems are being reduced or eliminated through continued evolution of existing systems supported by research, education, and legislation. Many countries now have versions of "good agricultural practice" (GAP), also called "best management practices", to assist management of environmental and food safety issues. Retail food companies, which account for a large majority of food sales in developed countries, increasingly require certification that production methods provide safe and healthy food. Now, socially and environmentally responsible production methods have been added to requirements for selling to these supermarkets. Vulnerability to litigation and large financial loss are primary motivations for such traceable, certified "farm to fork" programs. The impact of these trends also is felt in developing countries as they seek access to markets in developed countries.

In contrast to our argument that integrated agricultural systems, based on modern science and new technologies, can continue to evolve to successfully address current and future challenges, others believe that modern agriculture is inappropriate to the tasks ahead and must be redirected, perhaps radically, towards "alternative" methods.

18.4.2 Views of technology in agriculture

Alternative movements The range of technical, ecological, and social concerns over modern agriculture can be seen in the many groups that espouse changes. Membership of such groups is wide, including farmers and consumers, scientists and non-scientists, environmentalists, and industrialists. Their motives are equally diverse, ranging from reverence for nature, nostalgia for traditional methods of farming, dislike of science and technology (particularly of chemistry), and survivalist goals, to market opportunities and political change. Agriculture has always been blessed with theorists of new ways. Cato and Jethro Tull were each important in their time. What seems different now is that more of the criticism and advice directed at agriculture in developed countries is based in political movements and lacks practical experience in farming (Taverne 2005). In part, this reflects our encouragement of activism as a positive force in society and the increasing dominance of urban politics over rural affairs. A plethora of concepts about farming has emerged under labels such as "organic", "biodynamic", "alternative", "appropriate", and "sustainable". These concepts are only vaguely defined but less use of agrichemicals and inorganic fertilizers is usually involved.

Identification of links between pesticide exposure and cancer, even when such risks are very low, is a major cause of widespread public concerns about the use of agrichemicals. Ames *et al.* (1990a) have shown that over 99% of pesticides in US diets are natural and, of 52 that have been tested at high doses, 27 are carcinogenic to rodents. The authors also challenge procedures for testing carcinogenicity and the entire approach to setting tolerances (Ames *et al.* 1990b). At high dose rates most natural pesticides are also carcinogenic. Their point is that feeding toxic levels causes injury and stimulates cell division. Dividing cells are then susceptible to mutagenic defects arising from any source. In other words, the tests measure toxicity at massive doses rather than carcinogenicity. Humans, they point out, have many defenses for toxins that are equally effective against natural and synthetic compounds. Even though natural biochemicals are vastly more abundant in food, however, concerns over biocides have led some consumers to seek produce grown without them ("organic" food), and some farmers to provide those products, either for similar concerns and/or to take advantage of higher prices obtained for organic produce. Alternative movements sometimes argue that farming should be free of external inputs, which, they believe, represent an unsustainable drain on energy resources while destroying the natural biological order of agriculture. Some

Amish religious groups never adopted agrichemicals following a philosophy that farming should rely as much as possible upon human and animal labor. In their view, the natural order provides plants as fixers of energy for the biological system within which human activity ought also to be constrained.

Protagonists of organic agriculture (OA) are commonly at pains to explain that their system is unfairly described as one that simply prohibits the use of synthetic inputs, and more recently of transgenic plants and animals. They say OA is better described as a system that relies on natural processes of recycling and diversity to regenerate fertility and maintain productivity. To make the point, they portray conventional agriculture (non-OA) as a system that has forsaken those processes by substituting "industrial" inputs in their place. That is a serious misrepresentation. Modern agriculture "integrates" agrichemical inputs with biological systems, and does it productively and safely. Given that OA occupies such a small proportion of world agriculture (1%), readers can be assured that there is much more use of diverse cropping, rotations, residue management, manure utilization, conservation tillage and other soil building activities, and innovation outside OA than within.

Our perspective We see external inputs and sophisticated technology as essential in agriculture. Without them, productivity will spiral downward leading to poverty and starvation for many of the present world population and degradation of soil quality. The real issue for the future is provisioning and management of energy, nutrients, and agrichemicals in farming so that society can meet its needs with acceptable balance between use of land in agriculture and conservation of natural resources and biodiversity. Improvement of agricultural practice, of which one critical part is improvement of agrichemicals and fertilizers and their use, is a central challenge.

While we have no problem with alternative agriculturalists pursuing their own goals and methods to their financial benefit, we see significant dangers in attempts to malign conventional agriculture, or to impose poorly researched theories on the larger society through fear or government regulation without regard to consequences.

To be sufficient, agricultural systems must further intensify, which means increasing dependence on external sources of nutrients. Barring a collapse in the size of human populations, we see no alternative to that course. Our vision of farming is by no means a bleak industrial landscape saturated with agrichemicals, however. Rather, it is of a healthy landscape, well managed both ecologically and economically, with most current natural ecosystems and the wildlife they support spared from conversion to agriculture. Modern farming systems are sustainable as long as there is energy and inputs for their maintenance. Their weakness is not low sustainability but dependence on inputs (low autonomy). The same is true for most alternatives. High-yield organic agriculture, for example, generally requires large additions of organic material collected from outside the system. In this form, it also possesses low autonomy. Production from organic farms without external inputs is inevitably small either because all fields have low yields, determined by weathering rates of soil minerals and by other natural inputs, or because large portions of land must be assigned to legumes for N production. In those forms, organic farming can be sustainable but its productivity is sufficient only for a world population about half present size.

18.5   Improving technology

A recurring theme in this text is that agricultural technology, far from going away, must become increasingly sophisticated and integrated. Information is the basic element in future changes whether in the form of better understanding of the biology and ecology of pathogens, genetic traits for insect resistance, yield stability with variable weather, an IPM system, smart machines, or the dynamics of soil nutrient supply. Our discussion has identified three important tasks for the future: increase production; improve efficiency in the use of scarce resources; and resolve problems related to the use of agrichemicals and fertilizers. Accomplishing these tasks involves increasing the information content of agriculture, in particular improved understanding and management of energy, water, soil, genetic resources, weeds, pests, and diseases, and adapting to climate change. Each of these topics deserves brief attention.

18.5.1   Energy

Shortage of energy for agriculture is not a major issue in the foreseeable future because agriculture uses only a small portion (3–5%) of the total energy consumed in developed countries, and less in the world as a whole. Other aspects of current lifestyles such as home heating, lighting, refrigeration, and transport account for much larger fractions of energy use and offer far greater opportunities for conservation. Rising prices for energy will bring conservation in all of these sectors as well as in agriculture, seen most recently during the period of record oil prices ending in 2008. Agriculture's use of energy is embodied in three external inputs: labor, machines, and agrichemicals/fertilizers. These are closely linked so that energy expended in one form can substitute for either of the others (Chapter 15).

Reduction in use of machinery would require a return of human and animal labor to agriculture in disproportion to savings in liquid fuel. This arises because of high maintenance energy (food) requirements for animal and human workers and their slow work rate. Analyses of manual work (Table 15.1) emphasize that peak labor demands during critical procedures such as sowing and harvesting greatly restrict the overall efficiency of human work. Further restrictions arise from the large embodied energy cost of human labor with current Western standards of living ($c.$ 600 MJ d^{-1}). On the other hand, "smart labor", i.e., time spent in sophisticated management, can bring savings in total energy use. We do not expect that reduced supplies of petroleum will bring a return to horses and mules. Petroleum can be replaced by more expensive shale oil, or by H_2 from nuclear reactors; even biomass-fed steam engines would win in competition with animal power. Horses convert only about 7% of combustible energy in feed to work (Section 2.3) whereas efficiency of steam engines with similar biomass can exceed 30%.

Significant gains in energy efficiency were made in agriculture following the surge in energy prices in the 1970s. Greater use of diesel motors, larger tractors, reductions in tillage, and improved fertilizer use efficiency were the main elements. Larger tractors allowed improvements in yield through timeliness and uniformity as well as a significant reduction in embodied energy of labor (per unit production) through greater labor productivity. Avenues for additional improvements include further reductions in tillage and improvements in fertilizer practices. Conservation tillage systems, for example, are widespread (Section 12.5). World area of 106 Mha in 2007–8 is found mostly in the USA (27 Mha, or 38% of all cropland), followed by Brazil (26 Mha), Argentina (20 Mha), Canada (14 Mha), and Australia (12 Mha). Paraguay (2.4 Mha) and China (1.3 Mha), each having more area under conservation tillage than Europe (1.1 Mha) (Derpsch & Freidrich 2009). Moreover, only 7% of total cropland is under conservation tillage, which means it can be expanded considerably dependent on further improvements in herbicides and implements, especially for small-scale farming systems. "Smart" machines that sense needs for tillage, fertilizer, or herbicide on a microscale and apply them only to selected areas offer another avenue. Irrigation and drainage systems that depend upon pumping are particularly vulnerable to diminished supplies of energy. Farming below sea level in Dutch polders, and pump irrigation in the American Great Plains and the Indo-Gangetic plains of South Asia are examples of activities that will encounter problems if energy prices continue to rise.

Now, at the beginning of the twenty-first century, concerns about energy use in agriculture have receded. They have been replaced by concern, and mandated requirements in some countries, that agriculture should not just feed the population, but also play a significant role in fueling it, and at the same time mitigate climate change by sequestering carbon to offset society's other emissions of GHGs. Biofuels are readily made from sugar, starch, and oilseed crops, but at the expense of competition with resources for food production, and that competition is severe (Chapter 15). The Standard Nutritional Unit (SNU) of 500 kg grain, sufficient to feed one person for one year, produces just 200 L ethanol (140 L gasoline energy equivalent). Diversion of food products to biofuel production in amounts large enough to impact on energy supply must compete with food!

Growing non-edible crops, e.g., the oilseed crop Jatropa or the biomass crop switchgrass, is not a solution that avoids impacts on food production because those crops compete directly for land, water, nutrients, and investment. We have argued that biofuel made from crop residues, to the extent that they can be removed without deleterious effects on productivity, define the limit to biofuel production in a world that must seek to increase 2008 food production by 70% by 2050. To that end, three important tasks stand out. First, establishment of commercially viable conversion of cellulosic material to second generation fuels; second, design of highly integrated food and fuel production systems; and third, evaluation of where and how they can be managed for agronomic and ecological viability. The contribution will not be large on the world scale but there will be some countries with large agricultural resources relative to population size where contributions can be significant, including to agricultural development.

Manufacture of N fertilizer has a high energy cost (45 MJ kg^{-1} N). With limited energy, legumes will receive renewed emphasis because they can supply N with expenditure of only about 10% as much fossil fuel in their production. The extent that they may substitute for non-legumes, however, will continue to be limited by their requirement for fertile soil and by the problem of how we might use more forage. If energy really becomes scarce, it may be worthwhile bypassing forage legumes completely, provided non-legumes can be engineered to accept rhizobial symbiosis. That would, however, reduce their yields significantly due to higher energy costs associated with N fixation compared to reduction of nitrate (Section 8.6) and increase pressures for expansion of land under cultivation.

Despite the challenge that higher energy prices will impose on agriculture, we are heartened by the tremendous improvements in energy efficiency achieved during the past 30 years. In the first edition of this book, published in 1992, the data in Table 15.3 of that edition showed an output/input energy efficiency ratio for maize production in Indiana of 6.2 for a yield of 8.8 t ha^{-1}. In updating that table for this edition, data for an average maize crop in Iowa, now yielding 10.8 t ha^{-1}, revealed a net energy ratio of 10.9. This is an excellent example of the resource-conserving potential for ecological intensification (Section 1.2).

18.5.2   Water

Expansion of irrigation was a major contributor to increased agricultural production since the 1960s. It was a driving force for success of the "green revolution" in Asia that now has 60% of world irrigated area. Agriculture uses 69% of all extractions from rivers, lakes, and underground sources on 280 Mha of irrigated cropland. But society is also developing greater demands for urban, industrial, and recreational uses of water. Competition can only be resolved as far as greater extraction and efficiency of use allow.

Issues of extraction are difficult to assess. On the one hand there is plenty of water in the world. Fresh water accounts for about 1% with another 2% in polar ice caps and glaciers. Access to the major part, sea water, depends upon energy for desalination and pumping. Desalination, once only possible by vaporization and condensation, can now be achieved by the less energy demanding process of reverse osmosis. Osmosis, you recall (Chapter 9) maintains turgor in plant cells by attracting water through semi-permeable membranes that retain solutes. The industrial process works on the same principle, but in the opposite direction. Pressure is applied to a cell that has a saline solution flowing through it. The result is release of water leaving a more concentrated saline solution inside. Desalination is now providing fresh water for many cities around the world and in some areas, e.g., southern Spain, for agriculture also. There, it is argued to be a lower cost option to building infrastructure for transfers between rivers, such as one in neighboring France.

There are certainly many opportunities to improve the efficiency of water use in agriculture (Chapter 14) by changing crops, irrigation techniques, and rules and regulations that govern access to water. Aerobic rice can replace flooded rice, spray irrigation can replace flood and furrow irrigation, micro-irrigation can introduce the greatest water savings of all under field conditions. New technologies are also being evaluated vigorously as the cost of water increases. Information technology and computerized control systems are already in much use. There is a limit, however, to the efficiency with which plants use water that goes back to the essential process of plant growth. Plant leaves exchange water they absorb from soil for CO_2 from the atmosphere (Table 9.5). The terms of trade are not fixed but diminish rapidly with the drying power of the atmosphere. It is difficult, in practice, to obtain a mass ratio of water for biomass of less than 800. That inescapable physiological reality is the basis for "inefficient use" that competitors for water often attribute to agriculture.

Finally, irrigation is a major contributor to security and food supply. The 18% of crop-land that is irrigated provides 40% of food supplies. Some predict that the irrigated area might increase to 25% by 2050 and stabilize at that level. During that time some currently irrigated land will be lost to diminishing water supplies and salinization, perhaps aided by climate change. Hopefully, society will acknowledge the value of irrigation before then so that substantial efforts are directed to maintenance and renovation of exist-ing schemes and establishing systems of water trading that transfer water to the most efficient and responsible users. A major limitation in many old schemes is that systems of water rights isolate users from the real costs of water, which results in inefficient use, and contributes to environmental degradation.

18.5.3   Soil resources

Our improved understanding of soil chemistry and nutrient availability enables more crit-ical management of fertility. Previously, management was based solely upon enhance-ment of mineralization and weathering by tillage, return of crop residues and animal manures, and rotation with legumes. Manipulation of organic matter remains a cen-tral theme in crop nutrition but maintenance and improvement of fertility have been sharpened with fertilizers. The major part of fertilizer practice concerns provision of the macronutrients N, P, and K. The use of fertilizers allows all fields, rather than just a few, to be relieved of nutrient deficiencies. Deficiencies can be treated individually in relation to need (site-specific management, SSM, Section 12.1), a practice more difficult with organic materials such as manure, whereas fertilizers can be placed for ready access by crop roots and in synchrony with need.

Difficulties remain, however, because yields and nutrient use vary with weather, diseases, and pests, and because root systems may be inefficient in uptake. We require accurate predictions about those matters or about mineralization and losses by runoff, leaching, and volatilization, and fixation into unavailable forms. Those processes all vary spatially and temporally within and among fields, causing seemingly unavoidable inefficiencies in nutrient cycling and fertilizer practices. Inefficiencies resulting from poor placement and timing of fertilizer applications can be reduced. Better (and cheaper) methods for soil and plant analyses and greater attention to rooting traits also will help. But given the large amount of N that cycles through high-yield farming systems, such as a maize–soybean rotation (Chapter 17), losses of just a small fraction represent a large potential input to natural aquatic and terrestrial ecosystems that receive it. That problem occurs with any source of N from residues and manure to fertilizer. We can, however, hold farm effluents below levels that cause health or environmental damage by a variety of means. Dispersal of animal production, for example, would help through more uniform distribution of manure supplies, as would vegetative buffer strips to intercept runoff at field margins adjacent to streams and water bodies. Controlled release fertilizers also hold promise.

A popular concept for solutions to current dilemmas about surplus production and off-site effects is to reduce the intensity and thus the yield from farming through lower inputs of fertilizer. Proponents fail to realize that intensive systems have better energy efficiency per unit production than extensive ones, and that lower yielding systems would promote a massive expansion of agriculture at the expense of natural ecosystems. From an ecological perspective, it seems better to place marginal arable land into reserve programs than to adjust to demand through even lower levels of production. Among the reasons noted earlier: less land is then subject to erosion, water, biocide, and energy use per unit production are less, and production per unit capital and labor remain high.

18.5.4   Genetic resources

Considerable emphasis has been given in recent years to conservation of genetic resources and to molecular biology as important foundations for agriculture in the future. The emphasis on molecular biology is open to criticism, however, especially when claims are made for quantum leap gains in net primary productivity, drought resistance, and N use efficiency. Our view is that the easy gains in genetic advance have been in the application of modern genetics to plant breeding during the green revolution of the 1950s and 1960s. We judge that expected improvements from biotechnology will be mainly from simple, single-gene traits relating to matters such as grain quality and disease and insect resistance – at least during the next 10 to 20 years. While important, continued emphasis on "defect elimination" overlooks greater opportunities for progress through understanding and advancing integrative attributes such as rooting, canopy architecture, and root–shoot partitioning. We know of no crop for which we have clear ideotypes to direct such breeding efforts and efforts of crop physiologists who might contribute to progress have been diverted to problems of "stress" in extreme environments. Inability to better define objectives for plant breeders arises in part because our understanding of normal integrative controls of plant growth and function under field conditions is still rather primitive.

It is in the area of pest and disease control that molecular biology can continue to make its greatest contribution through more, and varied, insect-, disease-, and herbicide-resistant cultivars. Efforts should also be made to capitalize more on subtle beneficial interrelationships among soil microorganisms, or among predatory and phytophagous insects, but experience to date indicates that understanding and establishing reliable systems of that sort are difficult. Genetic engineering of intrinsically more productive plants remains a distant vision, and in considering the possibility of another "green revolution", we should be prepared to discover that the combination of genes, fertilizers, and irrigation that produced the first revolution was indeed unique. Instead, future progress will most likely come from numerous, small improvements that together sustain, or accelerate, productivity gains of our major crops and cropping systems.

18.5.5   Weeds, diseases, and pests

Problems arising from weeds, diseases, and pests have a number of elements in common. Species diversity of each class is very large requiring that we limit attention to a few key species that give most difficulty. Infestations with any of these organisms can cause large losses in yield but such events are highly variable. In some years or locations, unintended problems can cause loss of entire crops, in others, effects are absent or too small to be of concern. Estimates of average losses are large; even with significant efforts at control, yield losses due to weeds, diseases, and pests run to 10% or more.

Ecological methods are the principal means for controlling each of these groups of organisms. All are susceptible to their own arrays of diseases and predators and all generally do less damage to well managed, rapidly growing crops than, for example, to those deficient in nutrients. Sanitation efforts (cultivation of weeds, disposal of infested residues, crop rotation) are important aspects of management. Weeds can sometimes be overcome through competition by crops, and insects are subject to predation by beneficial insects. Pest organisms differ in that genetic resistance to insects, nematodes, and diseases has long been an important tool but genetic solutions to weed control are only now being explored. We might, for example, sacrifice some feature of an optimal canopy for crop photosynthesis to achieve a canopy that is more competitive with weeds.

Biocides are now also widely employed to control each of these groups. Control cannot and should not be directed at every organism in sight. Control action generally is justified only when an infestation reaches an "economic threshold" level that may eventually cause significant financial loss. In some cases, control can be achieved by biological methods relying upon predator–prey relationships; in others, understanding of the biology of the host–pest relationship enables more efficient and effective use of biocides.

Weed control Weeds have limited mobility and weed problems therefore have a residential character through soil seed banks to particular fields and to particular climates and farming systems. Movement from field to field is facilitated by machinery, grazing animals, and manure as well as by wind and other natural means. Weed management centers on limiting seed bank size through control of reproduction. Sanitation is important because small untended weed populations in one year can cause crop losses over a number of subsequent years. The economic threshold for weed control is therefore generally less than a density that would cause losses in the current year. Tillage, the traditional means for managing weed populations, has been rapidly replaced by herbicides, especially where conservation tillage is used. Herbicides enable the work to be done more quickly, more timely, and more cheaply with significantly less energy than with tillage or hand labor. Transgenic crops with resistance to broad spectrum herbicides provide further flexibility in the timing of application. Sole reliance on a single type of herbicide resistance, such as glyphosate, promotes rapid development of weeds with increasing resistance to it. Integrated weed management involving appropriate crop rotations and herbicides with different modes of action, and new transgenic cultivars with resistance to other types of herbicides, are needed.

Control of insects and disease Biocides are perhaps the most controversial management tools available to farmers. Objections to modern farming methods stem mostly from indiscriminate use of a few compounds. Synthesis and application of biocides require comparatively little energy, their major component is intellectual. Like fertilizer, biocides substitute chemical expertise for land and labor. An ideal biocide would be specific to target species and have no residual effect. Few have achieved that objective and some have been very wide of the mark. It is important that unsafe biocides be withdrawn from use and that registration procedures encourage improvement in quality of new compounds and direct the search for ideal biocides. Problems arise in defining "unsafe", however, and it is important to counter unreasonable criticism of their use. Investigation can establish if the chances of deleterious effects are negligible, but never better than that. To reject that analysis will place a large burden of inefficiency on farmers, consumers, and ultimately, the landscape.

Major progress has been made in developing integrated pest management programs (IPM) that base controls on careful monitoring of pest populations. This allows biocides to be used only at critical times and in optimal ways. Integrated pest management offers opportunities to optimize ecological controls and thus reduce biocide use, and also to slow development of resistance in target species. Critics of biocides sometimes challenge their use by citing evidence of pest adaptation. Given that success of biocides depends on exploitation of slight differences in response between species, this situation is unlikely to change. Pests pay a price, however. Those that adapt to biocides generally are then less effective as pests. The use of transgenic cultivars with insect resistance, such as Bt maize and cotton, has allowed tremendous reduction in pesticide use. Development and use of Bt cultivars of other major crops is expected, and we foresee substantial additional benefits from biotechnology in the effective management of insects and diseases.

18.5.6 Adapting to climate change

Agriculture is facing two key issues related to climate change. First, would be the need to adapt production systems to changing climate, whatever the cause. Second, is the call to mitigate AGW by both reducing emission of GHGs from agriculture and by developing systems to sequester CO_2 emissions from urban, industrial, and transport activities. Critical to this discussion is that progress towards both objectives must be achieved within the more certain requirement to increase food production by 70% by 2050.

Earlier analysis has shown great scope to adjust crop production systems to climate change by adjustments in sequence to planting dates, species combinations, and then locations (Chapters 5 and 16). The range of genetic adaptability in staple annual crops is large and expanding, as revealed by their current wide geographical distribution. For example, rice, maize, and soybean are now produced from 50° N latitude to the equator. Adaption of perennial crops could pose more serious problems, in part because of smaller genetic diversity, but also because of the time and cost required to establish new production systems, especially if climate change were rapid and continuous.

Agriculture's direct contribution to world GHG emissions, estimated at 6.6 Gt CO_2 equivalent, is 10 to 12% of world emissions, although about as much again is released by the current rate of land clearing for agricultural expansion (Section 6.10.4). Contribution of CO_2 from fossil fuel use in agriculture is small (1.5%), consistent with the small proportion of world energy (2–3%) used to support agricultural production to the farm gate (Chapter 15). The major contributions (69%) are by N_2O from soils and CH_4 from rice production and ruminants (Fig. 6.15). Again, our analyses have shown considerable scope to reduce GHG emissions from agriculture, especially through ecological intensification of cropping systems on existing farmland that increases yields and reduces emissions and negative environmental impact per unit of food production. We are optimistic that this can be achieved through expected improvements in nutrient use efficiency, expansion of conservation tillage, improved irrigation methods, and continued incremental improvement in stress resistance of commercial cultivars and hybrids. A confusion is found, however, in current policies of some governments on biofuel production. On one hand, biofuels compete with food production for land, water, nutrients, and investment, complicating the quest for greater food production, and on the other, they contribute to current rates of land clearing.

Analyses of biofuel production reveal that NER are small except for sugarcane and oil palm (Table 15.6), and those values ignore changes in SOM, and land clearing where required. Over the short term, however, expansion of biofuel production will contribute to higher commodity prices, which will help policy makers justify increased investment in agricultural research and development. Over the longer term, we see biofuel production, other than from residues, as a significant complication to meeting future world food demand. We are cautious about suggestions of a "win–win–win" solution to simultaneously produce bioenergy in large quantities, permanently sequester carbon, and improve soil and water quality. The scheme involves pyrolysis of cellulosic crops to make biofuel while applying the inert remainder ("biochar") to improve physical and chemical properties of soils (Laird 2008). Interestingly, however, the approach does embody the kind of innovative thinking required for success in ecological intensification and identifies the comprehensive analyses required to feed and fuel a large and expanding population. Fortunately, many new tools are available to measure the performance of production systems and to design and test new ones for any purpose. Present and future crop ecologists must apply and develop these tools to meet the immediate challenges facing agriculture.

Finally, what this book has hopefully shown is that our biosphere is a carbon-based world with strict interrelationships among resources such as sunlight and supplies of water and chemical nutrients. We have emphasized the importance of increasing efficiency of resource use by seeking higher productivity on suitable lands in order to preserve other land for nature. Given the efficiency gains that seem possible, it will be difficult to increase productivity to feed our expanding world population if serious restrictions are placed on areas available and/or access to water, energy, and fertilizer and if insufficient funds and incentives are provided for research and development. We do not foresee a major role for agriculture in providing fuel for a large population or a CO_2 sink to offset society's total GHG emissions. We again highlight the necessity that proposals for sustainable productivity goals be developed with a thorough understanding of crop ecology principles, and the tremendous opportunities for scientists to apply them to ensure global food security and conservation of natural resources for future generations.

18.6   Review of key concepts

Population, cropland, and food requirements

- Global population is projected to grow by 35% to 9.2 billion by 2050, stabilize after that, and decline toward the end of the century. Food production must increase by about

70% in 2050 to better feed a larger population that now includes almost 1 billion who are undernourished. Per capita consumption will increase most in developing countries where nearly all population growth will occur.

- Limited land is available for expansion of agriculture so greater food production must come mostly from greater productivity on existing farmland. Developing countries in Africa and Asia have much smaller cropland per capita than Europe and North America. Greatest potential for expansion exists in South America where international cooperation is required to conserve globally important rainforest and wetland ecosystems.

- Asia has achieved great increase in food production in the past 40 y, but endowments of land and climate, current population and growth rates make it unlikely they can achieve self-sufficiency by 2050. Small cropland per capita and increasing population present Africa with the greatest challenge of food security. North and South America, Europe, and Australia will continue to play critical roles in provisioning global food security.

Future limits, challenges, and opportunities

- Modern crop production methods, including inorganic fertilizers, biocides, mechanization and conservation tillage, modern cultivars (including transgenics), and maintenance of current irrigated area are essential to feed present and projected global populations.

- Current rates of yield gain are insufficient to meet future demand without a large expansion of crop area. Biotechnology solutions to re-engineer plant productivity are appealing and should be pursued as long-term goals, but not at the expense of research into the many other methods more certain to improve crop performance in the short term.

- Consumption of less meat and expansion of vegetarian diets in developed countries will have little impact on global grain requirements because savings there would be more than offset by increased meat and dairy consumption in the increasingly more populous developing countries.

- Agriculture will adapt to future climate change just as it has in the past. There is much scope for adjustments based on modifying crop growth periods by changing planting dates and using cultivars with later, or earlier, maturities. Greater atmospheric $[CO_2]$ will improve photosynthesis rates and water use efficiency of many crops. There will be winners and losers, however, as climate in some areas will become more, or less, favorable to support agricultural production.

- Biofuel production will compete with food crops for land, water, fertilizers, and investment. In some countries with small populations and large land resources, it can contribute to the development of agriculture and society. Sustainable solutions will require comprehensive application of crop ecology principles of efficient resource use to integrate biofuel and food production with conservation of the environment.

What will agriculture look like?

- Success in meeting production goals will depend on further refinement of timing and precision of all field operations, continued genetic advance, substantial improvements in crop tolerance of diseases and insects, and improvement in soil quality traits that support crop growth. Further innovation must come in remote sensing, geographic information systems, crop simulation models, access to high-quality, real-time, and historical weather data, weather forecasting, and equipment for planting, tillage, irrigation, and harvest. Holistic approaches that leverage all aspects of production are the key to success. Claims of silver bullet solutions should be viewed with caution.

- Modern integrated crop production systems raise a number of justifiable environmental concerns. They can be resolved, however, and at higher productivity, through focus on ecological intensification using current best management practices and new technologies. Public concerns, and pressures from large retailer food markets, will likely accelerate progress, through certification, to environmentally acceptable production practices.

- Greater investment in research and education, in public and private sectors, enlightened policies, and prioritization of activities at national and international levels to support agricultural development are essential. The projected population by 2050 establishes urgency for action so that food production can increase with demand. The major part of the 70% increase in production is needed soon, e.g., by 2030, to resolve the current problem of undernourishment and provide confidence for ultimate success.

Terms to remember: anthropogenic global warming (AGW); biofuel; climate change; conservation tillage; cropland per inhabitant; demand for food and energy; demographic shifts, developed and developing countries, urban and rural populations; desalination; dietary change; farming systems, intensive, integrated, low-input (traditional), organic; genetic engineering; global carrying capacity; good agricultural practices (GAP); grain-fed livestock; green revolution; irrigation water use efficiency; mechanization; molecular biology; pesticides and carcinogens; population growth and stabilization.

Species list

Common and scientific names of species mentioned in the text.

Common name	Scientific name
Alfalfa (lucerne)	*Medicago* spp.
Amaranth	*Amaranthus* spp.
Apple	*Malus* spp.
Asparagus	*Asparagus* spp.
Azolla	*Azolla anabaena*
Banana	*Musa* spp.
Barley	*Hordeum vulgare*
Bean	*Phaseolus, Vicia* spp.
dry	*P. vulgaris*
lima	*P. lunatus*
navy	*P. vulgaris*
broad, faba	*Vicia faba*
Birch	*Betula* spp.
Buckwheat	*Fagopyrum esculentum*
Buruti palm	*Mauritia flexuosa*
Cabbage	*Brassica oleracea*
Canary grass, annual	*Phalaris canariensis*
Canola	*Brassica napus*
Carrot	*Daucus carota*
Cassava	*Manihot esculenta*
Castor oil	*Ricinus communis*
Chard	*Beta vulgaris*
Chickpea	*Cicer arietinum*
Clover	*Trifolium, Melilotus* spp.
bcrseem	*T. alexandrinum*
red	*T. pratense*
subterranean	*T. subterraneum*
white	*T. repens*
sweet	*Melilotus* spp.
Coffee	*Coffea arabica*
Cotton	*Gossypium hirsutum*
Elephant grass	*Miscanthus hybrids*
Eucalypts	*Eucalyptus* spp.
Fodder beet	*Beta vulgaris*

(*cont.*)

Common name	Scientific name
Giant reed	*Arundo donax*
Grugri palm	*Acronomia aculeata*
Jatropa (Physic nut)	*Jatropa curcas*
Johnson grass	*Sorghum halepense*
Kentucky bluegrass	*Poa pratensis*
Lentil	*Lens culinaris*
Lettuce	*Latuca sativa*
Linseed	*Linum usitatissimum*
Lupin	*Lupinus* spp.
narrow leaf	*L. angustifolius*
Maize (corn)	*Zea mays*
Medic (annual)	*Medicago* spp.
Melon	*Cucumis* spp.
netted, muskmelon	*C. melo reticulatus*
Millet	*Pennisetum, Setaria*
pearl	*P. glaucum*
Mountain ash	*Eucalyptus regnans*
Neem	*Azadirachta indica*
Oat	*Avena* spp.
cultivated	*A. sativa*
wild	*A. fatua*
Oil radish	*Raphanus sativus*
Olive	*Olea europa*
Onion	*Allium* spp.
Orange	*Citrus sinensis*
Panic grass	*Panicum* spp.
green	*P. maximum*
Pea	*Pisum* spp.
field	*P. arvense*
Peach	*Prunus persica*
Peanut	*Arachis hypogaea*
Phalaris grass	*Phalaris aquatica*
Physic nut (Jatropa)	*Jatropa curcas*
Pineapple	*Ananas comosus*
Pongamia	*Pongamia* spp.
Poplar	*Populus* spp.
Potato	*Solanum tuberosum*
Quinoa	*Chenopodium quinoa*
Radish	*Raphanus sativus*
Rape	*Brassica* spp.
oilseed	*B. campestris*
winter	*B. napus*
Reed canary grass	*Phalaris arundinaceae*
Rhodes grass	*Chloris gayana*
Rice	*Oryza sativa*
Rutabaga	*Brassica napobrassica*
Rye	*Secale cereale*
Ryegrass	*Lolium* spp.
annual, italian	*L. multiflorum*

Common name	Scientific name
darnel	*L. temulentum*
perennial	*L. perenne*
wimmera	*L. rigidum*
Safflower	*Carthamus tinctorius*
Sisal	*Agave sisalana*
Sorghum	*Sorghum vulgare*
Spinach	*Spinacia oleracea*
Squash	*Curcurbita* spp.
Strawberry	*Fragaria* spp.
Sudangrass	*Sorghum bicolor*
Sugarbeet	*Beta vulgaris*
Sugarcane	*Saccharum officinarum*
Sunflower	*Helianthus annuus*
Sweet sorghum	*Sorghum bicolor*
Switch grass	*Panicim virgatum*
Timothy grass	*Phleum pratense*
Tobacco	*Nicotiana tobacum*
Tomato	*Lycopersicon esculentum*
Vetch	*Vicia sativa*
Wheat, bread	*Triticum* spp.
bread	*T. aestivum*
durum	*T. durum*
Wheat grass	*Agropyron* spp.

Conversions and constants useful in crop ecology

METRIC PREFIXES

Multiples of 1000 are preferred with SI units.

m	milli	10^{-3}
μ	micro	10^{-6}
n	nano	10^{-9}
p	pico	10^{-12}

k	kilo	10^3
M	mega	10^6
G	giga	10^9
T	tera	10^{12}
E	exa	10^{18}

CONVERSIONS

	Metric	*American*
Length:	1 m = 39.37 in	1 in = 25.4 mm
	1 km = 0.62 mi (statute)	1 foot = 0.305 m
Area:	1 ha = 10 000 m^2	1 acre = 43 560 $feet^2$
	1 ha = 2.47 acre	1 acre = 0.405 ha
		1 sq mi = 640 acre
Volume:	1 l = 1000 cm^3	1 acre-ft = 1233 m^3
	1 m^3 = 1000 L	1 cu ft = 28.3 L
	1 m^3 = 35.3 cu ft	1 bu (level) = 1.24 cu ft = 0.0352 m^3
	1 m^3 H_2O = 1000 kg	1 bbl (petroleum) = 42 gal
		1 gal (liq.) = 3.785 L
Mass:	1 kg = 2.205 lb	1 lb = 0.454 kg
	1 t = 1000 kg	1 ton = 2000 lb
	1 t = 2205 lb	
Yield:	1 kg ha^{-1} = 0.89 lb $acre^{-1}$	1 lb $acre^{-1}$ = 1.12 kg ha^{-1}
	1 kg ha^{-1} = 0.1 g m^{-2}	
	1 t ha^{-1} = 0.45 ton $acre^{-1}$	
Pressure:	1 bar = 10^6 dyne cm^2	1 atm = 1.013 bar
	1 bar = 0.987 atm	
	1 MPa = 10 bar	
Energy:	1 J = 10^7 erg (dyne cm)	1 BTU = 252 cal
	1 J = 0.239 cal	1 cal = 4.184 J
	1 J s^{-1} = 1 Watt	1 HP = 0.75 kW
		1 HP = 2542 BTU h^{-1}
Radiation:	1 W m^{-2} = 0.143 mcal cm^{-2} min^{-1}	
	1 einstein = 1 mol quanta	1 cal cm^{-2} min^{-1} = 697 W m^{-2}

CONSTANTS AND COEFFICIENTS

Avogadro's number		6.022×10^{23}
Gas constant		8.314 J mol^{-1} K^{-1}
RT (0°C)		2.27 L MPa mol^{-1}
RT (10°C)		2.35 L MPa mol^{-1}
RT (20°C)		2.44 L MPa mol^{-1}
RT (30°C)		2.52 L MPa mol^{-1}
Latent heat of fusion (0°C)		334 J g^{-1}
Latent heat of vaporization	(0°C)	2501 J g^{-1}
	(10°C)	2477 J g^{-1}
	(20°C)	2442 J g^{-1}
	(30°C)	2430 J g^{-1}
Planck's constant		6.63×10^{-34} J s
Saturation vapor pressure	(0°C)	0.611 kPa
(see Eq. 6.16)	(10°C)	1.227 kPa
	(20°C)	2.337 kPa
	(30°C)	4.243 kPa
Solar constant		1360 W m^{-2} (currently 1370 W m^{-2})
Specific heat of air	(0–40°C)	1.01 J g^{-1} C^{-1}
Specific heat of water	(17°C)	4.184 J g^{-1} C^{-1}
Speed of light (vacuum)		3.00×10^8 m s^{-1}
Stefan–Boltzmann constant		5.67×10^{-8} W m^{-2} K^{-4}

References

Abbate, P. E., J. L. Dardanelli, M. G. Cantarero, M. Maturano, R. J. M. Melchiori, and E. E. Suero. 2004. Climate and water availability effects on water-use efficiency in wheat. *Crop Sci.* **44**:474–83.

Adamchuk, V. I., A. Dobermann, and J. L. Ping. 2004a. *Listening to the story told by yield maps*. University of Nebraska Extension Circular 04–704. Lincoln, Nebraska.

Adamchuk, V. I., J. W. Hummel, M. T. Morgan, and S. K. Upadhyaya. 2004b. On-the-go soil sensors for precision agriculture. *Comput. Electron. Agr.* **44**:71–91.

Addiscott, T. M. 1988. Long-term leakage from bare unmanured land. *Soil Use Manage.* **4**:91–5.

Adviento-Borbe, M. A. A., M. L. Haddix, D. L. Binder, D. T. Walters, and A. Dobermann. 2007. Soil greenhouse gas fluxes and global warming potential in four high-yielding maize systems. *Global Change Biol.* **13**:1972–88.

Agricultural Research Council Working Party. 1980. *The Nutrient Requirements of Ruminant Livestock*. Commonwealth Agricultural Bureaux, Farnham Royal, Slough, UK.

Allen, R. G., L. S. Pereira, D. Raes, and M. Smith. 1998. *Crop Evapotranspiration. Guidelines for Computing Crop Water Requirements*. Irrigation and Drainage Paper 56, FAO, Rome.

American Society of Agricultural Engineers (ASAE). 1999. Manure production and characteristics. ASAE Standard ASAE D384.1. *ASAE Standards 1999*. St. Joseph, MI: ASAE, pp. 663–5.

Ames, B. N., M. Profet, and L. S. Gold. 1990a. Dietary pesticides (99.99% all natural). *Proc. Natl. Acad. Sci. USA*. **87**:7777–81.

Ames, B. N., M. Profet, and L. S. Gold. 1990b. Nature's chemicals and synthetic chemicals: comparative toxicology. *Proc. Natl. Acad. Sci. USA*. **87**:7782–6.

Amthor, J. S. 1989. *Respiration and Crop Productivity*. Springer-Verlag, New York.

Amthor, J. S. 2007. Improving photosynthesis and yield potential. In P. Ranalli (ed.) *Improvement of Crop Plants for Industrial End Uses*. Springer, pp. 27–58.

Anandakumar, K. 1999. Sensible heat flux over a wheat canopy: optical scintillometer measurements and surface renewal analysis estimations. *Agric. For. Meteorol.* **96**:145–56.

Annandale, J. G., N. Z. Jovanovic, G. S. Campbell, N. Du Sautoy, and P. Lobit. 2004. Two-dimensional solar radiation interception model for hedgerow fruit trees. *Agric. For. Meteorol.* **121**:207–25.

Angus, J. F. 2001. Nitrogen supply and demand in Australian agriculture. *Aust. J. Exp. Agric.* **41**:277–88.

Angus, J. F., M. W. Cunningham, M. W. Moncur, and D. J. Mackenzie. 1981. Phasic development in field crops. I. Thermal response in the seedling phase. *Field Crops Res.* **3**:365–78.

Anon. 1988. *Draft: Salinity and Drainage Strategy*. Discussion Paper No. 1. Murray–Darling Basin Ministerial Council, Canberra.

Anon. 2009. Supercharging the rice engine. *CSA News* **54**(7):4–6.

Anon. 2010. Biofuels and Land Use Change: A Science and Policy Review. Hart Consulting & CABI International, Wallingford, UK.

Arkley, R. J. 1963. Relationships between plant growth and transpiration. *Hilgardia* **34**:559–84.

Austin, R. B., J. Bingham, R. D. Blackwell, L. T. Evans, and **M. A. Ford**. 1980. Genetic improvements in winter wheat yields since 1900 and associated physiological changes. *J. Agric. Sci. (Camb.)* **94**:675–89.

Avlani, P. K. and **W. J. Chancellor**. 1977. Energy requirements for wheat production and use in California. *Trans. ASAE* **20**:429–37.

Ayers, R. S. and **D. W. Westcot**. 1985. *Water Quality for Agriculture*. Irrig. Drain. Paper No. 29(rev. 1). Rome, Italy: FAO.

Azevedo, J. and **P. R. Stout**. 1974. *Farm Animal Manures: An Overview of their Role in the Agricultural Environment*. Manual 44. Calif. Agric. Exp. Stn., Univ. California, Berkeley.

Badger, P. C. 2002. Ethanol from cellulose: a general review. In **J. Janick** and **A. Whipkey** (eds.) *Trends in New Crops and New Uses*. Alexandria, VA: ASHS Press, pp. 17–21.

Badgley, C., J. Moghtader, E. Quintero *et al.* 2007. Organic agriculture and the global food supply. *Renew. Agric. Food Syst.* **22**:86–108.

Bair, R. A. 1942. Growth rates of maize under field conditions. *Plant Physiol.* **17**:619–31.

Baker, D. N. and **R. E. Meyer**. 1966. Influence of stand geometry on light interception and net photosynthesis in cotton. *Crop Sci.* **6**:15–19.

Baker, D. N., J. R. Lambert, and **J. M. McKinion**. 1983. GOSSYM: *A Simulator of Cotton Crop Growth and Yield*. South Carolina Agric. Exp. Sm. Tech. Bull., 1089. Clemson Univ., Clemson.

Baker, J. M. and **T. J. Griffith**. 2005. Examining strategies to improve the carbon balance of corn/soybean agriculture using eddy covariance and mass balance techniques. *Agric. For. Meteorol.* **128**:163–77.

Baker, J. M., R. T. Ochsner, R. T. Venterea, and **T. J. Griffis**. 2007. Tillage and carbon sequestration – what do we really know? *Agric. Ecosys. Envir.* **118**:1–5.

Banziger, M., P. S. Setimela, D. Hodson, and **B. Vivek**. 2006. Breeding for improved abiotic stress tolerance in maize adapted to southern Africa. *Agric. Water Manage.* 2006: **80**:212–24.

Bartholomew, D. 1982. Environmental control of dry-matter production in pineapple. I. In **I. Ting** and **M. Gibbs** (eds.) *Crassulacean Acid Metabolism*. Rockville, Maryland: Am. Soc. Plant Physiol, pp. 278–94.

Bates, S. L., J. Z. Zhao, R. T. Roush, and **A. M. Shelton**. 2005. Insect resistance management in GM crops: past, present, and future. *Nature Biotech.* **23**:57–62.

Batjes, N. H. 2009. Harmonized soil profile data for applications at global and continental scales: updates to the WISE database. *J. Soil Use & Manage.* **25**:124–7.

Beale, P. E. 1974. Regeneration of *Trifolium subterraneum* cv Yarloop from seed reserves on Kangaroo Island. *J. Aust. Inst. Agric. Sci.* **40**:78–80.

Bellarby, J., B. Foereid, A. Hastings, and **P. Smith**. 2008. *Cool Farming*. Amsterdam: GreenPeace International.

Berni, J. A. J., P. J. Zarco-Tejada, G. Sepulcre-Cantó, E. Fereres, and **F. Villalobos**. 2009. Mapping canopy conductance and CWSI in olive orchards using high resolution thermal remote sensing imagery. *Remote Sensing Environ.* **113**:2380–8.

Betrán, F. J., D. Beck, M. Banziger, and **G. O. Edmeades**. 2003. Genetic analysis of inbred and hybrid grain yield under stress and nonstress environments in tropical maize. *Crop Sci.* **43**:807–17.

Bingham, J. 1972. Physiological objectives in breeding for grain yield in wheat. In **F. G. H. Lupton, G. Jenkins** and **R. Johnson** (eds.) *The Way Ahead in Plant Breeding* (Proc. 6th Eucarpia Congr.). Cambridge, UK: Plant Breeding Institute, pp. 15–29.

Biscoe, **P. V.**, **R. K. Scott**, and **J. L. Monteith**. 1975. Barley and its environment. III. Carbon budget of the stand. *J. Appl. Ecol.* **12**:269–91.

Björkman, **O.** (1981). Responses to different quantum flux densities. In **O. L. Lange**, **P. S. Nobel**, **C. B. Osmond** and **H. Zeigler** (eds.) *Physiological Plant Ecology. I. Responses to the Physical Environment (Encycl. Plant Physiol.*, new ser., vol. 12A). Berlin: Springer-Verlag, pp. 57–107.

Björkman, **O.**, **M. Nobs**, **R. Pearcy**, **J. Boynton**, and **J. Berry**. 1969. Characteristics of hybrids between C3 and C4 species of Atriplex. In **M. D. Hatch**, **C. B. Osmond**, and **R. O. Slayter** (eds.) *Photosynthesis and Photorespiration.* New York: Wiley-Interscience, pp. 105–19.

Blackmer, **A. M.**, **D. Pottker**, **M. E. Cerrato**, and **J. Webb**. 1989. Correlations between soil nitrate concentrations in late spring and corn yields in Iowa. *J. Prod. Agric.* **2**:103–9.

Blanco-Canqui, **H.** and **R. Lal**. 2008. No-tillage and soil-profile carbon sequestration: an on-farm assessment. *Soil Sci. Soc. Am. J.* **72**:693–701.

Bloom, **A. J.**, **F. S. Chapin**, and **H. A. Mooney**. 1985. Resource limitation in plants – an economic analogy. *Ann. Rev. Ecol. Syst.* **16**:363–92.

Bloom, **A. J.**, **R. M. Caldwell**, **J. Finazzo**, **R. L. Warner**, and **J. Weisshart**. 1989. Oxygen and carbon dioxide fluxes from barley shoots depend on nitrate assimilation. *Plant Physiol.* **91**:352–6.

Boddey, **R. M.**, **S. Urquiaga**, **V. Reis**, and **J. Döbereiner**. 1991. Biological nitrogen fixation associated with sugar cane. *Plant & Soil* **137**:111–17.

Boeringa, **R.** (ed.). 1980. *Alternative Methods of Agriculture.* Amsterdam: Elsevier Scientific.

Bolaños, **J.** and **G. O. Edmeades**. 1993. Eight cycles of selection for drought tolerance in lowland tropical maize. I. Response in grain yield, biomass, and radiation utilization. *Field Crops Res.* **31**:223–52.

Bolaños, **J.** and **G. O. Edmeades**. 1996. The importance of the anthesis–silking interval in breeding for drought tolerance in tropical maize. *Field Crops Res.* **48**:65–80.

Bolland, **M.** and **B. Gilkes**. 1990. The poor performance of rock phosphate fertilizers in Western Australia: part 1. The crop and pasture responses. *J. Aust. Inst. Agric. Sci. N.S.* **3**:43–8.

Bolton, **J. K.** and **R. H. Brown**. 1980. Photosynthesis of grass species differing in carbon dioxide fixation pathways. V. Response of *Panicum maximum*, *Panicum milioides* and tall fescue (*Festuca arundinacea*) to nitrogen nutrition. *Plant Physiol.* **66**:97–100.

Bonachela, **S.**, **F. Orgaz**, **F. J. Villalobos**, and **E. Fereres**. 2001. Soil evaporation from drip-irrigated olive orchards. *Irrig. Sci.* **20**:65–71.

Boote, **K. J.**, **J. W. Jones**, and **G. H. Hoogenboom**. 1998. Simulation of crop growth, CROPGRO model. In **R. M. Peart** and **R. B. Curry** (eds.) *Agricultural Systems Modeling and Simulation.* New York: Marcel Dekker, pp. 651–93.

Bouldin, **D. R.**, **S. D. Klausner**, and **W. S. Reid**. 1984. Use of nitrogen from manure. In **R. D. Hauck** (ed.) *Nitrogen in Crop Production.* Madison, Wisconsin: Am. Soc. Agron., pp. 221–45.

Bramley, **R. G. V.** 2009. Lessons from nearly 20 years of precision agriculture research, development, and adoption as a guide to its appropriate application. *Crop & Past. Sci.* **60**:197–217.

Broadbent, **F. E.** and **A. B. Carlton**. 1978. Field trials with isotopically labelled nitrogen fertilizer. In **D. R. Nielsen** and **J. G. MacDonald** (eds.) *Nitrogen in the Environment*, vol. 1. New York: Academic Press, pp. 1–41.

Brody, **S.** 1945. *Bioenergetics and Growth.* New York: Reinhold.

Brouder, **S. M.** and **K. G. Cassman**. 1994. Cotton root and shoot response to localized supply of nitrate, phosphate, and potassium: split-pot studies with nutrient solution and vermiculitic soil. *Plant & Soil* **161**:179–93.

Brouwer, **R.** 1983. Functional equilibrium: sense or nonsense? *Neth. J. Agric. Sci.* **31**:335–48.

Brown, D. L., J. T. Scott, E. J. de Peters, and **R. L. Baldwin.** 1989. Influence of sometribove, USAN (recombinant methionyl bovine somatotropin) on the body composition of lactating cattle. *Am. Inst. Nutr.* **1989**:633–8.

Brown, R. C. and **M. Wright.** 2009. Biomass conversion to fuels and electrical power. In **R. W. Howarth** and **S. Bringezu** (eds.) *Biofuels: Environmental Consequences and Interactions with Changing Land Use.* Proceedings of the Scientific Committee on Problems of the Environment (SCOPE) International Biofuels Project Rapid Assessment, September 22–25, 2008, Gummersbach Germany. Cornell University, Ithaca NY, USA (http://cip.cornell.edu/biofuels/), pp. 53–64.

Brown, R. H. 1978. A difference in N use efficiency in C3 and C4 plants and its implications in adaptation and evolution. *Crop Sci.* **18**:93–8.

Bryson, R. A. and **T. J. Murray.** 1977. *Climates of Hunger: Mankind and the World's Changing Weather.* Madison: University of Wisconsin Press.

Buol, S. W., F. D. Hole, and **R. J. McCracken.** 1989. *Soil Genesis and Classification,* 3rd edn. Ames: Iowa State University Press.

Buresh, R. J., K. R. Reddy, and **C. van Kessell.** 2008. Nitrogen transformations in submerged rice soils. In **J. S. Schepers** and **W. R. Raun** (eds.) *Nitrogen in Agricultural Soils.* Monograph 49. Madison, WI: Amer. Soc. Agron., pp. 401–36.

Buringh, P. and **H. D. van Heemst.** 1979. Potential world food production. In **H. Linnemann, J. de Hoogh, M. A. Keyzer,** and **H. D. van Heemst** (eds.) *MOIRA: Model of International Relations in Agriculture (Contr. Econ. Anal. No. 124).* Amsterdam, the Netherlands: Elsevier North-Holland, pp. 19–72.

Bywater, A. C. and **R. L. Baldwin.** 1980. Alternative strategies in food–animal production. In **R. J. Baldwin** (ed.) *Animals, Feed, Food and People: an Analysis of the Role of Animals in Food Production* (AAAS Sel. Symp. No. 42). Boulder, Colorado: Westview Press, pp. 1–29.

Cain, J. D., P. T. W. Rosier, W. Meijninger, and **H. A. R. De Bruin.** 2001. Spatially averaged sensible heat fluxes measured over barley. *Agric. For. Meteorol.* **107**:307–22.

Calvino, P., V. Sadras, M. Redolatti, and **M. Canepa.** 2004. Yield responses to narrow rows as related to interception of radiation and water decit in sunflower hybrids of varying cycle. *Field Crops Res.* **88**:261–7.

Cardwell, V. B. 1982. Fifty years of Minnesota corn production: sources of yield increase. *Agron. J.* **74**:984–90.

Cassman, K. G. 1999. Ecological intensification of cereal production systems: yield potential, soil quality, and precision agriculture. *Proc. National Acad. Sci. USA* **96**:5952–9.

Cassman, K. G. and **P. L. Pingali.** 1995. Intensification of irrigated rice systems: learning from the past to meet future challenges. *GeoJournal* **35**:299–305.

Cassman, K. G. and **S. Wood.** 2005. Cultivated systems. In *Millennium Ecosystem Assessment: Global Ecosystem Assessment Report on Conditions and Trends.* (www.maweb.org//en/products.global.aspx). Washington D.C.: Island Press, pp. 745–94.

Cassman, K. G., A. D. Dobermann, and **D. T. Walters.** 2002. Agroecosystems, N-use efficiency, and N management. *AMBIO* **31**:132–40.

Cassman, K. G., A. Dobermann, D. T. Walters, and **H. Yang.** 2003. Meeting cereal demand while protecting natural resources and improving environmental quality. *Annu. Rev. Environ. Resour.* **28**:315–58.

Cassman, K. G., S. Peng, D. C. Olk *et al.* 1998. Opportunities for increased nitrogen use efficiency from improved resource management in irrigated rice systems. *Field Crops Res.* **56**:7–39.

Cassman, K. G., B. A. Roberts, T. A. Kerby, D. C. Bryant, and S. L. Higashi. 1989. Soil potassium balance and cumulative cotton response to annual potassium additions on a vermiculitic soil. *Soil Sci. Soc. Am. J.* **53**:805–12.

Cassman, K. G., P. W. Singleton, and B. A. Linquist. 1993. Input/output analysis of the cumulative soybean response to phosphorus on an ultisol. *Field Crops Res.* **34**:23–36.

Cassman, K. G., A. S. Whitney, and K. R. Stockinger. 1980. Root growth and dry matter distribution of soybean as affected by phosphorus stress, nodulation, and N source. *Crop Sci.* **20**:239–44.

CAST (Council for Agriculture Science and Technology). 1999. *Animal Agriculture and Global Food Security*. Task Force Report no. 135. Ames, Iowa: Council for Agriculture Science and Technology.

Cerrato, M. E. and A. M. Blackmer. 1990. Comparison of models for describing corn yield response to nitrogen. *Agron. J.* **82**:138–43.

Cervinka, V. 1980. Fuel and energy efficiency. In D. Pimentel (ed.) *Handbook of Energy Utilization in Agriculture*. Boca Raton, Florida: CRC Press, pp. 15–21.

Chalmers, D. J., P. D. Mitchell, and L. van Heek. 1981. Control of peach tree growth and productivity by regulated water supply, tree density and summer pruning. *J. Am. Soc. Hort. Sci.* **106**:307–12.

Chang, C. C. 1963. *An Agricultural Engineering Analysis of Rice Farming Methods in Taiwan*. Conf. Paper No. 20. Los Baños, Philippines: IRRI.

Chapman, H. W., L. S. Gleason, and W. E. Loomis. 1954. The carbon dioxide content of field air. *Plant Physiol.* **29**:500–3.

Chapman, S. C. and G. O. Edmeades. 1999. Selection improves drought tolerance in tropical maize populations. II. Direct and correlated responses among secondary traits. *Crop Sci.* **39**:1315–24.

Chauhan, B. S., G. S. Gill, and C. Preston. 2006. Tillage system effects on weed ecology, herbicide activity and persistence: a review. *Aust. J. Exp. Agric.* **46**:1557–70.

Christopher, J. T., A. M. Manschadi, G. L. Hammer, and A. K. Borrell. 2008. Developmental and physiological traits associated with high-yield and stay-green phenotype in wheat. *Aust. J. Agric. Res.* **59**:354–64.

Clover, G. R. G., K. W. Jaggard, H. G. Smit, and S. N. Azam-Li. 2001. The use of radiation interception and transpiration to predict the yields of healthy, droughted and virus-infected sugar beet. *J. Agric. Sci.* **136**:169–78.

Cock, J. H., N. M. Riano, M. A. El-Sharkawy, Y. Lopez, and G. Bastidas. 1987. C3–C4 intermediate photosynthetic characteristics of cassava (*Manihot esculenta* Crantz). II. Initial products of $^{14}CO_2$ fixation. *Photosyn. Res.* **12**:237–41.

Cockroft, B. and W. Mason. 1987. Irrigated agriculture. In D. J. Connor and D. F. Smith (eds.) *Agriculture in Victoria*. Melbourne: Australian Institute of Agricultural Science, pp. 159–77.

Coehlo, D. T. and R. F. Dale. 1980. An energy-crop growth variable and temperature function for predicting corn growth and development: planting to silking. *Agron. J.* **72**:503–10.

Collino, D. J., J. L. Dardanelli, R. Sereno, and R. W. Racca. 2000. Physiological responses of argentine peanut varieties to water stress. Water uptake and water use efficiency. *Field Crops Res.* **68**:133–42.

Connor, D. J. 1983. Plant stress factors and their influence on production of agroforestry plant associations. In P. A. Huxley (ed.) *Plant Research and Agroforestry*. Nairobi, Kenya: ICRAF, pp. 401–26.

Connor, **D. J.** 2001. Optimizing crop diversity. In **J. Nosberger, H. H. Geiger** and **P. C. Struik** (eds.) *Crop Science – Progress and Prospects*. Proc. 4th Int. Crop Sci. Cong., Hamburg, 2000. Oxford, UK: CABI International, pp. 191–211.

Connor, **D. J.** 2004. Designing cropping systems for efficient use of limited water in southern Australia. *Europ. J. Agron*. **21**:419–31.

Connor, **D. J.** 2006. Towards optimal designs for hedgerow olive orchards. *Aust. J. Agric. Res*. **57**:1067–72.

Connor, **D. J.** 2008. Organic agriculture cannot feed the world. *Field Crops Res*. **106**:187–90.

Connor, **D. J.** and **J. H. Cock**. 1981. The response of cassava to water shortage. II. Canopy dynamics. *Field Crops Res*. **4**:285–96.

Connor, **D. J.** and **C. G. Hernández**. 2009. Crops for biofuels: current status and prospects for the future. In **R. W. Howarth** and **S. Bringezu** (eds.) *Biofuels: Environmental Consequences and Interactions with Changing Land Use*. Proceedings of the Scientific Committee on Problems of the Environment (SCOPE) International Biofuels Project Rapid Assessment, 22–25 September 2008, Gummersbach Germany (http://cip.cornell.edu/biofuels/). Ithaca NY, USA: Cornell University, pp. 65–80.

Connor, **D. J.** and **T. R. Jones**. 1985. Response of sunflower to strategies of irrigation. II. Morphological and physiological responses to water shortage. *Field Crops Res*. **12**:91–103.

Connor, **D. J.** and **R. S. Loomis**. 1991. Strategies and tactics for water-limited agriculture in low rainfall mediterranean climates. In **E. Acevedo, E. Fereres, C. Giménez** and **J. P. Srivastrava** (eds.) *Improvement and Management of Winter Cereals under Temperature, Drought and Salinity Stresses*. Proceedings of an International Symposium, October 26–29, 1987, Cordoba, Spain. Madrid: INIA, pp. 441–65.

Connor, **D. J.** and **J. A. Palta**. 1981. The response of cassava to water shortage. III. Stomatal control of plant water status. *Field Crops Res*. **4**:297–311.

Connor, **D. J., A. Centeno**, and **M. Gómez-del-Campo**. 2009. Yield determination in olive hedgerow orchards. II. Analysis of radiation and fruiting profiles in hedgerow olive orchards. *Aust. J. Crop & Past. Sci*. **60**:443–52.

Connor, **D. J., J. H. Cock**, and **G. H. Parra**. 1981. The response of cassava to water shortage. I. Growth and yield. *Field Crops Res*. **4**:181–200.

Connor, **D. J., T. J. Jones**, and **J. A. Palta**. 1985a. Response of sunflower to strategies of irrigation. I. Growth, yield and the efficiency of water use. *Field Crops Res*. **10**:15–36.

Connor, **D. J., J. A. Palta**, and **T. R. Jones**. 1985b. Response of sunflower to strategies of irrigation. III. Crop photosynthesis and transpiration. *Field Crops Res*. **12**:281–93.

Connor, **D. J., V. O. Sadras**, and **A. J. Hall**. 1995. Canopy nitrogen distribution and the photosynthetic performance of sunflower crops during grain filling – a quantitative analysis. *Oecol*. **101**:274–81.

Connor, **D. J., S. Theiveyanathan**, and **G. M. Rimmington**. 1992. Development, growth, water-use and yield of a spring and a winter wheat in response to time of sowing. *Aust. J. Agric. Res*. **43**:493–516.

Cook, **P. G., F. W. Leaney**, and **I. D. Jolly**. 2001. *Groundwater Recharge in the Mallee Region, and Salinity Implications for the Murray River. a Review*. CSIRO 45/01, Canberra.

Cooper, **J. P.** 1970. Potential production and energy conservation in temperate and tropical grasses. *Herbage Abst*. **40**:113–58.

Cooper, **P. J. M., P. J. Gregory, D. Tully**, and **H. C. Harris**. 1987. Improving water use efficiency of annual crops in the rainfed farming systems of west Asia and North Africa. *Exp. Agric*. **23**:113–58.

Cornish, P. S. 1987. Effects of residues and tillage on the water balance of a red earth soil. In **T. G. Reeves** (ed.) *Proceedings of the Fourth Australian Agronomy Conference*. Melbourne: Aust. Soc. Agron., p. 294.

Cornish, P. S. and **J. E. Pratley**. 1987. *Tillage: New Directions in Australian Agriculture*. Melbourne: Inkata Press.

Cowan, I. R. 1965. Transport of water in the soil–plant–atmosphere system. *J. Appl. Ecol.* **2**:221–39.

Cowan, I. R. 1982. Regulation of water use in relation to carbon gain in higher plants. In **O. L. Lange, P. S. Nobel, C. B. Osmond**, and **H. Ziegler** (eds.) *Physiological Plant Ecology 11. Water Relations and Carbon Assimilation (Encycl. Plant Physiol.*, new ser. vol. 12B). Berlin: Springer-Verlag, pp. 589–613.

Cowan, I. R. 1986. Economics of carbon fixation in higher plants. In **T. J. Givnish** (ed.) *On the Economy of Plant Form and Function*. Cambridge University Press, pp. 133–70.

DeDatta, S. K. 1981. *Principles and Practices of Rice Production*. New York: John Wiley.

Demment, M. W., M. M. Young, and **R. L. Sensenig**. 2003. Providing micronutrients through food-based solutions: a key to human and national development. *J. Nutr.* **133** supplement: 3879S–3885S.

Denison, R. F. 2007. When can intelligent design of crops by humans outperform natural selection? In **J. H. J. Spiertz, P. C. Struik**, and **H. H. van Laar** (eds.) *Scale and Complexity in Plant Systems Research, Gene–Plant–Crop Relations* (http://library.wur.nl/frontis/gene-plant-crop/24_struik.pdf). The Netherlands: Springer, pp. 287–302.

Denison, R. F. and **R. S. Loomis**. 1989. *An Integrative Physiological Model of Alfalfa Growth and Development*. Publ. No. 1926. Div. Agric. Nat. Res. Oakland, California: University of California.

Denison R. F., T. E. Kiers, and **S. A. West**. 2003. Darwinian agriculture: when can humans find solutions beyond the reach of natural selection? *Quart. Rev. Biol.* **78**:145–67.

Denmead, O. T. and **R. J. Shaw**. 1962. Availability of soil water to plants as affected by soil moisture content and meteorological conditions. *Agron. J.* **54**:385–90.

Derpsch, R. and **T. Freidrich**. 2009. *Development and Current Status of No-till Adoption in the World*. Proceedings on CD-ROM, 18th Triennial Conference of the International Soil Tillage Research Organization (ISTRO), June 15–19, 2009, Izmir, Turkey.

de Wit, C. T. 1958. Transpiration and crop yields. *Versl. Landbouwk. Onderz.* **64**:6.

de Wit, C. T. 1960. On competition. *Versl. Landbouwk. Onderz.* **66**:8.

de Wit, C. T. 1992. Resource use efficiency in agriculture. *Agric. Syst.* **40**:125–52.

de Wit, C. T. and **J. P. van den Bergh**. 1965. Competition between herbage plants. *Neth. J. Agric. Sci.* **13**:212–21.

de Wit, C. T. *et al.* 1978. *Simulation of Assimilation, Respiration and Transpiration of Crops*. Simulation Monographs. Wageningen, the Netherlands: Pudoc.

Díaz-Ambrona, C. G. H., G. J. O'Leary, V. O. Sadras, M. G. O'Connell, and **D. J. Connor**. 2005. Environmental risk analysis of farming systems in a semi-arid environment: effect of rotations and management practices on deep drainage. *Field Crops Res.* **94**: 257–71.

Diebert, E. J., M. Bijeriego, and **R. A. Olson**. 1979. Utilization of ^{15}N fertilizer by nodulating and non-nodulating soybean isolines. *Agron. J.* **71**:717–22.

Distelfeld, A., C. Li, and **J. Dubcovsky**. 2009. Regulation of flowering in temperate cereals. *Curr. Opin. Plant Biol.* **12**:1–7.

Dobermann, A. and **K. G. Cassman**. 2004. Cropping systems: irrigated continuous rice systems of tropical and subtropical Asia. In **R. M. Goodman** (ed.) *Encyclopedia of Plant and Crop Science*. New York: Marcel Dekker, pp. 349–54.

Dobermann, A. *et al.* (22 co-authors) 2002. Site-specific nutrient management for intensive rice cropping systems in Asia. *Field Crops Res.* **74**:37–66.

Dobos, R. R., H. R. Sinclair, and **K. W. Hipple**. 2008. *User Guide: National Commodity Crop Productivity Index*. Version 1.0. (ftp://ftp-fc.sc.egov.usda.gov/NSSC/NCCPI/NCCPI_user_guide.pdf). USDA-NRCS.

DOE (USA) 2006. *Breaking the Barriers to Cellulosic Ethanol. A Joint Research Agenda*. DOE/SC-0095.

DOE (USA) 2008. *Biomass: Multi Year Plan* 2008. Oak Ridge, Tennessee.

Docring, O. C. III 1977. *An Energy Based Analysis of Alternative Production Methods and Cropping Systems in the Corn Belt*. West Lafayette, Indiana: Purdue University, Agricultural Experiment Station.

Dolan, M. S., C. E. Clapp, R. R. Allmaras, J. M. Baker, and **J. A. E. Molina**. 2006. Soil organic carbon and nitrogen in a Minnesota soil as related to tillage, residue and nitrogen management. *Soil & Tillage Res.* **89**:221–31.

Donald, C. M. 1958. The interaction of competition for light and for nutrients. *Aust. J. Agric. Res.* **9**(4):421–35.

Donald, C. M. 1968. The breeding of crop ideotypes. *Euphytica* **17**:385–403.

Donald, C. M. and **J. Hamblin**. 1976. The biological yield and harvest index of cereals as agronomic and plant breeding criteria. *Adv. Agron.* **28**:361–405.

Donald, C. M. and **J. Hamblin**. 1983. The convergent evolution of annual seed crops in agriculture. *Adv. Agron.* **36**:97–143.

Dong, Z., M. J. Canny, M. E. McCully *et al.* 1994. A nitrogen-fixing endophyte of sugarcane stems. *Plant Physiol.* **105**:1139–47.

Dry, P. R. and **B. R. Loveys**. 1999. Grape shoot growth and stomatal conductance are reduced when part of the root system is dried. *Vitis* **38**:151–6.

Dry, P. R., B. R. Loveys, and **H. During**. 2000. Partial drying of the root-zone of grape. I. Transient changes in shoot growth and gas exchange. *Vitis* **39**:3–8.

Duncan, W. G., R. S. Loomis, W. A. Williams, and **R. Hanau**. 1967. A model for simulating photosynthesis in plant communities. *Hilgardia* **38**:181–205.

Duncan, W. G., D. E. McCloud, R. L. McGraw, and **K. J. Boote**. 1978. Physiological aspects of peanut yield improvement. *Crop Sci.* **18**:1015–20.

Dunin, F. X., C. J. Smith, S. J. Zeglin *et al.* 2001. Water balance changes in a crop sequence with lucerne. *Aust. J. Agric. Res.* **52**:247–61.

Duvick, D. N. and **K. G. Cassman**. 1999. Post-green-revolution trends in yield potential of temperate maize in the north-central United States. *Crop Sci.* **39**:1622–30.

EC (European Commission). (2000). *Organic Farming: Guide to Community Rules*. Brussels: Directorate General for Agriculture.

Economist. 2008. *The Power and the Glory. Special Report on Energy*. June 21.

Edmeades, G. O. 2008. Drought tolerance in maize: an emerging reality. In **J. Clive** (ed.) *Global Status of Commercialized Biotech/GM Crops: 2008*. ISAAA Brief No. 39. (http://www.isaaa.org). Ithaca, NY: ISAAA.

EEA (European Energy Agency). 2007a. *Estimating the Environmentally Compatible Bioenergy Potential from Agriculture*. Technical Report 12. Luxembourg: Office for Official Publications of the European Communities.

EEA (European Energy Agency). 2007b. *Environmentally Compatible Bioenergy Potential from European Forests*. Copenhagen: EEA.

Ehlers, W. and **M. Goss**. 2003. *Water Dynamics in Plant Production*. Wallingford: CABI.

Ehleringer, J. and **R. W. Pearcy**. 1983. Variation in quantum yield for CO_2 uptake among C3 and C4 plants. *Plant Physiol.* **73**:555–9.

Ekern, P. C. 1965. Evapotranspiration of pineapple in Hawaii. *Plant Physiol.* **40**:736–9.

Elliott, B. 1987. Field crops. In **D. J. Connor** and **D.F. Smith** (eds.) *Agriculture in Victoria*. Melbourne, Australia: Australian Institute of Agricultural Science, pp. 107–25.

El-Sharkawy, M. A. 2007. International research on cassava photosynthesis, productivity, eco-physiology, and responses to environmental stresses in the tropics. *Photosynthetica* **44**:481–512.

El-Sharkawy, M. A., Y. Lopez, and **L. M. Bernal**. 2008. Genotypic variations in activities of phosphoenolpyruvate carboxylase and correlations with leaf photosynthetic characteristics and crop productivity of cassava grown in low-land seasonally-dry tropics. *Photosynthetica* **46**:238–47.

EMBRAPA. 2006. *Brazilian Agroenergy Plan 2006–2011*. Brazilia DF: EMBRAPA.

Evans, J. R., S. von Caemmerer, and **W. W. Adams III** (eds.) 1988. *Ecology of Photosynthesis in Sun and Shade*. Australia: CSIRO.

Evans, L. T. 1971. Evolutionary, adaptive, and environmental aspects of the photosynthetic pathway: assessment. In **M. D. Hatch, C. B. Osmond**, and **R. O. Slayter** (eds.) *Photosynthesis and Photorespiration*. New York: Wiley-Interscience, pp. 130–6.

Evans, L. T., R. M. Visperas, and **B. S. Vergara**. 1984. Morphological and physiological changes among rice varieties used in the Philippines over the last seventy years. *Field Crops Res.* **8**:105–24.

FAO. 2000. *Land Resource Potential and Constraints at Regional and Country Levels*. World Soil Resources Report 90. Rome: FAO.

FAO. 2002. *World Agriculture: Towards 2015/2030*. Rome: FAO UN.

FAO. 2007. *AEZWIN. An Interactive Multiple Criteria Analysis Tool for Land Resources Appraisal*. World Soil Resources Report 87. (www.iiasa.ac.at/Research/LUC/SAEZ/index.html) Rome: FAO.

FAO. 2008. *The State of Food and Agriculture 2008. Biofuels: Prospects, Risks and Opportunities*. Rome: FAO.

FAO. 2009. *High-level Expert Forum: How to Feed the World in 2050, 12–13 October 2009*. Rome: FAO, Agricultural Economics Development Division.

FAOSTAT. (http://faostat.fao.org/).

FAO/WHO. 1973. *Energy and Protein Requirements. Report of Joint Expert Committee*. Rome: FAO UN.

FAO/WHO/UNU. 1985. *Energy and Protein Requirements. Joint Expert Consultation*. Tech. Rep. No. 724. Geneva: World Health Organization.

Farquhar, G. D. and **R. A. Richards**. 1984. Istopic composition of plant carbon correlates with water-use efficiency of wheat genotypes. *Aust. J. Plant Physiol.* **11**:539–52.

Farquhar, G. G., M. H. O'Leary, and **J. A. Berry**. 1982. On the relationship between carbon isotope discrimination and the intercellular carbon dioxide concentration in leaves. *Aust. J. Plant Physiol.* **9**:121–37.

Fereres, E. and **A. Soriano**. 2007. Deficit irrigation for reducing agricultural water use. *J. Exp. Bot.* **58**:147–59.

Fereres, E., D. A. Goldhammer, and **L. R. Parsons**. 2003. Irrigation water management of horticultural crops. *HortScience* **38**:1036–42.

Fick, G. W., W. A. Williams, and **R. S. Loomis**. 1971. Recovery from partial defoliation and root pruning in sugar beet. *Crop Sci.* **11**:718–21.

Finlay, K. W. and **G. N. Wilkinson**. 1963. The analysis of adaptation in a plant breeding programme. *Aust. J. Agric. Res.* **14**:742–54.

Fischer, G. and **G. K. Heilig**. 1997. Population momentum and the demand on land and water resources. *Phil. Trans. Roy. Soc. London B* **352**:869–89.

Fischer, G., H. van Velthuizen, M. Shah, and **F. Nachtergaele**. 2002. *Global Agro-ecological Assessment for Agriculture in the 21st century*. FAO: Rome; and IIASA: Laxenburg.

Fischer, R. A. 1979. Growth and water limitation to dryland wheat yield in Australia: a physiological framework. *J. Aust. Inst. Agric. Sci.* **45**:83–94.

Fischer, R. A. and **J. S. Armstrong**. 1987. Strategies and tactics with short fallows. In **T. G. Reeves** (ed.) *Proceedings of the Fourth Australian Agronomy Conference*. Melbourne: Aust. Soc. Agron., p. 300.

Fischer, R. A., D. Rees, K. D. Sayre *et al.* 1998. Wheat yield progress associated with higher stomatal conductance and photosynthetic rate of cooler canopies. *Crop Sci.* **38**: 1467–75.

Fitzpatrick, E. A. and **H. A. Nix**. 1970. The climatic factor in Australian grassland ecology. In **R. M. Moore** (ed.) *Australian Grasslands*. Canberra: Australian National University Press, pp. 3–26.

Fleagle, R. G. and **J. A. Businger**. 1980. *An Introduction to Atmospheric Physics*, 2nd edn. New York: Academic Press.

Fletcher, A. L., T. R. Sinclair, and **L. H. Allen Jr**. 2007. Transpiration responses to vapor pressure deficit of "slow-wilting" and commercial soybean. *Environ. Exp. Bot.* **61**:145–51.

Fluck, R. C. 1981. Net energy sequestered in agricultural labor. *Trans. ASAE* **24**:1449–55.

Forrester, J. W. 1961. *Industrial Dynamics*. Cambridge, Massachusetts: Massachusetts Institute of Technology Press.

Foyer, C. H. 1988. Feedback inhibition of photosynthesis through source-sink regulation in leaves. *Plant Physiol. Biochem.* **26**:483–92.

Francis, G., R. Edinger, and **K. Becker**. 2005. A concept for simultaneous wasteland reclamation, fuel production, and socio-economic development in degraded areas in India: need, potential and perspectives of Jatropha plantations. *Nat. Resour. Forum* **29**:12–24.

Frissel, M. (ed.) 1977. Cycling of mineral nutrients in agricultural systems. *Agro-Ecosys* (special issue) **4**:1–354.

Fröhlich, C. and **J. London**. 1985. *Radiation Manual*. Geneva: World Meteorological Organization.

Fronzek, S. and **T. R. Carter**. 2007. Assessing uncertainties in climate change impacts on resource potential for Europe based on projections from RCMs and GCMs. *Clim. Change* **81**(Supplement 1):357–37

Gaff, D. F. 1981. The biology of resurrection plants. In **J. S. Pate** and **A. J. McComb** (eds.) *The Biology of Australian Plants*. Nedlands: University of Western Australia Press, pp. 114–46.

Galloway, J. N., F. J. Dentener, D. G. Capone *et al.* 2004. Nitrogen cycles: past, present, future. *Biogeochem.* **70**:153–226.

Gardner, W. R. 1965. Dynamic aspects of soil-water availability to plants. *Ann. Rev. Plant Physiol.* **16**:323–42.

Gardner, W. R. and **R. H. Brooks**. 1957. A descriptive theory of leaching. *Soil Sci.* **83**:295–304.

Garner, W. W. and **H. A. Allard**. 1920. Effect of length of day and night and other factors of the environment on growth and reproduction in plants. *J. Agric. Res.* **18**:553–606.

Garrity, D. P., D. G. Watts, C. Y. Sullivan, and **J. R. Gilley**. 1982. Moisture deficits and grain sorghum performance: evapotranspiration–yield relationships. *Agron. J.* **74**: 815–20.

Gates, D. M. 1980. *Biophysical Ecology*. New York: Springer-Verlag.

Gifford, R. M. and **R. J. Millington**. 1975. *Energetics of Agriculture and Food Production*. Aust. Bull. No. 288. Melbourne, Australia: CSIRO.

Gilmartin, P. M. and **C. Bowler** (eds.) 2002. *Molecular Plant Biology Vol. 1*. Oxford and New York: Oxford University Press.

Gilmore, E. C. and **J. S. Rogers**. 1958. Heat units as a method of measuring maturity in corn. *Agron. J.* **50**:611–15.

Gimeno, V., J. M. Fernandez-Martinez, and **E. Fereres**. 1989. Winter planting as a means of drought escape in sunflower. *Field Crops Res.* **22**:307–16.

Givnish, T. J. (ed.) 1986. *On the Economy of Plant Form and Function*. Cambridge: Cambridge University Press.

Gladstones, J. S. 1967. Naturalized subterranean clover strains in Western Australia: a preliminary agronomic examination. *Aust. J. Agric. Res.* **18**:713–32.

Gleason, H. A. 1926. The individualistic concept of plant association. *Bull. Torrey Bot. Club* **53**:1–20.

Gómez-del-Campo, M., A. Centeno, and **D. J. Connor**. 2009. Yield determination in olive hedgerow orchards. I. Yield and profiles of yield components in north–south and east–west oriented hedgerows. *Aust. J. Crop & Past. Sci.* **60**:434–42.

Goyne, P. J. and **A. A. Schneiter**. 1987. Photoperiod influence on development in sunflower genotypes. *Agron. J.* **79**:704–9.

Graham, R. L., R. Nelson, J. Sheehan, R. D. Perlack, and **L. L. Wright**. 2007. Current and potential corn stover supplies. *Agron. J.* **99**:1–11.

Grassini, P., A. J. Hall, and **J. L. Mercau**. 2009a. Benchmarking sunflower water productivity in semiarid environments. *Field Crops Res.* **110**:251–62.

Grassini, P., H. S. Yang, and **K. G. Cassman**. 2009b. Limits to maize productivity in the Western Corn-Belt: a simulation analysis for fully irrigated and rainfed conditions. *Agric. Forest Meteorol.* **149**:1254–65.

Grassini, P., J. Thorburn, C. Burr, and **K. G. Cassman**. 2010. High-yield irrigated maize systems in Western U.S. Corn Belt: management factors that affect water productivity. *Field Crops Research.* **120**:133–44.

Greenland, D. 1971. Changes in the nitrogen status and physical conditions of soils under pastures, with special reference to the maintenance of fertility of Australian soils used for growing wheat. *Soils and Fert.* **34**:237–51.

Griffiths, J. B. and **D. N. Walsgott**. 1987. Water use of wheat in the Victorian Mallee. In **T. G. Reeves** (ed.) *Proceedings of the Fourth Australian Agronomy Conference*. Melbourne: Aust. Soc. Agron., p. 296.

Grimes, D. W. and **K. M. El-Zik**. 1982. *Water Management for Cotton*. Berkeley, California: Div. Agric. Sci., University of California.

Gulick, S. H., K. G. Cassman, and **S. R. Grattan**. 1989. Exploitation of soil potassium in layered profiles by root systems of cotton and barley. *Soil Sci. Soc. Am. J.* **53**:146–53.

Gutschick, V. P. 1987. *A Functional Biology of Crop Plants*. London: Croom Helm.

Guyol, N. D. 1977. *Energy Interrelationships. A Handbook of Tables and Conversion Factors for Combining and Comparing International Energy Data*. Publ. No. FEB/B-77/166. Washington, D.C.: Federal Energy Administration.

Haefele, S. M., J. D. L. C. Siopongco, A. A. Boling, B. A. M. Bouman, and **T. P. Tuong**. 2009. Transpiration efficiency of rice. *Field Crops Res.* **111**:1–10.

Haishun, Y., A. Dobermann, K. G. Cassman, and **D. T. Walters**. 2006. Features, applications, and limitations of the hybrid-maize simulation model. *Agron. J.* **98**:737–48.

Hall, A. J., A. Vilella, N. Trapani, and **C. Chimenti**. 1982. The effects of water stress and genotype on the dynamics of pollen-shedding and silking in maize. *Field Crops Res.* **5**:349–63.

Hammer, G. L., P. J. Goyne, and **D. R. Woodruff**. 1982. Phenology of sunflower cultivars. III. Models for prediction in field environments. *Aust. J. Agric. Res.* **33**:263–74.

Hammer, G. L., Z. Dong, G. McLean *et al.* 2009. Can changes in canopy and/or root system architecture explain historical yield trends in U.S. Corn Belt? *Crop Sci.* **49**:299–312.

Hanks, R. J. 1983. Yield and water use relationships. In **H. M. Taylor, W. R. Jordan**, and **T. R. Sinclair** (eds.) *Limitations to Efficient use of Water in Crop Production*. Madison, WI: Am. Soc. Agron., pp. 393–412.

Harlan, H. V. and **M. L. Martini**. 1938. The effect of natural selection in a mixture of barley varieties. *J. Agric. Res.* **57**:189–99.

Harper, J. L. 1977. *Population Biology of Plants*. New York: Academic Press.

Haun, J. R. 1973. Visual quantification of wheat development. *Agron. J.* **65**:116–19.

Hearn, A. B. 1994. OZCOT: a simulation model for cotton management. *Agric. Sys.* **44**:257–99.

Hearn, A. B. and **D. D. da Rosa**. 1985. A simple model for crop management applications for cotton (*Gossypium hirsutum* L.). *Field Crops Res.* **12**:49–69.

Heffer, P. 2009. *Assessment of Fertilizer Use by Crops at the Global Level, 2006/07–2007/08.* (www.fertilizer.org). Paris: International Fertilizer Association.

Heichel, G. H. 1978. Stabilizing agricultural energy needs: role of forages, rotations, and nitrogen fixation. *J. Soil Water Conserv.* **33**:279–82.

Heilman, J. L., K. J. McInnes, M. J. Savage, R. W. Gesch, and **R. J. Lascano**. 1994. Soil and canopy energy balances in a west Texas vineyard. *Agric. For. Meteorol.* **71**:99–114.

Heilman, J. L., K. J. McInnes, R. W. Gesch, R. J. Lascano, and **M. J. Savage**. 1996. Effects of trellising on the energy balance of a vineyard. *Agric. For. Meteorol.* **81**:79–93.

Helyar, K. R. and **W. M. Porter**. 1989. Soil acidification, its measurement and the processes involved. In **A. D. Robson** (ed.) *Soil Acidity and Plant Growth*. Merrickville, New South Wales: Academic Press Australia, pp. 61–101.

Helyar, K. R., Z. Hochman, and **J. P. Brennan**. 1988. The problem of acidity in temperate area soils and its management. In **J. Loveday** (ed.) *National Soils Conference Review Papers*. Nedlands, Western Australia: Australian Society of Soil Science, pp. 22–54.

Herridge, D. E., M. B. Peoples, and **R. M. Boddey**. 2008. Global inputs of biological nitrogen fixation in agricultural systems. *Plant & Soil* **311**:1–18.

Herridge, D. F. and **F. J. Bergersen**. 1988. Symbiotic nitrogen fixation. In **J. R. Wilson** (ed.) *Advances in Nitrogen Cycling in Agricultural Ecosystems*. Wallingford, Oxon, UK: CAB International, pp. 45–65.

Hiler, E. A., A. T. Howell, R. B. Lewis, and **R. P. Boos**. 1974. Irrigation timing by the stress day index method. *Trans. ASAE* **17**:393–8.

Hirose, T. and **M. J. A. Werger**. 1987. Maximizing daily canopy photosynthesis with respect to the leaf nitrogen allocation pattern in the canopy. *Oecologia* **72**:520–6.

Hochman, Z., D. Holzworth, and **J. R. Hunt**. 2009. Potential to improve on-farm wheat yield and WUE in Australia. *Crop & Past. Sci.* **60**:708–16.

Holaday, A. S. and **C. C. Black**. 1981. Comparative characterization of phosphoenolpyruvate carboxylase in C3, C3 and C3–C4 intermediate *Panicum* species. *Plant Physiol.* **67**:330–4.

Holland, E. A., Braswell, B. H., Sulzman, J., and **J.-F. Lamarque**. 2005. Nitrogen deposition onto the United States and western Europe: synthesis of observations and models. *Ecol. Applic.* **15**:38–57.

Hoskinson, P. E. and **C. O. Qualset**. 1967. Geographic variation in Balboa rye. *Tennessee Farm & Home Progress Report* **62**(2):8–9.

Houston, C. E. 1967. *Drainage of Irrigated Land*. University of California, Berkeley, California: Calif. Agric. Exp. Sta.

Hsiao, T. C., E. Fereres, E. Acevedo, and **D. W. Henderson**. 1976. Water stress and dynamics of growth and yield of crop plants. In **O. L. Lange, L. Kappen,** and **E.-D. Schulze** (eds.) *Water and Plant Life*. Ecological Studies Vol. 19. Berlin: Springer-Verlag, pp. 281–305.

Hsiao, T. C., P. Steduto, and **E. Fereres**. 2007. A systematic and quantitative approach to improve water use efficiency in agriculture. *Irrig. Sci.* **25**:209–31.

Huan, N. H., L. V. Thiet, H. V. Chien, and **K. L. Heong**. 2004. Farmers' evaluation of reducing pesticides, fertilizers and seed rates in rice farming through participatory research in the Mekong Delta, Vietnam. *Crop Prot.* **24**:457–64.

Huang, P. 1989. Feldspars, olivines, pyroxenes, and amphiboles. In **J. B Dixon** and **S. B. Weed** (eds.) *Minerals in the Soil Environment*. Madison, WI: Soil. Sci. Soc. Am., pp. 975–1050.

Hunt, D. 1983. *Farm Power and Machinery Management*. Ames, IA: Iowa State University Press.

Hunt, R. 1978. Plant growth analysis. *Studies in Biology, No. 96*. London: Edward Arnold.

Idso, C. and **S. F. Singer**. 2009. *Climate Change Reconsidered: 2009 Report of the Nongovernmental Panel on Climate Change (NIPCC)*. Chicago, Illinois: The Heartland Institute.

IEA (International Energy Agency). 2008. *World Energy Outlook, 2008*. Paris, France: IEA.

IFOAM (International Federation of Organic Agriculture Movements). 2002. *Norms and Accreditation Criteria for Organic Production and Processing*. Bonn, Germany: IFOAM.

Inman-Bamber, N. G. and **M. G. McGlinchey**. 2003. Crop coefficients and water-use estimates for sugarcane based on long-term Bowen ratio energy balance measurements. *Field Crops Res.* **83**:125–8.

IPCC-AR4. 2007. *Fourth Assessment Report of the Intergovernmental Panel on Climate Change*. **M. L. Parry, O. F. Canziani, J. P. Palutikof, P. J. van der Linden** and **C. E. Hanson** (eds.) Cambridge University Press.

Jackson, B. S., G. F. Arkin, and **A. B. Hearn**. 1988. The cotton simulation model "COTTAM": fruiting model calibration and testing. *Trans. Amer. Soc. Agric. Eng.* **31**:846–54.

Jacobs, M. R. 1955. *Growth Habits of Australian Eucalypts*. Canberra: Australian Government Printer.

James, C. 2008. *Global Status of Commercialized Biotech/GM Crops: 2008*. ISAAA Brief No. 39. (www.isaaa.org). Ithaca, NY: ISAAA.

Jamieson, P. D., I. R. Brooking, M. A. Semenov, and **J. R. Porter**. 1998. Making sense of wheat development: a critique of methodology. *Field Crops Res.* **55**:117–27.

Jamieson, P. D., I. R. Brooking, M. A. Semenov *et al.* 2007. Reconciling alternative models of phenological development in winter wheat. *Field Crops Res.* **103**:36–41.

Jenkinson, D. S. 1982. The nitrogen cycle in long term field experiments. *Phil. Trans. R. Soc. Lond.* B **296**:563–71.

Jenkinson, D. S. 1988. Determination of microbial biomass carbon and nitrogen in soil. In **J. R. Wilson** (ed.) *Advances in Nitrogen Cycling in Agricultural Ecosystems*. Wallingford, Oxon, UK: CAB International, pp. 368–86.

Jenny, H. 1930. *A Study on the Influence of Climate upon the Nitrogen and Organic Matter Content of the Soil*. University of Missouri Agric. Exp. Sta. Res. Bull. No. 152.

Jenny, H. 1941. *Factors of Soil Formation*. New York: McGraw-Hill Book Co.

Jenny, H. 1980. *The Soil Resource. Ecological Studies* Vol. 37. New York: Springer-Verlag.

Jensen, M. E. 1974. *Consumptive Use of Water and Irrigation Water Requirements*. New York: Am. Soc. Civil Eng.

Jensen, M. E. 2007. Beyond irrigation efficiency. *Irrig. Sci.* **25**:233–45.

Johnson, J. W., L. F. Welch, and **L. T. Kurtz**. 1975. Environmental implications of N fixation by soybeans. *J. Environ. Qual.* **4**:303–6.

Johnson, W. A., V. Stoltzfus, and **P. Craumer**. 1977. Energy conservation in Amish agriculture. *Science* **198**:373–8.

Johnson, W. G., V. M. Davis, G. R. Kruger, and **S. C. Weller**. 2009. Influence of glyphosate-resistant cropping systems on weed species shifts and glyphosate-resistant weed populations. *European J. Agron.* **31**:162–72.

Jones, C. A. and **J. R. Kiniry** (eds.). 1986. *CERES-MAIZE. A Simulation Model of Maize Growth and Development*. College Station: Texas A&M University Press.

Jones, J. W. 1983. Irrigation options to avoid critical stresses: optimization of on-farm water allocation to crops. In **H. M. Taylor, W. R. Jordan,** and **T. R. Sinclair** (eds.) *Limitations to Efficient Water Use in Crop Production*. Madison, WI: Am. Soc. Agron., pp. 507–616.

Jones, J. W., G. Hoogenboom, C. H. Porter *et al.* 2003. The DSSAT cropping system model. *Eur. J. Agron.* **18**:235–65.

Jones, M. R. 1989. Analysis of the use of energy in agriculture. *Agric. Syst.* **29**:339–55.

Jongschaap, R. E. E., W. J. Corre, P. S. Bindraban, and **W. A. Bandenburg**. 2007. *Claims and facts on* Jatropha curcas *L. Global* Jatropha curcas *Evaluation, Breeding and Propagation Programme*. Report 158. (www.jatropha-platform.org/Documents.html). Wagenignen: Plant Research International B.V.

Jordan, W. R. and **F. R. Miller**. 1980. Genetic variability in sorghum root systems: implications for drought tolerance. In **N. C. Turner** and **P. J. Kramer** (eds.) *Adaptation of Plants to Water and High Temperature Stress*. New York: John Wiley, pp. 383–99.

Jordan, W. R., W. A. Dugas, and **P. J. Shouse**. 1983. Strategies for crop improvement for drought-prone regions. *Agric. Water Manage.* **7**:281–99.

Kaddah, M. T. and **J. D. Rhoades**. 1976. Salt and water balance in Imperial Valley, California. *Soil Sci. Soc. Am. J.* **40**:93–100.

Kampmeijer, P. and **J. C. Zadoks**. 1977. *EPIMUL, a Simulator of Foci and Epidemics in Mixtures of Resistant and Susceptible Plants, Mosaics and Multilines. Simulation Monographs*. Wageningen, the Netherlands: Pudoc.

Katsura, K., S. Maeda, T. Horie, and **T. Shiraiwa**. 2009. Estimation of respiratory parameters for rice based on long-term and intermittent measurement of canopy CO_2 exchange rates in the field. *Field Crops Res.* **111**:85–91.

Keating, B. A., D. Gaydon, N. I. Huth *et al.* 2002. Use of modelling to explore the water balance of dryland farming systems in the Murray-Darling Basin, Australia. *Europ. J. Agron.* **18**:159–69.

Kemanian, A. R., C. O. Stockle, and **D. R. Huggins**. 2005. Transpiration-use efficiency of barley. *Agric. For. Meteorol.* **130**:1–11.

Kiehl, J. T. 2007. Twentieth century climate model response and climate sensitivity. *Geophys. Res. Lett.* **34**:L22710. DOI:10.1029/2007GL031383.

Kiesselbach, T. A. 1949. *The Structure and Reproduction of Corn*. Cold New York: Cold Spring Harbor Laboratory Press.

Kiniry, J. R., J. T. Ritchie, and **R. L. Musser**. 1983. Dynamic nature of the photoperiod response of maize. *Agron. J.* **75**:700–3.

Kira, T., H. Ogawa, and **K. Shinozaki**. 1953. Intraspecific competition among higher plants. 1. Competition-density-yield inter-relationships in regularly dispersed populations. *J. Inst. Polytech. Osaka City Univ.* **4**:1–16.

Kirby, E. J. M. and **M. Appleyard**. 1984. *Cereal Development Guide*, 2nd edn. Stoneleigh, UK: National Agricultural Centre.

Kirchmann, H. and **L. Bergstrom** (eds.) 2008. *Organic Crop Production: Ambitions and Limitations*. Springer.

Kirkegaard, J. A., P. A. Gardner, J. F. Angus, and **E. Koetz**. 1994. Effect of *Brassica* break crops on the growth and yield of wheat. *Aust. J. Agric. Res.* **45**:529–45.

Kirkegaard, J. A., G. N. Howe, and **P. M. Mele**. 1999. Enhanced accumulation of soil mineral-N following canola. *Aust. J. Exp Agric.* **39**:587–93.

Klingebiel, A. A. and **P. H. Montgomery**. 1961. *Land-use Capability Classification*. Agric. Handbook No. 210. Washington, D.C.: U.S. Government Printing Office.

Knutson, J. D., R. G. Curley, E. B. Roberts, R. M. Hagan, and **V. Cervinka**. 1977. Energy for irrigation. *Calif. Agric.* **31**(5):46–7.

Kreidemann, P. E. and **I. Goodwin**. 2003. *Regulated Deficit Irrigation and Partial Rootzone Drying. An Overview of Principles and Applications*. Irrigation Insights No. 3. Canberra: Land & Water Australia.

Kremer, C., C. O. Stockle, A. R. Kemanian, and **T. Howell**. 2008. A canopy transpiration and photosynthesis model for evaluating simple crop productivity models. In **Ahuja** *et al.* (eds.) *Response of Crops to Limited Water: Understanding and Modeling Water Stress Effects on Plant Growth Processes*. Madison, Wisconsin: ASA, CSSA, SSSA, pp. 301–55.

Laabs, V., W. Amelung, A. Pinto, A. Altstaedt, and **W. Zech**. 2000. Leaching and degradation of corn and soybean pesticides in an Oxisol of the Brasilian Cerrados. *Chemosphere* **41**:1441–9.

Lafitte, H. R. and **R. S. Loomis**. 1988a. Calculation of growth yield, growth respiration and heat content of grain sorghum from elemental and proximal analyses. *Ann. Bot.* **62**:353–61.

Lafitte, H. R. and **R. S. Loomis**. 1988b. The growth and composition of grain sorghum with limited nitrogen. *Agron. J.* **80**:492–3.

Laird, D. A. 2008. The charcoal vision: a win–win–win scenario for simultaneously producing bioenergy, permanently sequestering carbon, while improving soil and water quality. *Agron. J.* **100**:178–81.

Laing, D. R., P. J. Kretchmer, S. Zuluaga, and **P. G. Jones**. 1983. Field bean. In **W. H. Smith** and **S. J. Banta** (eds.) *Potential Productivity of Field Crops Under Different Environments*. Los Banos, Philippines: IRRI, pp. 227–48.

Lamb, H. H. 1977. *Climate: Past, Present and Future*. Vol. 2. *Climatic History and the Future*. London: Metheun.

Large, E. C. 1954. Growth stages in cereals. Illustrations of the Feekes' scale. *Plant Path.* **3**:128–9.

LaRue, T. A. and **T. G. Patterson**. 1981. How much nitrogen do legumes fix? *Adv. Agron.* **34**:15–38.

Latshaw, W. L. and **E. C. Miller**. 1924. Elemental composition of the corn plant. *J. Agric. Res.* **27**:845–60.

Lee, E. A. and **M. Tollenaar**. 2007. Physiological basis of successful breeding strategies for maize grain yield. *Crop Sci.* **47**(S3):S202–S215.

Lefroy, E. C. and **R. J. Stirzaker**. 1999. Agroforestry for water management in the cropping zone of southern Australia. *Agrofor. Sys.* **45**:277–302.

Legg, T. D., J. J. Fletcher, and **K. W. Easter**. 1989. Nitrogen budgets and economic efficiency: a case study of southeastern Minnesota. *J. Prod. Agric.* **2**:110–16.

Leitch, **I.** and **W. Godden**. 1953. *The Efficiency of Farm Animals in the Conversion of Feed-ingstuffs to Food for Man*. (Anim. Nut. Tech. Commun. No. 14.) Farnham Royal, Slough, UK: Commonwealth Agricultural Bureau.

Levi, **J.** and **M. L. Peterson**. 1972. Responses of spring wheats to vernalization and photoperiod. *Crop Sci.* **12**:487–90.

Lindsay, **W. L.**, **P. L. G. Vlek**, and **S. H. Chien**. 1989. Phosphate minerals. In **J. B. Dixon** and **S. B. Weed** (eds.) *Minerals in Soil Environments*. Madison, Wisconsin: Soil Sci. Soc. Am., pp. 1089–130.

Liska, **A. J.** and **K. G. Cassman**. 2008. Towards standardization of life-cycle metrics for biofuels: greenhouse gas emissions mitigation and net energy yield. *J. Biobased Materials* **2**:187–203.

Liska, **A. J.**, and **R. K. Perrin**. 2009. Indirect land use emissions in the life cycle of biofuels: regulations vs. science. *Biofuel Bioproduct Refining* **3**:318–28

Liska, **A. J.**, **H. S. Yang**, **V. R. Bremer** *et al.* 2009. Improvements in life cycle energy efficiency and greenhouse gas emissions of corn-ethanol. *J. Indust. Ecol.* **13**:58–74.

Lobell, **D. B.**, **K. G. Cassman**, and **C. B. Field**. 2009. Crop yield gaps: their importance, magnitudes, and causes. *Annu. Rev. Environ. Resour.* **34**:179–204.

Loffler, **C. M.**, **J. Wei**, **T. Fast** *et al.* 2005. Classification of maize environments using crop simulation and geographic information systems. *Crop Sci.* **45**:1708–16.

Loomis, **R. S.** 1978. Ecological dimensions of medieval agrarian systems: an ecologist responds. *Agric. Hist.* **52**:478–83.

Loomis, **R. S.** 1979. Ideotype concepts for sugarbeet improvement. *J. Am. Soc. Sugar Beet Technol.* **20**:323–41.

Loomis, **R. S.** 1983. Crop manipulations for efficient use of water: an overview. In **H. M. Taylor**, **W. R. Jordan**, and **T. R. Sinclair** (eds.) *Limitations to Efficient Use of Water in Crop Production*. Madison, Wisconsin: Am. Soc. Agron., pp. 345–74.

Loomis, **R. S.** 1985. Systems approaches for crop and pasture research. In **J. J. Yates** (ed.) *Proceedings of the Third Australian Agronomy Conference*, Melbourne, Aust. Soc. Agron., pp. 1–8.

Loomis, **R. S.** and **P. A. Gerakis**. 1975. Productivity of agricultural systems. In **J. P. Cooper** (ed.) *Photosynthesis and Productivity in Different Environments*. Cambridge: Cambridge University Press, pp. 145–72.

Loomis, **R. S.** and **H. R. Lafitte**. 1987. The carbon economy of a maize crop exposed to elevated CO_2 concentrations and water stress determined from elemental analysis. *Field Crops Res.* **17**:63–74.

Loomis, **R. S.** and **W. A. Williams**. 1963. Maximum crop productivity: an estimate. *Crop Sci.* **3**:67–72.

Loomis, **R. S.** and **W. A. Williams**. 1969. Productivity and the morphology of crop stands. In **J. D. Eastin**, **F. A. Haskin**, **C. Y. Sullivan**, and **C. H. M. Van Bavel** (eds.) *Physiological Aspects of Crop Yield*. Madison, Wisconsin: Am. Soc. Agron., pp. 27–47.

Loomis, **R. S.**, **J. H. Brickey**, **F. E. Broadbent**, and **G. F. Worker Jr**. 1960. Comparison of nitrogen source materials for midseason fertilization of sugar beets. *Agron. J.* **52**:97–101.

Loomis, **R. S.**, **Y. Luo**, and **P. Kooman**. 1990. Integration of activity in the higher plant. In **R. Rabbinge**, **J. Goudriaan**, **H. van Keulen**, **F. W. T. Penning de Vries**, and **H. H. van Laar** (eds.) *Theoretical Production Ecology: Reflections and Prospects*. Wageningen, the Nether-lands: Pudoc, pp. 105–24.

Loomis, **W. E.** 1932. Growth–differentiation balance vs. carbohydrate–nitrogen balance. *Proc. Am. Soc. Hort. Sci.* **29**:240–5.

Lopes, N. F. 1979. *Respiration Related to Growth and Maintenance in Radish* (Raphanus sativus L.). PhD dissertation. University of California, Davis.

López-Bellido, L., M. Fuentes, J. E. Castillo, F. J. López-Garrido, and **E. J. Fernández.** 1996. Long-term tillage, crop rotation, and nitrogen fertilizer effects on wheat yield under rainfed Mediterranean conditions. *Agron. J.* **88**:783–91.

Lorio., P. L. Jr. 1986. Growth-differentiation balance: a basis for understanding Southern Pine Beetle-tree interactions. *Forest Ecol. Manage.* **145**:259–73.

Ludlow, M. M. and **S. B. Powles.** 1988. Effects of photoinhibition induced by water stress on growth and yield of grain sorghum. In **J. R. Evans, S. von Caemmerer,** and **W. W. Adams III** (eds.) *Ecology of Photosynthesis in Sun and Shade.* Canberra, Australia: CSIRO, pp. 179–94.

Lush, W. M. and **L. T. Evans.** 1974. Translocation of photosynthetic assimilate from grass leaves, as influenced by environment and species. *Aust. J. Plant Physiol.* **1**:417–31.

Lyerly, P. J. and **D. E. Longeneker.** 1957. *Salinity Control in Irrigated Agriculture.* Texas Agric. Exp. Sta. Bull. No. 876.

Maas, E. V. 1984. Crop tolerance. *Calif. Agric.* **38**(10):20–l.

Maas, E. V. and **G. J. Hoffman.** 1977. Crop salt tolerance: current assessment. *J. Irrig. Drain.* **103**:115–34.

MacRae, R. J. and **G. R. Mehuys.** 1985. The effect of green manuring on the physical properties of temperate-area soils. *Adv. Soil Sci.* **3**:71–94.

Macedo, I. C. and **E. A. Seabra.** 2008. Mitigation of GHG emissions using sugarcane ethanol. In **P. Zuurbier** and **J. van de Vooren** (eds.) *Sugarcane Ethanol: Contributions to Climate Change Mitigation and the Environment.* The Netherlands: Wageningen Academic Publishers, pp. 95–111.

Major, D. J. 1980. Photoperiod response characteristics controlling flowering of nine crop species. *Can. J. Plant Sci.* **60**:777–84.

Marris, E. 2008. More crop per drop. *Nature* **452**:273–7.

Marten, G. G. 1988. Productivity, stability, sustainability, equitability and autonomy as properties for agroecosystem assessment. *Agric. Systems* **26**:291–316.

Mateos, L. 2008. Assessing a new paradigm for irrigation system performance. *Irrig. Sci.* **27**:25–34.

McBride, J. L. and **N. Nicholls.** 1983. Seasonal relationships between Australian rainfall and the southern oscillation index. *Month. Weath. Rev.* **111**:1998–2004.

McCallum, M. H., D. J. Connor, and **G. J. O'Leary.** 2001. Water use by lucerne and effect on crops in the Victoria Wimmera. *Aust. J. Agric. Res.* **52**:193–201.

McCree, K. J. 1970. An equation for the rate of respiration of white clover plants grown under controlled conditions. In **I. Setlik** (ed.) *Prediction and Measurement of Photosynthetic Productivity.* Proc. IBP/PP Tech. Mtg., Trebon. Pudoc, Wageningen, pp. 221–9.

McDermitt, D. K. and **R. S. Loomis.** 1981. Elemental composition of biomass and its relation to energy content, growth efficiency, and growth yield. *Ann. Bot.* **48**:275–90.

McGlasson, W. B. and **H. K. Pratt.** 1963. Fruit-set patterns and fruit growth in cantaloupe (*Cucumis melo* L., var. *reticulatis* Naud.). *Proc. Am. Soc. Hort. Sci.* **83**:495–505.

McGregor, K. C. and **C. K. Mutchler.** 1977. Status of the R factor in northern Mississippi. In **G. R. Foster** (ed.) *Soil Erosion: Prediction and Control.* Ankeny, Iowa: Soil Conservation Society, pp. 135–42.

McIlroy, I. C. 1971. An instrument for continuous recording of natural evaporation. *Agric Meteorol.* **9**:93–100.

McKinion, J. M., D. N. Baker, F. D. Whistler, and **J. R. Lambert**. 1989. Application of the GOSSYM/COMAX system to cotton crop management. *Agric. Sys.* **31**:55–65.

McLaughlin, N.B., C. F. Drury, W. D. Reynolds *et al.* 2008. Energy inputs for conservation and conventional tillage implements in a clay soil. *Amer. Soc. Agric. & Biol. Engineers* **51**:1153–63.

Meijninger, W. M. L. and **H. A. R. de Bruin**. 2000. The sensible heat fluxes over irrigated areas in western Turkey determined with a large aperture scintillometer. *J. Hydrol.* **229**:42–9.

Menz, K. M., D. N. Moss, R. Q. Cannell, and **W. A. Brun**. 1969. Screening for photosynthetic efficiency. *Crop Sci.* **9**:692–4.

Messina, C. D., J. W. Jones, K. J. Boote, and **C. E. Vallejos**. 2006. A gene-based model to simulate soybean development and yield responses to environment. *Crop Sci.* **46**:456–66.

Meyer, W. S. and **G. C. Green**. 1981. Plant indicators of wheat and soybean crop water stress. *Irrig. Sci.* **2**:167–76.

Midwest Plan Service. Livestock Wastes Subcommittee. 1985. *Livestock Waste Facilities Handbook*. Ames, Iowa State University Press: Midwest Plan Service.

Miller, E. E. and **A. Klute**. 1967. The dynamics of soil water. Part I. Mechanical forces. In **R. M. Hagan, H. R. Haise**, and **T. W. Edminster** (eds.) *Irrigation of Agricultural Lands*. Agronomy Monograph No. 11. Madison, Wisconsin: Am. Soc. Agron., pp. 209–44.

Mohanty, H. K., S. Mallik, and **A. Grover**. 2000. Prospects of improving flooding tolerance in lowland rice varieties by conventional breeding and genetic engineering. *Curr. Sci.* **78**:132–7.

Moncur, M. W. 1981. *Floral Initiation in Field Crops. An Atlas of Scanning Electron Micrographs*. Australia: CSIRO.

Monsi, M. and **T. Saeki**. 1953. Uber den Lichtfaktor in den Pflanzengesellschaften und seine Bedeutung fur die Stoffproduktion. *Jap. J. Bot.* **14**:22–52.

Monteith, J. L. 1964. Evaporation and environment. *Symp. Soc. Exp. Biol.* **19**:205–34.

Monteith, J. L. 1977. Climate and the efficiency of crop production in Britain. *Phil. Trans. R. Soc. Lond. B* **281**:277–94.

Monteith, J. L. 1978. Reassessment of maximum growth rates for C3 and C4 crops. *Exp. Agric.* **14**:1–5.

Monteith, J. L. and **M. H. Unsworth**. 1990. *Principles of Environmental Physics*, 2nd edn. London: Edward Arnold.

Mooney, H. A. and **E. L. Dunn**. 1970. Convergent evolution of Mediterranean climate evergreen sclerophyll shrubs. *Evolution* **24**:292–303.

Morgan, J. 1983. Osmoregulation as a selection criterion for drought tolerance in wheat. *Aust. J. Agric. Res.* **34**:607–14.

Morgan, J. M. 2000. Increases in grain yield of wheat by breeding for an osmoregulation gene: relationship to water supply and evaporative demand. *Aust. J. Agric. Res.* **51**:971–8.

Morgan, R. P. C. and **D. A. Davidson**. 1986. *Soil Erosion and Conservation*. Harlow, Essex, UK: Longman Scientific & Technical.

Morrison, M. J., H. D. Voldeng, and **E. R. Cober**. 1999. Physiological changes from 58 years of genetic improvement of short-season soybean cultivars in Canada. *Agron. J.* **91**:685–9.

Morton, D. C., R. S. DeFries, Y. E. Shimabukuro *et al.* 2006. Crop expansion changes deforestation dynamics in southern Brazilian Amazon. *Proc. Natl. Acad. Sci. USA* **103**:14637–41.

Myers, R. J. K. and **M. A. Foale**. 1981. Row spacing and population density in grain sorghum: a simple analysis. *Field Crops Res.* **4**:147–54.

National Research Council. Committee on Animal Nutrition. 1982. *United States-Canadian Tables of Feed Composition* (third revision). Washington, D.C.: National Academy of Sciences.

National Research Council. Committee on Animal Nutrition. 1984. *Nutrient Requirements of Beef Cattle*. Washington, D.C.: National Academy Press.

National Research Council. Subcommittee on the Tenth Edition of the RDAs. 1989. *Recommended Dietary Allowances*. Washington, D.C.: National Academy Press.

Nature. 2006. Biofuelling the future. Business feature. *Nature* **44**:669–78.

Naylor, R. L., A. J. Liska, M. B. Burk *et al.* 2007. The ripple effect: biofuels, food security, and the environment. *Environment* **49**:30–43.

Neales, T. F. and **L. D. Incoll**. 1968. The control of leaf photosynthesis rate by level of assimilate concentration in the leaf: a review of the hypothesis. *Bot. Rev.* **34**:107–25.

Neiburger, M., T. G. Edinger, and **W. D. Bonner**. 1982. *Understanding our Atmospheric Environment*. San Francisco: W. H. Freeman.

Nelson, D. E. *et al.* (18 co-authors) 2007. Plant nuclear factor Y (NF-Y) B subunits confer drought tolerance and lead to improved corn yields on water-limited acres. *Proc. Nat. Acad. Sci. USA* **104**:16451–5.

Nelson, D. W. 1982. Gaseous losses of nitrogen other than through denitrification. In **F. J. Stevenson** (ed.) *Nitrogen in Agricultural Soils*. Agronomy Monograph no. 22. Madison, Wisconsin: Am. Soc. Agron., pp. 327–63.

Ng, E. and **R. S. Loomis**. 1984. *Simulation of Growth and Yield of the Potato Crop*. Simulation Monographs. Wageningen, the Netherlands: Pudoc.

Nicholls, N. 1991. Advances in long-term weather forecasting. In **R. C. Muchow** and **J. A. Bellamy** (eds.) *Climatic Risk in Crop Production: Models and Management for the Semi-arid Tropics*. Wallingford, UK: CAB International, pp. 427–44.

NLWRA. 2000. *Dryland Salinity. Australia's Dryland Salinity Assessment 2000*. Canberra: National Land and Water Resources Audit.

Northcote, K. H., G. D. Hubble, R. F. Isbell, C. H. Thompson, and **E. Bettany**. 1975. *A Description of Australian Soils*. Melbourne: CSIRO.

Norton, J. M. 2008. Nitrification in agricultural soils. In **J. S. Schepers, W. R. Raun** (eds.) *Nitrogen in Agricultural Systems*. Agron. Monogr. 49. Madison, Wisconsin: American Society of Agronomy; Crop Science Society of America; Soil Science Society of America; pp. 173–99.

O'Connell, M. G., G. J. O'Leary, and **M. Incerti**. 1995. Potential groundwater recharge from fallowing in north-western Victoria, Australia. *Agric. Water Manage.* **29**:37–52.

O'Connell, M. G., G. J. O'Leary, and **D. J. Connor**. 2003. Drainage and change in soil water storage below the root zone under long fallow and continuous cropping sequences in the Victorian Mallee. *Aust. J. Agric. Res.* **54**:663–75.

Odum, H. T. 1967. Energetics of world food production. In *The President's Scientific Advisory Committee, Report of Problems of World Food Supply 3*. Washington, D.C.: The White House, pp. 55–94.

Ofori, F. and **W. R. Stern**. 1987. Cereal–legume intercropping systems. *Adv. Agron.* **41**:41–90.

O'Leary, G. J. 1996. The effects of conservation tillage on potential groundwater recharge. *Agric. Water Manage.* **31**:65–73.

O'Leary, G. J. and **D. J. Connor**. 1998. A simulation study of wheat crop response to water supply, nitrogen nutrition, stubble retention, and tillage. *Aust. J. Agric. Res.* **49**: 11–19.

O'Leary, G. J., D. J. Connor, and **D. H. White**. 1981. A simulation model of the development, growth and yield of the wheat crop. *Agric. Syst.* **17**:1–26.

O'Leary, G. J., D. Ormesher, and **M. Wells**. 2003. Detecting subsoil constraints on farms in the Murray Mallee. In *Solutions for a Better Environment*. Proceedings of the 11th

Australian Agronomy Conference, February 2–6, 2003, Geelong, Victoria. Australian Society of Agronomy. (www.regional.org.au/au/asa/2003/c/15/oleary.htm).

Olesen, T., S. Morris, and L. McFadyen. 2007. Modelling the interception of photosynthetically active radiation by evergreen subtropical hedgerows. *Aust. J. Agric. Res.* **58**:215–33.

Olk, D. C., K. G. Cassman, N. Mahieu, and E. W. Randall. 1998. Conserved chemical properties of young humic acid fractions in tropical lowland soil under intensive irrigated rice cropping. *Eur. J. Soil Sci.* **49**:337–49.

Orang, M. N., J. S. Matyac, and R. L. Snyder. 2008. Survey of irrigation methods in California. *J. Irrig. Drain. E. ASCE* **34**:96–100.

Orgaz, F., L. Mateos, and E. Fereres. 1992. Season length and cultivar determine the optimum evapotranspiration deficit in cotton. *Agron. J.* **84**:700–6.

Osmond, C. B., V. Oja, and A. Laisk. 1988. Regulation of carboxylation and photosynthetic oscillations during sunshade acclimation in *Helianthus annuus* measured with a rapid-response gas exchange system. In J. R. Evans, S. von Caemmerer, and W. W. Adams III (eds.) *Ecology of Photosynthesis in Sun and Shade*. Canberra, Australia: CSIRO, pp. 237–51.

Osmond, C. B., K. Winter, and H. Ziegler. 1982. Functional significance of different pathways of CO_2 fixation in photosynthesis. In O. L. Lange, P. S. Nobel, C. B. Osmond, and H. Zeigler (eds.) *Physiological Plant Ecology. II. Water Relations and Carbon Assimilation (Encyl. Plant Physiol.*, new ser. vol. 12B). Heidelberg: Springer-Verlag, pp. 479–547.

Oster, J. D., G. J. Hoffman, and F. E. Robinson. 1984. Dealing with salinity. Management alternatives: crop, water and soil. *Calif. Agric.* **38**(10):29–32.

Otegui, M. E., F. H. Andrade, and E. E. Suero. 1995. Growth, water use, and kernel abortion of maize subjected to drought at silking. *Field Crops Res.* **40**:87–94.

Oyarzun, R. A., C. O. Stockle, and M. D. Whiting. 2007. A simple approach to modeling radiation interception by fruit-tree orchards. *Agric. For. Meteorol.* **142**:12–24.

Paltridge, G. W. and J. V. Denholm. 1974. Plant yield and the switch from vegetative to reproductive growth. *J. Theor. Biol.* **44**:23–34.

Paltridge, G. W. and C. M. R. Platt. 1976. *Radiative Processes in Meteorology and Climatology.* (Developments in Atmospheric Science no. 5.) Amsterdam: Elsevier Scientific.

Paltridge, G. W., J. V. Denholm, and D. J. Connor. 1984. Determinism, senescence and the yield of plants. *J. Theor. Biol.* **10**:383–98.

Pannell, D. J. 1999. Social and economic challenges in the development of complex farming systems. *Agrofor. Syst.* **45**:393–409.

Passioura, J. B. 1983. Roots and drought resistance. *Agric. Water Manage.* **7**:265–80.

Paul, E. A. and F. E. Clark. 1989. *Soil Microbiology and Biochemistry*. San Diego, California: Academic Press.

Paul, E. A. and J. A. van Veen. 1978. The use of tracers to determine the dynamic nature of organic matter. *Trans, 11th Int. Congr. Soil Sci.* **3**:61–102.

Payne, P. R. 1978. Human protein requirements. In G. Norton (ed.) *Plant Proteins*. London: Butterworth, pp. 247–63.

Pearce, R. B., G. E. Carlson, D. K. Barnes, R. H. Hart, and C. H. Hanson. 1969. Specific leaf weight and photosynthesis in alfalfa. *Crop Sci.* **9**:423–6.

Peng, S., K. G. Cassman, S. S. Virmani, J. Sheehy, and G. S. Khush. 1999. Yield potential trends of tropical rice since the release of IR8 and the challenge of increasing rice yield potential. *Crop Sci.* **39**:1552–9.

Peng, S., F. V. Garcia, R. C. Laza *et al.* 1996. Increased N-use efficiency using a chlorophyll meter on high-yielding irrigated rice. *Field Crops Res.* **47**:243–52.

Peng, S., J. Huang, J. E. Sheehy *et al.* 2004. Rice yields decline with higher night temperature from global warming. *Proc. Natl. Acad. Sci. USA* **101**:9971–5.

Peng, S., G. S. Khush, P. Virk, Q. Tang, and **Y. Zo**. 2008. Progress in ideotype breeding to increase rice yield potential. *Field Crops Res.* **108**:32–8.

Penman, H. L. 1948. Natural evaporation from open water, bare soil, and grass. *Proc. R. Soc. Lond.* A **193**:120–45.

Penman, H. L., D. E. Angus, and **C. H. M. Van Bavel**. 1967. Microclimate factors affecting evaporation and transpiration. In **R. M. Hagen, H. R. Haise,** and **T. W. Edminister** (eds.) *Irrigation of Agricultural Lands.* Agronomy Monograph no. 11. Madison, Wisconsin: Am. Soc. Agron., pp. 483–505.

Penning de Vries, F. W. T. 1975. The cost of maintenance processes in plant cells. *Ann. Bot.* **39**:77–92.

Penning de Vries, F. W. T., A. H. M. Brunsting, and **H. H. van Laar**. 1974. Products, requirements and efficiency of biosynthesis: a quantitative approach. *J. Theor. Biol.* **45**:339–77.

Penning de Vries, F. W. T., H. H. van Laar, and **M. C. M. Chardon**. 1983. Bioenergetics and growth of fruits, seeds, and storage organs. In **W. H. Smith** and **S. J. Banta** (eds.) *Potential Productivity of Field Crops Under Different Environments.* Los Baños, Philippines: IRRI, pp. 37–59.

Perlack, R. D., L. L. Wright, A. F. Turhollow *et al.* 2005. *Biomass as Feedstock for a Bioenergy and Bioproducts Industry. The Technical Feasibility of a Billion Ton Annual Supply.* USDA and DOE.

Peterson, G. A., D. G. Westfall, and **C. V. Cole**. 1993. Agroecosystem approach to soil and crop management. *Soil Sci. Soc. Am. J.* **57**:1354–60.

Petheram, C., G. Walker, R. Grayson, T. Thierfelder, and **L. Zhang**. 2002. Towards a framework for predicting impacts of land-use on recharge: 1. A review of recharge studies in Australia. *Aust. J. Soil Res.* **40**:397–417.

Piha, M. I. and **D. N. Munns**. 1987. Nitrogen fixation potential of bean (*Phaseolus vulgaris* L.) compared with other grain legumes under controlled conditions. *Plant Soil* **98**:169–82.

Pimentel, D., L. E. Hurd, A. C. Belloti *et al.* 1973. Food production and the energy crisis. *Science* **206**:1277–80.

Pirt, S. J. 1965. The maintenance energy of bacteria in growing cultures. *Proc. R. Soc. Lond.* B **163**:224–31.

Plant and Soil Sciences eLibrary. http://plantandsoil.unl.edu/croptechnology2005/pages/index.jsp?what=topicsD&topicOrder=1&informationModuleId=1087488148.

Plant, R. E., R. D. Horrocks, D. W. Grimes, and **L. J. Zelinski**. 1992. CALEX/cotton: an integrated expert system application for irrigation scheduling. *Amer. Soc. Agric. Eng.* **35**:1833–9.

Pook, M., S. Lisson, J. Risbey *et al.* 2009. The autumn break for cropping in southeast Australia: trends, synoptic influences and impacts on wheat yield. *Int. J. Climatol.* **29**:2012–26.

Porter, J. R. 1993. AFRCWHEAT2: a model of the growth and development of wheat incorporating responses to water and nitrogen. *Eur. J. Agron.* **2**:69–82.

Portmann, F. T., S. Siebert, and **P. Döll**. 2010. MIRCA2000 – Global Monthly Irrigated and Rainfed Crop Areas around the year 2000: a new high-resolution data set for agricultural and hydrological modeling. *Global Biogeochem. Cy.* **24**, GB1011. doi:10.1029/2008GB003435.

Postel, S., P. Polak, F. Gonzales, and **J. Keller**. 2001. Drip irrigation for small farmers. A new initiative to alleviate hunger and poverty. *Water Int.* **26**:3–13.

Powlson, D. S., T. M. Addiscott, N. Benjamin *et al.* 2008. When does nitrate become a risk for humans? *J. Environ. Qual.* **37**:291–5.

Priestley, C. H. B. and R. J. Taylor. 1972. On the assessment of surface heat flux and evaporation using large scale parameters. *Mon. Weather Rev.* **100**:81–92.

Pruitt, W. O. 1964. Cyclic relations between evapotranspiration and radiation. *Trans. ASAE* **7**:271–5; 280.

Pruitt, W. O. 1986. *Prediction and Measurement of Crop Water Requirements: the Basis of Irrigation Scheduling.* University of Sydney, Sydney: Faculty of Agriculture.

Puckridge, D. W. and C. M. Donald. 1967. Competition among wheat plants sown at a wide range of densities. *Aust. J. Agric. Res.* **18**:193–211.

Pugsley, A. T. 1982. Additional genes inhibiting winter habit in wheat. *Euphytica* **21**:547–52.

Putman, A. and C. S. Tang (eds.) 1986. *The Science of Allelopathy.* New York: John Wiley.

Radford, P. J. 1967. Growth analysis formulae – their use and abuse. *Crop Sci.* **7**:171–5.

Radin, J. W. 1977. Contribution of the root system to nitrate assimilation in whole cotton plants. *Aust. J. Plant Physiol.* **4**:811–19.

Radin, J. W. 1983. Control of plant growth by nitrogen: differences between cereals and broadleaf species. *Plant Cell Environ.* **6**:65–8.

Rapoport, H. F. and R. S. Loomis. 1986. Structural aspects of root thickening in *Beta vulgaris* L.: comparative thickening in sugarbeet and chard. *Bot. Gaz.* **147**:270–7.

Rasmusson, D. C. 1987. An evaluation of ideotype breeding. *Crop. Sci.* **27**:1140–6.

Raun, W. R., J. B. Solie, G. V. Johnson *et al.* 2002. Improving nitrogen use efficiency in cereal grain production with optical sensing and variable rate application. *Agron. J.* **94**:815–20.

Rebetzke, G. J., A. G. Condon, R. A. Richards, and G. D. Farquhar. 2002. Selection for reduced carbon isotope discrimination increases aerial biomass and grain yield of rainfed bread wheat. *Crop Sci.* **42**:739–45.

Reddall, A. A., L. J. Wilson, P. C. Gregg, and V. O. Sadras. 2007. Photosynthetic response of cotton to spider mite damage: interaction with light and compensatory mechanisms. *Crop Sci.* **47**:2047–57.

Reeve, R. C. and M. Fireman. 1967. Salt problems in relation to irrigation. In R. M. Hagan, H. R. Haise and T. W. Edminister (eds.) *Irrigation of Agricultural Lands.* Agronomy Monographs. Madison, Wisconsin: Am. Soc. Agron., pp. 988–1008.

Rengasamy, P. 2002. Transient salinity and subsoil constraints to dryland farming in Australian sodic soils: an overview. *Aust. J. Exp. Agric.* **42**:351–61.

Richards, R. A. 2000. Selectable traits to increase crop photosynthesis and yield of grain crops. *J. Exp. Bot.* **51**:447–58.

Richards, R. A. 2006. Physiological traits used in the breeding of new cultivars for water-scarce environments. *Agr. Water Manage.* **80**:197–211.

Richards, R. A. and J. B. Passioura. 1989. A breeding program to reduce the diameter of the major xylem vessel in the seminal roots of wheat and its effect on grain yield in rain-fed environments. *Aust. J. Agric. Res.* **40**:943–50.

Richards, Q. D., M. P. Bange, and S. B. Johnston. 2008. HydroLogic: an irrigation management system for Australian cotton. *Agric. Sys.* **98**:40–9.

Ridley, A. M., B. Christy, F. X. Dunin *et al.* 2001. Lucerne in crop rotations on the Riverine Plains. I. The soil water balance. *Aust. J. Agric. Res.* **52**:263–77.

Ridley, A. M., R. J. Simpson, and R. E. White. 1999. Nitrate leaching under phalaris, cocksfoot, and annual ryegrass pastures and implications for soil acidification. *Aust. J. Agric. Res.* **50**:55–64.

Rimmington, G. M. and **N. Nicholls**. 1993. Forecasting wheat yields in Australia with the southern oscillation index. *Aust. J. Agric. Res.* **44**:625–32.

Ritchie, J. T. 1971. Dryland evaporative flux in a subhumid climate: I. Micrometeorological influences. *Agron. J.* **63**:51–5.

Ritchie, J. T. 1972. Model for predicting evaporation from a row crop with incomplete cover. *Water Resour. Res.* **8**:1204–13.

Ritchie, J. T. and **E. Burnett**. 1971. Dryland evaporative flux in a subhumid climate: II. Plant influences. *Agron. J.* **63**:56–62.

Ritchie, J. T. and **S. Otter**. 1985. *Description and Performance of CERES-Wheat. A User-oriented Wheat Yield Model. ARS Wheat Yield Project, ARS* **38**:159–76.

Roberts, A. M., M. J. Helmers, and **I. R. P. Fillery**. 2009. The adoptability of perennial-based farming systems for hydrologic and salinity control in dryland farming systems in Australia and the United States of America. *Crop & Past. Sci.* **60**:83–99.

Roberts, E. H. and **R. J. Summerfield**. 1987. Measurement and prediction of flowering in annual crops. In **J. G. Atherton** (ed.) *Manipulation of Flowering*. Proc. 45th Easter School Agricultural Science, Nottingham University. London: Butterworth, pp. 17–50.

Roberts, H. A. 1981. Seed banks in the soil. *Adv. Appl. Biol.* **6**:1–55.

Robertson, G. W. 1973. Development of simplified agroclimatic procedures for assessing temperature effects on crop development. In **R. O. Slayter** (ed.) *Plant Responses to Climatic Factors, Proc. Uppsala Symp. 1970*. Paris: UNESCO, pp. 327–43.

Robertson, M. J., I. R. Brooking, and **J. T. Ritchie**. 1996. Temperature response of vernalization in wheat: modelling the effect on the final number of mainstem leaves. *Ann. Bot.* **78**:371–81.

Robson, M. J. 1982. The growth and carbon economy of selection lines of *Lolium perenne* cv. S23 with "differing" rates of dark respiration. I. Grown as simulated swards during a regrowth period. *Ann. Bot.* **49**:321–9.

Rolston, D. E., D. L. Hoffman, and **D. W. Toy**. 1978. Field measurement of denitrification: I. Flux of N_2 and N_2O. *Soil Sci. Soc. Am. J.* **42**:863–9.

Rose, C. W. 1985. Developments in soil erosion and deposition models. *Adv. Soil Sci.* **2**:1–63.

Rosegrant, M. W., X. Cai, and **S. A. Cline**. 2002. *World Water and Food to 2025: Dealing with Scarcity*. Washington, DC: Intern. Food Policy Res. Inst.

Rosenberg, N. J., B. L. Blad, and **S. B. Verma**. 1983. *Microclimate, the Biological Environment*, 2nd edition. New York: John Wiley.

Rossiter, R. C. and **W. T. Collins**. 1988. Genetic diversity in old subterranean clover (*Trifolium subterranean* L.) populations in Western Australia. II. Pastures sown initially to the Mt. Barker strain. *Aust. J. Agric. Res.* **39**:1063–74.

Rovira, A. D. 1992. Dryland Mediterranean farming systems of Australia. *Aust. J. Exp. Agric.* **32**:808–9.

Rovira, A. D. 1993. *Sustainable Farming Systems in the Cereal-livestock Areas of the Mediterranean Region of Australia*. Glen Osmond, South Australia: Cooperative Research Centre for Soil and Land Management.

Running, S. W. 1980. Relating plant capacitance to water relations of *Pinus contorta. Forest Ecol. Manage.* **2**:237–52.

Russell, J. S. 1967. Nitrogen fertilizer and wheat in a semi-arid environment. I. Effect on yield. *Aust. J. Agric. Anim. Husb.* **7**:453–62.

Ruthenberg, H. 1980. *Farming Systems in the Tropics*. Clarendon Press, Oxford.

Ruttan, V. W. 1982. *Agricultural Research Policy*. Minneapolis: University of Minnesota Press.

Ryle, **G. J. A.**, **C. E. Powell**, and **A. J. Gordon**. 1979. The respiratory costs of nitrogen fixation in soyabean, cowpea, and white clover. I. Nitrogen fixation and the respiration of the nodulated root. *J. Exp. Bot.* **30**:135–44.

Sadras, **V. O.** 2002. Canopy management. In **D. Pimentel** (ed.) *Encyclopedia of Pest Management*. New York: Marcel Dekker, pp. 112–14.

Sadras, **V. O.** 2003. Influence of size of rainfall events on water-driven processes. I. Water budget of wheat crops in south-eastern Australia. *Aust. J. Agric. Res.* **54**:341–51.

Sadras, **V. O.** 2009. Does partial root-zone drying improve irrigation water productivity in the field? A meta analysis. *Irrig Sci.* **27**:183–90.

Sadras, **V. O.** and **J. F. Angus**. 2006. Benchmarking water-use efficiency of rainfed wheat in dry environments. *Aust. J. Agric. Res.* **57**:847–56.

Sadras, **V. O.** and **D. J. Connor**. 1991. Physiological basis of the response of harvest index to the fraction of water transpired after anthesis: a simple model to estimate harvest index for determinate species. *Field Crop Res.* **26**:227–39.

Sadras, **V. O.**, **D. K. Roget**, and **M. Krause**. 2003. Dynamic cropping strategies for risk management in dry-land farming systems. *Agric. Syst.* **76**:929–48.

Saeki, **T.** 1963. Light relations in plant communities. In **L. T. Evans** (ed.) *Environmental Control of Plant Growth*. New York: Academic Press, pp. 79–94.

Sage, **R. F.** and **C. D. Reid**. 1994. Photosynthetic response mechanisms to environmental change in C3 plants. In **R. E. Wilkinson** (ed.) *Plant–Environment Interactions*. New York: Marcel Dekker, pp. 413–19.

Sale, **P. J. M.** 1974. Productivity of vegetable crops in a region of high solar input. III. Carbon balance of potato crops. *Aust. J. Plant Physiol.* **1**:283–96.

Sale, **P. J. M.** 1975. Productivity of vegetable crops in a region of high solar input. IV. Field chamber measurements of french beans (*Phaseolus vulgaris* L.) and cabbages (*Brassica oleracea* L.). *Aust. J. Plant Physiol.* **2**:461–70.

Sale, **P. J. M.** 1977. Net carbon exchange rates of field grown crops in relation to irradiance and dry weight accumulation. *Aust. J. Plant Physiol.* **4**:555–69.

Salisbury, **F. B.** 1963. *The Flowering Process*. New York: Pergamon Press.

Salisbury, **F. B.** 1981. Response to photoperiod. In **O. L. Lange**, **P. S. Nobel**, **C. B. Osmond**, and **H. Zeigler** (eds.) *Physiological Plant Ecology. I. Responses to the Physical Environment*, (*Encyl. Plant Physiol*, New Ser. Vol. 12A). Heidelberg: Springer-Verlag, pp. 135–67.

Sanchez, **P. A.** 1976. *Properties and Management of Soils in the Tropics*. New York: John Wiley.

Schneider, **S. H.**, **P. H. Gleick**, and **L. O. Mearns**. 1990. Prospects for climate change. In **P. E. Waggoner** (ed.) *Climate Change and U.S. Water Resources*. New York: John Wiley, pp. 41–73.

Schouten, **H.** 1986. Low-input farming. In **H. van Keulen** and **J. Wolf** (eds.) *Modelling of Agricultural Production: Weather, Soils and Crops*. Simulation Monographs. Wageningen, the Netherlands: Pudoc, pp. 263–76.

Schubert, **K. R.** (ed.) 1982. *The Energetics of Biological Nitrogen Fixation*. Rockville, Maryland: Am. Soc. Plant Physiol.

Schultz, **J. E.** 1971. Soil water changes under fallow crop treatments in relation to soil type, rainfall, and yield. *Aust. J. Exp. Agric. Anim. Husb.* **11**:236–42.

Searchinger, **T.**, **R. Heimlich**, **R. Houghton** *et al.* 2008. Use of U.S. croplands for biofuels increases greenhouse gases through emissions from land use change. *Science* **113**:2380–8.

Shannon, **M. C.** 1984. Breeding, selection and the genetics of salt tolerance. In **R. C. Staples** and **G. H. Toenniessen** (eds.) *Salinity Tolerance in Plants: Strategies for Crop Improvement*. New York: Wiley-Interscience, pp. 231–54.

Shantz, H. L. and **L. N. Piemeisel**. 1927. Water requirement of plants at Akron, Colorado. *J. Agric. Res.* **34**:1093–190.

Sheehy, J. E., P. L. Mitchell, and **A. B. Ferrer**. 2006. Decline in rice grain yields with temperature. Models and correlations can give different estimates. *Field Crop Res.* **98**:151–6.

Sheldrick, W., J. K. Syers, and **J. Lingard**. 2003. Contribution of livestock excreta to nutrient balances. *Nutr. Cycl. Agroecosys.* **66**:119–31.

Shibles, R. M. and **C. R. Weber**. 1965. Leaf area, solar radiation interception and dry matter production by soybeans. *Crop Sci.* **5**:575–7.

Shouse, P., W. A. Jury, L. H. Stolzy, and **S. Dasburg**. 1982. Field measurement and modelling of cowpea water use and yield under stressed and well-watered conditions. *Hilgardia* **50**:1–25.

Silsbury, J. H. 1977. Energy requirements of symbiotic nitrogen fixation. *Nature* **267**:1149–50.

Simmonds, N. W. 1979. *Principles of Crop Improvement*. London: Longman.

Sinclair, T. R., G. L. Hammer, and **E. J. van Oosterom**. 2005. Potential yield and water-use efficiency benefits in sorghum from limited maximum transpiration rate. *Funct. Plant Biol.* **32**:945–52.

Slatyer, R. O. 1967. *Plant–water Relations*. New York: Academic Press.

Slicher van Bath, B. H. 1963. *The Agrarian History of Western Europe: A.D. 500–1500*. London: Edward Arnold.

Smika, D. E. 1970. Summer fallow for dryland winter wheat in the semiarid Great Plains. *Agron. J.* **62**:15–17.

Smil, V. 1999. Crop residues, agriculture's largest harvest. *Bioscience* **49**:299–308.

Smil, V. 2001. *Feeding the World. A Challenge for the Twenty-first Century*. Cambridge, Massachusetts: MIT Press.

Smil, V. 2004. *Enriching the Earth. Fritz Haber, Carl Bosch, and the Transformation of World Food Production*. Cambridge, Massachusetts: MIT Press.

Smith, B. D. 1990. Origins of agriculture in eastern North America. *Science* **246**:1566–71.

Smith, D. F. 2000. *Natural Gain: in the Grazing Lands of Southern Australia*. Sydney: UNSW Press.

Smith, W. H. and **S. J. Banta** (eds.) 1983. *Potential Productivity of Field Crops under Different Environments*. Los Baños, Philippines: IRRI.

Snyder, C. S., T. W. Bruulsema, T. L. Jensen, and **P. E. Fixen**. 2009. Review of greenhouse gas emissions from crop production systems and fertilizer management effects. *Agric. Ecosys. & Environ.* **133**:247–66.

Soil Survey Staff. 1988. *Keys to Soil Taxonomy*. Soil Management Support Services, AID, USDA, Cornell University, Ithaca, New York.

Sommer, S. G., J. K. Schjoerring, and **O. T. Denmead**. 2004. Ammonia emission from mineral fertilizers and fertilized crops. *Adv. Agron.* **82**:558–622.

Soriano, M. A., F. Orgaz, F. J. Villalobos, and **E. Fereres**. 2004. Efficiency of early plantings of sunflower. *Eur. J. Agron.* **21**:465–76.

Specht, J. E., J. H. Williams, and **C. J. Weidenbenner**. 1986. Differential responses of soybean genotypes subjected to a seasonal soil water gradient. *Crop Sci.* **26**:922–34.

Specht, J. E., D. J. Hume, and **S. V. Kumudini**. 1999. Soybean yield potential – a genetic and physiological perspective. *Crop Sci.* **39**:1560–70.

Spitters, C. J. T. 1980. Competition effects within mixed stands. In **R. G. Hurd, P. V. Biscoe**, and **C. Dennis** (eds.) *Opportunities for Increasing Crop Yields*. London: Pitman, pp. 219–31.

Squire, G. R. 1990. *The Physiology of Tropical Crop Production*. Wallingford: CABI.

Staff of the L. H. Bailey Hortorium. 1976. *Hortus Third*. New York: Macmillan.

Stahl, R. S. and **K. J. McCree**. 1988. Ontogenetic changes in the respiration coefficients of grain sorghum. *Crop Sci.* **28**:111–13.

Stainforth, D. A., T. Aina1, C. Christensen *et al.* 2005. Uncertainty in predictions of the climate response to rising levels of greenhouse gases. *Nature* **433**:403–6.

Stapper, M. 1984. *SIMTAG: A Simulation Model of Wheat Genotypes.* University of New England and ICARDA, Armidale, Australia, and Aleppo, Syria.

Steduto, P. and **R. Albrizio**. 2005. Resource use efficiency of field-grown sunflower, sorghum and chickpea II. Water use efficiency and comparison with radiation use efficiency. *Agric. For. Meteorol.* **130**:269–81.

Steele, K. W. and **I. Vallis**. 1988. The nitrogen cycle in pastures. In **J. R. Wilson** (ed.) *Advances in Nitrogen Cycling in Agricultural Ecosystems.* Wallingford, Oxon, UK: CAB International, pp. 274–91.

Stern, W. R. and **C. M. Donald**. 1962. Light relations in grass-clover swards. *Aust. J. Agric Res.* **13**:599–614.

Stevenson, F. J. (ed.). 1982. *Nitrogen in Agricultural Soils.* (Agronomy Monograph no. 22.) Madison, Wisconsin: Am. Soc. Agron.

Stewart, J. I. 1988. *Response Farming in Rainfed Agriculture.* Davis, California: WHARF Foundation Press.

Stewart, J. I. and **C. T. Hash**. 1982. Impact of weather analysis on agricultural production and planning decisions for semiarid areas of Kenya. *J. Appl. Meteorol.* **21**:477–94.

Stout, B. A., C. A. Myers, A. Hurrand, and **L. W. Faidley**. 1979. *Energy for World Agriculture.* Rome: UN FAO.

Struik, P. C., Cassman, K. G., and **M. Koornneef**. 2007. A dialogue on interdisciplinary collaboration to bridge the gap between plant genomics and crop science. In **J. H. J. Spiertz, P. C. Struik** and **H. H. van Laar** (eds.) *Scale and Complexity in Plant Systems Research: Gene–Plant–Crop Relations.* Springer. pp. 319–28.

Suyker, A. E. and **S. B. Verma**. 2009. Evapotranspiration of irrigated and rainfed maize–soybean cropping systems. *Agric. For. Meteorol.* **149**:443–52.

Tanaka, A. and **K. Fujita**. 1979. Growth, photosynthesis and yield components in relation to grain yield of the field bean. *J. Fac. Agric. Hokkaido Univ.* **59**(2):146–238.

Tanner, C. B. 1981. Transpiration efficiency of potato. *Agron. J.* **73**:59–64.

Tanner, C. B. and **T. R. Sinclair**. 1983. Efficient water use in crop production: Research or research? In **H. M. Taylor, W. R. Jordan**, and **T. R. Sinclair** (eds.) *Limitations to Efficient Water Use in Crop Production.* Madison, Wisconsin: Am. Soc. Agron., pp. 1–27.

Taverne, D. 2005. *The March of Unreason. Science, Democracy, and the New Fundamentalism.* Oxford: Oxford University Press.

Tennant, D. and **D. Hall**. 2001. Improving water use of annual crops and pastures – limitations and opportunities in Western Australia. *Aust. J. Agric. Res.* **52**:171–82.

Testi, L., F. J. Villalobos, F. Orgaz, and **E. Fereres**. 2006. Water requirements of olive orchards. 1. Simulation of daily evapotranspiration for scenario analysis. *Irrig. Sci.* **24**:69–76.

Tetio-Kagho, F. and **F. P. Gardner**. 1988. Responses of maize to plant population density. II. Reproductive development, yield and yield adjustments. *Agron. J.* **80**:935–40.

Thomas, M. D. and **G. R. Hill**. 1949. Photosynthesis under field conditions. In **J. Franck** and **W. E. Loomis** (eds.) *Photosynthesis in Plants.* Ames: Iowa State College Press, pp. 19–51.

Thompson, P. B. 2008. The agricultural ethics of biofuels: a first look. *J. Agr. Environ. Ethic.* **21**:183–98.

Thornley, J. H. M. (1976). *Mathematical Models in Plant Physiology: a Quantitative Approach to Problems in Plant and Crop Physiology*. London: Academic Press.

Timmermann, C., R. Gerhards, and **W. Kühbauch**. 2003. The economic impact of site-specific weed control. *Precis. Agric.* **4**:249–60.

Tollenaar, M. 1989. Genetic improvement in grain yield of commercial maize hybrids grown in Ontario from 1959 to 1988. *Crop Sci.* **29**:1365–71.

Tollenaar, M., and **M. A. Lee**. 2002. Yield potential, yield stability, and stress tolerance in maize. *Field Crop Res.* **75**:161–9.

Tottmann, D. R., R. J. Makepeace, and **H. Broad**. 1979. An explanation of the decimal code for the growth stages of cereals, with illustrations. *Ann. Appl. Biol.* **93**:221–34.

Trabalka, J. R. and **D. E. Reichle** (eds.). 1986. *The Changing Carbon Cycle*. New York: Springer-Verlag.

Trenbath, B. R. 1974. Biomass productivity of mixtures. *Adv. Agron.* **26**:177–210.

Trevaskis, B., M. N. Hemming, E. S. Dennis, and **W. J. Peacock**. 2007. The molecular basis of vernalization-induced flowering in cereals. *Trends Plant Sci.* **12**:352–7.

Trewartha, G. T. and **L. H. Horn**. 1980. *An Introduction to Climate*. New York: McGraw-Hill.

Trewavas, A. 1986. Resource allocation under poor growth conditions. A major role for growth substances in developmental plasticity. In **D. H. Jennings** and **A. J. Trewavas** (eds.) *Plasticity in Plants*. Symp. Soc Exp. Biol. no. 40. Cambridge University, Cambridge, UK: Company of Biologists, pp. 31–76.

Trumble, H. C. 1939. Climatic factors in relation to agricultural regions of southern Australia. *Trans. Roy. Soc. S. Aust.* **63**:36–43.

Turner, N. C. 1986. Crop water deficits: a decade of progress. *Adv. Agron.* **39**:1–51.

Turner, N. C. 1997. Further progress in crop water relations. *Adv. Agron.* **58**:293–338.

Turner, N. C. and **J. E. Begg**. 1981. Plant–water relations and adaptation to drought. *Plant Soil* **58**:97–113.

Turner, R. E., N. N. Rabalais, and **D. Justic**. 2008. Gulf of Mexico hypoxia: alternate states and a legacy. *Environ. Sci. Technol.* **42**:2323–7.

Ulrich, A. 1961. Variety climate interactions of sugar beet varieties in simulated climates. *J. Am. Soc. Sugar Beet Technol.* **11**:376–87.

Ulrich, A., D. Ririe, F. J. Hills, A. George, and **M. D. Morse**. 1959. *Plant Analysis. A Guide for Sugar Beet Fertilization*. (Bull. no. 766, part 1.) University of California, Berkeley, California: Calif. Agric. Exp. Sta.

Ummenhofer, C. C., M. H. England, P. C. McIntosh *et al.* 2009. What causes south east Australias's worst droughts? *Geophys. Res. Lett.* **36** L04706, doi:10.1029/2008GL036801.

United Nations Population Division. www.un.org/esa/population.

Unger, P. W. 1984. *Tillage Systems for Soil and Water Conservation*. (FAO Soils Bulletin no. 54.) Rome: FAO.

Unkovich, M., D. Herridge, M. Peoples *et al.* 2008. *Measuring Plant-associated Nitrogen Fixation in Agricultural Systems*. ACIAR Monograph No. 136. Canberra, Australia.

USDA (United States Department of Agriculture). 1974. *Summer Fallow in the Western United States. Conservation Research Report*. Agric. Res. Ser. Washington, DC: Government Printing Office, USDA, US.

USDA. 2009. Manure use for fertilizer and energy: report to Congress. Economic Research Service. U.S. Washington, D.C.: Department of Agriculture.

USDA-ERS (USDA Economic Research Service). 2005. *Agricultural Resource Management Survey*. Washington, D.C.: USDA-ERS.

USDA-ERS website. http://www.ers.usda.gov/Data/FertilizerUse/.

USDA-NASS (USDA National Agricultural Statistics Service) website. www.nass.usda.gov/.

van Alphen, B. J., H. W. G. Booltink, and **J. Bouma**. 2001. Combining pedotransfer functions with physical measurements to improve estimation of soil hydraulic properties. *Geoderma* **103**:133–47.

Vandenbygaart, A. J. and **D. A. Angers**. 2006. Towards accurate measurements of soil organic carbon stock change in agroecosystems. *Can. J. Soil Sci.* **86**:465–71.

Van Der Meer, H. G., R. J. Unwin, T. A. van Dijk, and **G. C. Ennik** (eds.) 1987. Animal manure on grassland and fodder crops. Fertilizer or waste? (*Proc. Intl. Symp. European Grassland Fed.*) Dordrecht, the Netherlands: Martinus Nijhoff.

Van Der Plank, J. E. 1963. *Plant Disease: Epidemics and Control*. New York: Academic Press.

van Heemst, H. D., J. J. Nerkelijn, and **H. van Keulen**. 1981. Labour requirements in various agricultural systems. *Quart. J. Int. Agric.* **120**:178–201.

van Heerwarden, A. F., G. D. Farquhar, J. F. Angus, R. A. Richards, and **G. N. Howe**. 1998. "Haying off", the negative grain yield response of dryland wheat to nitrogen fertilizer. I. Biomass, grain yield, and water use. *Aust. J. Agric. Res.* **49**:1067–81.

van Keulen, H. 1982. Graphical analysis of annual crop response to fertilizer application. *Agric. Syst.* **9**:113–26.

van Keulen, H. and **H. D. J. van Heemst**. 1982. Crop supply of macronutrients. *Agric. Res. Rep.* **916**:46.

Van Noordwijk, M. and **B. Lusiana**. 1999. WaNuLCAS, a model of water, nutrient and light capture in agroforestry systems. *Agrofor. Syst.* **43**:217–42.

Van Soest, P. J. 1982. *Nutritional Ecology of the Ruminant*. Corvallis, Oregon: O & B Books.

van Wijk, W. R. and **D. A. de Vries**. 1963. Periodic temperature variations in a homogeneous soil. In **W. R. van Wijk** (ed.) *Physics of the Plant Environment*. Amsterdam: North-Holland, pp. 102–43.

Venterea, R. D., Burger, M., and **K. A. Spokas**. 2005. Nitrogen oxide and methane emissions under varying tillage and fertilizer management. *J. Environ. Qual.* **34**:1467–77.

Verma, S. B., A. Dobermann, K. G. Cassman *et al.* 2005. Annual carbon dioxide exchange in irrigated and rainfed maize-based agroecosystems. *Agric. For. Meteorol.* **131**:77–96.

Vertregt, N. and **F. W. T. Penning de Vries**. 1987. A rapid method for determining the efficiency of biosynthesis of plant biomass. *J. Theor. Biol.* **128**:109–19.

Vertregt, N. and **B. Rutgers**. 1988. *Ammonia Volatilization from Grazed Pastures*. (CABO- Report 84. Report 64–2.) Wageningen, the Netherlands: Centre for Agrobiological Research.

Vince-Prue, D. 1975. *Photoperiodism in Plants*. New York: McGraw-Hill.

Wafula, B. M. 1995. Applications of crop simulation in agricultural extension and research in Kenya. *Agric. Sys.* **49**:399–412.

Wallace, J. S. 2000. Increasing agricultural water use efficiency to meet future food production. *Agric. Ecosyst. Environ.* **82**:105–19.

Wang, Y. P. and **D. J. Connor**. 1996. Simulation of optimal development for spring wheat at two locations in southern Australia under present and changed climate conditions. *Agric. For. Meteorol.* **79**:9–28.

Ward, P. R., F. X. Dunin, and **S. F. Micin**. 2001. Water balance of annual and perennial pastures on a duplex soil in a Mediterranean environment. *Aust. J. Agric. Res.* **52**:203–9.

Warren Wilson, J. 1959. Analysis of the spatial distribution of foliage by two-dimensional point quadrats. *New Phytol.* **58**:92–101.

Warren Wilson, J. 1967. Stand structure and light penetration. III. Sunlit foliage area. *J. Appl. Ecol.* **4**:159–65.

Warrington, I. J. and **E. T. Kanemasu**. 1983. Corn growth response to temperature and photoperiod. I. Seedling emergence, tassel initiation, and anthesis. *Agron. J.* **75**:749–54.

Weir, A. H., P. L. Bragg, J. R. Porter, and **J. H. Rayner**. 1984. A winter wheat crop simulation model without water or nutrient limitations. *J. Agric. Sci. Camb.* **102**:371–82.

Weldon, C. W. and **W. L. Slauson**. 1986. The intensity of competition versus its importance: an overlooked distinction and some implications. *Quart. Rev. Biol.* **61**:23–44.

Wellings, S. R. and **P. Bell**. 1980. Movement of water and nitrate in the unsaturated zone of the upper Chalk near Winchester, Hants. *J. Hydrol.* **48**:119–36.

White, J. W. 1981. *A Quantitative Analysis of the Growth and Development of Bean Plants* (Phaseolus vulgaris L.) PhD Dissertation, University of California. Berkeley.

White, J. W., M. Herndl, L. A. Hunt, T. S. Payne, and **G. Hoogenboom**. 2008. Simulation-based analysis of effects of *Vrn* and *Ppd* loci on flowering in wheat. *Crop Sci.* **48**:678–87.

Whitfield, D. M. 1990. Canopy conductance, carbon assimilation and water use in wheat. *Agr. Forest Meterol.* **53**:1–18.

Whitfield, D. M., D. J. Connor, and **P. J. M. Sale**. 1980. Carbon dioxide exchanges in response to change of environment and to defoliation in a tobacco crop. *Aust. J. Plant Physiol.* **7**: 473–85.

Whitfield, D. M., G. C. Wright, O. A. Gyles, and **A. J. Taylor**. 1986. Effects of stage growth, irrigation frequency, and gypsum treatment on CO_2 assimilation of lucerne (*Medicago sativa* L.) grown on a heavy clay soil. *Irrig. Sci.* **7**:169–81.

Whitson, R. E., R. D. Kay, W. A. LePori, and **E. M. Rister**. 1981. Machinery and crop selection with weather risk. *Trans. ASAE* **24**:288–91; 295.

Wilcke, B. 2008. Energy costs for corn drying and cooling. *Minnesota Crop eNews*. (www. extension.umn.edu/cropenews).

Wilhelm, W. W., J. M. F. Johnson, D. L. Karlen, and **D. T. Lightle**. 2007. Corn stover to sustain soil organic carbon further constrains biomass supply. *Agron. J.* **99**:1665–7.

Williams, C. H. 1980. Soil acidification under clover pasture. *Aust. J. Exp. Agric. Anim. Husb.* **20**:531–67.

Wilson, D. and **J. G. Jones**. 1982. Effect of selection for dark respiration rate of mature leaves on crop yields of *Lolium perenne* cv. S23. *Ann. Bot.* **49**:313–20.

Wischmeier, W. H. and **D. D. Smith**. 1978. *USDA Handbook no. 537*. Washington D.C.: U.S. Government Printing Office.

Witt, C., K. G. Cassman, D. C. Olk *et al.* 2000. Crop rotation and residue management effects on carbon sequestration, nitrogen cycling, and productivity of irrigated rice systems. *Plant & Soil* **225**:263–78.

Woodruff, N. P. and **F. H. Siddoway**. 1965. A wind erosion equation. *Soil Sci. Soc. Am. Proc.* **29**:602–8.

Wrage, N., G. L. Velthof, M. L. van Beusichem, and **O. Oenema**. 2001. Role of nitrifier denitrification in the production of nitrous oxide. *Soil Biol. & Biochem.* **33**:1723–32.

World Food Prize website. www.worldfoodprize.org/Laureates/laureates.htm

WWF. 2008. *Global Market Study on Jatropa. Final Report*. London: GEXSI LLP.

Xu, K., X. Xu, T. Fukao *et al.* 2006. Sub1A encodes an ethylene responsive-like factor that confers submergence tolerance to rice. *Nature* **442**:705–8.

Xue, C., X. Yang, B. A. M. Bouman *et al.* 2008. Optimizing yield, water requirements, and water productivity of aerobic rice for the North China Plain. *Irrig. Sci.* **26**:459–74.

Yang, **H.**, **A. Dobermann**, **K. G. Cassman**, and **D. T. Walters**. 2006. Features, applications, and limitations of the Hybrid-Maize Simulation Model. *Agron. J.* **98**:737–48.

Yorinori, **J. T.**, **W. M. Paiva**, **R. D. Frederick** *et al.* 2005. Epidemics of Soybean Rust (*Phakopsora pachyrhizi*) in Brazil and Paraguay from 2001 to 2003. *Plant Dis.* **89**:675–7.

Yoshida, **S.** 1981. *Fundamentals of Rice Crop Science*. Los Baños, Philippines: International Rice Research Institute.

Zadoks, **J. C.**, **T. T. Chang**, and **C. F. Konzak**. 1974. A decimal code for the growth stages of cereals. *Weed Res.* **14**:415–21.

Zhang, **D.-Y.**, **G.-J. Sun**, and **X.-H. Jiang**. 1999. Donald's ideotype and growth redundancy: a game theoretical analysis. *Field Crop Res.* **61**:179–87.

Index

Abbreviations used in this index: BNF; biological nitrogen fixation, Bt; *Bacillus thurigensis*, C/N; carbon-to-nitrogen ratio, C3 and C4; photosynthetic pathways, CAM; crassulacean acid metabolism, CEC; cation exchange capacity, CNCp; critical nutrient concentration (plant), CNCs; critical nutrient concentration (soil), Es; soil evaporation, ET; evapotranspiration, GHG; greenhouse gas, IPCC; International Panel on Climate Change, IPM; integrated pest management, LAI; leaf area index, N; nitrogen, NER; net energy ratio, NUE; nitrogen-use efficiency, P; phosphorus, PAR; photosynthetically active radiation, RUE; radiation-use efficiency, SNU; standard nutritional unit, SOM; soil organic matter, TE; transpiration efficiency, TU; thermal unit, WUE; water-use efficiency.

Except as a ratio in the abbreviation "C/N", the symbol "/" may be read as *and* or *or*, or *both*.